Apollo By The Numbers
A Statisical Reference

Apollo By The Numbers
A Statisical Reference

Richard W. Orloff

GOVERNMENT REPRINTS PRESS
An imprint of
Ross & Perry, Inc.
Washington, D.C.

Printed in The United States of America
Ross & Perry, Inc. Publishers
717 Second St., N.E., Suite 200
Washington, D.C. 20002
(202) 675-8300

SAN 253-8555

Government Reprints Press Edition 2001
Government Reprints Press is an Imprint of Ross & Perry, Inc.

Library of Congress Control Number: 2001091288

http://www.GPO.reprints.com

ISBN 1-931641-00-5

Foreword

In a spring 1999 poll of opinion leaders sponsored by leading news organizations in the United States, the 100 most significant events of the 20th century were ranked. The Moon landing was a very close second to the splitting of the atom and its use during World War II. "It was agonizing," CNN anchor and senior correspondent Judy Woodruff said of the selection process. Probably historian Arthur M. Schlesinger, Jr., best summarized the position of a large number of individuals polled. "The one thing for which this century will be remembered 500 years from now was: This was the century when we began the exploration of space." He noted that Project Apollo gave many a sense of infinite potential. "People always say: If we could land on the Moon, we can do anything," said Maria Elena Salinas, co-anchor at Miami-based Spanish-language cable network Univision, who also made it her first choice.

Perhaps because of his long life, Schlesinger has looked toward a positive future, and that prompted him to rank the lunar landing first. "I put DNA and penicillin and the computer and the microchip in the first 10 because they've transformed civilization. Wars vanish," Schlesinger said, and many people today cannot even recall when the Civil War took place. "Pearl Harbor will be as remote as the War of the Roses," he said, referring to the English civil war of the 15th century. And there's no need to get hung up on the ranking, he said. "The order is essentially very artificial and fictitious," he said. "It's very hard to decide the atomic bomb is more important than getting on the Moon."

There have been many detailed historical studies of Project Apollo completed in the more than thirty years since the first lunar landing in 1969. The major contours of the American sprint to the Moon during the 1960s have been told and retold many times, notably in several books in the NASA History Series, and by William Burroughs, Andrew Chaikin, and Charles Murray and Catherine Bly Cox. All provide he end of the decade through the first lunar landing on July 20, 1969, on to the last of six successful Moon landings with Apollo 17 in December 1972, NASA carried out Project Apollo with enthusiasm and aplomb. With the passage of time, the demise of the Soviet Union, the end of the Cold War, and the subsequent opening of archives on both sides of the space race, however, there are opportunities not present before to reconsider Project Apollo anew.

While there have been many studies recounting the history of Apollo, this new book in the NASA History Series seeks to draw out the statistical information about each of the flights that have been long buried in numerous technical memoranda and historical studies. It seeks to recount the missions, measuring results against the expectations for them.

This work appears in the NASA History Series as a Special Publication (SP) in the Reference Works section, SP-4000, of the series. Works in this section provide information, usually in dictionary, encyclopedic, or chronological form, for use by NASA personnel, scholars, and the public. This new publication captures for the use of all detailed information about Apollo and its unfolding during the 1960s and early 1970s.

Roger D. Launius
Chief Historian
National Aeronautics and Space Administration
October 2, 2000

Introduction

The purpose of this work is to provide researchers, students, and space enthusiasts with a comprehensive reference for facts about Project Apollo, America's effort to put humans on the Moon.

Research for this work started in 1988, when the author discovered that, despite the number of excellent books that focused on the drama of events that highlighted Apollo, there were none that focused on the drama of the numbers.

It may be impossible to produce the perfect Apollo fact book. For a program of the magnitude of Apollo, many NASA Centers and contractors maintained data files for each mission. As a result, the same measurements from different sources vary, sometimes significantly. In addition, there are notable errors and conflicts even within official NASA and contractor documents. In order to minimize conflicts, the author sought original documents to create this work. Some documents were previously unavailable to the public, and were released only following the author's petitions through the Freedom of Information Act.

This book is separated into two parts. The first part contains narratives for the Apollo 1 fire and the 11 flown Apollo missions. Included after each narrative is a series of data tables, followed by a comprehensive timeline of events from just before liftoff to just after crew and spacecraft recovery. The second part contains more than 50 tables. These tables organize much of the data from the narratives in one place so they can be compared among all missions. The tables offer additional data as well. The reader can select a specific mission narrative or specific data table by consulting the Table of Contents.

Event times in this work are expressed mostly as GMT (Greenwich Mean Time) and GET (Ground Elapsed Time). Local U.S. Eastern time, in which all missions were launched, is included only for significant events. In regular usage, GMT does not use a colon between the hours and minutes; however for the convenience of readers of this work, most of whom are in the United States, where time is expressed as "00:00", the colon is included.

The term "GET" (Ground Elapsed Time), used for manned U.S. spaceflights prior to the Space Shuttle, was referenced to "Range Zero," the last integral second before liftoff. With the first launch of the Shuttle, NASA began using the term "MET" (Mission Elapsed Time), which begins at the moment of solid rocket booster ignition. The format for GET used here is hhh:mm:ss.sss (e.g., hours:minutes:seconds). Example: 208:23:45.343, with "GET" excluded and assumed in order to avoid confusion with GMT.

Some other abbreviations used frequently in this work include:

B.S.: Bachelor of Science degree
CM: Command Module
CMP: Command Module Pilot
CSM: Command and Service Module(s) (combined structure)
GH_2: Gaseous Hydrogen
LH_2: Liquid Hydrogen
LM: Lunar Module
LMP: Lunar Module Pilot
LOX: Liquid Oxygen
LRV: Lunar Rover Vehicle (used on Apollos 15, 16, and 17)

M.S.: Master of Science degree
MET: Modular Equipment Transport (used only on Apollo 14)
NASA: National Aeronautics and Space Administration
Ph.D.: Doctor of Philosophy degree
Sc.D.: Doctor of Science degree
S-IB: Saturn IB launch vehicle
S-IVB: Saturn IV-B launch vehicle
SM: Service Module
SPS: Service Propulsion System

Trivia buffs will have a field day with the data published here, and it's a sure bet that a few readers will disagree with some of it. However, it is a start. Enjoy!

Comments and documented potential corrections are welcomed. Mail inquiries should be sent to Richard Orloff, Apollo by the Numbers, c/o NASA History Division, NASA Headquarters, Mail Code ZH, Washington, DC 20546, U.S.A.

Richard W. Orloff
October 2000

Acknowledgments

The information contained in the mission narratives in this work was derived primarily from uncopyrighted NASA and contractor mission reports, and, in some cases, is quoted verbatim from the original text without attribution. Readers interested in specific sources will find them listed in the bibliography which appears at the end of this work. In a few cases, it was necessary to include information from copyrighted works, and the author acknowledges those cases as follows:

The source for some of the **astronaut biographical data** is *Who's Who In Space: The International Space Year Edition*, by Michael Cassutt, although most information was derived from NASA biographies.

The primary source for descriptions of the **mission emblems** is the official NASA text that accompanied each emblem. However, additional information has been used from *Space Patches From Mercury to the Space Shuttle*, written by Judith Kaplan and Robert Muniz. Another source is Dick Lattimer's unpublished draft of *Astronaut Mission Patches and Spacecraft Callsigns*, available at the time of this writing at Rice University's Fondren Library.

The source for the **COSPAR designations** for the various Apollo spacecraft and launch vehicle stages once on orbit is the **R.A.E. Table of Earth Satellites 1957-1986.**

The author gratefully acknowledges the assistance of the following people for helping to locate original NASA documents, images and other information, and for checking the transcript for errors.

Becky Fryday, formerly Media Services, Lyndon B. Johnson Space Center; Bunda L. Dean, formerly Lyndon B. Johnson Space Center; Dale Johnson, George C. Marshall Space Flight Center; Daryl L. Bahls, The Boeing Company; David Ransom, Jr., Rancho Palos Verdes, CA; J.L. Pickering, Normal, IL; Ricky Lanclos, Nederland, TX; Dr. Eric M. Jones, editor of the Apollo Lunar Surface Journal Internet Web site; Dr. John B. Charles, Lyndon B. Johnson Space Center; Florastela Luna, Lyndon B. Johnson Space Center; Gary Evans, TRW; Gordon Davie, Edinburgh, Scotland; Janet Kovacevich, Lyndon B. Johnson Space Center; Joan Ferry and Lois Morris, Woodsen Research Center, Rice University; Joey Pellarin Kuhlman, formerly Lyndon B. Johnson Space Center; Kenneth Nail, formerly John F. Kennedy Space Center; Kipp Teague, Lynchburg, VA; Lee Saegesser, formerly NASA Headquarters; Lisa Vazquez, formerly Lyndon B. Johnson Space Center; Mike Gentry, Lyndon B. Johnson Space Center; Margaret Persinger, Kennedy Space Center; Oma Lou White, formerly George C. Marshall Space Flight Center; Paulo D'Angelo, Rome, Italy; Philip N. French and Jonathan Grant, NASA Center for Aerospace Information; Robert Sutton, Chantilly, VA; Robert W. Fricke, Jr., Lockheed Martin; Ruud Kuik, Amsterdam, The Netherlands; Dr. David R. Williams, National Space Data Center, GSFC; Hayes M. Harper, Downers Grove, IL; Lt. Col. George H. Orloff USA-RET, Oakhurst, NJ; Harald Kucharek, Karlsruhe, Germany; Kay Grinter, Kennedy Space Center; and Louise Alstork, Stanley Artis, Steve Garber, Hope Kang, Roger Launius, Warren Owens, and Michael Walker, NASA Headquarters, Washington, DC.

This book is dedicated to

ROBERT W. FRICKE, JR.
Lockheed Martin/Lyndon B. Johnson Space Center, Houston, Texas

Bob started working in the space program during Project Mercury. He's seen it all, and his insights have been invaluable in making this book come to life. In fact, it was Bob's gift to me of a copy of the Apollo Program Summary Report more than a decade ago that helped give birth to the concept of *Apollo by the Numbers*. During those years, Bob has continued to be a source of information, inspiration, and above all, a dear friend.

In recognition of the fact that he has worked on post-mission reports for more than 100 U.S. piloted spaceflights, NASA presented Bob with the coveted "Silver Snoopy" award for his outstanding achievement.

Richard W. Orloff
October 2000

Table of Contents

APOLLO 1

The Fire

Apollo 1 Fire Summary

(27 January 1967)

The Apollo 1 crew (l. to r.): Ed White, Gus Grissom, Roger Chaffee (NASA S66-30236).

Author's Note: None of the crew member photos in this chapter were taken on the day of the fire. These photos are used strictly to provide examples of training activities.

Background

The first piloted Apollo mission was scheduled for launch on 21 February 1967 at Cape Kennedy Launch Complex 34. However, the death of the prime crew in a command module fire during a practice session on 27 January 1967 put America's lunar landing program on hold.

The crew consisted of Lt. Colonel Virgil Ivan "Gus" Grissom (USAF), command pilot; Lt. Colonel Edward Higgins White, II (USAF), senior pilot; and Lt. Commander Roger Bruce Chaffee (USN), pilot.

Selected in the astronaut group of 1959, Grissom had been pilot of MR-4, America's second and last suborbital flight, and command pilot of the first two-person flight, Gemini 3. Born on 3 April 1926 in Mitchell, Indiana, Grissom was 40 years old on the day of the Apollo 1 fire. Grissom received a B.S. in mechanical engineering from Purdue University in 1950. His backup for the mission was Captain Walter Marty "Wally" Schirra (USN).

White had been pilot for the Gemini 4 mission, during which he became the first American to walk in space. He was born 14 November 1930 in San Antonio, Texas, and was 36 years old on the day of the Apollo 1 fire. He

received a B.S. from the U.S. Military Academy at West Point in 1952, an M.S. in aeronautical engineering from the University of Michigan in 1959, and was selected as an astronaut in 1962. His backup was Major Donn Fulton Eisele (USAF).

Chaffee was training for his first spaceflight. He was born 15 February 1935 in Grand Rapids, Michigan, and was 31 years old on the day of the Apollo 1 fire. He received a B.S. in aeronautical engineering from Purdue University in 1957, and was selected as an astronaut in 1963. His backup was Ronnie Walter "Walt" Cunningham.

The Accident

The accident occurred during the Plugs Out Integrated Test. The purpose of this test was to demonstrate all space vehicle systems and operational procedures in as near a flight configuration as practical and to verify systems capability in a simulated launch.

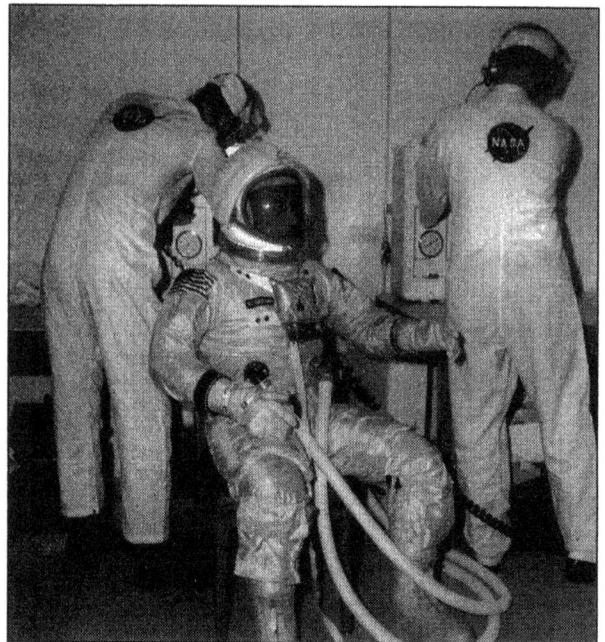

Grissom being checked out in Apollo 1 pressure suit (NASA S66-58023).

The test was initiated at 12:55 GMT on 27 January 1967. After initial system tests were completed, the flight crew entered the command module at 18:00 GMT. The command pilot noted an odor in the spacecraft environmental control system suit oxygen loop and the count was held at 18:20 GMT while a sample of the oxygen in this system was taken. The count was resumed at 19:42 GMT with hatch installation and subsequent cabin purge with oxygen beginning at 19:45 GMT. (The odor was later determined not to be related to the fire.)

Communication difficulties were encountered and the count was held at approximately 22:40 GMT to troubleshoot the problem. The problem consisted of a continuously live microphone that could not be turned off by the crew. Various final countdown functions were still performed during the hold as communications permitted.

By 23:20 GMT, all final countdown functions up to the transfer to simulated fuel cell power were completed and the count was held at T-10 minutes pending resolution of the communications problems.

Grissom, Chaffee, and White during Apollo 1 training (NASA S66-49181).

From the start of the T-10 minute hold at 23:20 GMT until about 23:30 GMT, there were no events that appear to be related to the fire. The major activity during this period was routine troubleshooting of the communications problem; all other systems were operating normally. There were no voice transmissions from the spacecraft from 23:30:14 GMT until the transmission reporting the fire, which began at 23:31:04.7 GMT.

During the period beginning about 30 seconds before the report, there were indications of crew movement. These indications were provided by the data from the biomedical sensors, the command pilot's live microphone, the guidance and navigation system, and the environmental control system. There was no evidence as to what this movement was or that it was related to the fire.

The biomedical data indicated that just prior to the fire report the senior pilot was performing essentially no activity until about 23:30:21 GMT, when a slight increase in pulse and respiratory rate was noted. At 23:30:30 GMT, the electrocardiogram indicated some muscular activity for several seconds. Similar indications were noted at 23:30:39 GMT. The data show increased activity but are not indicative of an alarm type of response. By 23:30:45 GMT, all of the biomedical parameters had reverted to the baseline "rest" level.

Apollo 1 commander Grissom (l.) inspects the CM during a visit to North American Aviation in 1966 (NASA S66-40760).

Beginning at about 23:30 GMT, the command pilot's live microphone transmitted brushing and tapping noises which were indicative of movement. The noises were similar to those transmitted earlier in the test by the live microphone when the command pilot was known to have been moving. These sounds ended at 23:30:58.6 GMT.

Any significant crew movement would result in minor motion of the command module as detected by the guidance and navigation system; however, the type of movement could not be determined. Data from this system indicated a slight movement at 23:30:24 GMT, with more intense activity beginning at 23:30:39 GMT and ending at 23:30:44 GMT. More movement began at 23:31:00 GMT and continued until loss of data transmission during the fire.

Increases of oxygen flow rate to the crew suits also indicated movement. All suits had some small leakage, and this leakage rate varied with the position of each crew member in the spacecraft. Earlier in the Plugs Out Integrated Test, the crew reported that a particular movement, the nature of which was unspecified, provided increased flow rate.

This was also confirmed from the flow rate data records. The flow rate showed a gradual rise at 23:30:24 GMT which reached the limit of the sensor at 23:30:59 GMT.

At 23:30:54.8 GMT, a significant voltage transient was recorded. The records showed a surge in the AC Bus 2 voltage. Several other parameters being measured also showed anomalous behavior at this time.

Beginning at 23:31:04.7 GMT, the crew gave the first verbal indication of an emergency when they reported a fire in the command module.

Emergency procedures called for the senior pilot, occupying the center couch, to unlatch and remove the hatch while retaining his harness buckled. A number of witnesses who observed the television picture of the command module hatch window discerned motion that suggested that the senior pilot was reaching for the inner hatch handle. The senior pilot's harness buckle was found unopened after the fire, indicating that he initiated the standard hatch-opening procedure. Data from the Guidance and Navigation System indicated considerable activity within the command module after the fire was discovered. This activity was consistent with movement of the crew prompted by proximity of the fire or with the undertaking of standard emergency egress procedures.

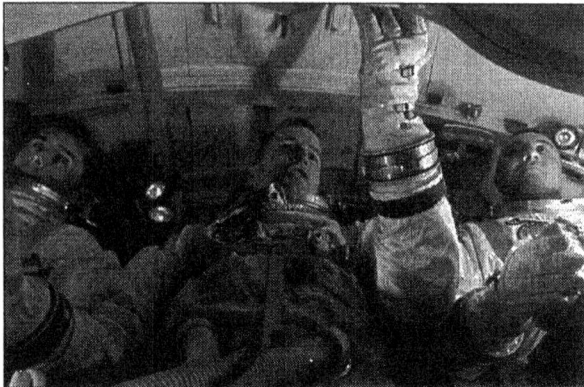

Apollo 1 crew training (NASA 57-HC-21).

Personnel located on adjustable level 8 adjacent to the command module responded to the report of the fire. The pad leader ordered crew egress procedures to be started and technicians started toward the White Room which surrounded the hatch and into which the crew would step upon egress. Then, at 23:31:19 GMT, the command module ruptured.

All transmission of voice and data from the spacecraft terminated by 23:31:22.4 GMT, three seconds after rupture. Witnesses monitoring the television showing the hatch window report that flames spread from the left to the right side of the command module and shortly thereafter covered the entire visible area.

Flames and gases flowed rapidly out of the ruptured area, spreading flames into the space between the command module pressure vessel and heat shield through access hatches and into levels A-8 and A-7 of the service structure. These flames ignited combustibles, endangered pad personnel, and impeded rescue efforts. The burst of fire, together with the sounds of rupture, caused several pad personnel to believe that the command module had exploded or was about to explode.

The immediate reaction of all personnel on level A-8 was to evacuate the level. This reaction was promptly followed by a return to effect rescue. Upon running out on the swing arm from the umbilical tower, several personnel obtained fire extinguishers and returned along the swing arm to the White Room to begin rescue efforts. Others obtained fire extinguishers from various areas of the service structure and rendered assistance in fighting the fires.

Three hatches were installed on the command module. The outermost hatch, called the boost protective cover (BPC) hatch, was part of the cover which shielded the command module during launch and was jettisoned prior to orbital operation. The middle hatch was termed the ablative hatch and became the outer hatch when the BPC was jettisoned after launch. The inner hatch closed the pressure vessel wall of the command module and was the first hatch to be opened by the crew in an unaided crew egress.

The day of the fire, the outer or BPC hatch was in place but not fully latched because of distortion in the BPC caused by wire bundles temporarily installed for the test. The middle hatch and inner hatch were in place and latched after crew ingress.

Although the BPC hatch was not fully latched, it was necessary to insert a specially-designed tool into the hatch in order to provide a hand-hold for lifting it from the command module. At this time the White Room was filling with dense, dark smoke from the command module interior and from secondary fires throughout level A-8. While some personnel were able to locate and don operable gas masks, others were not. Some proceeded without masks while others attempted without success to render masks operable. Even operable masks were unable to cope with the dense smoke present because they were designed for use in toxic rather than dense smoke atmospheres.

Visibility in the White Room was virtually nonexistent. It was necessary to work essentially by touch since visual observation was limited to a few inches at best. A hatch removal tool was in the White Room. Once the small fire near the BPC hatch had been extinguished and the tool located, the pad leader and an assistant removed the BPC hatch. Although the hatch was not latched, removal was difficult.

The personnel who removed the BPC hatch could not remain in the White Room because of the smoke. They left the White Room and passed the tool required to open each hatch to other individuals. A total of five individuals took part in opening the three hatches and each made several trips into the White Room and out for breathable air.

The middle hatch was removed with less effort than was required for the BPC hatch.

The inner hatch was unlatched and an attempt was made to raise it from its support and to lower it to the command module floor. The hatch could not be lowered the full distance to the floor and was instead pushed to one side. When the inner hatch was opened intense heat and a considerable amount of smoke issued from the interior of the command module.

Apollo 1 crew members inspect equipment before fire (NASA S66-40472).

When the pad leader ascertained that all hatches were open, he left the White Room, proceeded a few feet along the swing arm, donned his headset and reported this fact. From a voice tape it has been determined that this report

came approximately 5 minutes 27 seconds after the first report of the fire. The pad leader estimates that his report was made no more than 30 seconds after the inner hatch was opened. Therefore, it was concluded that all hatches were opened and the two outer hatches removed approximately five minutes after the report of fire or at about 23:36 GMT.

Medical opinion, based on autopsy reports, concluded that chances of resuscitation decreased rapidly once consciousness was lost (about 15 to 30 seconds after the first suit failed) and that resuscitation was impossible by 23:36 GMT. Cerebral hypoxia due to cardiac arrest resulting from myocardial hypoxia caused a loss of consciousness. Factors of temperature, pressure, and environmental concentrations of carbon monoxide, carbon dioxide, oxygen, and pulmonary irritants were changing extremely rapidly. It was impossible to integrate these variables on the basis of available information with the dynamic physiological and metabolic conditions they produced in order to arrive at a precise time when consciousness was lost and death supervened. The combined effect of these environmental factors dramatically increased the lethal effect of any factor by itself.

Visibility within the command module was extremely poor. Although the lights remained on, they could be perceived only dimly. No fire was observed. Initially, the crew was not seen. The personnel who had been involved in removing the hatches attempted to locate the crew without success.

Throughout this period, other pad personnel were fighting secondary fires on level A-8. There was considerable fear that the launch escape tower, mounted above the command module, would be ignited by the fires below and destroy much of the launch complex.

Shortly after the report of the fire, a call was made to the fire department. From log records, it appeared that the fire apparatus and personnel were dispatched at about 23:32 GMT. After hearing the report of the fire, the doctor monitoring the test from the blockhouse near the pad proceeded to the base of the umbilical tower.

The exact time at which firefighters reached Level A-8 is not known. Personnel who opened the hatches unanimously stated that all hatches were open before any firefighters were seen on the level or in the White Room. The first firefighters who reached Level A-8 stated that all hatches were open, but that the inner hatch was inside the command module when they arrived. This placed arrival of the firefighters after 23:36 GMT. It was estimated on the basis of tests that seven to eight minutes were required to travel from the fire station to the launch complex and to ride the elevator from the ground to Level A-8. Thus, the estimated time the firefighters arrived at level A-8 was shortly before 23:40 GMT.

When the firefighters arrived, the positions of the crew couches and crew could be perceived through the smoke but only with great difficulty. An unsuccessful attempt was made to remove the senior pilot from the command module.

Initial observations and subsequent inspection revealed the following facts. The command pilot's couch (the left couch) was in the "170 degree" position, in which it was essentially horizontal throughout its length. The foot restraints and harness were released and the inlet and outlet oxygen hoses were connected to the suit. The electrical adapter cable was disconnected from the communications cable. The command pilot was lying supine on the aft bulkhead or floor of the command module, with his helmet visor closed and locked and with his head beneath the pilot's head rest and his feet on his own couch. A fragment of his suit material was found outside the command module pressure vessel five feet from the point of rupture. This indicated that his suit had failed prior to the time of rupture (23:31:19.4 GMT), allowing convection currents to carry the suit fragment through the rupture.

The senior pilot's couch (the center couch) was in the "96 degree" position in which the back portion was horizontal and the lower portion was raised. The buckle releasing the shoulder straps and lap belts was not opened. The straps and belts were burned through. The suit oxygen outlet hose was connected but the inlet hose was disconnected. The helmet visor was closed and locked and all electrical connections were intact. The senior pilot was lying transversely across the command module just below the level of the hatchway.

The pilot's couch (the couch on the right) was in the "264 degree" position in which the back portion was horizontal and the lower portion dropped toward the floor. All restraints were disconnected, all hoses and electrical connections were intact and the helmet visor was closed and locked. The pilot was supine on his couch.

From the foregoing, it was determined that in all probability the command pilot left his couch to avoid the initial fire, the senior pilot remained in his couch as planned for emergency egress, attempting to open the hatch until his restraints burned through. The pilot remained in his couch

to maintain communications until the hatch could be opened by the senior pilot as planned. With a slightly higher pressure inside the command module than outside, opening the inner hatch was impossible because of the resulting force on the hatch. Thus the inability of the pressure relief system to cope with the pressure increase due to the fire made opening the inner hatch impossible until after cabin rupture. After rupture, the intense and widespread fire, together with rapidly increasing carbon monoxide concentrations, further prevented egress.

Whether the inner hatch handle was moved by the crew cannot be determined because the opening of the inner hatch from the White Room also moves the handle within the command module to the unlatched position.

Immediately after the firefighters arrived, the pad leader on duty was relieved to allow treatment for smoke inhalation. He had first reported over the headset that he could not describe the situation in the command module. In this manner he attempted to convey the fact that the crew was dead to the Test Conductor without informing the many people monitoring the communication channels. Upon reaching the ground the pad leader told the doctors that the crew was dead. The three doctors proceeded to the White Room and arrived there shortly after the arrival of the firefighters. The doctors estimate their arrival to have been at 23:45 GMT. The second pad leader reported that medical support was available at approximately 23:43 GMT. The three doctors entered the White Room and determined that the crew had not survived the heat, smoke, and thermal burns. The doctors were not equipped with breathing apparatus, and the command module still contained fumes and smoke. It was determined that nothing could be gained by immediate removal of the crew. The firefighters were directed to stop removal efforts.

When the command module had been adequately ventilated, the doctors returned to the White Room with equipment for crew removal. It became apparent that extensive fusion of suit material to melted nylon from the spacecraft would make removal very difficult. For this reason it was decided to discontinue removal efforts in the interest of accident investigation and to photograph the command module with the crew in place before evidence was disarranged.

Photographs were taken and the removal efforts resumed at approximately 00:30 GMT, 28 January. Removal of the crew took approximately 90 minutes and was completed about seven and one-half hours after the accident.

Chronology of the Fire

It was most likely that the fire began in the lower forward portion of the left equipment bay, to the left of the command pilot, and considerably below the level of his couch.

Once initiated, the fire burned in three stages. The first stage, with its associated rapid temperature rise and increase in cabin pressure, terminated 15 seconds after the verbal report of fire. At this time, 23:31:19 GMT, the command module cabin ruptured. During this first stage, flames moved rapidly from the point of ignition, traveling along debris traps installed in the command module to prevent items from dropping into equipment areas during tests or flight. At the same time, Velcro strips positioned near the ignition point also burned.

The fire was not intense until about 23:31:12 GMT. The slow rate of buildup of the fire during the early portion of the first stage was consistent with the opinion that ignition occurred in a zone containing little combustible material. The slow rise of pressure could also have resulted from absorption of most of the heat by the aluminum structure of the command module.

The original flames rose vertically and then spread out across the cabin ceiling. The debris traps provided not only combustible material and a path for the spread of the flames, but also firebrands of burning molten nylon. The scattering of these firebrands contributed to the spread of the flames.

By 23:31:12 GMT, the fire had broken from its point of origin. A wall of flames extended along the left wall of the module, preventing the command pilot, occupying the left couch, from reaching the valve that would vent the command module to the outside atmosphere.

Although operation of this was the first step in established emergency egress procedures, such action would have been to no avail because the venting capacity was insufficient to prevent the rapid buildup of pressure due to the fire. It was estimated that opening the valve would have delayed command module rupture by less than one second.

The command module was designed to withstand an internal pressure of approximately 13 pounds per square inch above external pressure without rupturing. Data recorded during the fire showed that this design criterion was exceeded late in the first stage of the fire and that rupture occurred at about 23:31:19 GMT. The point of rupture was where the floor or aft bulkhead of the command module joined the wall, essentially opposite the point of origin of the fire. About three seconds before rupture, at 23:31:16.8 GMT, the final crew communication began. This communication ended shortly after rupture at 23:31:21.8 GMT, followed by loss of telemetry at 23:31:22.4 GMT.

Apollo 1 CM after the fire (NASA S90-35348).

Rupture of the command module marked the beginning of the brief second stage of the fire. This stage was characterized by the period of greatest conflagration due to the forced convection that resulted from the outrush of gases through the rupture in the pressure vessel. The swirling flow scattered firebrands throughout the crew compartment, spreading fire. This stage of the fire ended at approximately 23:31:25 GMT. Evidence that the fire spread from the left side of the command module toward the rupture area was found on subsequent examination of the module and crew suits. Evidence of the intensity of the fire includes burst and burned aluminum tubes in the oxygen and coolant systems at floor level.

This third stage was characterized by rapid production of high concentrations of carbon monoxide. Following the loss of pressure in the command module and with fire now throughout the crew compartment, the remaining atmosphere quickly became deficient in oxygen so that it could not support continued combustion. Unlike the earlier stages where the flame was relatively smokeless, heavy smoke now formed and large amounts of soot were deposited on most spacecraft interior surfaces as they cooled. The third stage of the fire could not have lasted more than a few seconds because of the rapid depletion of oxygen. It was estimated that the command module atmosphere was lethal by 23:31:30 GMT, five seconds after the start of the third stage.

External view of fire damage to the Apollo 1 CM (NASA S67-21295).

Although most of the fire inside the command module was quickly extinguished because of a lack of oxygen, a localized, intense fire lingered in the area of the environmental control unit. This unit was located in the left equipment bay, near the point where the fire was believed to have started. Failed oxygen and water/glycol lines in this area continued to supply oxygen and fuel to support the localized fire that melted the aft bulkhead and burned adjacent portions of the inner surface of the command module heat shield.

The Investigation

Immediately after the accident, additional security personnel were positioned at Launch Complex 34 and the complex was impounded. Prior to disturbing any evidence, numerous external and internal photographs were taken of

the spacecraft. After crew removal, two experts entered the command module to verify switch positions. Small groups of NASA and North American Aviation management, Apollo 204 Review Board members, representatives, and consultants inspected the exterior of Spacecraft 012.

Internal view of fire damage to the Apollo 1 CM (NASA S67-21294).

A series of close-up stereo photographs of the command module was taken to document the as-found condition of the spacecraft systems. After the couches were removed, a special false floor with removable 18-inch transparent squares was installed to provide access to the entire inside of the command module without disturbing evidence. A detailed inspection of the spacecraft interior was then performed, followed by the preparation and approval by the Board of a command module disassembly plan.

Command module 014 was shipped to NASA Kennedy Space Center (KSC) on 1 February 1967 to assist the Board in the investigation. This command module was placed in the Pyrotechnics Installation Building and was used to develop disassembly techniques for selected components prior to their removal from command module 012. By 7 February 1967, the disassembly plan was fully operational. After the removal of each component, photographs were taken of the exposed area. This step-by-step photography was used throughout the disassembly of the spacecraft. Approximately 5,000 photographs were taken.

All interfaces such as electrical connectors, tubing joints, physical mounting of components, etc. were closely inspected and photographed immediately prior to, during, and after disassembly. Each item removed from the command module was appropriately tagged, sealed in clean plastic containers, and transported under the required security to bonded storage.

On 17 February 1967, the Board decided that removal and wiring tests had progressed to a point which allowed moving the command module without disturbing evidence. The command module was moved to the Pyrotechnics Installation Building at KSC, where better working conditions were available.

With improved working conditions, it was found that a work schedule of two eight-hour shifts per day for six days a week was sufficient to keep pace with the analysis and disassembly planning. The only exception to this was a three-day period of three eight-hour shifts per day used to remove the aft heat shield, move the command module to a more convenient workstation and remove the crew compartment heat shield. The disassembly of the command module was completed on 27 March 1967.

Cause of the Apollo I Fire

Although the Board was not able to determine conclusively the specific initiator of the Apollo 204 fire, it identified the conditions that led to the disaster. These conditions were:

1. A sealed cabin, pressurized with an oxygen atmosphere.

2. An extensive distribution of combustible materials in the cabin.

3. Vulnerable wiring carrying spacecraft power.

4. Vulnerable plumbing carrying a combustible and corrosive coolant.

5. Inadequate provisions for the crew to escape.

6. Inadequate provisions for rescue or medical assistance.

Having identified these conditions, the Board addressed the question of how these conditions came to exist. Careful consideration of this question led the Board to the conclusion that in its devotion to the many difficult problems of space travel, the Apollo team failed to give adequate attention to certain mundane but equally vital questions of crew safety. The Board's investigation revealed many deficiencies in design and engineering, manufacture, and quality control.

As a result of the investigation, major modifications in design, materials, and procedures were implemented. The two-piece hatch was replaced by a single quick-operating, outward opening crew hatch made of aluminum and fiberglass. The new hatch could be opened from inside in seven seconds and by a pad safety crew in 10 seconds. Ease of opening was enhanced by a gas-powered counterbalance mechanism. The second major modification was the change in the launch pad spacecraft cabin atmosphere for pre-launch testing from 100 percent oxygen to a mixture of 60 percent oxygen and 40 percent nitrogen to reduce support of any combustion. The crew suit loops still carried 100 percent oxygen. After launch, the 60/40 mix was gradually replaced with pure oxygen until cabin atmosphere reached 100 percent oxygen at 5 pounds per square inch. This "enriched air" mix was selected after extensive flammability tests in various percentages of oxygen at varying pressures.

Other changes included: substituting stainless steel for aluminum in high-pressure oxygen tubing, armor plated water-glycol liquid line solder joints, protective covers over wiring bundles, stowage boxes built of aluminum, replacement of materials to minimize flammability, installation of fireproof storage containers for flammable materials, mechanical fasteners substituted for gripper cloth patches, flameproof coating on wire connections, replacement of plastic switches with metal ones, installation of an emergency oxygen system to isolate the crew from toxic fumes, and the inclusion of a portable fire extinguisher and fire-isolating panels in the cabin.

Safety changes were also made at Launch Complex 34. These included structural changes to the White Room for the new quick-opening spacecraft hatch, improved firefighting equipment, emergency egress routes, emergency access to the spacecraft, purging of all electrical equipment in the White Room with nitrogen, installation of a hand-held water hose and a large exhaust fan in the White Room to draw smoke and fumes out, fire-resistant paint, relocation of certain structural members to provide easier access to the spacecraft and faster egress, addition of a water spray system to cool the launch escape system (the solid propellants could be ignited by extreme heat), and the installation of additional water spray systems along the egress route from the spacecraft to ground level.

Apollo 1 Spacecraft History

EVENT	DATE
Fabrication of spacecraft 012 at North American Aviation, Downey, CA.	Aug 1964
Basic structure completed.	Sept 1965
Installation and final assembly of subsystems completed. Critical design reviews completed. Checkout of all subsystems initiated, followed by integrated testing of all spacecraft subsystems.	Mar 1966
Customer acceptance readiness review completed. NASA issued certificate of flightworthiness and authorized spacecraft to be shipped to KSC.	Aug 1966
Command module received at KSC.	26 Aug 1966
CM-012 mated with service module in altitude chamber. Alignment, subsystems and system certification tests and functional checks performed.	Sept 1966
First combined systems tests completed.	1 Oct 1966
Design certification document issued which certified design as flightworthy, pending satisfactory resolution of open items.	7 Oct 1966
First piloted test at sea level pressure to verify total spacecraft system operation completed.	13 Oct 1966
Unpiloted test at altitude pressures using oxygen to verify spacecraft system operation.	15 Oct 1966
Piloted test with flight crew completed.	19 Oct 1966
Second piloted altitude test with backup crew initiated, but discontinued when failure occurred in oxygen system regulator in spacecraft environmental control system. Regulator removed and found to have design deficiency.	21 Oct 1966
Apollo program director conducted recertification review which closed out majority of open items remaining from previous reviews.	21 Dec 1966
Sea level and unpiloted altitude tests completed.	21 Dec 1966
Piloted altitude test with backup flight crew completed.	30 Dec 1966
Command module removed from altitude chamber.	3 Jan 1967
Spacecraft mated to launch vehicle at Cape Kennedy Launch Complex 34. Various tests and equipment installations and replacements performed.	6 Jan 1967

Apollo I Fire Timeline

Event	GMT Date	GMT Time
Plugs Out Integrated Test initiated when power applied to spacecraft.	27 Jan 1967	12:55
Following completion of initial verification tests of system operation, command pilot entered spacecraft, followed by pilot and senior pilot.		18:00
Count held when command pilot noted odor in spacecraft environmental control system suit oxygen. Sample taken.		18:20
Count resumed after hatch installed.		19:42
Cabin purged with oxygen.		19:45
Open microphone first noted by test crew.		22:25
Count held while communication difficulties checked. Various final countdown functions performed during hold as communications permitted.		22:40
From this time until about 23:53 GMT, flight crew interchanged equipment related to communications systems in effort to isolate communications problem. During troubleshooting period, problems developed with ability of various ground stations to communicate with one another and with crew.		22:45
Final countdown functions up to transfer to simulated fuel cell power completed and count held at T-10 minutes pending resolution of communications problems. For next 10 minutes, no events related to fire. Major activity was routine troubleshooting of communications problem. All other systems operated normally during this period.		23:20
First indication by either cabin pressure or battery compartment sensors of a pressure increase.		23:21:11
Command pilot live microphone transmitted brushing and tapping noises, indicative of movement. Noises similar to those transmitted earlier in test by live microphone when command pilot was known to be moving.		23:30
No voice transmissions from spacecraft from this time until transmission reporting fire.		23:30:14
Slight increase in pulse and respiratory rate noted from senior pilot.		23:30:21
Data from guidance and navigation system indicated undetermined type of crew movement. Gradual rise in oxygen flow rate to crew suits began, indicating movement. Earlier in Plugs Out Integrated Test, crew reported that an unspecified movement caused increased flow rate.		23:30:24
Senior pilot's electrocardiogram indicated muscular activity for several seconds.		23:30:30
Additional electrocardiogram indications from senior pilot. Data show increased activity but were not indicative of alarm type of response. More intense crew activity sensed by guidance and navigation system.		23:30:39
Crew movement ended.		23:30:44
All of senior pilot's biomedical parameters reverted to "rest" level.		23:30:45
Variation in signal output from gas chromatograph.		23:30:50
First voice transmission ended.		23:31:10
Fire broke from its point of origin. Evidence suggests a wall of flames extended along left wall of module, preventing command pilot, occupying left couch, from reaching valve which would vent command module to outside atmosphere. Original flames rose vertically and spread out across cabin ceiling. Scattering of firebrands of molten burning nylon contributed to spread of flames. It was estimated that opening valve would have delayed command module rupture by less than one second.		23:31:12
Cabin pressure exceeded range of transducers, 17 pounds per square inch absolute (psia) for cabin and 21 psia for battery compartment transducers. Rupture and resulting jet of hot gases caused extensive damage to exterior.		23:31:16

Apollo 1 Fire Timeline

Event	GMT Date	GMT Time
Beginning of final voice transmission from crew. Entire transmission garbled. Sounded like, "They're fighting a bad fire—let's get out. Open 'er up." Or, "We've got a bad fire—let's get out. We're burning up." Or, "I'm reporting a bad fire. I'm getting out." Transmission ended with cry of pain, perhaps from pilot.	27 Jan 1967	23:31:16.8
Command module ruptured, start of second stage of fire. First stage marked by rapid temperature rise and increase in cabin pressure. Flames had moved rapidly from point of ignition, traveling along net debris traps installed to prevent items from dropping into equipment areas. At same time, Velcro strips positioned near ignition point also burned.		23:31:19
End of final voice transmission.		23:31:21.8
All spacecraft transmissions ended. Television monitors showed flames spreading from left to right side of command module and shortly covered entire visible area. Telemetry loss made determination of precise times of subsequent occurrences impossible.		23:31:22.4
Third stage of fire characterized by greatest conflagration due to forced convection from outrush of gases through rupture in pressure vessel. Swirling flow scattered firebrands, spreading fire. Pressure in command module dropped to atmospheric pressure five or six seconds after rupture.		23:31:25
Command module atmosphere reached lethal stage, characterized by rapid production of high concentrations of carbon monoxide. Following loss of pressure, and with fire throughout crew compartment, remaining atmosphere quickly became deficient in oxygen and could not support continued combustion. Heavy smoke formed and large amounts of soot deposited on most spacecraft interior surfaces. Although oxygen leak extinguished most of fire, failed oxygen and water/glycol lines supplied oxygen and fuel to support localized fire that melted aft bulkhead and burned adjacent portions of inner surface of command module heat shield.		23:31:30
Fire apparatus and firefighting personnel dispatched.		23:32
Attempts to remove hatches.		23:32:04
Pad leader reported that attempts had started to remove hatches.		23:32:34
Hatches opened, outer hatches removed. Resuscitation of crew impossible.		23:36
Pad leader ascertained all hatches open, left White Room, proceeded a few feet along swing arm, donned headset and reported this fact.		23:36:31
Firefighters arrived at Level A-8. Positions of crew couches and crew could be perceived through smoke but only with great difficulty. Unsuccessful attempt to remove senior pilot from command module.		23:40
Doctors arrived.		23:43
Photographs taken, and removal efforts started.	28 Jan 1967	00:30
Removal of crew completed, about seven and one-half hours after accident.		07:00
Command module 014 shipped to KSC to develop disassembly techniques for selected components prior to their removal from command module 012.	1 Feb 1967	
Disassembly plan fully operational.	7 Feb 1967	
Command module moved to pyrotechnics installation building at KSC, where better working conditions available.	17 Feb 1967	
Disassembly of command module completed.	27 Mar 1967	

APOLLO 7

The First Mission:
Testing the CSM in Earth Orbit

Apollo 7 Summary

(11 October–22 October 1968)

The Apollo 7 crew (l. to. r.): Donn Eisele, Wally Schirra, Walt Cunningham (NASA S68-33744).

Background

Twenty-one months after the Apollo 1 fire, the United States was ready to begin the piloted phase of the Apollo program. The primary objectives of the first mission were:

• to demonstrate CSM and crew performance;

• to demonstrate crew, space vehicle, and mission support facilities performance; and

• to demonstrate CSM rendezvous capability.

The crew members were Captain Walter Marty "Wally" Schirra, Jr. [shi-RAH] (USN), commander; Major Donn Fulton Eisele [EYES-lee] (USAF), command module pilot; and Ronnie Walter "Walt" Cunningham, lunar module pilot.

Selected in the original astronaut group in 1959, Schirra had been pilot of the fifth (third orbital) Mercury mission (MA-8) and command pilot of Gemini 6-A. With Apollo 7, Schirra would become the first person to make three trips into space. Born 12 March 1923 in Hackensack, New Jersey, Schirra was 45 years old at the time of the Apollo 7 mission. Schirra received a B.S. degree from the U.S. Naval Academy in 1945. His backup for the mission was Colonel Thomas Patten Stafford (USAF).

Eisele and Cunningham were each making their first space-flight. Born 23 June 1930 in Columbus, Ohio, Eisele was

38 years old at the time of the Apollo 7 mission. He received a B.S. in astronautics in 1952 from the U.S. Naval Academy, and an M.S. in astronautics in 1960 from the U.S. Air Force Institute of Technology, and was selected as an astronaut in 1963.[1] His backup was Commander John Watts Young (USN).

Born 16 March 1932 in Creston, Iowa, Cunningham was 36 years old at the time of the Apollo 7 mission. He received a B.A. in physics in 1960 and an M.A. in physics in 1961 from the University of California at Los Angeles. He was selected as an astronaut in 1963. His backup was Commander Eugene Andrew "Gene" Cernan (USN).

The capsule communicators (CAPCOMs) for the mission were Stafford, Lt. Commander Ronald Ellwin Evans (USN), Major William Reid Pogue (USAF)[2], John Leonard "Jack" Swigert, Jr. [SWY-girt], Young, and Cernan. The support crew were Swigert, Evans, and Pogue. The flight directors were Glynn S. Lunney (first shift), Eugene F. Kranz (second shift), and Gerald D. Griffin (third shift).

The Apollo 7 launch vehicle was a Saturn IB, an "uprated" Saturn, designated SA-205. The mission also carried the designation Eastern Test Range #66. The CSM combination was designated CSM-101 and formed the first block II configuration spacecraft flown, that is, with the capability to accommodate the LM and other systems advancements.

Launch Preparations

The countdown began at 19:00 GMT on 6 October 1968. There were three planned holds. The first two, at T-72 hours for six hours and at T-33 hours for three hours, allowed sufficient time to fix any spacecraft problems. The final hold, at T-6 hours, provided a rest period for the launch crew. Six hours later, the clock resumed at 09:00 GMT, 11 October 1968.

The final countdown proceeded smoothly until T-10 minutes when thrust chamber jacket chilldown was initiated for the launch vehicle S-IVB stage. The procedure took longer than necessary and would have required a recycling of the clock to T-15 minutes if the proper temperature were not reached in time for initiation of the automatic countdown sequence. As a result, a hold was called at T-6 minutes 15 seconds, and lasted for 2 minutes 45 seconds. Postlaunch analysis determined that chilldown would have occurred without the hold, but the hold was advisable in real-time to meet revised temperature requirements. At 14:56:30 GMT, the countdown resumed and continued to liftoff without further problems.

---

[1] Eisele died of a heart attack 1 December 1987 in Tokyo, Japan (*Houston Chronicle*, 2 Dec 1987, p. 6).

[2] Pogue replaced Major Edward Galen Givens, Jr. (USAF), who died in an automobile accident in Pearland, TX, on 6 June 1967. Givens had been selected in the astronaut class of 1966 (*Houston Chronicle*, 8 Jun 1967).

A large high pressure system centered over Nova Scotia caused high easterly surface winds at launch time. The upper winds, above 30,000 feet, were light from the west. Surface wind speeds were the highest observed for any Saturn vehicle to date. A few scattered clouds were in the area. Cumulonimbus clouds covered 30 percent of the sky with a base at 2,100 feet, visibility 10 statute miles, temperature 82.9° F, relative humidity 65 percent, dew point 70.0° F, barometric pressure 14.765 lb/in², and winds 19.8 knots at 90° from true north measured by the anemometer on the light pole 59.4 feet above ground at the launch site.

Ascent Phase

Apollo 7 was launched from Launch Complex 34 at Cape Kennedy, Florida (USAF Eastern Test Range). Liftoff occurred at Range Zero time of 15:02:45 GMT (11:02:45 a.m. EDT) on 11 October 1968, well within the planned launch window of 15:00:00 to 19:00:00 GMT.

The ascent phase was nominal. Moments after liftoff, the vehicle rolled from a launch pad azimuth of 100° to a flight azimuth of 72° east of north. The first stage provided continuous thrust until center engine cutoff at 000:02:20.65. The outboard engine shut down 3.67 seconds later at an Earth-fixed velocity of 6,479.1 ft/sec. Cutoff conditions were very close to prediction.

The S-IB was separated from the upper stage at 000:02:25.59, followed by S-IVB engine ignition at 000:02:26.97. Cutoff occurred at 00:10:16.76, with deviations from the planned trajectory of only 2.3 ft/sec in velocity and 0.054 n mi in altitude. The S-IVB burn time of 469.79 seconds was within one second of prediction, and all structural load limits were well within design tolerances during ascent.

The maximum wind conditions encountered during ascent were 81 knots at 172,000 feet. Wind shear in the high dynamic pressure region reached 0.0113 sec⁻¹ in the pitch plane at 48,100 feet. The maximum wind speed in the high dynamic pressure region was 30.3 knots from 309° at 44,500 feet.

The probable impact of the spent S-IB was determined from a theoretical, tumbling, free flight trajectory. Assuming the booster remained intact during entry, the impact occurred in the Atlantic Ocean at latitude 29.76° north and longitude 75.72° west, 265.01 n mi from the launch site.

At 000:10:26.76, the spacecraft entered Earth orbit, defined as S-IVB cutoff plus 10 seconds to account for engine tailoff and other transient effects. At insertion, conditions

were: apogee and perigee 153.7 by 123.3 n mi, inclination 31.58°, period 89.70 minutes, and velocity 25,538.6 ft/sec.

Apollo 7's Saturn IB lifts off from Cape Canaveral Pad 34 (NASA S68-48778).

The international designation for the spacecraft upon achieving orbit was 1968-089A and the S-IVB was designated 1968-089B.[3]

Inflight Activities

The crew adapted quickly and completely to the weightless environment. There were no disorientation problems associated with movement inside the CM nor looking out the windows at Earth. In fact, an attempt by the lunar module pilot to induce vertigo or motion sickness by movement of the head in all directions at rapid rates met with negative results. Early in the mission, however, the crew reported some soreness of their back muscles in the kidney area. The soreness was relieved by exercise and hyperextension of the back.

Prior to separation from the S-IVB, a 2-minute 56-second manual takeover of attitude control from the launch vehicle stage was performed at 002:30:48. The crew exercised the

[3] *RAE Table of Earth Satellites 1957-1986*, pps. vii, and viii. The international Committee on Space Research (COSPAR) has given all satellites a designation based on the year of launch (first four digits) and number of successful launches during that year (next three digits). In COSPAR terminology, the letter A usually refers to the instrumented spacecraft, B to the rocket, and C, D, E, etc. to fragments.

manual S-IVB/IU orbital attitude control capability. This consisted of a test of the closed loop spacecraft/launch vehicle control system by performing manual pitch, roll, and yaw maneuvers. The control system responded properly. After completion of the test, the crew switched attitude control back to the automatic launch vehicle system which resumed the normal attitude timeline. By the time the CSM/S-IVB separated at 002:55:02, venting of S-IVB propellants had raised the orbit to 167.0 by 125.3 n mi.

One objective of Apollo 7 was to perform a "safing" of the S-IVB stage by lowering pressure in the propellant tanks and high-pressure bottles to a level that would permit safe rendezvous and simulated docking maneuvers. The safing was scheduled to take place in several stages. First, the LH_2 tank safing was to be performed by three pre-programmed ventings; however, four additional ventings were required because the pre-programmed ones did not adequately safe the tank under the orbital conditions experienced. The first venting occurred at 000:10:17, and the final one ended at 005:11:15. The seven ventings totaled 3,274.1 seconds. Second, a liquid oxygen dump was initiated at 001:34:28 and lasted 721.00 seconds. Third, a cold helium dump was performed at 001:42:28 and again at 004:30:16, lasting 2,868.00 and 1,199.99 seconds, respectively. Finally, a stage control sphere helium dump occurred at 003:17:33, but was terminated by ground command after 2,967 seconds to save the remaining helium for control of the LH_2 tank vent-and-relief valve. Safing, however, was adequately accomplished.

During the second revolution the crew observed that one of the spacecraft/LM adapter panels on the S-IVB was deployed only 25° instead of the normal 45°. It had opened fully, but a retention cable designed to prevent the panel from closing had become stuck and the panel had partially closed. This was not a problem because the panels would be jettisoned on future missions. By the 19th revolution, the panel had moved to the full open position.

In order to establish conditions required for rendezvous with the S-IVB, a 16.3-second phasing maneuver was performed at 003:20:09 using the service module reaction control system. This resulted in an orbit of 165.2 by 124.8 n mi.

The phasing burn was intended to place the spacecraft 76.5 n mi ahead of the S-IVB. However, the S-IVB orbit decayed more rapidly than anticipated during the six subsequent revolutions. An additional phasing maneuver of 17.6 seconds was performed at 015:52:00 to obtain the desired conditions. The resulting orbit was 164.7 by 120.8 n mi.

Rendezvous operations with the S-IVB stage (NASA AS07-03-1541).

At 014:46, it was reported that the commander had developed a bad head cold, which had begun about one hour after liftoff, and that he had taken two aspirins. The next day, the other two crew members also experienced head cold symptoms. This condition, which continued throughout the mission, caused extreme discomfort because it was very difficult to clear the ears, nose, and sinuses in "zero g" conditions. Medication was taken, but the symptoms persisted.

At 023:33, the spacecraft commander canceled the first television transmission, scheduled to begin in 20 minutes. Annoyed that mission control had added two burns and a urine dump to the crew's workload while they were testing a new vehicle, and still suffering from a cold, Schirra reported that, "...TV will be delayed without further discussion..."

Two service propulsion system firings were required for rendezvous with the S-IVB. The first firing, a 9.26-second corrective combination maneuver at 026:24:55, was necessary to achieve the desired 1.32° phase and 8.0 nautical mile altitude offset so that the second firing would produce an orbit coelliptic with that of the S-IVB. The result was an orbit of 194.1 by 123.0 n mi. During this period, the sextant was used to track the S-IVB, which was visible in reflected sunlight. The 7.76-second firing at 028:00:56 occurred when the spacecraft was 80 n mi behind and 7.8 n mi below the S-IVB, and created a more circularized orbit of 153.6 by 113.9 n mi.

The two firings achieved the desired conditions for the 46-second rendezvous terminal phase initiation, which occurred at 029:16:33, about four and one half minutes earlier than planned because of a minor variation in the orbit. A small midcourse correction was made at 029:37:48, followed by a 708-second braking maneuver at 029:43:55, and final closure to within 70 feet of the tumbling S-IVB. Stationkeeping was performed for 25 minutes starting at 029:55:43 in an orbit of 161.0 by 122.1 n mi, after which a 5.4-second service module reaction control system posigrade maneuver removed the CSM from the vicinity of the S-IVB stage. The crew maneuvered the CSM around the S-IVB in order to inspect and photograph it.

The rendezvous maneuver was important because it demonstrated the ability of the spacecraft to rendezvous with the LM (represented by the S-IVB) if the ascent stage became disabled after leaving the lunar surface. However, the crew reported that the manually-controlled braking maneuver was frustrating because no reliable backup ranging information was available, as would be the case during an actual rendezvous with the LM.

The next 24-hour period was devoted to a sextant calibration test at 041:00, two attitude control tests at 049:00 and 050:40, and two primary evaporator tests at 049:50 and 050:40. In addition, the crew performed a rendezvous navigation test, using the sextant to track the S-IVB visually to a distance of 160 n mi at 044:40 and to 320 n mi at 053:20. The crew later reported sighting the S-IVB at a range of nearly 1,000 n mi.

To ensure maximum return from Apollo 7, it was planned to complete as many primary and secondary objectives as possible early in the flight, and, by the end of the second day, more than 90 percent had been accomplished.

Three tests of the rendezvous radar transponder were performed. This system would be essential for docking the LM ascent stage to the CM after liftoff from the lunar surface. The first two tests occurred at 061:00 and 071:40. The third was performed during revolution 48 at 076:27, when the ground radar at White Sands Missile Range, New Mexico, acquired and locked onto the spacecraft transponder at a range of 390 n mi and tracked it to 415 n mi.

At 071:43, the first of seven television transmissions began and lasted for seven minutes. It was the first live television transmission from a piloted American spacecraft. The crew opened the telecast with a sign that read "From the lovely Apollo room high atop everything." They then aimed the camera out the window as the spacecraft passed over New Orleans and then over the Florida peninsula. The orbital motion of the spacecraft was evident.

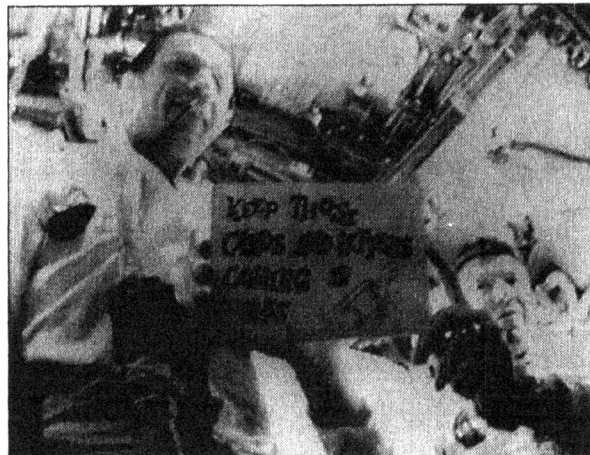

Message to the world from the Apollo 7 crew during first live television transmission (NASA S68-50713).

The service propulsion system was fired six additional times during the mission. The third firing, at 075:48:00 (advanced 16 hours from the original plan), was a 9.10-second maneuver controlled by the stabilization and control system. The maneuver was performed early to increase the backup deorbit capability of the service module reaction control system by lowering the perigee to 90 n mi and placing it in the northern hemisphere. The resulting orbit was 159.7 by 89.5 n mi.

After the third firing, a three-hour cold soak of the service propulsion thermal control system was performed. The cold soak stabilized the spacecraft and exposed one side away from the Sun for a period of time to lower the temperature and monitor the effects of the cold space environment. The thermal characteristics of the system were better than anticipated for random, drifting flight, because the temperature decrease was less than predicted.

A test to determine whether the environmental control system radiator surface coating had degraded was conducted between 092:37 and 097:00. Results indicated that the solar absorptivity of the radiator panel tested was within predicted limits, and validated the system for lunar flight.

The second television transmission started at 095:25 and lasted about 11 minutes. The program included a tour of the CM including various controls, a demonstration of the exercise device, and an attempt to show water condensation inside the spacecraft.

Condensation was a major problem associated with the cabin and suit circuits. This problem was anticipated in the cabin because the cold coolant lines from the radiator to the environment control unit and from the environment control unit to the inertial measurement unit were not insulated. Each time excessive condensation was noted on the coolant lines or in a puddle on the aft bulkhead after service propulsion system maneuvers, the crew vacuumed the water overboard. Experiment S005 (Synoptic Terrain Photography) began at 098:40, using a hand-held modified 70 mm Hasselblad 500C camera. The photographs were used to study the origin of the Carolina bays in the United States, wind erosion in desert regions, coastal morphology, and the origin of the African rift valley. Near-vertical, high-sun-angle photographs of Baja California, other parts of Mexico, and parts of the Middle East were useful for geologic studies. Photographs of New Orleans and Houston were generally better for geographic urban studies than those available from previous programs.

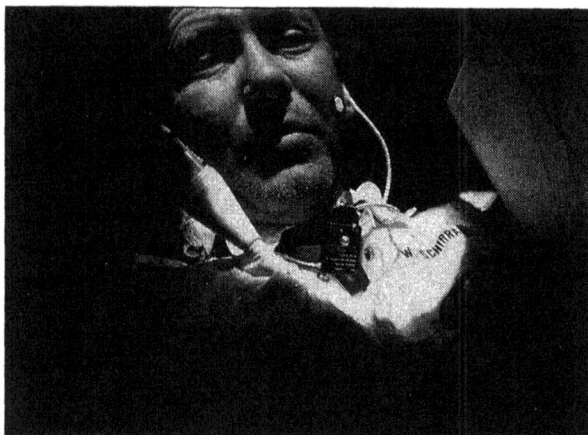

Mission commander Wally Schirra (NASA AS07-04-1582).

Areas of oceanographic interest, particularly islands in the Pacific Ocean, were photographed for the first time. In addition, the mission obtained the first extensive photographic coverage of northern Chile, Australia, and other areas. Of the 500 photographs taken of land and ocean areas, approximately 200 were usable, and, in general, the color and exposure were excellent. The need to change the film magazines, filters, and exposure settings hurriedly when a target came into view, and to hold the camera steady, accounted for the improper exposure of many frames.

The purpose of Experiment S006 (Synoptic Weather Photography) was to photograph as many as possible of 27 basic categories of weather phenomena, and began at 099:10. The camera was the same used for Experiment S005. Of the 500 photographs taken, approximately 300 showed clouds or other items of meteorological interest, and approximately 80 contained features of interest in oceanography. Categories considered worthy of additional interest included

weather systems, winds and their effects on clouds, ocean surfaces, underwater zones of Australian reefs, the Pacific atolls, the Bahamas and Cuba, landform effects, climactic zones, and hydrology. Oceanographic surface features were revealed more clearly than in any of the preceding piloted flights. The photographs of Hurricane Gladys and Typhoon Gloria, taken on 17 October and 20 October 1968, respectively, were the best-to-date views of tropical storms. Image sharpness of photographs for this experiment ranged from fair to excellent, again affected by the difficulty in holding the camera steady. Regardless, ocean swells could be resolved from altitudes near 100 n mi.

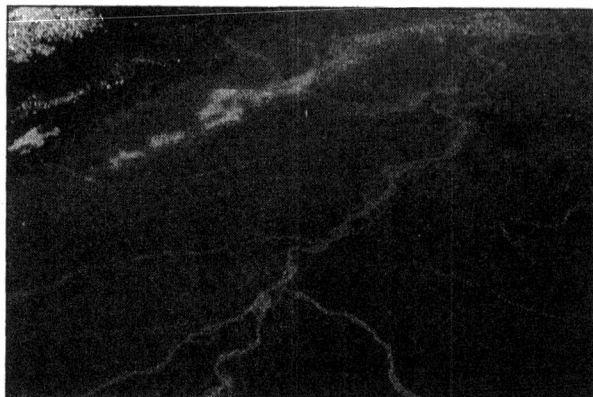

Example of Synoptic Terrain Photography: India, Nepal, Tibet, and Himalayas from 126 n mi altitude (NASA AS07-11-1980).

The third television transmission began at 119:08 and lasted about ten minutes. It featured a demonstration of how to prepare food in space, in particular a package of dried fruit juice reconstituted with water. The telecast also showed the process of vacuuming water that had accumulated on the cold glycol lines. Various controls at the commander's workstation were also viewed.

The fourth service propulsion system firing, at 120:43:00.44, was performed to evaluate the minimum-impulse capability of the service propulsion engine. It lasted only 0.48 seconds and produced an orbit of 156.7 by 89.1 n mi.

A tour of the CM, the fourth television transmission, began at 141:11. The crew trained their camera on deposits on window 1 and on optical site markings used to measure pitch angle on window 2. Panning around the spacecraft, the camera gave viewers a look at sleep stations, stowage areas, helmet bags and pressure suit hoses. The commander also demonstrated weightlessness by blowing on a floating pen to control its motion. By 141:27, the crew had signed off and the transmission signal had faded.

Example of Synoptic Weather Photography: a view of Hurricane Gladys over the Pacific Ocean at an altitude of 99 n mi (NASA AS07-07-1877).

During this time, the S-IVB stage continued to orbit the Earth. It impacted the Indian Ocean at 09:30 GMT on 18 October. The estimated impact point was latitude 8.9° south and longitude 81.6° east.

A fifth service propulsion system firing was performed to position the spacecraft for an optimum deorbit maneuver at the end of the planned orbital phase by allowing at least two minutes of tracking by the Hawaii ground station if another orbit were required. This occurred at 165:00:00.42. To ensure verification of the propellant gauging system, the firing duration was increased from the original plan.

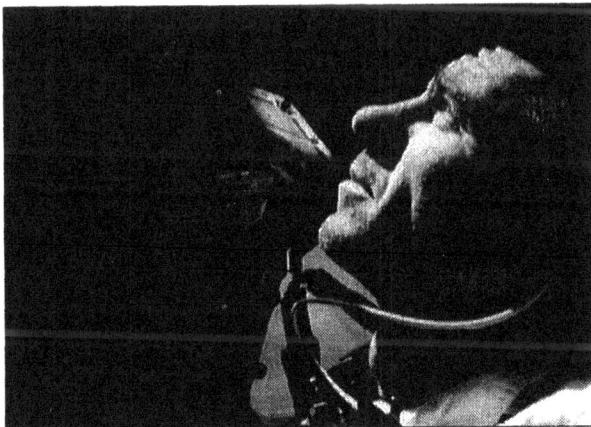

CMP Walt Cunningham peers out the spacecraft window (NASA AS07-04-1584).

The 66.95-second maneuver produced the largest velocity change of the mission, 1,691.3 ft/sec, and incorporated a manual thrust-vector-control takeover halfway through the maneuver. The resulting orbit was 244.2 by 89.1 n mi.

During translunar and transearth flight on future missions, it would be necessary to put the spacecraft into a slow "barbecue" roll to maintain an even external temperature. This maneuver, called passive thermal control, was tested twice on Apollo 7, first at 167:00 and next at 212:00.

The fifth television transmission, starting at 189:04, featured another spacecraft tour. The program began with a view of the instrument panel including attitude thruster switches and the display keyboard, and cryogenic controls, and ended with the crew performing a military "close order drill." An attempt to show scenes of Earth was unsuccessful.

The sixth SPS firing was performed during the eighth day, at 210:07:59, and was the second minimum-impulse maneuver. At the time, the apogee was 234.6 n mi and the perigee was 88.4 n mi. This firing lasted 0.50 seconds and was directed out-of-plane because no change in orbit was desired.

For the sixth television transmission, starting at 213:10, the crew aimed the camera out the window and gave ground controllers a view of the Florida peninsula. They then turned the camera inside the spacecraft to show off the beards they had grown during the mission.

At 231:08, the Solar Particle Alert Network facility at Carnarvon, Australia, detected a Class 1B solar flare. Analysis of data confirmed the flare would have no effect on the spacecraft or crew. However, this exercise proved to be an excellent checkout of the systems and procedures that would be used in the event of a solar flare during a lunar mission. This event was followed by the seventh service propulsion system firing, a 7.70-second maneuver at 239:06:11, which placed the spacecraft perigee at the proper longitude for entry and recovery, and lowered the orbit to 229.8 by 88.5 n mi.

For the final television transmission, starting at 236:18 and lasting for about 11 minutes, the crew showed off their beards again, and reported seeing several jet contrails far below them over the Gulf Coast. They also described the bands of color created by the day air glow above the Earth.

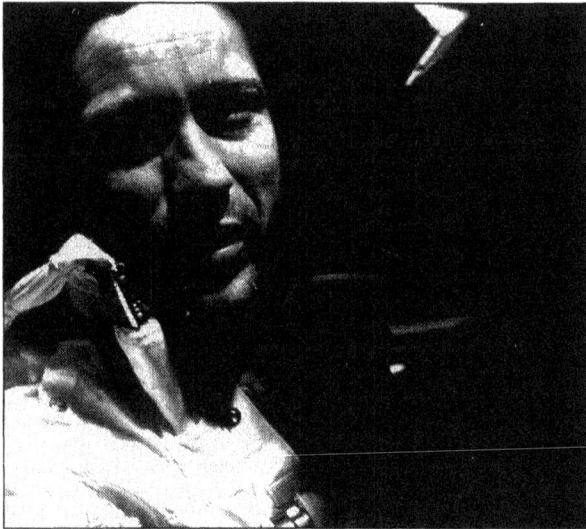

LMP Donn Eisele poses for a photo (NASA AS07-04-1583).

The midcourse navigation program, using the Earth horizon and a star, could not be accomplished because the Earth horizon was indistinct and variable. The air glow was about three degrees wide and had no distinct boundaries or lines when viewed through the sextant. This problem seemed to be associated with the spacecraft being in a low Earth orbit. Using this same program on lunar landmarks and a star, however, the task was very easy to perform. Lunar landmarks showed up nearly as well as Earth landmarks. Stars could be seen at 10° and 15°, and greater, from the Moon.

Sextant/star counts and star checks and star/horizon sightings were made throughout the mission; lunar landmark/star sightings were attempted at 147:00.

Recovery

The final day of the mission was devoted primarily to preparations for the deorbit maneuver. This was accomplished by the eighth SPS firing, an 11.79-second eighth service maneuver at 259:39:16 over Hawaii, during the 163rd orbit. During the final orbit, the apogee was 225.3 n mi, the perigee was 88.2 n mi, the period was 90.39 minutes, and the inclination 29.88°.

Because of their cold symptoms, there was a considerable amount of discussion about whether the crew should wear helmets and gloves during entry. With helmets on, it might be impossible to properly clear the throat and ears as increasing gravity drew mucus down from the head area, where it remained during zero gravity conditions. It was

decided 48 hours prior to entry, and at the crew's insistence, that helmets and gloves would not be worn.

The service module was jettisoned at 259:43:33, and the CM entry followed both automatic and manually guided profiles. The command module reentered the Earth's atmosphere (400,000 feet altitude) at 259:53:26 at a velocity of 25,846 ft/sec. Trajectory reconstruction indicated that the service module impacted the Atlantic Ocean at 260:03:00 at a point estimated to be latitude 29° north and longitude 72° west. During entry, three objects—the CM, the service module, and a 12-foot insulation disk between the two—were tracked simultaneously and were also sighted visually.

The parachute system effected a soft splashdown of the CM in the Atlantic Ocean southeast of Bermuda at 11:11:48 GMT (07:11:48 a.m. EDT) on 22 October 1968. Mission duration was 260:09:03. The impact point was 1.9 n mi from the target point and 7 n mi from the recovery ship U.S.S. *Essex*. The splashdown site was estimated to be latitude 27.63° north and longitude 64.15° west. After splashdown, the CM assumed an apex-down flotation attitude, but was successfully returned to the normal flotation position within 13 minutes by the inflatable bag uprighting system. During this period, the recovery beacon was not visible and voice communication with the crew was interrupted.

The crew was retrieved by helicopter and was aboard the recovery ship 56 minutes after splashdown. The CM was recovered 55 minutes later. The estimated CM weight at splashdown was 11,409 pounds, and the estimated distance traveled for the mission was 3,953,842 n mi.

After splashdown, Wally Schirra exits the command module with the aid of a Navy support team member (NASA S68-49529).

Apollo 7 crew safely aboard recovery ship U.S.S. *Essex* following successful mission (NASA S68-49744).

At CM retrieval, the weather recorded onboard the *Essex* showed light rain showers, 600-foot ceiling; visibility 2 n mi; wind speed 16 knots from 260° true north; air temperature 74° F; water temperature 81° F; with waves to 3 feet from 260° true north.

The CM was offloaded from the *Essex* on 24 October at the Norfolk Naval Air Station, Norfolk, Virginia, and the Landing Safing Team began the evaluation and deactivation procedures at 14:00 GMT. Deactivation was completed at 01:30 GMT on 27 October 1968. The CM was then flown to Long Beach, California and trucked to the North American Rockwell Space Division facility at Downey, California for postflight analysis.

Conclusions

The Apollo 7 mission was successful in every respect. All spacecraft systems operated satisfactorily, and all but one of the detailed test objectives were met. As an engineering test flight, Apollo 7 demonstrated the performance of the orbital safing experiment, the adequacy of attitude control in both the manual and automatic modes, and that the vehicle systems could perform for extended periods in orbit. For the first time, a mixed cabin atmosphere consisting of 65 percent oxygen and 35 percent nitrogen was used aboard an American piloted spacecraft. All previous flights had used 100 percent oxygen, a procedure changed as a result of recommendations made by the Apollo 1 fire investigation board. Another "first" was the availability of hot and cold drinking water for the crew as a by-product of the service module fuel cells, an important element for piloted lunar excursions. Consumables usage was maintained at safe levels, and permitted the introduction of additional flight activities toward the end of the mission.

The most significant aerodynamic effect encountered was the unexpected phenomenon noted as "perigee torquing," a rotation of the CSM most noticeable when the perigee was at 90 n mi.

The following conclusions were made from an analysis of post-mission data:

1. The results of the Apollo 7 mission, when combined with results of previous missions and ground tests, demonstrated that the CSM was qualified for operation in the Earth orbital environment and was ready for tests in the cislunar and lunar orbital environments.

2. The concepts and operational functioning of the crew/spacecraft interfaces, including procedures, provisioning, accommodations, and displays and controls, were acceptable.

3. The overall thermal balance of the spacecraft, for both active and passive elements, was more favorable than predicted for the near-Earth environment.

4. The endurance required for systems operation on a lunar mission was demonstrated.

5. The capability of performing rendezvous using the CSM, with only optical and onboard data, was demonstrated; however, it was determined that ranging information would be extremely desirable for the terminal phase.

6. Navigation techniques in general were demonstrated to be adequate for lunar missions. Specifically:

   a. Onboard navigation using the landmark tracking technique proved feasible in Earth orbit.

   b. The Earth horizon was not usable for optics measurements in low Earth orbit with the available optics design and techniques.

   c. Although a debris cloud of frozen liquid particles following venting obscured star visibility with the scanning telescope, it could be expected to dissipate rapidly in Earth orbit without significantly contaminating the optical surfaces.

   d. Star visibility data with the scanning telescope indicated that in cislunar space, with no venting and with proper spacecraft orientation to shield the optics from the Sun and Earth or Moon light, constellation recognition would be adequate for platform inertial orientation.

   e. Sextant star visibility was adequate for platform realignments in daylight using Apollo navigation stars as close as 30° from the Sun line-of-sight.

7. The rendezvous radar acquisition and tracking test demonstrated the capability of performance at ranges required for rendezvous between the CSM and the LM.

8. Mission support facilities, including the Piloted Space Flight Network and the recovery forces were satisfactory for an Earth orbital mission.

Apollo 7 Objectives[4]

Launch Vehicle Primary Detailed Objectives

1. To demonstrate the adequacy of the launch vehicle attitude control system for orbital operation. *Achieved.*

2. To demonstrate S-IVB orbital safing capability. *Achieved.*

3. To evaluate S-IVB J-2 engine augmented spark igniter line modifications. *Achieved.*

Launch Vehicle Secondary Detailed Test Objectives

1. To evaluate the S-IVB/instrument unit orbital coast lifetime capability. *Achieved.*

2. To demonstrate command and service module piloted launch vehicle orbital attitude control. *Achieved.*

Spacecraft Primary Objectives

1. To demonstrate command and service module and crew performance. *Achieved.*

2. To demonstrate crew, space vehicle, and mission support facilities performance. *Achieved.*

3. To demonstrate command and service module rendezvous capability. *Achieved.*

Spacecraft Primary Detailed Test Objectives

1. P1.6: To perform inertial measurement unit alignments using the sextant. *Achieved.*

2. P1.7: To perform an internal measurement unit orientation determination and a star pattern daylight visibility check. *Achieved.*

3. P1.8: To perform onboard navigation using the technique of the scanning telescope landmark tracking. *Achieved.*

4. P1.10: To perform optical tracking of a target vehicle using the sextant. *Achieved, during rendezvous.*

5. P1.12: To demonstrate guidance navigation control system automatic and manual attitude controlled reaction control system maneuvers. *Partially achieved, by the automatic mode prior to the service propulsion system burns and the manual mode. Although all required modes were demonstrated, all rates were not checked.*

6. P1.13: To perform guidance navigation control system controlled service propulsion system and reaction control system velocity maneuvers. *Achieved, at various times during the mission.*

7. P1.14: To evaluate the ability of the guidance navigation control system to guide the entry from Earth orbit. *Achieved, during entry.*

8. P1.15: To perform star and Earth horizon sightings to establish an Earth horizon model. *Not achieved. On the two occasions attempted, the Earth horizon was indistinct and variable, with no defined boundaries or lines, thus precluding obtaining the necessary data.*

9. P1.16: To obtain inertial measurement unit performance data in the flight environment. *Achieved, in conjunction with the inertial measurement unit alignment checks. Two pulse integrating pendulous accelerometer bias tests were also performed.*

10. P2.3: To monitor the entry monitoring system during service propulsion velocity changes and entry. *Achieved, during the first service propulsion service burn and entry.*

11. P2.4: To demonstrate the stabilization control system automatic and manual attitude controlled reaction control system maneuvers. *Achieved, except for testing the high and auto rate modes.*

12. P2.5: To demonstrate the command and service module stabilization control system velocity control capability. *Achieved.*

13. P2.6: To perform a manual thrust vector control takeover. *Achieved.*

14. P2.7: To obtain data on the stabilization control systems capability to provide a suitable inertial reference in a flight environment. *Achieved, during the zero-g phase of the mission prior to the fourth service propulsion system burn and prior to the S-IVB separation. Desired data during the boost phase was not obtained.*

15. P2.10: To accomplish the backup mode of the gyro display coupler-flight director attitude indicator alignment using the scanning telescope in preparation for an increment velocity maneuver. *Achieved, although there was a problem with the flight director attitude indicator in the latter part of the mission.*

16. P3.14: To demonstrate the service propulsion system minimum impulse burns in a space environment. *Achieved, during the fourth and sixth service propulsion burns.*

[4] Apollo objectives and their level of achievement for all flights are derived from mission reports and from Boeing's final flight evaluation reports for Apollo 7, 8, 9, and 10.

17. P3.15: To perform a service propulsion system performance burn in the space environment. *Achieved, during the fifth service propulsion burn.*

18. P3.16: To monitor the primary and auxiliary gauging system. *Achieved, during the fifth service propulsion burn.*

19. P3.20: To verify the adequacy of the propellant feed line thermal control system. *Achieved, by the demonstration of normal operation and the cold soak test.*

20. P4.4: To verify the life support functions of the environmental control system. *Achieved.*

21. P4.6: To obtain data on operation of the waste management system in the flight environment. *Achieved.*

22. P4.8: To operate the secondary coolant loop. *Achieved, and included daily redundant component tests.*

23. P4.9: To demonstrate the water management subsystems operation in the flight environment. *Achieved, throughout the mission, despite a problem with the chlorination procedure and some hardware problems.*

24. P4.10: To demonstrate the postlanding ventilation circuit operation. *Achieved.*

25. P5.8: To obtain data on thermal stratification with and without the cryogenic fans of the cryogenic gas storage system. *Achieved. Although only two of the three stratification tests were successful and part of the third test was accomplished (the rest was deleted), sufficient data were obtained.*

26. P5.9: To verify automatic pressure control of the cryogenic tank systems in a zero-g environment. *Achieved.*

27. P5.10: To demonstrate fuel cell water operations in a zero-g environment. *Achieved.*

28. P6.7: To demonstrate S-band data uplink capability. *Achieved.*

29. P6.8: To demonstrate a simulated command and service module overpass of the lunar module rendezvous radar during the lunar stay. *Achieved, during the 48th revolution.*

30. P7.19: To obtain data on the environmental control system primary radiator thermal coating degradation. *Achieved, from 092:37 to 097:00.*

31. P7.20: To obtain data on the block II forward heat shield thermal protection system. *Achieved, during entry.*

32. P20.8: To perform a command and service module/S-IVB separation, transposition, and simulated docking. *Achieved.*

33. P20.10: To demonstrate the performance of the command and service module/Piloted Space Flight Network S-band communication system. *Achieved.*

34. P20.11: To obtain data on all command and service module consumables. *Achieved.*

35. P20.13: To perform a command and service module active rendezvous with the S-IVB. *Achieved.*

36. P20.15: To obtain crew evaluation of intravehicular activity in general. *Achieved.*

Spacecraft Secondary Detailed Test Objectives

1. S1.11: To monitor the guidance navigation control systems and displays during launch. *Achieved.*

2. S3.17: To obtain data on the service module reaction control subsystem pulse and steady state performance. *Achieved.*

3. S7.24: To obtain data on initial coning angles when in the spin mode as used during transearth flight. *Partially achieved. The first of three tests was accomplished. A pitch control mode was also accomplished but was not planned prior to launch. The third test was deleted (the crew objected because they expected excessive cross-coupling).*

4. S7.28: To obtain command and service module vibration data. *Achieved, during boost, powered flight, and deorbit.*

5. S20.9: To perform manual out-of-window command and service module attitude orientation for retrofire. *Achieved, by two tests.*

6. S20.12: To perform crew controlled manual S-IVB attitude maneuvers in three axes. *Achieved.*

7. S20.14: To verify that the launch vehicle propellant pressure displays are adequate to warn of a common bulkhead reversal. *Achieved.*

8. S20.16: To obtain photographs of the command module rendezvous windows during discrete phases of the mission. *Achieved, although the second and third of four scheduled tests were deleted.*

9. S20.17: To obtain data on propellant slosh damping following service propulsion system cutoff and following reaction control subsystem burns. *Achieved, by three tests.*

10. S20.18: To obtain data via the command and service module/Apollo range instrumentation aircraft communication subsystems. *Achieved.*

11. S20.19: To demonstrate command and service module VHF voice communications with the Manned Space Flight Network. *Achieved, throughout the mission and during recovery.*

12. S20.20: To evaluate the crew optical alignment sight for docking, rendezvous, and proper attitude verification. *Achieved, throughout the mission and in conjunction with deorbit attitude.*

13. S7.21: To obtain data on the service module lunar module adapter deployment system operation. *Achieved.*

Experiments

1. S005 (Synoptic Terrain Photography): To obtain elective, high quality photographs with color and panchromatic film of selected land and ocean areas. *Achieved. Of the more than 500 photographs obtained, approximately 200 were usable for the purposes of the experiment. The objective of comparing color with black-and-white photography of the same areas was not successful because of problems with focus, exposure, and filters.*

2. S006 (Synoptic Weather Photography): To obtain selective, high quality color cloud photographs to study the fine structure of Earth's weather system. *Achieved. In particular, excellent views of Hurricane Gladys and Typhoon Gloria were obtained. The color photographs enabled meteorologists to ascertain much more accurately the types of clouds involved than with black-and-white satellite photographs. Oceanographic surface features were also revealed more clearly than in any of the preceding piloted flights.*

3. M006: To establish the occurrence and degree of bone demineralization during long spaceflights. *Achieved, by preflight and postflight x-ray studies of selected bones of crew members.*

4. M011: To determine if the space environment fosters any cellular changes in human blood. *Achieved, by comparison of preflight and postflight crew blood samples.*

5. M023: To measure changes in lower body negative pressure as evidence of cardiovascular deconditioning resulting from prolonged weightlessness. *Achieved, by preflight and postflight medical examinations.*

Test Objectives Added During Mission

1. Pitch about Y axis. *Achieved.*

2. Optics degradation evaluation. *Achieved.*

3. Sextant/horizon sightings. *Not achieved. Erroneous procedures were given to the crew.*

4. Three additional S-band communication modes. *Achieved.*

Apollo 7 Spacecraft History[5]

EVENT	DATE
Individual and combined CM and SM systems test completed at factory.	18 Mar 1968
Saturn IB stage delivered to KSC.	28 Mar 1968
Saturn IV-B stage delivered to KSC.	7 Apr 1968
Saturn IB instrument unit delivered to KSC.	11 Apr 1968
Integrated CM and SM systems test completed at factory.	29 Apr 1968
CM #101 and SM #101 ready to ship from factory to KSC.	29 May 1968
CM #101 and SM #101 delivered to KSC.	30 May 1968
CM #101 and SM #101 mated.	11 Jun 1968
CSM #101 combined systems test completed.	19 Jun 1968
CSM #101 altitude tests completed.	29 Jul 1968
Space vehicle moved to Cape Kennedy Launch Complex 34.	9 Aug 1968
CSM #101 integrated systems test completed.	27 Aug 1968
CSM #101 electrically mated to launch vehicle.	30 Aug 1968
Space vehicle overall test completed.	4 Sep 1968
Space vehicle countdown demonstration test completed.	17 Sep 1968
Space vehicle flight readiness test completed.	25 Sep 1968

Apollo 7 Ascent Phase

Event	GET (hhh:mm:ss)	Altitude (n mi)	Range (n mi)	Earth Fixed Velocity (ft/sec)	Space Fixed Velocity (ft/sec)	Event Duration (deg E)	Geocentric Latitude (deg N)	Longitude (deg E)	Space Fixed Flight Path Angle (deg)	Space Fixed Heading Angle (E of N)
Liftoff[6]	000:00:00.36	0.019	0.000	0.0	1,341.7		28.3608	-80.5611	0.06	90.01
Mach 1 achieved	000:01:02.15	4.120	0.753	1,039.1	1,960.1		28.3649	-80.5477	29.63	86.70
Maximum dynamic pressure	000:01:15.5	6.567	1.933	1,459.4	2,408.8		28.3708	-80.5264	31.64	83.65
S-IB center engine cutoff	000:02:20.65	30.626	29.184	6,264.7	7,394.5	123.64	28.5090	-80.0349	27.09	75.87
S-IB outboard engine cutoff	000:02:24.32	32.678	32.418	6,479.1	7,616.8	147.31	28.5252	-79.9765	26.55	75.78
S-IB/S-IVB separation[7]	000:02:25.59	33.389	33.561	6,472.1	7,612.6		28.5310	-79.9558	26.32	75.79
S-IVB engine cutoff	000:10:16.76	123.167	983.290	24,181.2	25,525.9	469.79	31.3633	-61.9777	0.00	85.91
Earth orbit insertion	000:10:26.76	123.177	1,121.743	24,208.5	25,553.2		31.4091	-61.2293	0.005	86.32

---

[5] There are conflicts in NASA literature regarding the history of Apollo hardware. Where conflicts exist, the author has used the dates that appear to be most logical. The sources for these events are: *Apollo Program Summary Report* (JSC-09423); *Stages To Saturn: A Technological History of Saturn/Apollo Launch Vehicles* (SP-4206); and the *Saturn V Flight Evaluation Report* for each mission.

[6] Altitude on the launch pad is measured at the instrument unit for all Apollo missions.

[7] Only the commanded time is available for this event.

Apollo 7 Earth Orbit Phase

Event	GET (hhh:mm:ss)	Space Fixed Velocity (ft/sec)	Event Duration (sec)	Velocity Change (ft/sec)	Apogee (n mi)	Perigee (n mi)	Perigee (mins)	Inclination (deg)
Earth orbit insertion	000:10:26.76	25,553.2			152.34	123.03	89.55	31.608
Separation of CSM from S-IVB	002:55:02.40	25,499.5			170.21	123.01	89.94	31.640
1st rendezvous phasing ignition	003:20:09.9	25,531.7			167.0	125.3	89.99	31.61
1st rendezvous phasing cutoff	003:20:26.2	25,525.0	16.3	5.7	165.2	124.8	89.95	31.62
2nd rendezvous phasing ignition	015:52:00.9	25,283.1			165.1	124.7	89.95	31.62
2nd rendezvous phasing cutoff	015:52:18.5	25,277.4	17.6	7.0	164.7	120.8	89.86	31.62
1st SPS ignition	026:24:55.66	25,289.9			164.6	120.6	89.86	31.62
1st SPS cutoff	026:25:05.02	25,354.0	9.36	204.1	194.1	123.0	90.57	31.62
2nd SPS ignition	028:00:56.47	25,446.5			194.1	123.0	90.57	31.62
2nd SPS cutoff	028:01:04.23	25,357.2	7.76	173.8	153.6	113.9	89.52	31.63
Terminal phase initiation ignition	029:16:33	25,327.1			153.6	113.9	89.52	31.63
Terminal phase initiation cutoff	029:17:19		46	17.7				
Terminal phase finalize (braking)	029:43:55				154.1	121.6	89.68	31.61
Terminal phase end	029:55:43	25,546.1	708	49.1	161.0	122.1	89.82	31.61
Separation ignition	030:20:00.0	25,514.1			161.0	122.1	89.82	31.61
Separation cutoff	030:20:05.4	25,515.1	5.4	2.0	161.0	122.2	89.82	31.61
3rd SPS ignition	075:48:00.27	25,326.1			159.4	121.3	89.77	31.61
3rd SPS cutoff	075:48:09.37	25,273.9	9.10	209.7	159.7	89.5	89.17	31.23
4th SPS ignition	120:43:00.44	25,661.2			149.4	87.5	88.94	31.25
4th SPS cutoff	120:43:00.92	25,670.6	0.48	12.3	156.7	89.1	89.11	31.24
5th SPS ignition	165:00:00.42	25,519.3			146.5	87.1	88.88	31.25
5th SPS cutoff	165:01:07.37	25,714.9	66.95	1,691.3	244.2	89.1	90.77	30.08
6th SPS ignition	210:07:59.99	25,354.7			234.8	88.5	90.59	30.08
6th SPS cutoff	210:08:00.49	25,354.6	0.50	14.2	234.6	88.4	90.58	30.07
7th SPS ignition	239:06:11.97	25,864.6			228.3	88.4	90.24	30.07
7th SPS cutoff	239:06:19.67	25,866.4	7.70	220.1	229.8	88.5	90.48	29.87
8th SPS ignition (deorbit)	259:39:16.36	25,155.3			225.3	88.2	90.39	29.88
8th SPS cutoff	259:39:28.15	24,966.5	11.79	343.6				

Apollo 7 Timeline

Event	GET (hhh:mm:ss)	GMT Time	GMT Date
Countdown started at T-101 hours.	-101:00:00	19:00:00	06 Oct 1968
Scheduled 6-hour hold at T-72 hours.	-072:00:00	00:00:00	08 Oct 1968
Countdown resumed at T-72 hours.	-072:00:00	06:00:00	08 Oct 68
Scheduled 3-hour hold at T-33 hours.	-033:00:00	21:00:00	09 Oct 1968
Countdown resumed at T-33 hours.	-033:00:00	00:00:00	10 Oct 1968
Terminal countdown started.	-018:00:00	14:30:00	10 Oct 1968
Scheduled 6-hour hold at T-6 hours.	-006:00:00	03:00:00	11 Oct 1968
Terminal countdown started.	-006:00:00	09:00:00	11 Oct 1968
Crew ingress.	-002:27	12:35	11 Oct 1968
Unscheduled 2-minute 45-second hold to complete propellant chilldown.	-000:06:15	14:53:45	11 Oct 1968
Countdown resumed at T-6 minutes 15 seconds.	-000:06:15	14:56:30	11 Oct 1968
Guidance reference release.	-000:00:04.972	15:02:40	11 Oct 1968
S-IB engine start command.	-000:00:02.988	15:02:42	11 Oct 1968
Range zero.	000:00:00.00	15:02:45	11 Oct 1968
All holddown arms released (1st motion) (1.21 g).	000:00:00.17	15:02:45	11 Oct 1968
Liftoff (umbilical disconnected).	000:00:00.36	15:02:45	11 Oct 1968
Pitch and roll maneuver started.	000:00:10.31	15:02:55	11 Oct 1968
Roll maneuver ended.	000:00:38.46	15:03:23	11 Oct 1968
Mach 1 achieved.	000:01:02.15	15:03:47	11 Oct 1968
Maximum bending moment achieved (7,546,000 lbf-in).	000:01:13.1	15:03:58	11 Oct 1968
Maximum dynamic pressure (665.60 lb/ft^2).	000:01:15.5	15:04:00	11 Oct 1968
Pitch maneuver ended.	000:02:14.26	15:04:59	11 Oct 1968
S-IB maximum total inertial acceleration (4.28 g).	000:02:20.10	15:05:05	11 Oct 1968
S-IB center engine cutoff.	000:02:20.65	15:05:05	11 Oct 1968
S-IB outboard engine cutoff.	000:02:24.32	15:05:09	11 Oct 1968
S-IB maximum Earth-fixed velocity.	000:02:24.6	15:05:09	11 Oct 1968
S-IB/S-IVB separation command.	000:02:25.59	15:05:10	11 Oct 1968
S-IVB engine ignition command.	000:02:26.97	15:05:12	11 Oct 1968
S-IVB ullage case jettisoned.	000:02:37.58	15:05:22	11 Oct 1968
Launch escape tower jettisoned.	000:02:46.54	15:05:31	11 Oct 1968
Iterative guidance mode initiated.	000:02:49.76	15:04:54	11 Oct 1968
S-IB apex.	000:04:19.4	15:06:54	11 Oct 1968
S-IB impact in the Atlantic Ocean (theoretical).	000:09:20.2	15:12:05	11 Oct 1968
S-IVB engine cutoff.	000:10:16.76	15:13:01	11 Oct 1968
S-IVB maximum total inertial acceleration (2.55 g).	000:10:16.9	15:12:45	11 Oct 1968
S-IVB safing experiment—Start 1st LH$_2$ tank vent.	000:10:17.37	15:13:02	11 Oct 1968
S-IVB safing experiment—Tank passivization valve open.	000:10:17.56	15:13:02	11 Oct 1968
S-IVB maximum Earth-fixed velocity.	000:10:19.3	15:12:54	11 Oct 1968
Earth orbit insertion.	000:10:26.76	15:13:11	11 Oct 1968
Orbital navigation started.	000:10:32.2	15:13:17	11 Oct 1968
S-IVB safing experiment—Start LOX tank vent.	000:10:47.17	15:13:32	11 Oct 1968
S-IVB safing experiment—End LOX tank vent.	000:11:17.17	15:14:02	11 Oct 1968
S-IVB safing experiment—End 1st LH$_2$ tank vent (approximate due to data dropout).	000:31:17:36	15:34:02	11 Oct 1968
S-IVB safing experiment—Start 2nd LH$_2$ tank vent.	000:54:06.95	15:56:52	11 Oct 1968
S-IVB safing experiment—End 2nd LH$_2$ tank vent.	000:59:06.95	16:01:52	11 Oct 1968
Start of two-minute power failure in Mission Control Center started. No loss of communications.	001:18:34	16:21:19	11 Oct 1968
S-IVB safing experiment—LOX dump started.	001:34:28.96	16:37:14	11 Oct 1968
S-IVB safing experiment—LOX tank non-propulsive vent valve open (until end of mission).	001:34:38.95	16:37:24	11 Oct 1968
S-IVB safing experiment—Start 3rd LH$_2$ tank vent.	001:34:42.95	16:37:28	11 Oct 1968
S-IVB safing experiment—Start 1st cold helium dump.	001:42:28.95	16:45:14	11 Oct 1968
S-IVB safing experiment—End 3rd LH$_2$ tank vent.	001:44:42.95	16:47:28	11 Oct 1968

Apollo 7 Timeline

Event	GET (hhh:mm:ss)	GMT Time	GMT Date
S-IVB safing experiment—LOX dump ended.	001:46:29.96	16:49:15	11 Oct 1968
S-IVB safing experiment—End 1st cold helium dump.	002:30:16.95	17:33:02	11 Oct 1968
Manual takeover of S-IVB attitude control started.	002:30:48.80	17:32:45	11 Oct 1968
Manual takeover—Pitch maneuver started.	002:31:22	17:34:07	11 Oct 1968
Manual takeover—Pitch maneuver ended.	002:32:15	17:35:00	11 Oct 1968
Manual takeover—Roll maneuver started.	002:32:22	17:35:07	11 Oct 1968
Manual takeover—Roll maneuver ended.	002:32:51	17:35:36	11 Oct 1968
Manual takeover—Yaw maneuver started.	002:33:01	17:35:46	11 Oct 1968
Manual takeover—Yaw maneuver ended.	002:33:31	17:36:16	11 Oct 1968
Manual takeover of S-IVB attitude control ended.	002:33:44.80	17:35:45	11 Oct 1968
Window photography.	002:45	12:17	11 Oct 1968
Separation of CSM from S-IVB.	002:55:02.40	17:57:47	11 Oct 1968
S-IVB safing experiment—Start 4th LH_2 tank vent.	003:09:14.48	18:11:59	11 Oct 1968
S-IVB safing experiment—End 4th LH_2 tank vent.	003:15:56.11	18:18:41	11 Oct 1968
S-IVB safing experiment—Start stage control sphere helium dump.	003:17:33.95	18:20:19	11 Oct 1968
1st rendezvous phasing maneuver ignition.	003:20:09.9	18:22:54	11 Oct 1968
1st rendezvous phasing maneuver cutoff.	003:20:26.2	18:23:11	11 Oct 1968
S-IVB safing experiment—Start 5th LH_2 tank vent.	004:05:47.27	19:08:32	11 Oct 1968
S-IVB safing experiment—End stage control sphere helium dump.	004:07:01.27	19:09:46	11 Oct 1968
S-IVB safing experiment—End 5th LH_2 tank vent.	004:10:08.43	19:12:53	11 Oct 1968
S-IVB safing experiment—Start 2nd cold helium dump.	004:30:16.96	19:33:02	11 Oct 1968
S-IVB safing experiment—Start 6th LH_2 tank vent.	004:43:55.85	19:46:40	11 Oct 1968
S-IVB safing experiment—End 6th LH_2 tank vent	004:49:01.73	19:51:46	11 Oct 1968
S-IVB safing experiment—End 2nd cold helium dump.	004:50:16.95	19:53:02	11 Oct 1968
S-IVB safing experiment—Start 7th LH_2 tank vent.	005:08:58.99	20:11:44	11 Oct 1968
S-IVB safing experiment—End 7th LH_2 tank vent.	005:11:15.43	20:14:00	11 Oct 1968
Hydrogen stratification test.	013:28:00	04:30:45	12 Oct 1968
2nd rendezvous phasing maneuver ignition.	015:52:00.9	06:54:45	12 Oct 1968
2nd rendezvous phasing maneuver cutoff.	015:52:18.5	06:55:03	12 Oct 1968
Y-Pulse Integrating Pendulum Accelerometer test.	022:30	13:32	12 Oct 1968
S-IVB optical tracking.	025:10	16:12	12 Oct 1968
Oxygen stratification test.	025:14:00	16:16:45	12 Oct 1968
1st SPS ignition (NCC/corrective combination maneuver—initiation of rendezvous sequence).	026:24:55.66	17:27:40	12 Oct 1968
1st SPS cutoff.	026:25:05.02	17:27:50	12 Oct 1968
2nd SPS ignition (NSR/coelliptic maneuver).	028:00:56.47	19:03:41	12 Oct 1968
2nd SPS cutoff.	028:01:04.23	19:03:49	12 Oct 1968
S-IVB optical tracking.	028:20	19:22	12 Oct 1968
Terminal phase initiation ignition.	029:16:33	20:19:18	12 Oct 1968
Terminal phase initiation cutoff.	029:17:19	20:20:04	12 Oct 1968
Midcourse correction.	029:30:42	20:33:27	12 Oct 1968
Terminal phase finalize (braking).	029:43:55	20:46:40	12 Oct 1968
Terminal phase end/start station-keeping.	029:55:43	20:58:28	12 Oct 1968
Separation maneuver ignition.	030:20:00.0	21:22:45	12 Oct 1968
Separation maneuver cutoff.	030:20:05.4	21:22:50	12 Oct 1968
Sextant calibration test.	041:00	08:02	13 Oct 1968
Sextant tracking of S-IVB started.	044:40	11:42	13 Oct 1968
Sextant tracking of S-IVB ended at 160 n mi.	045:30	12:32	13 Oct 1968
Attitude hold test.	049:00	16:02	13 Oct 1968
Primary evaporator test.	049:50	16:52	13 Oct 1968
Primary evaporator test.	050:30	17:32	13 Oct 1968
Attitude hold test.	050:40	17:42	13 Oct 1968

Apollo 7 Timeline

Event	GET (hhh:mm:ss)	GMT Time	GMT Date
Sextant tracking of S-IVB started.	052:10	19:12	13 Oct 1968
Sextant tracking of S-IVB ended at 320 n mi.	053:20	20:22	13 Oct 1968
Rendezvous radar transponder test.	061:00	04:02	14 Oct 1968
Rendezvous radar transponder test.	071:40	14:42	14 Oct 1968
1st television transmission started.	071:43	14:45	14 Oct 1968
1st television transmission ended.	071:50	14:52	14 Oct 1968
3rd SPS ignition (to position and size orbital ellipse).	075:48:00.27	18:50:45	14 Oct 1968
3rd SPS cutoff.	075:48:09.37	18:50:54	14 Oct 1968
Rendezvous radar transponder test.	076:27	19:29	14 Oct 1968
Radiator degradation test started.	092:37:00	11:39:45	15 Oct 1968
2nd television transmission started.	095:25	14:27	15 Oct 1968
2nd television transmission ended.	095:36	14:38	15 Oct 1968
Radiator surface coating degradation test ended.	097:00	16:02	15 Oct 1968
Hydrogen stratification test.	098:11	17:13	15 Oct 1968
Experiment S005 photography.	098:40	17:42	15 Oct 1968
Experiment S006 photography.	099:10	18:12	15 Oct 1968
Window photography.	101:10	20:12	15 Oct 1968
3rd television transmission started.	119:08	14:10	16 Oct 1968
3rd television transmission ended.	119:18	14:20	16 Oct 1968
4th SPS ignition (minimum impulse burn).	120:43:00.44	15:45:45	16 Oct 1968
4th SPS cutoff.	120:43:00.92	15:45:45	16 Oct 1968
Star/horizon sightings.	124:00	19:02	16 Oct 1968
Oxygen stratification test.	131:52	02:54	16 Oct 1968
4th television transmission started.	141:11	12:13	17 Oct 1968
4th television transmission ended.	141:27	12:29	17 Oct 1968
Lunar landmark star sightings.	147:00	18:02	17 Oct 1968
S-IVB impact (theoretical).	162:27:15	09:30:00	18 Oct 1968
5th SPS ignition (to position and size orbital ellipse).	165:00:00.42	12:02:45	18 Oct 1968
5th SPS cutoff.	165:01:07.37	12:03:52	18 Oct 1968
Passive thermal control test started.	167:00	14:02	18 Oct 1968
Passive thermal control test ended.	167:50	14:52	18 Oct 1968
Service propulsion cold soak test started.	168:00	15:02	18 Oct 1968
Service propulsion cold soak test ended.	171:10	18:12	18 Oct 1968
5th television transmission.	189:04	12:06	19 Oct 1968
Morse code emergency keying test started.	190:36:06	13:38:51	19 Oct 1968
Morse code emergency keying test ended.	190:43:01	13:45:46	19 Oct 1968
Oxygen stratification test.	198:27:00	21:29:45	19 Oct 1968
6th SPS ignition (minimum impulse burn).	210:07:59.99	09:10:45	20 Oct 1968
6th SPS cutoff.	210:08:00.49	09:10:45	20 Oct 1968
Passive thermal control test (pitch procedure) started.	212:00	11:02	20 Oct 1968
Passive thermal control test ended.	212:50	11:52	20 Oct 1968
6th television transmission.	213:10	12:12	20 Oct 1968
Star/horizon sightings.	213:30	12:32	20 Oct 1968
Hydrogen stratification test.	227:12	02:14	21 Oct 1968
Optics degradation test started.	228:30	03:32	21 Oct 1968
Solar Particle Alert Network Facility at Carnarvon reported class 1B solar flare.	231:08	06:10	21 Oct 1968
7th television transmission started.	236:18	11:20	21 Oct 1968
7th television transmission ended.	236:29	11:31	21 Oct 1968
7th SPS ignition (time anomaly adjust for deorbit burn).	239:06:11.97	14:08:57	21 Oct 1968
7th SPS cutoff.	239:06:19.67	14:09:04	21 Oct 1968
Window photography.	242:30	17:32	21 Oct 1968

Apollo 7 Timeline

Event	GET (hhh:mm:ss)	GMT Time	GMT Date
8th SPS ignition (deorbit burn).	259:39:16.36	10:42:01	22 Oct 1968
8th SPS cutoff.	259:39:28.15	10:42:13	22 Oct 1968
CM/SM separation.	259:43:33.8	10:46:18	22 Oct 1968
Entry.	259:53:26	10:56:11	22 Oct 1968
Communication blackout started.	259:54:58	10:57:43	22 Oct 1968
Communication blackout ended.	259:59:46	11:02:31	22 Oct 1968
Maximum entry g force (3.33 g).	260:01:09	11:03:54	22 Oct 1968
SM impact in the Atlantic Ocean. S-band contact with CM by recovery aircraft.	260:03	11:055	22 Oct 1968
Drogue parachute deployed.	260:03:23	11:06:08	22 Oct 1968
Main parachute deployed. VHF voice contact with CM established by recovery forces.	260:04:13	11:06:58	22 Oct 1968
Splashdown (went to apex-down).	260:09:03	11:11:48	22 Oct 1968
Inflation of flotation bags started.	260:18	11:20	22 Oct 1968
CM returned to apex-up position.	260:22	11:24	22 Oct 1968
VHF recovery beacon signal received by recovery aircraft.	260:23	11:25	22 Oct 1968
VHF voice communication with CM reestablished.	260:24	11:26	22 Oct 1968
CM sighted by recovery helicopter.	260:30	11:32	22 Oct 1968
Swimmers and flotation collar deployed.	260:32	11:34	22 Oct 1968
Flotation collar inflated.	260:41	11:43	22 Oct 1968
CM hatch opened.	260:45	11:47	22 Oct 1968
Crew aboard recovery helicopter.	260:58	12:00	22 Oct 1968
Recovery ship at CM. Crew aboard recovery ship.	261:06	12:08	22 Oct 1968
CM aboard recovery ship.	262:01	13:03	22 Oct 1968
Crew departed recovery ship.	285:54	12:56	23 Oct 1968
Crew arrived at Cape Kennedy.	288:43	15:45	23 Oct 1968
CM offloaded at Norfolk Naval Air Station.			24 Oct 1968
Safing team started CM deactivation.	310:58	14:00	24 Oct 1968
Deactivation of CM completed.	370:28	01:30	27 Oct 1968

APOLLO 8

The Second Mission:
Testing the CSM in Lunar Orbit

Apollo 8 Summary

(21 December–27 December 1968)

The Apollo 8 crew (l. to. r.): Bill Anders, Jim Lovell, Frank Borman (NASA S68-53187).

Background

Apollo 8 was a Type "C prime" mission, a CSM piloted flight demonstration in lunar orbit instead of Earth orbit like Apollo 7. It was the first mission to take humans to the vicinity of the Moon, a bold step forward in the development of a lunar landing capability.

The mission was originally designated SA-503, an unpiloted Earth orbital mission to be launched in May 1968 with boilerplate payload BP-30 instead of an operational spacecraft. The success of Apollo 6 (AS-502), however, led to the decision on 27 April that AS-503 would be a piloted mission with a CSM and LM instead of BP-30.

The change to a piloted flight required that the S-II stage be returned to the Mississippi Test Facility for "man-rating." Additional tests for a piloted flight continued at KSC. The Mississippi tests were successfully completed on 30 May 1968 and the stage returned to the Kennedy Space Center on 27 June.

After two months of testing, which started 11 June 1968, it was determined that the LM would not be ready for the projected early December launch. Therefore, the decision was made on 19 August that a 19,900-pound LM test article would be installed in the spacecraft/launch vehicle adapter for mass loading purposes, replacing the LM. It was also on this date that the crew was instructed to train for a mission to the Moon, officially designated "Apollo 8."

The possibility of conducting a lunar mission was first discussed with the crew on 10 August, and the results of Apollo 7, to be launched in October, would determine whether the mission would be lunar orbital, circumlunar, or Earth orbital. All training immediately focused on the lunar orbital mission, the most difficult of the three, and ground support preparations were accelerated. The first simulation exercise was conducted on 9 September, and the space vehicle was transferred to the launch site on 9 October.

Following the successful completion of Apollo 7 on 22 October, the official decision to conduct a lunar orbit mission was made 12 November, just five weeks before the scheduled launch. The decision was made after a thorough evaluation of spacecraft performance during Apollo 7's ten days in Earth orbit and an assessment of the risks involved in a lunar orbit mission. These risks included the total dependency upon the service propulsion engine for propelling the spacecraft from lunar orbit, and a lunar orbit return time of three days, compared to an Earth orbit return of just 30 minutes to three hours. Also considered was the value of the flight in furthering the goal of landing a human on the Moon before the end of 1969. The principal gains from a lunar mission would include experience in deep space navigation, communications, and tracking; greater knowledge of spacecraft thermal response to deep space; and crew operational experience—all directly applicable to lunar landing missions.

Apollo 8 was the first piloted mission launched with the three-stage Saturn V vehicle; the two previous Saturn V flights had been unpiloted. The spacecraft was a Block II CSM, and the spacecraft/launch vehicle adapter was the first to incorporate a mechanism to jettison the panels that would cover the LM on future missions.

The primary objectives of Apollo 8 were:

• to demonstrate the combined performance of the crew, space vehicle, and mission support team during a piloted Saturn V mission with the CSM; and

• to demonstrate the performance of nominal and selected backup lunar orbit rendezvous procedures.

The crew members were Colonel Frank Frederick Borman II (USAF), commander; Captain James Arthur Lovell, Jr. (USN), command module pilot; and Major William Alison Anders (USAF), lunar module pilot.

Selected in the astronaut group of 1962, Borman had been command pilot of Gemini 7. Born 14 March 1928 in Gary, Indiana, he was 40 years old at the time of the Apollo 8 mission. Borman received a B.S. from the U.S. Military Academy in 1950 and an M.S. in Aeronautical Engineering in 1957 from the California Institute of Technology. His backup for the mission was Neil Alden Armstrong.

Lovell had been pilot for the Gemini 7 mission and command pilot for Gemini 12. Born 25 March 1928 in Cleveland, Ohio, he was 35 years old at the time of the Apollo 8 mission. Lovell received a B.S. in 1952 from the U.S. Naval Academy, and was selected as an astronaut in 1962. His backup was Colonel Edwin Eugene "Buzz" Aldrin, Jr. (USAF).

Anders was making his first spaceflight. Born 17 October 1933 in Hong Kong, he was 35 years old at the time of the Apollo 8 mission. Anders received a B.S. in Electrical Engineering in 1955 from the U.S. Naval Academy and an M.S. in Nuclear Engineering in 1962 from the U.S. Air Force Institute of Technology, and was selected as an astronaut in 1963. His backup was Fred Wallace Haise, Jr.

The capsule communicators (CAPCOMs) for the mission were Lt. Col. Michael Collins (USAF), Lt. Commander Thomas Kenneth "Ken" Mattingly II (USN), Major Gerald Paul Carr (USMC), Armstrong, Aldrin, Vance DeVoe Brand, and Haise. The support crew were Brand, Mattingly, and Carr. The flight directors were Clifford E. Charlesworth (first shift), Glynn S. Lunney (second shift), and Milton L. Windler (third shift).

The Apollo 8 launch vehicle was a Saturn V, designated SA-503. The mission also carried the designation Eastern Test Range #170. The CSM combination was designated CSM-103. The lunar module test article was designated LTA-B.

Because this was a lunar mission, it was necessary for the vehicle to be launched within a particular daily launch "window", or time period, within a monthly launch window. Part of the constraints were dictated by the desire to pass over selected lunar sites with lighting conditions simi- lar to those planned for the later landing missions. Lunar orbit inclination, inclination of the free return trajectory, and spacecraft propellant reserves were other primary factors considered in the mission planning.

The first monthly window was in December 1968, with launch dates of 20–27 December, and January 1969 as a backup. It was decided to make the first attempt on 21 December to have the total available daily window during daylight. Targeting for this day would allow the flight to pass over a future lunar landing site at latitude 2.63° and longitude 34.03° with a sun elevation angle of 6.74°. The window for 21 December lasted from 12:50:22 to 17:31:40 GMT, with liftoff scheduled for 12:51:00 GMT.

Launch Preparations

The terminal countdown sequence (T-28 hours) began at 13:51 GMT on 19 December. At that time, space vehicle operations were functionally ahead of the clock. Later in the count, it was discovered that the onboard liquid oxygen supply for the spacecraft environmental control system and fuel cell systems was contaminated with nitrogen. Preparations were made to replace the liquid oxygen, the reservicing operations were completed, and the tanks were pressurized at T-10 hours.

During the planned six-hour hold period at T-9 hours, virtually all of the countdown tasks, delayed by the liquid oxygen detanking and retanking operations, were brought back in line. When the count was picked up again at T-9 hours, space vehicle operations were essentially on schedule. At T-8 hours, S-IVB liquid oxygen loading operations began. The cryogenic loading operations were completed at 08:29 GMT on December 21, eight minutes into the one-hour scheduled hold. The count was picked up at T-3 hours 30 minutes at 09:21 GMT, and the crew entered the spacecraft at T-2 hours 53 minutes.

A cold front passed through the launch area the afternoon before launch and became a stationary front about launch time, laying through the Miami area. At launch time, surface winds were from the north but changed to westerly at 4,900 feet and remained generally from the west above that region. Cirrus clouds covered 40 percent of the sky (cloud base not recorded), visibility was 10 statute miles, the temperature was 59.0° F, relative humidity was 88 percent, dew point was 56 percent, barometric pressure was 14.804 lb/in² and winds were 18.7 ft/sec at 348° from true north measured by the anemometer on the light pole 60.0 feet above ground at the launch site.

Ascent Phase

Apollo 8 was launched from Launch Complex 39, Pad A, at the Kennedy Space Center, Florida. Liftoff occurred at a Range Zero time of 12:51:00 GMT (07:51:00 a.m. EST) on 21 December 1968, well within the planned launch window.

The ascent phase was nominal. Moments after liftoff, the vehicle rolled from a launch pad azimuth of 90° to a flight azimuth of 72.124° east of north. The S-IC engine shut down at 000:02:33.82, followed by S-IC/S-II separation, and S-II engine ignition. The S-II engine shut down at 000:08:44.04 followed by separation from the S-IVB, which ignited at 000:08:48.29. The first S-IVB engine cutoff occurred at 000:11:24.98, with deviations from the planned trajectory of only +1.44 ft/sec in velocity and only -0.01 n mi in altitude.

Apollo 8, the first piloted Saturn V flight and humankind's first trip to the Moon, lifts off from Kennedy Space Center Pad 39A (NASA S68-56050).

The S-IC stage impacted at 000:09:00.410 in the Atlantic Ocean at latitude 30.2040° north and longitude 74.1090° west, 353.462 n mi from the launch site. The S-II stage impacted at 000:19:25.106 in the Atlantic Ocean at latitude 31.8338° north and longitude 37.2774° west, 2,245.913 n mi from the launch site.

Four recoverable film camera capsules were carried aboard the S-IC stage. Two were located in the forward interstage looking forward to view S-IC/S-II separation and S-II engine start. The other two were mounted on top of the S-IC stage LOX tank and contained pulse cameras which viewed aft into the LOX tank through fiber optics bundles. One of the LOX tank capsules was recovered by helicopter at 00:19:30 at latitude 30.22° north and longitude 73.97° west. Despite film damage caused by sea water and dye marker which had leaked into the camera compartment, the film provided usable data. It was not known if the other three capsules were ejected. There were also two television cameras on the S-IC to view propulsion and control system components. Both provided good quality data.

The maximum wind conditions encountered during ascent were 114.1 ft/sec at 284° from true north at 49,900 feet (high dynamic pressure region). Component wind shears were of low magnitude at all altitudes. The largest wind shear was a pitch plane shear of 0.0103 sec^{-1} at 52,500 feet.

At 000:11:34.98, the spacecraft entered Earth orbit, defined as S-IVB cutoff plus 10 seconds to account for engine tail-off and other transient effects. At insertion, conditions were: apogee and perigee 99.99 by 99.57 n mi, inclination 32.509°, period 89.19 minutes, and velocity 25,567.06 ft/sec. The apogee and perigee were based upon a spherical Earth with a radius of 3,443.934 n mi.

The international designation for the spacecraft upon achieving orbit was 1968-118A and the S-IVB was designated 1968-118B.

Earth Orbit Phase

At 000:42:05, the optics cover was jettisoned and the crew performed star checks over the Carnarvon, Australia, tracking station to verify platform alignment. During the second revolution, at 001:56:00, all spacecraft systems were approved for translunar injection.

Because of the risks involved, the mission had been structured with three commit points: launch, Earth parking orbit, and translunar coast preceding the point where the CSM was to brake into lunar orbit. Had any problems been detected at these points, the plan was to shift to alternate missions, which provided for maximum crew safety and maximum scientific and engineering benefit. Had there been reason for not committing to the third point, the CSM would have continued on its "free-return" trajectory, looping behind the Moon and returning directly to Earth.

After inflight systems checks, it was determined that liquid oxygen venting through the J-2 engine had increased the apogee by 6.4 n mi, a condition that was only 0.7 n mi greater than predicted.

The 317.72-second translunar injection maneuver (second S-IVB firing) was performed at 002:50:37.79. The S-IVB engine shut down at 002:55:55.51 and translunar injection occurred ten seconds later, at a velocity of 35,504.41 ft/sec, after 1.5 Earth revolutions lasting 2 hours 44 minutes 30.53 seconds.

Translunar Phase

The spacecraft was separated from the S-IVB at 003:20:59.3 by a small maneuver of the service module reaction control system, and the high-gain antenna was deployed (later used for the first time at 006:33:04). After spacecraft turn-around, the crew observed and photographed the S-IVB and practiced station-keeping. At 003:40:01, a 1.1 ft/sec maneuver was performed using the service module reaction control system to increase the distance between the spacecraft and the S-IVB. The distance did not increase as rapidly as desired, and a second, 7.7 ft/sec maneuver was performed at 004:45:01.

View of Saturn V stage following separation from the CSM (NASA AS08-16-2583).

One objective of the mission was to place the S-IVB into solar orbit. The "slingshot" maneuver required to accom-

plish this objective included a continuous LH_2 vent, a LOX dump, and an auxiliary propulsion system ullage burn. At 004:55:56.02, the LH_2 vent valve was opened, and the remaining liquid oxygen and the auxiliary propulsion system propellant in the S-IVB were used to change the trajectory of the S-IVB stage. The liquid oxygen was expelled through the J-2 engine starting at 005:07:55.82 and ended five minutes later.

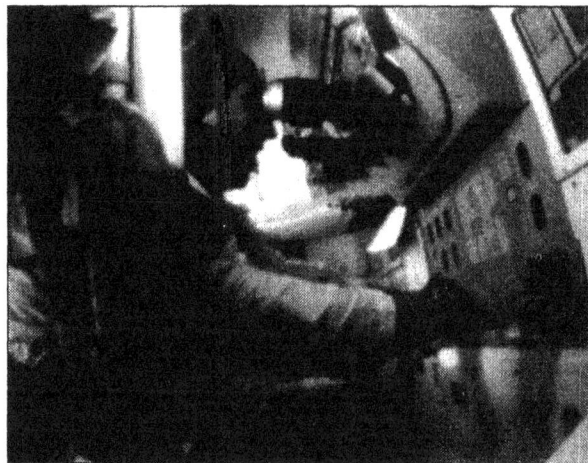

CMP Jim Lovell navigates during the first trip to the Moon (NASA S69-35097).

The auxiliary propulsion motors were fired from 005:25:55.85 to depletion at 005:38:34.00. The resulting velocity increment targeted the S-IVB to go past the trailing edge of the Moon. The closest approach of the S-IVB to the Moon was 682 n mi at 069:58:55.2. The point of closest approach was latitude 19.2 north by longitude 88.0 east. The trajectory after passing from the lunar sphere of influence resulted in a solar orbit with a semi-major axis of 77.130 million n mi, an aphelion and perihelion of 79.770 million by 74.490 million n mi, an inclination of 23.47°, and a period of 340.8 days.

The translunar injection maneuver was so accurate that only one small midcourse correction would have been sufficient to achieve the desired lunar orbit insertion altitude of 65 n mi. However, the second of the 2 maneuvers that separated the spacecraft from the S-IVB altered the trajectory so that a 2.4-second midcourse correction of 20.4 ft/sec at 010:59:59.2 was required to achieve the desired trajectory.[1] For this midcourse correction, the service propulsion system was used to reduce the altitude of closest approach to the Moon from 458.1 to 66.3 n mi. An additional 11.9-second midcourse correction of only 1.4 ft/sec was performed at 060:59:55.9 to refine the lunar insertion conditions further.

---

[1] The maneuver at 010:59:59.2 was targeted for a velocity change of 24.8 ft/sec. Only 20.4 ft/sec was achieved because thrust was less than expected. The firing time of 2.4 seconds was correct for the constants loaded into the computer, but was approximately 0.4 second too short for the actual engine performance.

Earth view following translunar injection (NASA AS08-16-2596).

During the translunar coast, the crew made systems checks and navigation sightings, and tested the spacecraft high-gain antenna, a four-dish unified S-band antenna that swung out from the service module after separation from the S-IVB.

Apollo 8 was the first piloted U.S. mission in which the crew members experienced symptoms of a mild motion sickness, identical to incipient mild seasickness. Soon after leaving their couches, all three experienced nausea as a result of rapid body movements. The duration of symptoms varied between 2 and 24 hours but did not interfere with operational effectiveness. After waking from a fitful rest period at 016:00:00, the commander experienced a headache, nausea, vomiting, and diarrhea. These symptoms were diagnosed inflight as a possible viral gastroenteritis, an epidemic that was noted in the Cape Kennedy area prior to the mission. During the post mission medical debriefing, the commander reported that the symptoms may have been a side effect of a sleeping tablet he had taken at 011:00:00, which had produced similar symptoms during pre-mission testing of the drug (Seconal™).

Two of the six live television transmissions were also made during translunar flight. The first was a 23-minute 37-second transmission at 031:10:36. The wide-angle lens was used to obtain excellent pictures of the inside of the spacecraft and Lovell preparing a meal; however the telephoto lens passed too much light and pictures of Earth were very poor. A procedure for taping certain filters from the still camera to the television camera improved later transmissions. A 25-minute 38-second transmission at 55:02:45 provided scenes of Earth's western hemisphere.

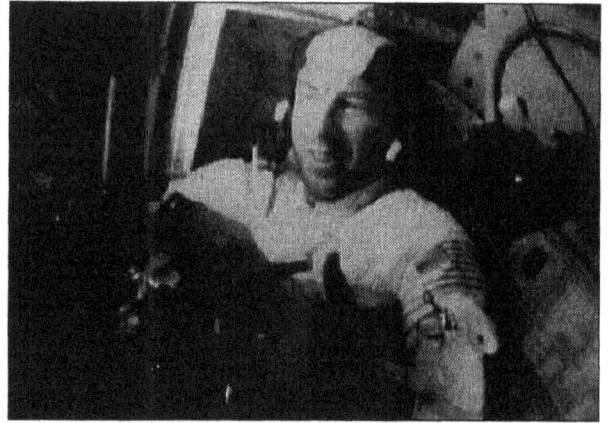

CMP Jim Lovell with camera (NASA S68-56533).

At 055:38:40 the crew were notified that they had become the first humans to travel to a place where the pull of Earth's gravity was less than that of another body. The spacecraft was 176,250 n mi from Earth, 33,800 n mi from the Moon, and their velocity had slowed to 3,261 ft/sec. Gradually, as it moved farther into the Moon's gravitational field, the spacecraft picked up speed.

Ignition for lunar orbit insertion was performed with the service propulsion system at 069:08:20.4, at an altitude of 76.6 n mi above the Moon. The 246.9-second burn resulted in an orbit of 168.5 by 60.0 n mi and a velocity of 5,458 ft/sec. The translunar coast had lasted 66 hours 16 minutes 21.79 seconds.

View of the Moon from Apollo 8 (NASA AS08-14-2506).

Lunar Orbit Phase

As the spacecraft passed behind the Moon for the first time, and communications were interrupted, the Apollo 8 crew became the first humans to see the far side of the Moon. After four hours of navigation checks, ground-based determination of the orbital parameters, and a 12-minute television transmission of the lunar surface at 071:40:52, a 9.6-second lunar orbit circularization maneuver was performed at 073:35:06.6, which resulted in an orbit of 60.7 by 59.7 n mi.

The next 12 hours of crew activity in lunar orbit involved photography of both the near and far sides of the Moon and landing-area sightings. The principal photographic objectives were to obtain vertical and oblique overlapping (stereo strip) photographs during at least two revolutions, photographs of specified targets of opportunity, and photographs through the spacecraft sextant of a potential landing site. The purpose of the overlapping photography was to determine elevation and geographical position of lunar far side features. The targets of opportunity were areas recommended for photography if time and circumstances permitted. They were selected to provide either detailed coverage of specific features or broad coverage of areas not adequately covered by satellite photography. Most were proposed to improve knowledge of areas on the Earth-facing hemisphere. The sextant photography was included to provide image comparisons for landmark evaluation and navigation training purposes. A secondary objective was to photograph one of the certified Apollo landing sites.

The Apollo 8 photography afforded the first opportunity to analyze the intensity and spectral distribution of lunar surface illumination free from the atmospheric modulation present in Earth telescopic photography and without the electronic processing losses present in satellite photography.

The crew completed photographic exercises in an excellent manner. Over 800 70 mm still photographs were obtained. Of these, 600 were good-quality reproductions of lunar surface features, and the remainder were of the S-IVB during separation and venting, and long-distance Earth and lunar photography.

Over 700 feet of 16 mm film were also exposed during the S-IVB separation, lunar landmark photography through the sextant, lunar surface sequence photography, and documentation of intravehicular activity.

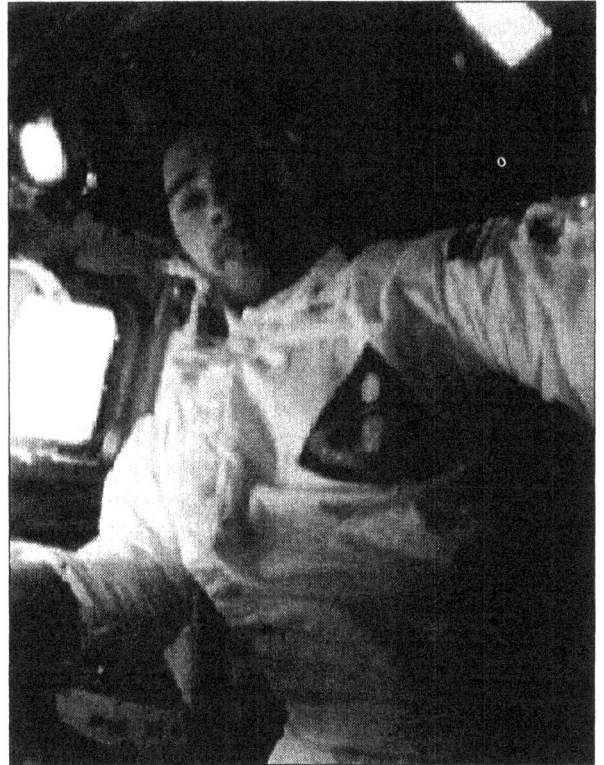

LMP Bill Anders during CM activities (NASA S69-56532).

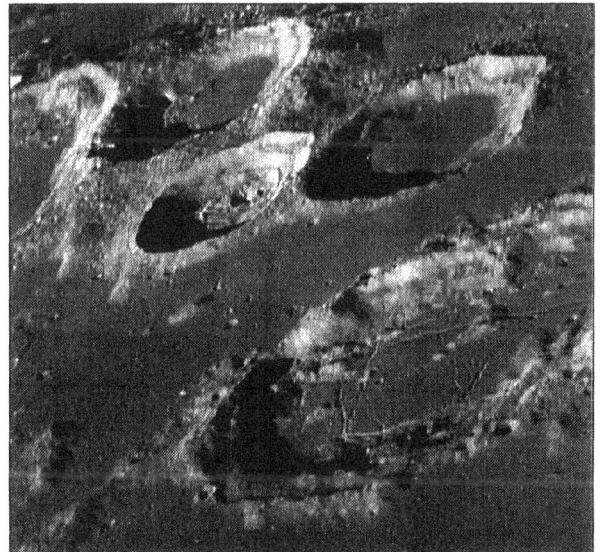

Craters Goclenius (foreground), Columbo A, and Maegelhans (background) (NASA AS08-13-2224).

The still photography contributed significantly to knowledge of the lunar environment. In addition, many valuable observations were made by the crew. Their initial comments during the lunar orbit phase included descriptions of the color of the lunar surface as "black-and-white," "absolutely no color" or "whitish gray, like dirty beach sand." As expected, the crew could recognize surface features in shadow zones and extremely bright areas of the lunar surface, but these features were not well delineated in the photographs.

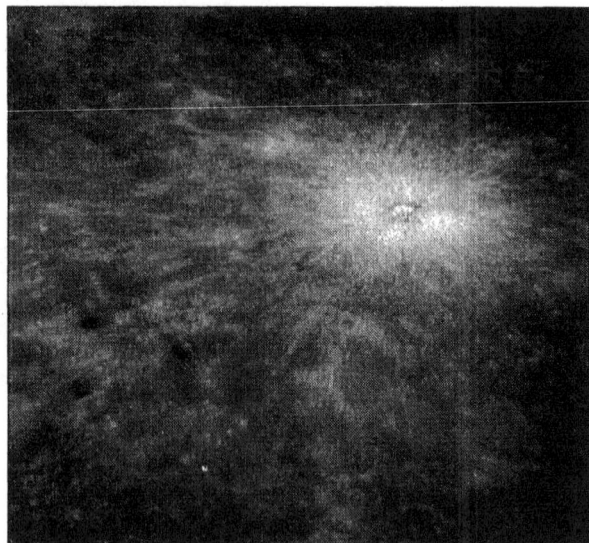

Brightly rayed crater on far side of the Moon (NASA AS08-13-2327).

This recognition combined with the photographic information enabled new interpretations of lunar surface features and phenomena. As a result, lunar-surface lighting constraints for the lunar landing missions were widened.

Prior to Apollo 8, the lower limit for lunar lighting was believed to be 6°. The Apollo 8 crew observed surface detail at sun angles in the vicinity of 2° or 3° and stated that these low angles should present no problem for a lunar landing, but landing sites in long shadow areas, however, were to be avoided. At the higher limit, an upper bound of 16° would still provide very good definition of surface features for most of the critical landing phase near touchdown. Between 16° and 20°, lighting was judged acceptable for viewing during final descent. A sun angle above 20° was considered unsatisfactory for a manual landing maneuver.

The crew report of the absence of sharp color boundaries was significant. The lack of visible contrast from an altitude of 60 n mi reduced the probability that a crew would be able to use color to distinguish geologic units while operating near or on the lunar surface.

View of the Sea of Tranquility, target site for the first piloted lunar landing attempt during Apollo 11 seven months later (NASA AS08-13-2344).

Just prior to sunrise on one of the early lunar orbit revolutions, the command module pilot observed what was believed to be zodiacal light and solar corona through the telescope. The lunar module pilot observed a cloud or bright area in the sky during lunar darkness on two successive revolutions. The identification, if correct, indicated that one of the Magellanic clouds had been observed.

Long-distance Earth photography of general interest highlighted global weather and terrain features. Lunar photography had not been accomplished during translunar coast because of rigid spacecraft attitude constraints. However, good quality photography of most of the Moon disk was accomplished during transearth coast.

The crew initially followed the lunar orbit mission plan and performed all scheduled tasks. However, because of crew fatigue, the commander made the decision at 084:30 to cancel all activities during the final four hours in lunar orbit to allow the crew to rest. The only activities during this period were a required platform alignment and preparation for transearth injection. A planned 26-minute 43-second television transmission of the Moon and Earth was made at 085:43:03, on Christmas eve. It was during this transmission that the crew read from the Bible the first ten verses of Genesis, and then wished viewers "Good night, good luck, a Merry Christmas, and God bless all of you, all of you on the good Earth." An estimated one billion people in 64 countries heard or viewed the live reading and greeting; delayed broadcasts reached an additional 30 countries that same day.

The Earth rising over the lunar surface as seen by the crew of Apollo 8 (NASA AS08-14-2383).

Orbit analysis indicated that previously unknown mass concentrations or "mascons" were perturbing the orbit. As a result, the final lunar orbit had an apogee and perigee of 63.6 by 58.6 n mi. The 203.7-second transearth injection maneuver was performed with the service propulsion system at an altitude of 60.2 n mi at 089:19:16.6 after ten revolutions and 20 hours 10 minutes 13.0 seconds in lunar orbit. The velocity at transearth injection was 8,842 ft/sec. During the mission, the spacecraft reached a maximum distance from Earth of 203,752.37 n mi.

Transearth Phase

After emerging from lunar occlusion following transearth injection, Apollo 8 experienced the only significant communications difficulty of the mission. Although two-way phaselock was established at 089:28:47, two-way voice contact and telemetry synchronization were not achieved until 089:33:28 and 089:43:00, respectively. Data indicated that high-gain antenna acquisition may have been attempted while line-of-sight was within the service module reflection region and that the reflections may have caused the antenna to track on a side lobe. In addition, the spacecraft was erroneously configured for high-bit-rate transmission; therefore the command at 089:29:29 that configured the spacecraft for normal voice and subsequent playback of the data storage equipment, selected an S-band signal combi-

nation that was not compatible with the received carrier power.

The transearth coast activities included star/horizon navigation sightings using both Moon and Earth horizons. Passive thermal control, using a roll rate of one revolution per hour, was used during most of the translunar and transearth coast phases to maintain nearly stable onboard temperatures. Only one small transearth midcourse correction, a 15.0-second maneuver using the service module reaction control system, was required at 104:00:00, and changed the velocity by 4.8 ft/sec.

Because of a crew procedural error, the onboard state vector and platform alignment were lost at 106:00:26. Realignment was performed at 106:45.

A special test of the automatic acquisition mode of the high-gain antenna was performed at 110:16:55. Results indicated that the antenna performed as predicted.

View of Earth on the way home (NASA AS08-15-2561).

The final two television transmissions were made during transearth coast. The fifth was a 9-minute 31-second transmission of the spacecraft interior at 104:24:04. The sixth transmission was for 19 minutes 54 seconds at 127:45:33 and featured views of Earth, particularly of the western hemisphere.

The service module was jettisoned at 146:28:48, and the CM entry followed an automatically guided entry profile. No radar tracking data for the service module were available during entry, but photographic coverage information correlated well with the predicted trajectory in altitude, latitude, longitude, and time.

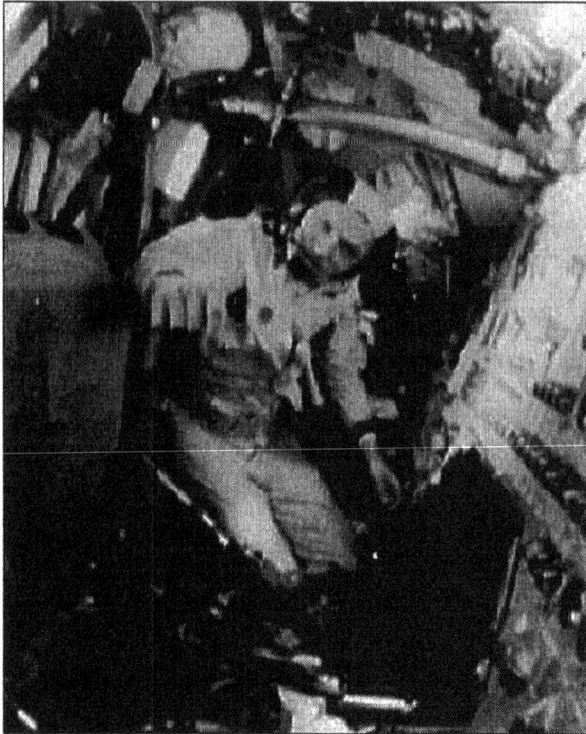

Apollo 8 commander Frank Borman (NASA S68-56531).

Recovery

The command module reentered Earth's atmosphere (400,000 feet altitude) at 146:46:12.8 at a velocity of 36,221.1 ft/sec, following a transearth coast of 57 hours 23 minutes 32.5 seconds. The ionization became so bright during entry that the CM interior was bathed in a cold blue light as bright as daylight. At 180,000 feet, as expected, the lift of the CM bounced it to 210,000 feet, where it then resumed its downward course.

Apollo 8 crew safely aboard the recovery ship U.S.S Yorktown (NASA S69-15737).

The parachute system effected splashdown of the CM in the Pacific Ocean at 10:51:42 GMT (05:51:42 a.m. EST) on 27 December. Mission duration was 147:00:42.0. The impact point was 1.4 n mi from the target point and 2.6 n mi from the recovery ship U.S.S. *Yorktown*. The splashdown site was estimated to be latitude 8.10° north and longitude 165.00° west. Due to the splashdown impact, the CM assumed an apex-down flotation attitude, but was successfully returned to the normal flotation position 6 minutes and 3 seconds later by the inflatable bag uprighting system.

As planned, helicopters and aircraft hovered over the spacecraft and pararescue personnel were not deployed until local sunrise, 43 minutes after splashdown. At dawn, the crew was retrieved by helicopter and were aboard the recovery ship 88 minutes after splashdown. The spacecraft was recovered 60 minutes later. Estimated distance traveled for the mission was 504,006 n mi.

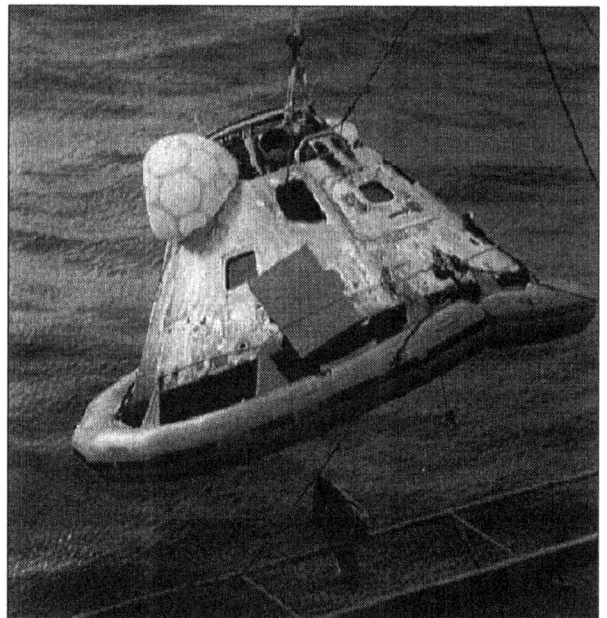

Apollo 8 CM is hoisted aboard the recovery ship (NASA S68-56304).

At the time the recovery swimmers were deployed, the weather recorded onboard the *Yorktown* showed scattered clouds at 2,000 feet and overcast at 9,000 feet, visibility ten n mi, wind speed 19 knots from 70° true north, water temperature 82° F, and waves to six feet from 110° true north.

The CM was offloaded from the *Yorktown* on December 29 at Ford Island, Hawaii. The Landing Safing Team began the evaluation and deactivation procedures at 09:00 GMT, and completed them on 1 January 1969. The CM was then flown to

Long Beach, California, and trucked to the North American Rockwell Space Division facility at Downey, California for postflight analysis. It arrived on 2 January 1969 at 09:00 GMT.

Conclusions

With only minor problems, all Apollo 8 spacecraft systems operated as intended, and all primary mission objectives were successfully accomplished. Crew performance was admirable throughout the mission. Approximately 90 percent of the photographic objectives were accomplished and 60 percent of the additional lunar photographs requested as "targets of opportunity" were also taken, despite fogging of three of the spacecraft windows due to exposure of the window sealant to the space environment and early curtailment of crew activities due to fatigue. Many smaller lunar features, previously undiscovered, were photographed. These features were located principally on the far side of the Moon in areas which had been photographed only at much greater distances by automated spacecraft. In addition, the heat shield system was not adversely affected by exposure to cislunar space or to the lunar environment and performed as expected. The following conclusions were made from an analysis of post-mission data:

1. The CSM systems were operational for a piloted lunar mission.

2. All system parameters and consumable quantities were maintained well within their design operating limits during both cis-lunar and lunar orbit flight.

3. Passive thermal control, a slow rolling maneuver perpendicular to the Sun line, was a satisfactory means of maintaining critical spacecraft temperatures near the middle of the acceptable response ranges.

4. The navigation techniques developed for translunar and lunar orbit flight were proved to be more than adequate to maintain required lunar orbit insertion and transearth injection guidance accuracies.

5. Non-simultaneous sleep periods adversely affected the normal circadian cycle of each crew member and provided a poor environment for undisturbed rest. Mission activity scheduling for the lunar orbit coast phase also did not provide adequate time for required crew rest periods.

6. Communications and tracking at lunar distances were excellent in all modes. The high-gain antenna, flown for the first time, performed exceptionally well and withstood dynamic structural loads and vibrations which exceeded anticipated operating levels.

7. Crew observations of the lunar surface showed the "washout" effect (surface detail being obscured by backscatter) to be much less severe than anticipated. In addition, smaller surface details were visible in shadow areas at low sun angles, indicating that lighting for lunar landing should be photometrically acceptable.

8. To accommodate the change in Apollo 8 from an Earth orbital to a lunar mission, pre-mission planning, crew training, and ground support reconfigurations were completed in a time period significantly shorter than usual. The required response was particularly demanding on the crew and, although not desirable on a long-term basis, exhibited a capability which had never before been demonstrated.

Apollo 8 Objectives

Spacecraft Primary Objectives

1. To demonstrate crew/space vehicle/mission support facilities performance during a piloted Saturn V mission with the command and service module. *Achieved.*

2. To demonstrate the performance of nominal and selected back-up lunar orbit rendezvous mission activities, including:

   a. Saturn targeting for translunar injection. *Achieved.*

   b. Long-duration service propulsion burns and midcourse corrections. *Achieved.*

   c. Pre-translunar injection procedures. *Achieved.*

   d. Translunar injection. *Achieved.*

   e. Command and service module orbital navigation. *Achieved.*

Primary Detailed Test Objectives

1. P1.31: To perform a guidance and navigation control system controlled entry from a lunar return. *Achieved.*

2. P1.33: To perform star-lunar horizon sightings during the translunar and transearth phases. *Achieved, although the field of view in the scanning telescope was obscured by what appeared to be particles whenever the telescope optics were repositioned.*

3. P1.34: To perform star-Earth horizon sightings during translunar and transearth phases. *Achieved, although the field of view in the scanning telescope was obscured by what appeared to be particles whenever the telescope optics were repositioned.*

4. P6.11: To perform manual and automatic acquisition, tracking, and communication with the Manned Space Flight Network using the high-gain command and service module S-band antenna during a lunar mission. *Achieved.*

5. P7.31: To obtain data on the passive thermal control system during a lunar orbit mission. *Achieved.*

6. P7.32: To obtain data on the spacecraft dynamic response. *Achieved.*

7. P7.33: To demonstrate spacecraft lunar module adapter panel jettison in a zero-g environment. *Achieved.*

8. P20.105: To perform lunar orbit insertion service propulsion system guidance and navigation control system controlled burns with a fully loaded command and service module. *Achieved.*

9. P20.106: To perform a transearth insertion guidance and navigation control system controlled service propulsion system burn. *Achieved.*

10. P20.107: To obtain data on the command module crew procedures and timeline for lunar orbit mission activities. *Achieved.*

11. P20.109: To demonstrate command service module passive thermal control modes and related communication procedures during a lunar orbit mission. *Achieved.*

12. P20.110: To demonstrate ground operational support for a command and service module lunar orbit mission. *Achieved.*

13. P20.111: To perform lunar landmark tracking in lunar orbit from the command and service module. (The intent of this objective was to establish that an onboard capability existed to compute relative position data for the lunar landing mission. This mode was to be used in conjunction with the Manned Space Flight Network state-vector update). *Partially achieved. All portions of the objective were satisfied except for the functional test, which required the use of onboard data to determine the error uncertainties in the landing site location. A procedural error caused the time intervals between the mark designations to be too short; thus, the data may have been correct but may not have been representative. The accuracy of the onboard capability was not determined because the data analysis was not complete at the time the mission report was published. Sufficient data were obtained to determine that no constraint existed for subsequent missions. A demonstration of this technique was planned for the next lunar mission.*

14. P20.112: To prepare for translunar injection and monitor the guidance and navigation control system and launch vehicle tank pressure displays during the translunar injection burn. *Achieved.*

15. P20.114: To perform translunar and transearth midcourse corrections. *Achieved, although the service propulsion system engine experienced a momentary drop in chamber pressure from 94 psi to 50 psi during the service propulsion system burn for midcourse correction, and the entry monitoring system velocity counter counted through zero at the termination of the transearth midcourse correction.*

Secondary Detailed Test Objectives

1. S1.27: To monitor the guidance and navigation control system and displays during launch. *Achieved.*

2. S1.30: To obtain inertial measurement unit performance data in the flight environment. *Achieved.*

3. S1.32: To perform star-Earth landmark sighting navigation during translunar and transearth phases. *Partially achieved. The three sets of sightings required at less than 50,000 n mi altitude were not obtained. The accuracy of other navigation modes was sufficient to preclude the necessity of using star-Earth landmarks for midcourse navigation. No constraint on subsequent missions resulted from this problem.*

4. S1.35: To perform an inertial measurement unit alignment and a star pattern visibility check in daylight. *Achieved.*

5. S3.21: To perform service propulsion system lunar orbit injection and transearth injection burns and monitor the primary and auxiliary gauging systems. *Achieved.*

6. S4.5: To obtain data on the block II environmental control system performance during piloted lunar return entry conditions. *Achieved, although the #2 cabin fan was noisy.*

7. S6.10: To communicate with the Manned Space Flight Network using the command and service module S-band omni antennas at lunar distance. *Achieved.*

8. S7.30: To demonstrate the performance of the block II thermal protection system during a piloted lunar return entry. *Achieved.*

9. S20.104: To perform a command and service module/S-IVB separation and a command and service module transposition on a lunar mission timeline. *Achieved.*

10. S20.108: To obtain data on command and service module consumables for a command and service module lunar orbit mission. *Achieved.*

11. S20.115: To obtain photographs during the transearth, translunar and lunar orbit phases for operational and scientific pur-

poses. *Achieved, although the hatch and side windows were obscured by fog or frost throughout the mission.*

12. S20.116: To obtain data to determine the effect of the tower jettison motor, S-II retro and service module reaction control system exhausts, and other sources of contamination on the command module windows. *Achieved. The hatch and side windows were obscured by fog or frost throughout the mission.*

Functional Tests Added to Primary Detailed Test Objectives During the Mission

1. P1.34: Star/earth horizon photography through the sextant. *Achieved.*

2. P1.34: Midcourse navigation with helmets on. *Achieved.*

3. P1.34: Navigation with long eyepiece. *Achieved.*

4. P6.11: High-gain antenna, automatic reacquisition. *Achieved.*

5. P20.109: Passive thermal control, roll rate of 0.3° per second. *Achieved.*

Launch Vehicle Primary Detailed Test Objectives

1. To verify that modifications incorporated in the S-IC stage since the Apollo 6 flight suppress low-frequency longitudinal oscillations (POGO). *Achieved.*

2. To confirm the launch vehicle longitudinal oscillation environment during the S-IC stage burn. *Achieved.*

3. To verify the modifications made to the J-2 engine since the Apollo 6 flight. *Achieved.*

4. To confirm the J-2 engine environment in the S-II and S-IVB stages. *Achieved.*

5. To demonstrate the capability of the S-IVB to restart in Earth orbit. *Achieved.*

6. To demonstrate the operation of the S-IVB helium heater repressurization system. *Achieved.*

7. To demonstrate the capability to safe the S-IVB stage in orbit. *Achieved.*

8. To verify the capability to inject the S-IVB/instrument unit/lunar module test article "B" into a lunar "slingshot" trajectory. *Achieved.*

9. To verify the capability of the launch vehicle to perform a free-return translunar injection. *Achieved.*

Launch Vehicle Secondary Detailed Test Objective

To verify the onboard command and communications system and ground system interface and the operation of the command and communications system in the deep space environment. *Achieved.*

Apollo 8 Spacecraft History

EVENT	DATE
Saturn S-II stage #3 delivered to KSC.	26 Dec 1967
Saturn S-IC stage #3 delivered to KSC.	27 Dec 1967
Saturn S-IC stage #3 erected on MLP #1.	30 Dec 1967
Saturn S-IVB stage #503 delivered to KSC.	30 Dec 1967
Saturn V instrument unit #503 delivered to KSC.	04 Jan 1968
BP-30 delivered to KSC.	06 Jan 1968
Lunar test article B delivered to KSC.	09 Jan 1968
Lunar test article B mated to spacecraft/LM adapter.	19 Jan 1968
Saturn S-II stage #3 erected.	31 Jan 1968
Saturn S-IVB stage #503 erected.	01 Feb 1968
Saturn V instrument unit #503 erected.	01 Feb 1968
Boilerplate payload (BP-30) and summary launch escape system erected.	05 Feb 1968
Launch vehicle electrically mated.	12 Feb 1968
Space vehicle overall test #1 completed (for unpiloted mission).	11 Mar 1968
Space vehicle pull test completed (for unpiloted mission).	25 Mar 1968
Space vehicle overall test #2 completed (for unpiloted mission).	08 Apr 1968
Decision made to de-erect boilerplate payload (BP-30) for service propulsion system skirt modifications.	10 Apr 1968
C mission changed to C prime mission.	27 Apr 1968
Spacecraft/LM adapter #11, instrument unit #503 and Saturn S-IVB stage #503 de-erected.	28 Apr 1968
Saturn S-II stage #3 de-erected.	29 Apr 1968
Saturn S-II stage #3 departed for Mississippi Test Facility for man-rating tests.	01 May 1968
Individual and combined CM and SM systems test completed at factory.	02 Jun 1968
LM descent stage #3 delivered to KSC.	09 Jun 1968
LM ascent stage #3 delivered to KSC.	14 Jun 1968
Saturn S-II stage #3 delivered to KSC from Mississippi Test Facility.	27 Jun 1968
Integrated CM and SM systems test completed at factory.	21 Jul 1968
Saturn S-II stage #3 re-erected.	06 Aug 1968
CSM #103 quads delivered to KSC.	11 Aug 1968
CM #103 and SM #103 ready to ship from factory to KSC.	11 Aug 1968
Service module #103 delivered to KSC.	12 Aug 1968
CM #103 delivered to KSC.	14 Aug 1968
Saturn S-IVB stage #503 erected.	14 Aug 1968
Saturn V instrument unit #503 erected.	15 Aug 1968
Facility verification vehicle erected.	16 Aug 1968
AS-503 designated Apollo 8. Decision made to replace LM with spacecraft/LM adapter and lunar test article B.	19 Aug 1968
CM #103 and SM #103 mated.	22 Aug 1968
Launch vehicle electrical systems test completed.	23 Aug 1968
CSM #103 combined systems test completed.	05 Sep 1968
Facility verification vehicle de-erected.	14 Sep 1968
BP-30 erected for service arm checkout.	15 Sep 1968
Spacecraft/LM adapter #11 delivered to KSC.	18 Sep 1968
CSM #103 altitude tests completed.	22 Sep 1968
Lunar test article B mated with spacecraft/LM adapter.	29 Sep 1968
Service arm overall test completed.	02 Oct 1968
BP-30 de-erected.	04 Oct 1968
CSM #103 moved to VAB.	07 Oct 1968
Space vehicle and MLP #1 transferred to launch complex 39A.	09 Oct 1968
Mobile service structure transferred to launch complex 39A.	12 Oct 1968
Space vehicle cutoff and malfunction test completed.	22 Oct 1968
CSM #103/Mission Control Center Houston test completed.	29 Oct 1968
CSM #103 integrated systems test completed.	02 Nov 1968
CSM #103 electrically mated to launch vehicle.	04 Nov 1968

Apollo 8 Spacecraft History

EVENT	DATE
Space vehicle electrically mated.	05 Nov 1968
Space vehicle overall test completed.	06 Nov 1968
Space vehicle overall test #1 (plugs in) completed.	07 Nov 1968
Launch vehicle/Mission Control Center Houston test completed.	11 Nov 1968
Launch umbilical tower/pad water system test completed.	12 Nov 1968
Space vehicle flight readiness test completed.	19 Nov 1968
Space vehicle hypergolic fuel loading completed.	30 Nov 1968
Saturn S-IC stage #3 RP-1 fuel loading completed.	02 Dec 1968
Space vehicle countdown demonstration test (wet) completed.	10 Dec 1968
Space vehicle countdown demonstration test (dry) completed.	11 Dec 1968

Apollo 8 Ascent Phase

Event	GET (hhh:mm:ss)	Altitude (n mi)	Range (n mi)	Earth Fixed Velocity (ft/sec)	Space Fixed Velocity (ft/sec)	Event Duration (deg E)	Geocentric Latitude (deg N)	Longitude (deg E)	Space Fixed Flight Path Angle (deg)	Space Fixed Heading Angle (E of N)
Liftoff	000:00:00.67	0.032	0.000	2.2	1,340.7		28.4470	-80.6041	0.00	90.00
Mach 1 achieved	000:01:01.45	3.971	1.297	1,076.3	2,078.4		28.4526	-80.5805	26.79	85.21
Maximum dynamic pressure	000:01:18.9	7.252	3.545	1,735.4	2,754.7		28.4645	-80.5398	29.56	82.43
S-IC center engine cutoff[2]	000:02:05.93	22.398	22.704	5,060.1	6,213.78	132.52	28.5581	-80.1934	24.527	76.572
S-IC outboard engine cutoff	000:02:33.82	35.503	48.306	7,698.0	8,899.77	160.41	28.6856	-79.7302	20.699	75.387
S-IC/S-II separation[2]	000:02:34.47	35.838	49.048	7,727.36	8,930.15		28.6893	-79.7168	20.605	75.384
S-II engine cutoff	000:08:44.04	103.424	812.267	21,055.6	22,379.1	367.85	31.5492	-65.3897	0.646	81.777
S-II/S-IVB separation[2]	000:08:44.90	103.460	815.159	21,068.14	22,391.60		31.5565	-65.3338	0.636	81.807
S-IVB 1st burn cutoff	000:11:24.98	103.324	1,391.631	24,238.3	25,562.43	156.69	32.4541	-54.0565	-0.001	88.098
Earth orbit insertion	000:11:34.98	103.326	1,430.363	24,242.9	35,532.41		32.4741	-53.2923	-2.072	87.47

Apollo 8 Earth Orbit Phase

Event	GET (hhh:mm:ss)	Space Fixed Velocity (ft/sec)	Event Duration (sec)	Velocity Change (ft/sec)	Apogee (n mi)	Perigee (n mi)	Period (mins)	Inclination (deg)
Earth orbit insertion	000:11:34.98	25,567.06			99.99	99.57	88.19	32.509
S-IVB 2nd burn ignition	002:50:37.79	25,558.6						
S-IVB 2nd burn cutoff	002:55:55.51	35,532.41	317.72	9,973.81				30.639

[2] Only the commanded time is available for this event.

Apollo 8 Translunar Phase

Event	GET (hhh:mm:ss)	Altitude (n mi)	Space Fixed Velocity (ft/sec)	Event Duration (sec)	Velocity Change (ft/sec)	Space Fixed Flight Path Angle (deg)	Space Fixed Heading Angle (E of N)
Translunar injection	002:56:05.51	187.221	35,505.41			7.897	67.494
CSM separated from S-IVB	003:20:59.3	3,797.775	24,974.90			45.110	107.122
Midcourse correction ignition	010:59:59.2	52,768.4	8,187			73.82	120.65
Midcourse correction cutoff	011:00:01.6	52,771.7	8,172	2.4	20.4	73.75	120.54
Midcourse correction ignition	060:59:55.9	21,064.5	4,101			-84.41	-86.90
Midcourse correction cutoff	061:00:07.8	21,059.2	4,103	11.9	1.4	-84.41	-87.01

Apollo 8 Lunar Orbit Phase

Event	GET (hhh:mm:ss)	Altitude (n mi)	Space Fixed Velocity (ft/sec)	Event Duration (sec)	Velocity Change (ft/sec)	Apogee (n mi)	Perigee (n mi)
Lunar orbit insertion ignition	069:08:20.4	75.6	8,391				
Lunar orbit insertion cutoff	069:12:27.3	62.0	5,458	246.9	2,997	168.5	60.0
Lunar orbit circularization ignition	073:35:06.6	59.3	5,479				
Lunar orbit circularization cutoff	073:35:16.2	60.7	5,345	9.6	134.8	60.7	59.7

Apollo 8 Transearth Phase

Event	GET (hhh:mm:ss)	Altitude (n mi)	Space Fixed Velocity (ft/sec)	Event Duration (sec)	Velocity Change (ft/sec)	Space Fixed Flight Path Angle (deg)	Space Fixed Heading Angle (E of N)
Transearth injection ignition	089:19:16.6	60.2	5,342			-0.16	-110.59
Transearth injection cutoff	089:22:40.3	66.1	8,842	203.7	3,519.0	5.10	-115.00
Midcourse correction ignition	104:00:00.00	165,561.5	4,299			-80.59	52.65
Midcourse correction cutoff	104:00:15.00	167,552.0	4,298	15.00	4.8	-80.60	52.65

Apollo 8 Timeline

Event	GET (hhh:mm:ss)	GMT Time	GMT Date
Terminal countdown started.	-028:00:00	01:51:00	20 Dec 1968
Scheduled 6-hour hold at T-9 hours.	-009:00:00	20:51:00	20 Dec 1968
Countdown resumed at T-9 hours.	-009:00:00	02:51:00	21 Dec 1968
Scheduled 1-hour hold at T-3 hours 30 minutes.	-003:30:00	08:21:00	21 Dec 1968
Countdown resumed at T-3 hours 30 minutes.	-003:30:00	09:21:00	21 Dec 1968
Crew ingress.	-002:53	09:58	21 Dec 1968
Guidance reference release.	-000:00:16.970	12:50:43	21 Dec 1968
S-IC engine start command.	-000:00:08.89	12:50:51	21 Dec 1968
S-IC engine ignition (#5).	-000:00:06.585	12:50:53	21 Dec 1968
All S-IC engines thrust OK.	-000:00:01.387	12:50:58	21 Dec 1968
Range zero.	000:00:00.00	12:51:00	21 Dec 1968
All holddown arms released.	000:00:00.27	12:51:00	21 Dec 1968
1st motion (1.16 g).	000:00:00.33	12:51:00	21 Dec 1968
Liftoff (umbilical disconnected).	000:00:00.67	12:51:00	21 Dec 1968
Tower clearance yaw maneuver started.	000:00:01.76	12:51:01	21 Dec 1968
Yaw maneuver ended.	000:00:09.72	12:51:09	21 Dec 1968
Pitch and roll maneuver started.	000:00:12.11	12:51:12	21 Dec 1968
Roll maneuver ended.	000:00:31.52	12:51:31	21 Dec 1968
Mach 1 achieved.	000:01:01.45	12:52:01	21 Dec 1968
Maximum bending moment achieved (60,000,000 lbf-in).	000:01:14.7	12:52:14	21 Dec 1968
Maximum dynamic pressure (776.938 lb/ft^2).	000:01:18.9	12:52:18	21 Dec 1968
S-IC center engine cutoff command.	000:02:05.93	12:53:05	21 Dec 1968
Pitch maneuver ended.	000:02:25.50	12:53:25	21 Dec 1968
S-IC outboard engine cutoff.	000:02:33.82	12:53:33	21 Dec 1968
S-IC maximum total inertial acceleration (3.96 g).	000:02:33.92	12:53:33	21 Dec 1968
S-IC maximum Earth-fixed velocity; S-IC/S-II separation command.	000:02:34.47	12:53:34	21 Dec 1968
S-II engine start command.	000:02:35.19	12:53:35	21 Dec 1968
S-II ignition.	000:02:36.19	12:53:36	21 Dec 1968
S-II aft interstage jettisoned.	000:03:04.47	12:54:04	21 Dec 1968
Launch escape tower jettisoned.	000:03:08.6	12:54:08	21 Dec 1968
Iterative guidance mode initiated.	000:03:16.22	12:54:16	21 Dec 1968
S-IC apex.	000:04:26.54	12:55:26	21 Dec 1968
S-II engine cutoff.	000:08:44.04	12:59:44	21 Dec 1968
S-II maximum total inertial acceleration (1.86 g).	000:08:44.14	12:59:44	21 Dec 1968
S-II maximum Earth-fixed velocity; S-II/S-IVB separation command.	000:08:44.90	12:59:44	21 Dec 1968
S-IVB 1st burn start command.	000:08:45.00	12:59:45	21 Dec 1968
S-IVB 1st burn ignition.	000:08:48.29	12:59:48	21 Dec 1968
S-IVB ullage case jettisoned.	000:08:56.8	12:59:56	21 Dec 1968
S-IC impact (theoretical).	000:09:00.41	13:00:00	21 Dec 1968
S-II apex.	000:09:20.34	13:00:20	21 Dec 1968
S-IVB 1st burn cutoff.	000:11:24.98	13:02:25	21 Dec 1968
S-IVB 1st burn maximum total inertial acceleration (0.72 g).	000:11:25.08	13:02:25	21 Dec 1968
S-IVB 1st burn maximum Earth-fixed velocity.	000:11:25.50	13:02:25	21 Dec 1968
Earth orbit insertion.	000:11:34.98	13:02:35	21 Dec 1968
Maneuver to local horizontal attitude started.	000:11:45.19	13:02:45	21 Dec 1968
Orbital navigation started.	000:13:05.19	13:04:05	21 Dec 1968
S-II impact (theoretical).	000:19:25.106	13:10:25	21 Dec 1968
Optics cover jettisoned.	000:42:05	13:33:05	21 Dec 1968
All spacecraft systems approved for translunar injection.	001:56:00	14:47:00	21 Dec 1968
CAPCOM (Collins): "All right, Apollo 8. You are go for TLI."	002:27:22	15:18:22	21 Dec 1968
S-IVB 2nd burn restart preparation.	002:40:59.54	15:31:59	21 Dec 1968

Apollo 8 Timeline

Event	GET (hhh:mm:ss)	GMT Time	GMT Date
S-IVB 2nd burn restart command.	002:50:29.51	15:41:29	21 Dec 1968
S-IVB 2nd burn ignition.	002:50:37.79	15:41:37	21 Dec 1968
S-IVB 2nd burn cutoff.	002:55:55.51	15:46:55	21 Dec 1968
S-IVB 2nd burn maximum total inertial acceleration (1.55 g).	002:55:55.61	15:46:55	21 Dec 1968
S-IVB LH$_2$ tank latch relief valve open.	002:55:55.91	15:46:55	21 Dec 1968
S-IVB 2nd burn maximum Earth-fixed velocity.	002:55:56.00	15:46:56	21 Dec 1968
S-IVB LH$_2$ tank CVS valve open.	002:55:56.19	15:46:56	21 Dec 1968
S-IVB safing procedures started.	002:55:56.19	15:46:56	21 Dec 1968
S-IVB LOX tank non-propulsive vent valve open.	002:55:56.42	15:46:56	21 Dec 1968
Translunar injection.	002:56:05.51	15:47:05	21 Dec 1968
Maneuver to local horizontal attitude and orbital navigation started.	002:56:15.77	15:47:15	21 Dec 1968
S-IVB LOX tank non-propulsive vent valve closed.	002:58:26.39	15:49:26	21 Dec 1968
S-IVB LH$_2$ tank CVS valve and tank relief valve closed.	003:10:55.71	16:01:55	21 Dec 1968
Maneuver to transposition and docking attitude started.	003:10:58.40	16:01:58	21 Dec 1968
Sequence to separate CSM from S-IVB/LTA started. High-gain antenna deployed.	003:20:56.3	16:11:56	21 Dec 1968
CSM separated from S-IVB.	003:20:59.3	16:11:59	21 Dec 1968
1st CSM evasive maneuver from S-IVB (RCS).	003:40:01	16:31:01	21 Dec 1968
S-IVB LH$_2$ tank latch relief valve open.	003:55:56.16	16:46:56	21 Dec 1968
S-IVB LH$_2$ tank latch relief valve closed.	004:10:55.77	17:01:55	21 Dec 1968
Last reported VHF uplink reception.	004:39:54	17:30:54	21 Dec 1968
S-IVB lunar slingshot attitude maneuver initiated.	004:44:56.63	17:35:56	21 Dec 1968
2nd CSM evasive maneuver from S-IVB (RCS).	004:45:01	17:36:01	21 Dec 1968
Last reported VHF downlink reception.	004:48	17:39	21 Dec 1968
S-IVB LH$_2$ tank CVS valve open.	004:55:56.02	17:46:56	21 Dec 1968
S-IVB lunar slingshot maneuver—LH$_2$ vent valve open command.	004:55:56.02	17:46:56	21 Dec 1968
S-IVB LOX dump start.	005:07:55.82	17:58:55	21 Dec 1968
S-IVB lunar slingshot maneuver—LOX dump started.	005:07:55.82	17:58:55	21 Dec 1968
S-IVB lunar slingshot maneuver—Apply velocity change.	005:07:56.03	17:58:56	21 Dec 1968
S-IVB start bottle vent dump start.	005:08:25.82	17:59:25	21 Dec 1968
S-IVB start bottle vent dump end.	005:10:55.83	18:01:55	21 Dec 1968
S-IVB pneumatic sphere dump start.	005:12:25.83	18:03:25	21 Dec 1968
S-IVB LOX dump end.	005:12:55.82	18:03:55	21 Dec 1968
S-IVB lunar slingshot maneuver—LOX dump ended.	005:12:56.03	18:03:56	21 Dec 1968
S-IVB LOX tank non-propulsive vent valve open.	005:12:59.0	18:03:59	21 Dec 1968
S-IVB LH$_2$ tank latch relief valve open.	005:13:01.23	18:04:01	21 Dec 1968
S-IVB cold helium dump start.	005:13:03.6	18:04:03	21 Dec 1968
S-IVB lunar slingshot maneuver—APS ignition.	005:25:55.85	18:16:55	21 Dec 1968
S-IVB lunar slingshot maneuver—APS cutoff.	005:38:08.56	18:29:08	21 Dec 1968
S-IVB lunar slingshot maneuver—APS depletion.	005:38:34.00	18:29:34	21 Dec 1968
S-IVB cold helium dump end.	006:03:03.5	18:54:03	21 Dec 1968
S-IVB pneumatic sphere dump end.	006:11:05.88	19:02:05	21 Dec 1968
1st use of high gain antenna.	006:33:04	19:24:04	21 Dec 1968
Midcourse correction ignition.	010:59:59.2	23:50:59	21 Dec 1968
Midcourse correction cutoff.	011:00:01.6	23:51:01	21 Dec 1968
Data processing by missions operations computer and backup computer lost for ten minutes due to undesirable instruction sequence.	011:51:00	00:42:00	22 Dec 1968
S-band mode testing started.	012:03:01	00:54:01	22 Dec 1968
1st television transmission started.	031:10:36	20:01:36	22 Dec 1968
1st television transmission ended.	031:34:13	20:15:13	22 Dec 1968
2nd television transmission started.	055:02:45	19:53:45	23 Dec 1968
2nd television transmission ended.	055:28:23	20:19:23	23 Dec 1968

Apollo 8 Timeline

Event	GET (hhh:mm:ss)	GMT Time	GMT Date
Equigravisphere.	055:38	20:29	23 Dec 1968
Midcourse correction ignition.	060:59:55.9	01:50:55	24 Dec 1968
Midcourse correction cutoff.	061:00:07.8	01:51:07	24 Dec 1968
CAPCOM: "Apollo 8, this is Houston. At 68:04, you are go for LOI."	068:04:07	08:55:07	24 Dec 1968
Lunar orbit insertion ignition.	069:08:20.4	09:59:20	24 Dec 1968
CAPCOM: "Apollo 8, Houston. One minute to LOS. All systems go."	068:57:06	09:48:00	24 Dec 1968
CAPCOM: "Safe journey, guys."	068:57:19	09:48:19	24 Dec 1968
LMP (Anders): "Thanks a lot, troops."	068:57:24	09:48:24	24 Dec 1968
CMP (Lovell): "We'll see you on the other side."	068:57:26	09:48:26	24 Dec 1968
CAPCOM: "Apollo 8, 10 seconds to go. You're go all the way."	068:57:54	09:48:54	24 Dec 1968
CDR (Borman): "Roger."	068:58:00	09:49:00	24 Dec 1968
CAPCOM: "Apollo 8, Houston. Over."	069:33:44	10:24:44	24 Dec 1968
CMP: "Go ahead, Houston. This is Apollo 8. Burn complete..."	069:33:52	10:24:52	24 Dec 1968
CAPCOM: "Apollo 8, this is Houston. Roger...good to hear your voice."	069:34:07	10:25:07	24 Dec 1968
Lunar orbit insertion cutoff.	069:12:27.3	10:03:27	24 Dec 1968
S-IVB closest approach to lunar surface.	069:58:55.2	10:49:55	24 Dec 1968
Control point sightings.	071:00	11:51	24 Dec 1968
16 mm camera photography started.	071:10	12:01	24 Dec 1968
3rd television transmission started.	071:40:52	12:31:52	24 Dec 1968
3rd television transmission ended.	071:52:52	12:43:52	24 Dec 1968
Pseudo-landing site sightings.	071:55	12:46	24 Dec 1968
16 mm photography stopped.	072:20	13:11	24 Dec 1968
Lunar orbit circularization ignition.	073:35:06.6	14:26:06	24 Dec 1968
Lunar orbit circularization cutoff.	073:35:16.2	14:26:16	24 Dec 1968
Training photography.	074:00	14:51	24 Dec 1968
CSM landmark tracking and photography.	074:15	15:06	24 Dec 1968
Stereo photography started.	075:20	16:11	24 Dec 1968
Stereo photography ended.	076:00	16:51	24 Dec 1968
Landmark lighting evaluation.	076:15	17:06	24 Dec 1968
Control point sightings.	079:20	20:11	24 Dec 1968
Control point sightings.	077:20	18:11	24 Dec 1968
Pseudo-landing site sightings.	078:00	18:51	24 Dec 1968
Pseudo-landing site sightings.	080:00	20:51	24 Dec 1968
Control point sightings.	081:20	22:11	24 Dec 1968
Pseudo-landing site sightings.	082:00	22:51	24 Dec 1968
4th television transmission started.	085:43:03	02:34:03	25 Dec 1968
LMP: "We are now approaching the lunar sunrise, and for all the people back on Earth, the crew of Apollo 8 has a message that we would like to send to you. In the beginning (reading from the Bible)..."	086:06:56	02:57:56	25 Dec 1968
CMP: "And God called the light 'Day'..."	086:07:29	02:58:29	25 Dec 1968
CDR: "And from the crew of Apollo 8, we close with good night, good luck, a Merry Christmas, and God bless all of you, all of you on the good Earth."	086:08:36	02:59:36	25 Dec 1968
4th television transmission ended.	086:09:46	03:00:46	25 Dec 1968
Maneuver to transearth injection attitude.	087:15	04:06	25 Dec 1968
CAPCOM: "Okay, Apollo 8...you have a go for TEI."	088:03:36	04:54:36	25 Dec 1968
Transearth injection ignition (SPS).	089:19:16.6	06:10:16	25 Dec 1968
Transearth injection cutoff.	089:22:40.3	06:13:40	25 Dec 1968
Two-way communication phaselock established, but no voice or telemetry.	089:28:47	06:19:47	25 Dec 1968
Two-way voice synchronization established.	089:33:28	06:24:28	25 Dec 1968
CMP: "Houston, Apollo 8, over."	089:34:16	06:25:16	25 Dec 1968
CAPCOM: "Hello, Apollo 8. Loud and clear."	089:34:19	06:25:19	25 Dec 1968

Apollo 8 Timeline

Event	GET (hhh:mm:ss)	GMT Time	GMT Date
CMP: "Roger. Please be informed there IS a Santa Claus."	089:34:25	06:25:25	25 Dec 1968
CAPCOM: "That's affirmative. You are the best ones to know."	089:34:31	06:25:31	25 Dec 1968
Two-way telemetry synchronization established.	089:43:00	06:34:00	25 Dec 1968
Midcourse correction ignition.	104:00:00.00	20:51:00	25 Dec 1968
Midcourse correction cutoff.	104:00:15.00	20:51:15	25 Dec 1968
5th television transmission started.	104:24:04	21:15:04	25 Dec 1968
5th television transmission ended.	104:33:35	21:24:35	25 Dec 1968
Onboard state vector and platform alignment data corrupted due to crew error.	106:26	23:17	25 Dec 1968
State vector and platform alignment data corrected.	106:45	23:36	25 Dec 1968
Test of high-gain antenna automatic acquisition.	110:16:55	3:07:55	26 Dec 1968
6th television transmission started.	127:45:33	20:36:33	26 Dec 1968
6th television transmission ended.	128:05:27	20:56:27	26 Dec 1968
1st reception of ground VHF during transearth coast.	142:16:00	11:07:00	27 Dec 1968
CM/SM separation.	146:28:48.0	15:19:48	27 Dec 1968
Entry.	146:46:12.8	15:37:12	27 Dec 1968
Communication blackout started.	146:46:37	15:37:37	27 Dec 1968
Maximum entry g force (6.84 g).	146:47:38.4	15:38:38	27 Dec 1968
Recovery aircraft received direction-finding signals from CM and established visual contact.	146:49	15:4	27 Dec 1968
Radar contact with CM established by recovery ship at 270 nautical miles.	146:50	15:41	27 Dec 1968
Radar contact with CM established by recovery ship at 109 nautical miles.	146:51	15:42	27 Dec 1968
Communication blackout ended.	146:51:42.0	15:42:42	27 Dec 1968
Radar contact with CM established by recovery ship at 60 nautical miles.	146:52	15:43	27 Dec 1968
Drogue parachute deployed.	146:54:47.8	15:45:47	27 Dec 1968
Main parachute deployed.	146:55:38.9	15:46:38	27 Dec 1968
Voice contact established with CM by recovery helicopter. Recovery beacon signal contact established with CM by recovery aircraft.	146:56:01	14:52	27 Dec 1968
Recovery beacon contact with CM established.	146:57:05	15:48:05	27 Dec 1968
Splashdown (went to apex-down).	147:00:42.0	15:51:42	27 Dec 1968
CM went to apex down position. Voice contact lost.	147:00:50	15:51:50	27 Dec 1968
CM returned to apex-up position.	147:07:45	15:58:45	27 Dec 1968
Crew aboard recovery ship.	148:29	17:20	27 Dec 1968
Recovery ship arrived at CM.	149:22	18:13	27 Dec 1968
Crew in life raft.	148:15	17:06	27 Dec 1968
Swimmers deployed to CM.	147:44	16:35	27 Dec 1968
Flotation collar inflated.	148:07	16:58	27 Dec 1968
CM hatch opened.	148:12	17:03	27 Dec 1968
Crew aboard recovery helicopter.	148:23	17:14	27 Dec 1968
CM aboard recovery ship.	149:29	18:20	27 Dec 1968
Deactivation of CM started at Ford Island, Hawaii.	200:09	21:00	29 Dec 1968
CM arrived at contractor's facility in Downey, CA.	296:09	21:00	02 Jan 1969

APOLLO 9

The Third Mission:
Testing the LM in Earth Orbit

Apollo 9 Summary

(3 March–13 March 1969)

Apollo 9 crew (l. to r.): Jim McDivitt, Dave Scott, Rusty Schweickart (NASA S69-17590).

Background

Apollo 9 was a Type D mission, a lunar module piloted flight demonstration in Earth orbit. It was the first piloted test of the "lunar ferry" that would put astronauts on the Moon. A lunar module first had been flown without a crew aboard Apollo 5 on 22 January 1968.

Many of the LM tests on Apollo 9 exceeded conditions expected in a lunar landing. To ensure that major objectives would be accomplished if Apollo 9 ended early, the schedule for the first half of the mission also included more work for the crew than the schedule of either Apollo 7 or Apollo 8.

Apollo 9 lunar module being prepared for altitude chamber testing (NASA S68-44471).

The primary objectives were as follows:

• to demonstrate crew, space vehicle, and mission support facilities performance during a piloted Saturn V mission with command and service modules and lunar module;

• to demonstrate lunar module crew performance;

• to demonstrate performance of nominal and selected backup lunar orbit rendezvous mission activities; and

• to assess command and service module and lunar module consumables.

To meet these objectives, the lunar module was evaluated during three separate piloting periods that required multiple activation and deactivation of systems, a situation unique to this mission.

The crew members were Colonel James Alton McDivitt (USAF), commander; Colonel David Randolph Scott (USAF), command module pilot; and Russell Louis "Rusty" Schweickart, lunar module pilot.

Selected in the astronaut group of 1962, McDivitt had been command pilot of Gemini 4. Born 10 June 1929 in Chicago, Illinois, he was 39 years old at the time of the Apollo 9 mission. McDivitt received a B.S. in aeronautical engineering from the University of Michigan in 1959. His backup for the mission was Commander Charles "Pete" Conrad, Jr. (USN).

Scott had been pilot of Gemini 8. Born 6 June 1932 in San Antonio, Texas, he was 36 years old at the time of the Apollo 9 mission. Scott received a B.S. from the U.S. Military Academy in 1954 and an M.S. in aeronautics and astronautics from the Massachusetts Institute of Technology in 1962. He was selected as an astronaut in 1963. His backup was Commander Richard Francis "Dick" Gordon, Jr. (USN).

Schweickart, a civilian, was making his first spaceflight. Born 25 October 1935 in Neptune, New Jersey, he was 33 years old at the time of the Apollo 9 mission. Schweickart received a B.S. in aeronautical engineering in 1956 and an M.S. in aeronautics and astronautics in 1963 from the Massachusetts Institute of Technology. His backup was Commander Alan LaVern Bean (USN).

The capsule communicators (CAPCOMs) for the mission were Major Stuart Allen Roosa (USAF), Lt. Commander Ronald Ellwin Evans (USN), Major Alfred Merrill Worden

(USAF), Conrad, Gordon, and Bean. The support crew were Major Jack Robert Lousma (USMC), Lt. Commander Edgar Dean Mitchell (USN/Sc.D.), and Worden. The flight directors were Eugene F. Kranz (first shift), Gerald D. Griffin (second shift), and M.P. "Pete" Frank (third shift).

The Apollo 9 launch vehicle was a Saturn V, designated SA-504. The mission also carried the designation Eastern Test Range #9025. The CSM was designated CSM-104 and had the call-sign "Gumdrop," derived from the appearance of the command module when it was transported on Earth. During shipment, it was covered in blue wrappings that gave it the appearance of a wrapped gumdrop. The lunar module was designated LM-3 and had the call-sign "Spider," derived from its arachnid-like configuration.

Launch Preparations

The launch was originally scheduled for 28 February 1969, and the terminal countdown had actually begun for that launch at 03:00:00 GMT on 27 February at T-28 hours. However, one-half hour into the scheduled 3-hour hold at T-16 hours, the countdown was recycled to T-42 hours to allow the crew to recover from a mild viral respiratory illness. The count was picked up at 19:30:00 GMT on 1 March.

A low-pressure disturbance southwest of Cape Kennedy in the Gulf of Mexico was the principal cause of overcast conditions. At launch time, stratocumulus clouds covered 70 percent of the sky (base 3,500 feet) and altostratus clouds covered 100 percent (base 9,000 feet); the temperature was 67.3° F; the relative humidity was 61 percent; and the barometric pressure was 14.642 lb/in². The winds, as measured by the anemometer on the light pole 60.0 feet above ground at the launch site, measured 13.4 knots at 160° from true north.

Ascent Phase

Apollo 9 was launched from Kennedy Space Center Launch Complex 39, Pad A, at a Range Zero time of 16:00:00 GMT (11:00:00 a.m. EST) on 3 March 1969. The planned launch window for Apollo 9 extended to 19:15:00 GMT.

Between 000:00:13.3 and 000:00:33.0, the vehicle rolled from a launch pad azimuth of 90° to a flight azimuth of 72°. The S-IC engine shut down at 000:02:42.76, followed by S-IC/S-II separation and S-II engine ignition. The S-II engine shut down at 000:08:56.22, followed by separation from the S-IVB, which ignited at 000:09:00.82. The first S-IVB engine cutoff occurred at 000:11:04.66, with devia-

tions from the planned trajectory of +2.86 ft/sec in velocity and -0.17 n mi in altitude.

Apollo 9, the first piloted flight with a lunar module, lifts off from Kennedy Space Center Pad 39A to test the LM in Earth orbit (NASA S1969-25863).

The S-IC stage impacted at 000:08:56.44 in the Atlantic Ocean at latitude 30.183° north and longitude 74.238° west, 346.64 n mi from the launch site. The S-II stage impacted at 000:20:25.35 in the Atlantic Ocean at latitude 31.462° north and longitude 34.041° west, 2,413.2 n mi from the launch site.

The maximum wind conditions encountered during ascent were 148.1 knots at 264° from true north at 38,480 feet, and a maximum wind shear of 0.0254 sec-¹ at 48,160 feet.

Parking orbit conditions at insertion, 000:11:14.65 (S-IVB cutoff plus 10 seconds to account for engine tailoff and other transient effects), showed an apogee and perigee of 100.74 by 99.68 n mi, an inclination of 32.552°, a period of 88.20 minutes, and a velocity of 25,569.78 ft/sec. The apogee and perigee were based upon a spherical Earth with a radius of 3,443.934 n mi.

The international designation for the CSM upon achieving orbit was 1969-018A; the S-IVB was designated 1969-018B. After undocking, the LM ascent stage would be designated 1969-018C and the descent stage 1969-018D.

Earth Orbit Phase

After post-insertion checkout, the CSM was separated from the S-IVB stage at 002:41:16.0. The adapter that housed the LM and shielded it from the rigors of launch was then jettisoned. The CM was turned so its apex, holding the docking probe, faced the LM. Docking with the LM was completed at 003:01:59.3.

Lunar module inside S-IVB stage following separation (NASA AS09-19-2919).

Once docking was complete, the commander and lunar module pilot started preparations for their eventual entry into the LM. They pressurized the tunnel between the two spacecraft, and with the aid of the CMP, removed the CM hatch and checked the latches on the docking ring to verify the seal. Then they connected the electrical umbilical lines that would provide power to the LM while docked to the CM. The hatch was then replaced.

At 004:08:06, an ejection mechanism, used for the first time, ejected the docked spacecraft from the S-IVB.

Following a separation maneuver, the S-IVB was restarted at 004:45:55.54 and burned for 62.06 seconds. Ten seconds later, the S-IVB entered a 1,671.58 by 105.75 n mi intermediate coasting orbit that would allow the engine to cool down sufficiently prior to a restart within one revolution. The period of the orbit was 119.22 seconds, the inclination was 32.302°, and the velocity at insertion was 27,753.61 ft/sec.

At 005:59:01.07, the crew performed the first of eight service propulsion firings, a 5.23-second maneuver that raised the CSM/LM orbit to 127.6 by 113.4 n mi.

The third and final S-IVB ignition at 006:07:19.26 was a 242.06-second maneuver to demonstrate restart capability after the 80-minute coast and to test the engine performance under "out-of-specification" conditions. It also provided better ground tracking lighting conditions for the upcoming rendezvous. The escape orbit was achieved 10 seconds after S-IVB engine cutoff, and the velocity was 31,619.85 ft/sec. S-IVB performance was not as predicted due to various anomalies, including the failure of an LH_2 and LOX dump. The LH_2 dump through the engine could not be accomplished due to loss of pneumatic control of the engine valves. The LOX dump was not performed due loss of engine pneumatic control during the third burn. The LOX tank was satisfactorily safed by utilizing the LOX non-propulsive venting system.

The third ignition also served to place the S-IVB into a solar orbit with an aphelion and perihelion of 80,280,052 by 69,417,732 n mi, an inclination of 24.390°, an eccentricity of 0.07256, and a period of 325.8 days.

Crew activity on the second day was devoted to systems checks, pitch and roll yaw maneuvers, and the second, third, and fourth service propulsion system burns while docked to the LM. The second burn, a 110.29-second maneuver at 022:12:04.07, raised the orbit to 192.5 by 110.76 n mi. The third burn, at 025:17:39.27, lasted 279.88 seconds. It raised the orbit to 274.9 by 112.4 n mi and lightened the spacecraft so that it could be controlled by the reaction control system engines later in the mission and be in a better rescue position for rendezvous activities. During these two burns, tests were made to measure the oscillatory response of a docked spacecraft to provide data to improve the autopilot response for this configuration. The fourth burn, at 028:24:41.37, was a 27.87-second phasing maneuver to shift the node east and put the spacecraft in a better position later for lighting, braking, and docking.

On the third day, at 043:15, the lunar module pilot transferred to the LM to activate and check out the systems. The commander followed at 044:05. The LM landing gear was deployed at 045:00.

At 045:40, the commander reported that the lunar module pilot had been sick on two occasions and that the crew was behind in the timeline. For these reasons, the extravehicular activity was restricted to one daylight pass and would include only the opening of the hatches of the CM and LM. It was also decided to keep the lunar module pilot connected to the environmental control system hoses.

After communication checks for both vehicles, a five-minute television transmission was broadcast at 046:25 from inside the LM. The camera was trained on the instrument displays, other features of the LM interior, and the crew. The picture was good, but the sound was unsatisfactory.

The descent engine was fired for 371.51 seconds at 49:41:34.46 with the vehicles still docked. Attitude control with the digital autopilot and manual throttling of the descent engine to full thrust were also demonstrated. Transfer back to the CM began at 050:15, and the LM was deactivated at 051:00. The fifth service propulsion system firing, 43.26 seconds in duration, occurred at 054:26:12.27 to circularize the orbit for the LM active rendezvous. The resulting orbit was 131.0 by 125.9 n mi, compared to a desired circular orbit of 130.0 n mi, but it was considered acceptable for the rendezvous sequence.

Extravehicular operations were demonstrated on the fourth day of the mission. The plan was for the lunar module pilot to exit the LM, transfer to the open hatch in the CM, and then return. This plan was abbreviated from 2 hours 15 minutes to 39 minutes because of several bouts of nausea experienced by the lunar module pilot on the preceding day and because of the many activities required for rendezvous preparation.

The LM was depressurized at 072:45 and the forward hatch opened at 072:53. The lunar module pilot began his egress to the forward platform at 72:59:02, feet first and face up, and completed egress at 073:07. He was wearing the extravehicular mobility unit backpack, which provided communications and oxygen; it also circulated water through the suit to keep him cool. His only connection to the LM was a 25-foot nylon rope to keep him from drifting into space. He secured his feet in the "golden slippers," the gold-painted restraints affixed to the surface outside the hatch, called the "front porch" by the astronauts, where he remained while outside the LM.

During this same period, the command module pilot, dependent on CSM systems for life support, depressurized the CM and opened the side hatch at 073:02:00. He partially exited the hatch for observation, photography, and retrieval of thermal samples from the side of the CM. The samples were missing, so he retrieved the service module thermal samples at 073:26. The lunar module pilot retrieved the LM thermal samples at 073:39. Three minutes later, he began an abbreviated evaluation of translation and

body-attitude-control capability using the extravehicular transfer handrails. The initially planned hand-over-hand trip from the LM to the CM was not made. During this period, the lunar module pilot also completed 16 mm and 70 mm photography of the command module pilot's activities and the exterior of both spacecraft.

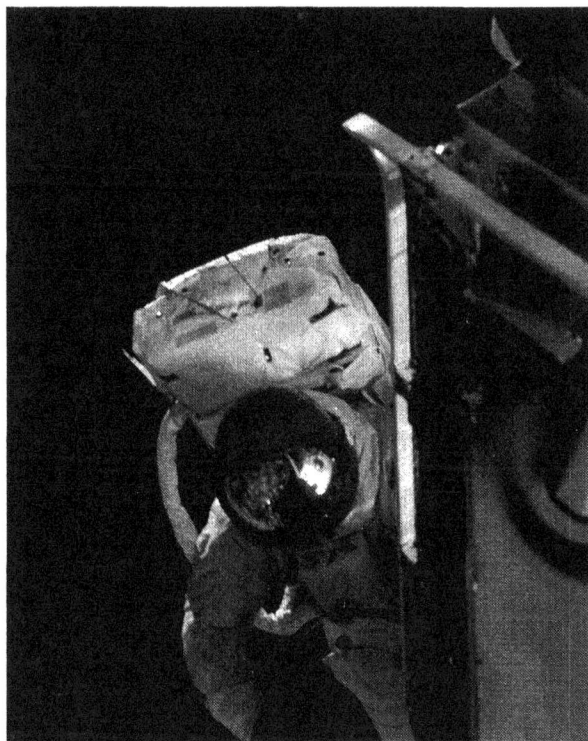

Schweickart on "porch" of LM during EVA activities (NASA AS09-19-2994).

The lunar module pilot began his ingress at 073:45 and completed it at 073:46:03. By 073:53, the forward hatch was closed and locked and the LM was repressurized. The CM hatch was then closed and locked at 073:49, and the CM was repressurized by 074:02. The second television transmission was made at 074:55. The commander returned to the CM at 075:15, followed by the lunar module pilot at 076:55.

After the lunar module pilot came back inside, both spacecraft were repressurized, and a second and final 10-minute television broadcast was telecast from inside the LM. Voice and pictures were both good, an improvement over the previous day's transmission.

On the fifth day, the lunar module pilot transferred to the LM at 088:05, followed by the commander at 088:55, to prepare for the first LM free flight and active rendezvous.

The CSM was maneuvered to the inertial undocking attitude at 092:22. Undocking was attempted at 092:38:00, but the capture latches did not release immediately. Undocking occurred at 092:39:36, and the LM was rolled on its axis so that the CMP could make a visual inspection. A small separation maneuver at 093:02:54, using the service module reaction control system, placed the LM 2.0 n mi behind the CSM 45 minutes later. The maximum range between the LM and CSM was 98 n mi, achieved about halfway between the coelliptical sequence initiation and constant differential height maneuver.

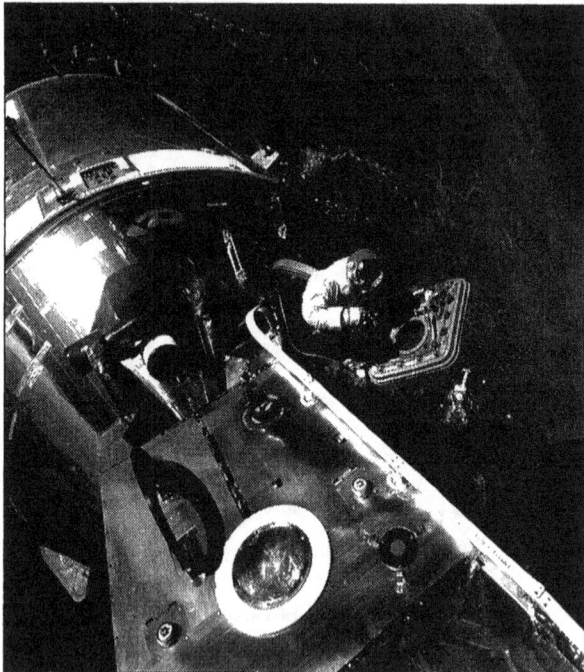

Scott opens the CM hatch during EVA activities (NASA AS09-20-3064).

Schweickart during EVA (NASA AS09-19-2983).

During this maneuver, the LM engine ran smoothly until throttled to 20 percent, at which time it chugged noisily. The commander stopped throttling and waited. Within seconds, the chugging stopped. He accelerated to 40 percent before shutting down and had no more problems. The LM crew then checked their systems and fired the descent engine again to 10 percent. It ran evenly.

LM in first free flight following separation from CSM (NASA AS09-21-3199).

The first LM rendezvous phasing maneuver was executed at 093:47:35.4 with the descent propulsion system under abort guidance control. This maneuver placed the LM in a near equiperiod orbit with apogee and perigee altitudes 12.2 n mi above and below the CSM. The second maneuver was not applied; it was a computation to be used only in case of a contingency requiring an LM abort. The solution time was 094:57:53. The third rendezvous maneuver was executed at 095:39:08.06 and resulted in an LM orbit of 138.9 by 133.9 n mi.

Coelliptic sequence initiation was performed at 096:16:06.54, and the descent stage was jettisoned immediately after the start of reaction control system thrusting. The maneuver left the LM 10 n mi below and 82 n mi behind the CSM. The descent stage remained in Earth orbit until 03:45 GMT on 23 March, when it impacted the Indian Ocean off the coast of eastern Africa.

The resulting ascent stage orbit was 116 by 111 n mi. After coelliptic sequence initiation using the CSM reaction control system, rendezvous radar tracking was reestablished, but the CM was unable to acquire the ascent stage tracking light, which had failed. The constant differential height maneuver was performed at 096:58:15.0, using the ascent stage engine for the first time. The onboard solution for terminal phase initiation was executed at 097:57:59, creating an ascent stage orbit of about 126 by 113 n mi. Two small midcourse corrections were performed at 10 and 22 minutes after terminal phase initiation. Terminal phase braking began at 098:30:03, followed by stationkeeping, formation flying, photography, and docking at 099:02:26. The ascent stage had been separated from the CSM for 6 hours 22 minutes 50 seconds.

LM ascent stage following separation from descent stage, preparing to redock with CM (NASA AS09-21-3236).

After docking, the crew transferred back to the CSM by 101:00. The ascent stage was jettisoned at 101:22:45.0, and the ascent engine fired for 362.4 seconds at 101:53:15.4 until oxidizer depletion. The final orbit for the ascent stage was 3,760.9 by 126.6 n mi, with an expected orbital lifetime of five years; however, entry occurred on 23 October 1981.

The sixth service propulsion burn, a 1.43-second maneuver at 123:25:06.97, had been postponed for one revolution because the reaction control translation required prior to ignition for propellant settling was improperly programmed. The maneuver, originally scheduled for 121:48:00, was an orbit-shaping retrograde maneuver to lower the perigee so that the reaction control system deorbit capability would be enhanced in the event of a contingency.

During the final four days in orbit, the crew conducted Earth resources and multispectral terrain photography experiments over the southern United States, Mexico, Brazil, and Africa. One objective, designated experiment S065, was to determine the extent to which multiband photography in the visible and near-infrared regions from orbit may be effectively applied to the Earth resources disciplines.

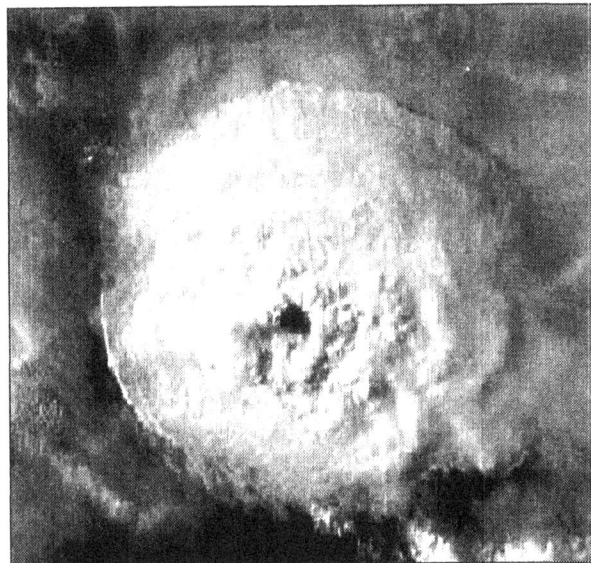

Thunderhead over South America as seen in nearly vertical view from Apollo 9 (NASA AS09-22-3374).

The other objective was to obtain simultaneous photographs with four different film/filter combinations from orbit to assist in defining future multispectral photographic systems. The results were excellent. The quality and subject material exceeded that of any previous orbital mission and would aid in future program planning. The reasons for the excellent results were the amount of time available (four days so the crew could wait for cloud cover to pass); the orbital inclination of 33.6°, which permitted vertical and near-vertical coverage of areas never photographed before; sufficient reaction control propellants which allowed the crew to orient the spacecraft whenever necessary; the lack of contamination on the spacecraft windows; and the continuous assistance and evaluation of the science support room at the Manned Spacecraft Center.

The crew also made an inertial measurement unit alignment with a sighting of the planet Jupiter (the first time a planet had been used) and performed a number of daylight star sightings, landmark sightings, and star sextant sightings. During two successive revolutions, at 192:43 and 194:13, the crew successfully tracked the Pegasus III satellite at a range of 1,000 n mi. Pegasus III had been launched on 30 July 1965.

While over Hawaii, the crew made a sighting of the ascent stage from 222:38:40 to 222:45:40.

The service propulsion system was fired for the seventh time at 169:30:00.36, a 24.90-second maneuver that raised the apogee to 253.2 by 100.7 n mi and established the desired conditions for the nominal deorbit point. If the service propulsion system had failed at deorbit, the reaction control system could have conducted a deorbit maneuver from this apogee condition and still landed near the primary recovery area. The deorbit maneuver was accomplished after 151 orbits with the eighth service propulsion firing, an 11.74-second maneuver at 240:31:14.84. It was performed one revolution later than planned because of unfavorable weather in the planned recovery area.

Recovery

The service module was jettisoned at 240:36:03.8, and the CM entry followed a primary guidance system profile. The command module reentered Earth's atmosphere (400,000 feet altitude) at 240:44:10.2 at a velocity of 25,894 ft/sec. Although the service module could not survive entry intact, radar tracking data predicted impact in the Atlantic Ocean at a point estimated to be latitude 22.0° north and longitude 65.3° west, 175 n mi downrange from the CM.

The parachute system effected splashdown of the CM in the Atlantic Ocean at 17:00:54 GMT (12:00:54 p.m. EST) on 13 March. Mission duration was 241:00:54. The impact point was about 2.7 n mi from the target point and 3 n mi from the recovery ship U.S.S. Guadalcanal. The splashdown site was estimated to be latitude 23.22° north and longitude 67.98° west. After splashdown, the CM assumed an apex-up flotation attitude. The crew was retrieved by helicopter and was aboard the recovery ship 49 minutes after splashdown. The CM was recovered 83 minutes later. The estimated CM weight at splashdown was 11,094 pounds, and the estimated distance traveled for the mission was 3,664,820 n mi.

At CM retrieval, the weather recorded onboard the Guadalcanal showed scattered clouds at 2,000 feet and broken clouds at 9,000, visibility 10 n mi, wind speed 9 kn from 200° true north, air temperature 79° F, and water temperature 76° F, with waves to seven feet.

The crew left the Guadalcanal by helicopter at 15:00 GMT on 14 March and arrived at Eleuthera, Bahamas, at 16:30 GMT. From there, they were flown to Houston.

The CM was offloaded from the Guadalcanal on 16 March at the Norfolk Naval Air Station, Norfolk, Virginia, and the Landing Safing Team began the evaluation and deactivation procedures at 16:00 GMT. Deactivation was completed on 19 March. The CM was then flown to Long Beach, California, and trucked to the North American Rockwell Space Division facility at Downey, California, for postflight analysis, where it arrived on 21 March.

Apollo 9 CM on parachute system just before splashdown (NASA S69-20364).

Apollo 9 crew aboard recovery ship U.S.S. Guadalcanal (l. to r.: Schweickart, Scott, McDivitt) (NASA S69-27921).

Apollo 9 CM onboard recovery ship (NASA S69-20239).

Conclusions

The following conclusions were made from an analysis of post-mission data:

1. The onboard rendezvous equipment and procedures in both spacecraft provided the required precision for rendezvous operations to be conducted during a lunar landing mission. The CSM computations and preparations for mirror-image maneuvers were completed on time by the command module pilot.

2. The functional operation of the docking process of the two spacecraft was demonstrated. However, the necessity for proper lighting conditions for the docking alignment aids was illustrated.

3. The performance of all systems in the extravehicular mobility unit was excellent throughout the entire extravehicular operation. The results of this mission, plus satisfactory results from additional qualification tests of minor design changes, provided verification of the operation of the extravehicular mobility unit on the lunar surface.

4. The extent of the extravehicular activity indicated the practicality of extravehicular crew transfer in the event of a contingency. Cabin depressurization and normal repressurization were demonstrated in both spacecraft.

5. Performance of the lunar module systems demonstrated the operational capability to conduct a lunar mission, except for the steerable antenna, which was not operated, and the landing radar, which could not be fully evaluated in Earth orbit. None of the anomalies adversely affected the mission. The concepts and operational functioning of the crew/spacecraft interfaces, including procedures, provisioning, restraints, displays, and controls, were satisfactory for piloted lunar module functions. The interfaces between the two spacecraft, while both docked and undocked, were also verified.

6. The lunar module consumable expenditures were well within predicted values, thus demonstrating adequate margins to perform the lunar mission.

7. Gas in the CM potable water supply interfered with proper food rehydration and therefore had some effect on food taste and palatability. Lunar module water was acceptable.

8. Orbital navigation of the CSM, using the yaw-control technique for landmark tracking, was demonstrated and reported to be adequate. The star visibility threshold of the CM scanning telescope was not definitely established for the docked configuration; therefore, platform orientation using the sun, the Moon, and planets may be required if inertial reference is inadvertently lost during translunar flight.

9. Mission support, including the Manned Space Flight Network, adequately provided simultaneous ground control of two piloted spacecraft.

Apollo 9 Objectives

Spacecraft Primary Objectives

1. To demonstrate crew/space vehicle/mission support facilities performance during a piloted Saturn V mission with command, service, and lunar modules. *Achieved.*

2. To demonstrate lunar module/crew performance. *Achieved.*

3. To demonstrate performance of nominal and selected backup lunar orbit rendezvous mission activities, including the following:

   a. Transposition, docking, and lunar module withdrawal. *Achieved.*

   b. Intravehicular and extravehicular crew transfer. *Achieved.*

   c. Extravehicular capability. *Achieved.*

d. Service propulsion system and descent propulsion system burns. *Achieved.*

e. Lunar module active rendezvous and docking. *Achieved.*

4. To assess command, lunar, and service module consumables. *Achieved.*

Mandatory Detailed Test Objectives

1. M11.6: To perform a medium-duration descent propulsion system firing to include manual throttling with command and service module and lunar module docked, and a short-duration descent propulsion system firing with an undocked lunar module and approximately half-full descent propulsion system propellant tanks. *Achieved; the primary guidance and navigation control system/digital auto-pitch performance was monitored and found acceptable during the first and second descent propulsion system burns.*

2. M13.11: To perform a long-duration ascent propulsion system burn. *Achieved; a burn to depletion was performed by the ascent propulsion system for an extended period.*

3. M13.12: To perform a long-duration descent propulsion system burn and obtain data to determine that no adverse interactions exist between propellant slosh, vehicle engine vibration, and descent propulsion system performance during a burn. *Achieved; data were collected during the docked descent propulsion system burn and the rendezvous.*

4. M14: To demonstrate the performance of the environmental control system during lunar module activity periods. *Achieved, although minor problems occurred in the system.*

5. M15.3: To determine the performance of the lunar module electrical power subsystem in the primary and backup modes. *Achieved, despite some problems in the fuel cells.*

6. M16.7: To operate the landing radar during the descent propulsion system burns. *Achieved.*

7. M17.9: To deploy the lunar module landing gear and obtain data on landing gear temperatures resulting from descent propulsion system operation. *Achieved.*

8. M17.17: To verify the performance of the passive thermal subsystems (thermal blanket, plume protection, ascent and descent stage base heat shields, and thermal control coatings) to provide adequate thermal control when the spacecraft is exposed to the natural and propulsion-induced thermal environments. *Achieved; lunar module environmental and thermal effect data were collected during the docked descent propulsion system burn, extravehicular activity, and post-rendezvous inspection.*

9. M17.18: To demonstrate the structural integrity of the lunar module during Saturn V launch and during descent propulsion system and ascent propulsion system burn in an orbital environment. *Achieved.*

Primary Detailed Test Objectives

1. P1.23: To demonstrate block II command and service module attitude control during service propulsion system thrusting with the command and service module and lunar module docked. *Achieved during the first, second, and third service propulsion burns.*

2. P1.24: To perform inertial measurement unit alignments using the sextant while docked. *Achieved.*

3. P1.25: To perform an inertial measurement unit and a star pattern visibility check in daylight while docked. *Achieved; many daytime sightings were made with visible star patterns, although reflective light hindered some tests.*

4. P2.9: To perform manual thrust vector control takeover of a guidance navigation control system initiated service propulsion docked burn. *Achieved during the third service propulsion system burn.*

5. P7.29: To obtain data on the effects of the tower jettison motor, S-II retrorockets, and service module reaction control system exhaust on the command and service module. *Achieved. Spacecraft exhaust effects data were collected following Earth orbital insertion, and lunar module/command and service module ejection during the revised extravehicular period and during the post-rendezvous inspection; however, the revised extravehicular activity permitted recovery of only part of the thermal samples.*

6. P11.5: To perform lunar module inertial measurement unit alignments using the alignment optical telescope and calibrate the coarse optical alignment sight. *Achieved; lunar module inflight inertial measurement unit alignment data were collected at various times during lunar module activity periods.*

7. P11.7: To demonstrate reaction control system translation and attitude control of the staged lunar module using automatic and manual primary guidance and navigation control system controls. *Achieved.*

8. P11.10: To obtain data to verify inertial measurement unit performance in the flight environment. *Achieved; lunar module primary guidance and navigation control system and command and service module guidance navigation control system inertial measurement unit performance data were collected throughout the mission.*

9. P11.14: To perform a primary guidance and navigation control system/digital autopilot controlled long-duration ascent propulsion burn. *Achieved.*

10. P12.2: To demonstrate an abort guidance system calibration and obtain abort guidance system performance data in the flight environment. *Achieved during docked descent propulsion system burn and the rendezvous phasing burn.*

11. P12.3: To demonstrate reaction control system translation and attitude control of unstaged lunar module using automatic and manual abort guidance system/control electric section control modes. *Achieved.*

12. P12.4 To perform an abort guidance system/control electric section controlled descent propulsion system burn with a heavy descent stage. *Achieved.*

13. P16.4: To demonstrate tracking of command and service module rendezvous radar transponder at various ranges between the command and service module and the lunar module. *Achieved.*

14. P16.6: To perform a landing radar self-test. *Achieved.*

15. P16.19: To obtain data on rendezvous radar corona susceptibility during lunar module -X translation reaction control system engine firings while undocked and during -X reaction control system engine firings while docked. *Partially Achieved. Data were obtained, but the rendezvous radar failed to lock.*

16. P20.21: To demonstrate the lunar module/Manned Space Flight Network operational S-band communication subsystem capability. *Achieved, despite intermittent discrepancies.*

17. P20.22: To demonstrate lunar module/command and service module/ Manned Space Flight Network/extravehicular activity operational S-band and VHF communication compatibility. *Achieved, despite sporadic failures.*

18. P20.24: To demonstrate command and service module docking with the S-IVB/spacecraft/lunar module adapter/lunar module. *Achieved.*

19. P20.25: To demonstrate lunar module separation and ejection of the command and service module/lunar module from the spacecraft/lunar module adapter. *Achieved.*

20. P20.26: To demonstrate the technique to be employed for the undocking of the lunar module from the command and service module prior to lunar descent. *Achieved.*

21. P20.27: To perform a lunar module active rendezvous with a passive command and service module. *Achieved.*

22. P20.28: To demonstrate lunar module active docking capability with the passive command and service module. *Achieved.*

23. P20.29: To perform a pyrotechnic separation of the lunar module and command and service module in flight. *Achieved.*

24. P20.31: To demonstrate mission support facilities performance during an Earth orbital mission. *Achieved.*

25. P20.33: To perform procedures required to prepare for a command and service module active rendezvous with the lunar module. *Achieved; the command and service modules were maintained in a recovery mode during the lunar module simulated descent.*

26. P20.34: To demonstrate crew capability to transfer themselves and equipment from the command and service module to the lunar module and return. *Achieved; the crew was successful in making the transfer in the time allotted.*

27. P20.35: To demonstrate extravehicular transfer and obtain extravehicular activity data. *Achieved, although the program was modified during the mission.*

Secondary Detailed Test Objectives

1. S1.26: To perform onboard navigation using the technique of scanning telescope landmark tracking. *Achieved.*

2. S13.10: To perform an unpiloted ascent propulsion burn to depletion. *Achieved.*

3. S20.32: To evaluate one-person lunar module operation capability and obtain data on crew maneuverability, crew compartmentation, and propulsive venting. *Achieved.*

4. S20.37: To obtain data on descent propulsion plume effects on astronauts' visibility. *Achieved; the descent propulsion system did not affect the crew's visibility during the two burns.*

5. S20.120: To obtain data on the electromagnetic compatibility of the command and service module, lunar module, and portable life support system. *Achieved; the command and service module, lunar module, and portable life support system were electromagnetically compatible with respect to any conducted or radiated electromagnetic interference.*

Functional Tests Added During The Mission

1. Command and service module intravehicular transfer, unsuited. *Achieved.*

2. Tunnel clearing, unsuited. *Achieved.*

3. Command module platform alignment in daylight. *Achieved.*

4. Command module platform alignment, using a planet (Jupiter). *Achieved.*

5. Digital autopilot orbital rate, pitch and roll. *Achieved.*

6. Backup gyro display coupler alignment of stabilization and control system. *Achieved.*

7. Window degradation photography. *Achieved.*

8. Satellite tracking, ground inputs. *Achieved.*

9. Command and service module high-gain S-band antenna reacquisition test. *Achieved.*

10. Passive thermal control cycling at 0.1°/second at three deadbands: +/-10°, +/-20°, and +/-25°. *Achieved.*

Experiment

S-065: To obtain selective, simultaneous multispectral photographs, with four different film/filter combinations, of selected land and ocean areas. *Achieved.*

Launch Vehicle Primary Objective

To demonstrate S-IVB/instrument unit control capability during transposition, docking, and lunar module ejection maneuver. *Achieved.*

Launch Vehicle Secondary Objectives

1. To demonstrate S-IVB restart capability. *Achieved.*

2. To verify J-2 engine modifications. *Achieved.*

3. To confirm J-2 environment in S-II stage. *Achieved.*

4. To confirm launch vehicle longitudinal oscillation environment during S-IC stage burn period. *Achieved.*

5. To demonstrate helium heater repressurization system operation. *Achieved.*

6. To demonstrate S-IVB propellant dump and safing with a large quantity of residual S-IVB propellants. *Partially achieved. The S-IVB stage was adequately safed; however, propellant dump was not achieved due to loss of engine helium control regulator discharge pressure.*

7. To verify that modifications incorporated in the S-IC stage suppress low-frequency longitudinal oscillations. *Achieved.*

8. To demonstrate 80-minute restart capability. *Partially achieved. The experimental start was achieved, and it accomplished the planned S-IVB third burn. However, rough combustion, a gas generator spike at ignition, and control oscillations resulted in a low performance at start, performance loss during the burn, and loss of engine helium control regulator discharge pressure.*

9. To demonstrate dual repressurization capability. *Achieved.*

10. To demonstrate helium heater restart capability. *Achieved.*

11. To verify the onboard command and communications system/ground system interface and operation in the deep space environment. *Achieved.*

Apollo 9 Spacecraft History

EVENT	DATE
LM #3 integrated test at factory.	31 Jan 1968
Saturn S-II stage #4 delivered to KSC.	15 May 1968
LM #3 final engineering evaluation acceptance test at factory.	17 May 1968
LM descent stage #3 ready to ship from factory to KSC.	04 Jun 1968
LM descent stage #3 delivered to KSC.	09 Jun 1968
LM ascent stage #3 ready to ship from factory to KSC.	12 Jun 1968
LM ascent stage #3 delivered to KSC.	14 Jun 1968
LM ascent stage #3 and descent stage #3 mated.	30 Jun 1968
LM #3 combined systems test completed.	01 Jul 1968
Individual and combined CM and SM systems test completed at factory.	20 Jul 1968
LM #3 reassigned to Apollo 9.	19 Aug 1968
Integrated CM and SM systems test completed at factory.	31 Aug 1968
Saturn S-IVB stage #504 delivered to KSC.	12 Sep 1968
LM #3 altitude tests completed.	27 Sep 1968
Saturn S-IC stage #4 delivered to KSC.	30 Sep 1968
Saturn V instrument unit #504 delivered to KSC.	30 Sep 1968
CM #104 and SM #104 ready to ship from factory to KSC.	05 Oct 1968
CM #104 and SM #104 delivered to KSC.	05 Oct 1968
CM #104 and SM #104 mated.	08 Oct 1968
CSM #104 combined systems test completed.	24 Oct 1968
CSM #104 altitude tests completed.	18 Nov 1968
CSM #104 mated to space vehicle.	03 Dec 1968
CSM #104 moved to VAB.	03 Dec 1968
LM #3 combined systems test completed.	07 Dec 1968
CSM #104 integrated systems test completed.	11 Dec 1968
CSM #104 electrically mated to launch vehicle.	26 Dec 1968
Space vehicle overall test completed.	27 Dec 1968
Space vehicle and MLP #2 transferred to launch complex 39A.	03 Jan 1969
Space vehicle flight readiness test completed.	18 Jan 1969
LM #3 flight readiness test completed.	19 Jan 1969
Space vehicle countdown demonstration test (wet) completed.	11 Feb 1969
Space vehicle countdown demonstration test (dry) completed.	12 Feb 1969
Terminal countdown initiated.	26 Feb 1969
Terminal countdown interrupted due to illness of crew.	27 Feb 1969
Terminal countdown reinitiated following crew medical clearance.	01 Mar 1969

Apollo 9 Ascent Phase

Event	GET (hhh:mm:ss)	Altitude (n mi)	Range (n mi)	Earth Fixed Velocity (ft/sec)	Space Fixed Velocity (ft/sec)	Event Duration (sec)	Geocentric Latitude (deg N)	Longitude (deg E)	Space Fixed Flight Path Angle (deg)	Space Fixed Heading Angle (E of N)
Liftoff	000:00:00.67	0.032	000.0	1.8	1,340.7		28.4470	-80.6041	0.08	90.00
Mach 1 achieved	000:01:08.2	4.243	1.383	1,088.4	2,100.7		28.4545	-80.5794	26.35	84.50
Maximum dynamic pressure	000:01:25.5	7.429	3.789	1,737.7	2,783.2		28.4666	-80.5369	28.08	81.87
S-IC center engine cutoff[1]	000:02:14.34	22.459	24.602	5,154.1	6,329.49	140.64	28.5720	-80.1602	22.5766	76.420
S-IC outboard engine cutoff	000:02:42.76	34.808	51.596	7,793.3	9,013.71	169.06	28.7071	-79.6718	18.5394	75.335
S-IC/S-II separation[1]	000:02:43.45	35.144	52.410	7,837.89	9,059.28		28.7111	-79.6571	18.449	75.337
S-II engine cutoff	000:08:56.22	100.735	830.505	21,431.9	22,753.54	371.06	31.6261	-65.0422	0.9177	81.872
S-II/S-IVB separation[1]	000:08:57.18	100.794	833.794	21,440.5	22,762.27		31.6343	-64.9786	0.906	81.907
S-IVB 1st burn cutoff	000:11:04.66	103.156	1,296.775	24,240.6	25,563.98	123.84	32.4266	-55.9293	-0.0066	86.979
Earth orbit insertion	000:11:14.66	103.154	1,335.515	24,246.39	25,569.78		32.4599	-55.1658	-0.0058	87.412

[1] Only the commanded time is available for this event.

Apollo 9 Earth Orbit Phase

Event	GET (hhh:mm:ss)	Space Fixed Velocity (ft/sec)	Event Duration (sec)	Velocity Change (ft/sec)	Apogee (n mi)	Perigee (n mi)	Perigee (mins)	Inclination (deg)
Earth orbit insertion	000:11:14.66	25,569.78			100.74	99.68	88.20	32.552
CSM separated from S-IVB	002:41:16.0	25,553						
CSM/LM ejected from S-IVB	004:08:09	25,565.3						
S-IVB 2nd burn restart[2]	004:45:47.20	25,556.1						
S-IVB 2nd burn cutoff	004:46:57.60	27,742.03	62.06					32.303
S-IVB intermediate orbit insertion	004:47:07.60	27,753.61			1,671.58	105.75	119.22	32.302
1st SPS ignition	005:59:01.07	25,549.8						
1st SPS cutoff	005:59:06.30	25,583.8	5.23	36.6	127.6	111.3	88.8	32.56
S-IVB 3rd burn restart[2]	006:06:27.35	20,766.0						
S-IVB 3rd burn cutoff	006:11:21.32	31,589.17	242.06					33.824
S-IVB escape orbit insertion	006:11:31.32	31,619.85						33.825
2nd SPS ignition	022:12:04.07	25,588.2		31.8				
2nd SPS cutoff	022:13:54.36	25,701.7	110.29	850.5	192.5	110.7	90.0	33.46
3rd SPS ignition	025:17:39.27	25,692.4						
3rd SPS cutoff	025:22:19.15	25,794.3	279.88	2567.9	274.9	112.6	91.6	33.82
4th SPS ignition	028:24:41.37	25,807.7						
4th SPS cutoff	028:25:09.24	25,798.9	27.87	300.5	275.0	112.4	91.6	33.82
DPS docked ignition	049:41:34.46	25,832.7						
DPS docked cutoff	049:47:45.97	25,783.0	371.51	1737.5	274.6	112.1	91.5	33.97
5th SPS ignition	054:26:12.27	25,700.8						
5th SPS cutoff	054:26:55.53	25,473.2	43.26	572.5	131.0	125.9	89.2	33.61
CSM/LM separation ignition	093:02:54	25,480.5						
CSM/LM separation cutoff	093:03:03.5	25,480.5	9.5		127	122		
LM descent phasing ignition	093:47:35.4	25,518.9						
LM descent phasing cutoff	093:47:54.4	25,518.2	19.0		137	112		
LM descent insertion ignition	095:39:08.06	25,412.6						
LM descent insertion cutoff	095:39:30.43	25,453.0	22.37		138.9	133.9		
LM coelliptic sequence ignition	096:16:06.54	25,452.0						
LM coelliptic sequence cutoff	096:16:38.25	25,412.0	31.71		138	113		
LM constant differential height ignition	096:58:15.0	25,592.0						
LM constant differential height cutoff	096:58:17.9	25,550.6	2.9		116	111		
LM terminal phase initiation ignition	097:57:59	25,540.8						
LM terminal phase initiation cutoff	097:58:36.6	25,560.5	37.6		126	113		
LM ascent engine depletion ignition	101:53:15.4	25,480.3						
LM ascent engine depletion cutoff	101:59:17.7	29,415.4	362.3	5,373.4	3,760.9	126.6	165.3	28.95
6th SPS ignition	123:25:06.97	25,522.2						
6th SPS cutoff	123:25:08.40	25,489.0	1.43	33.7	123.1	108.5	88.7	33.62
7th SPS ignition	169:39:00.36	25,589.6						
7th SPS cutoff	169:39:25.26	25,825.9	24.90	650.1	253.2	100.7	90.9	33.51
8th SPS ignition	240:31:14.84	25,318.4						
8th SPS cutoff	240:31:26.58	25,142.8	11.74	322.7	240.0	-4.2	88.8	33.52

[2] Only the commanded time is available for this event.

Event	GET (hhh:mm:ss)	GMT Time	GMT Date
Terminal countdown started.	-028:00:00	03:00:00	27 Feb 1969
Scheduled 3-hour hold at T-16 hours.	-016:00:00	15:00:00	27 Feb 1969
Decision made to recycle countdown to T-42 hours due to health of crew.	-016:30:00	15:30:00	27 Feb 1969
Countdown resumed at T-42 hours.	-042:00:00	07:30:00	01 Mar 1969
Scheduled 5-hour 30-minute hold at T-28 hours.	-028:00:00	21:30:00	01 Mar 1969
Countdown resumed at T-28 hours.	-028:00:00	03:00:00	02 Mar 1969
Scheduled 3-hour hold at T-16 hours.	-016:00:00	15:00:00	02 Mar 1969
Countdown resumed at T-16 hours.	-016:00:00	18:00:00	02 Mar 1969
Scheduled 6-hour hold at T-9 hours.	-009:00:00	01:00:00	03 Mar 1969
Countdown resumed at T-9 hours.	-009:00:00	07:00:00	03 Mar 1969
Guidance reference release.	-000:00:16.97	15:59:43	03 Mar 1969
S-IC engine start command.	-000:00:08.9	15:59:51	03 Mar 1969
S-IC engine ignition (#5).	-000:00:06.3	15:59:53	03 Mar 1969
All S-IC engines thrust OK.	-000:00:01.3	15:59:58	03 Mar 1969
Range zero.	000:00:00.00	16:00:00	03 Mar 1969
All holddown arms released (1st motion) (1.10 g).	000:00:00.26	16:00:00	03 Mar 1969
Liftoff (umbilical disconnected).	000:00:00.67	16:00:00	03 Mar 1969
Tower clearance yaw maneuver started.	000:00:01.7	16:00:01	03 Mar 1969
Yaw maneuver ended.	000:00:09.7	16:00:09	03 Mar 1969
Pitch and roll maneuver started.	000:00:13.3	16:00:13	03 Mar 1969
Roll maneuver ended.	000:00:33.0	16:00:33	03 Mar 1969
Mach 1 achieved.	000:01:08.2	16:01:08	03 Mar 1969
Maximum bending moment (86,000,000 lbf-in).	000:01:19.4	16:01:19	03 Mar 1969
Maximum dynamic pressure (630.740 lb/ft^2).	000:01:25.5	16:01:25	03 Mar 1969
S-IC center engine cutoff command.	000:02:14.34	16:02:14	03 Mar 1969
Pitch maneuver ended.	000:02:38.0	16:02:38	03 Mar 1969
S-IC outboard engine cutoff.	000:02:42.76	16:02:42	03 Mar 1969
S-IC maximum total inertial acceleration (3.85 g).	000:02:42.84	16:02:42	03 Mar 1969
S-IC/S-II separation command and S-IC maximum Earth-fixed velocity.	000:02:43.45	16:02:43	03 Mar 1969
S-II engine start command.	000:02:44.17	16:02:44	03 Mar 1969
S-II ignition.	000:02:45.16	16:02:45	03 Mar 1969
S-II aft interstage jettisoned.	000:03:13.5	16:03:13	03 Mar 1969
Launch escape tower jettisoned.	000:03:18.3	16:03:18	03 Mar 1969
Iterative guidance mode initiated.	000:03:24.6	16:03:24	03 Mar 1969
S-IC apex.	000:04:26.03	16:04:26	03 Mar 1969
S-II engine cutoff.	000:08:56.22	16:08:56	03 Mar 1969
S-II maximum total inertial acceleration (2.00 g).	000:08:56.31	16:08:56	03 Mar 1969
S-IC impact (theoretical).	000:08:56.436	16:08:56	03 Mar 1969
S-II maximum Earth-fixed velocity.	000:08:56.45	16:08:56	03 Mar 1969
S-II/S-IVB separation command.	000:08:57.18	16:08:57	03 Mar 1969
S-IVB 1st burn start command.	000:08:57.28	16:08:57	03 Mar 1969
S-IVB 1st burn ignition.	000:09:00.82	16:09:00	03 Mar 1969
S-IVB ullage case jettisoned.	000:09:09.0	16:09:09	03 Mar 1969
S-II apex.	000:09:53.58	16:09:53	03 Mar 1969
S-IVB 1st burn cutoff.	000:11:04.66	16:11:04	03 Mar 1969
S-IVB 1st burn maximum total inertial acceleration (0.80 g).	000:11:04.74	16:11:04	03 Mar 1969
Earth orbit insertion. S-IVB 1st burn maximum Earth-fixed velocity.	000:11:14.66	16:11:14	03 Mar 1969
Maneuver to local horizontal attitude started.	000:11:24.9	16:11:24	03 Mar 1969
Orbital navigation started.	000:12:47.7	16:12:47	03 Mar 1969
S-II impact (theoretical).	000:20:25.346	16:20:25	03 Mar 1969
Maneuver to transposition and docking attitude.	002:34:01.0	18:34:01	03 Mar 1969

Apollo 9 Timeline

Event	GET (hhh:mm:ss)	GMT Time	GMT Date
CSM separated from S-IVB (command).	002:41:16.0	18:41:16	03 Mar 1969
Formation flying. CSM docked with LM/S-IVB.	003:01:59.3	19:01:59	03 Mar 1969
CSM/LM ejected from S-IVB.	004:08:09	20:08:09	03 Mar 1969
Maneuver to local horizontal attitude started.	004:25:05.1	20:25:05	03 Mar 1969
S-IVB 2nd burn restart preparation.	004:36:17.24	20:36:17	03 Mar 1969
S-IVB 2nd burn restart command.	004:45:47.20	20:45:47	03 Mar 1969
S-IVB 2nd burn ignition (for intermediate orbit insertion).	004:45:55.54	20:45:55	03 Mar 1969
S-IVB 2nd burn cutoff.	004:46:57.60	20:46:57	03 Mar 1969
S-IVB 2nd burn maximum total inertial acceleration (1.24 g).	004:46:57.68	20:46:57	03 Mar 1969
S-IVB 2nd burn maximum Earth-fixed velocity.	004:46:58.20	20:46:58	03 Mar 1969
S-IVB intermediate orbit insertion.	004:47:07.60	20:47:07	03 Mar 1969
Orbital navigation started.	004:47:14.2	20:47:14	03 Mar 1969
Maneuver to local horizontal attitude started.	004:47:18.6	20:47:18	03 Mar 1969
1st SPS ignition.	005:59:01.07	21:59:01	03 Mar 1969
1st SPS cutoff.	005:59:06.30	21:59:06	03 Mar 1969
Powered flight navigation started.	005:59:39.0	21:59:39	03 Mar 1969
S-IVB 3rd burn restart preparation.	005:59:40.98	21:59:41	03 Mar 1969
S-IVB 3rd burn restart command.	006:06:27.35	22:06:27	03 Mar 1969
S-IVB 3rd burn ignition (Earth escape trajectory).	006:07:19.26	22:07:19	03 Mar 1969
S-IVB 3rd burn maximum total inertial acceleration (1.69 g).	006:08:53.00	22:08:53	03 Mar 1969
S-IVB 3rd burn cutoff.	006:11:21.32	22:11:21	03 Mar 1969
S-IVB safing procedures started.	006:11:21.92	22:11:21	03 Mar 1969
S-IVB 3rd burn maximum Earth-fixed velocity.	006:11:23.50	22:11:23	03 Mar 1969
S-IVB escape orbit insertion.	006:11:31.32	22:11:31	03 Mar 1969
Orbital navigation started.	006:11:38.0	22:11:38	03 Mar 1969
Maneuver to local horizontal attitude started.	006:11:42.0	22:11:42	03 Mar 1969
S-IVB safing—LOX dump started (failed due to loss of engine pneumatic control during 3rd burn).	006:12:15.5	22:12:15	03 Mar 1969
S-IVB safing—LH$_2$ dump started (failed due to loss of pneumatic control of engine valves).	006:24:11.3	22:24:11	03 Mar 1969
S-IVB safing—LOX NPV valve latched open to safe LOX tank.	006:24:02	22:24:02	03 Mar 1969
S-IVB safing—APS depletion firing ignition.	007:34:04.6	23:34:04	03 Mar 1969
S-IVB safing—APS depletion firing cutoff.	007:41:53	23:41:53	03 Mar 1969
2nd SPS ignition.	022:12:04.07	14:12:04	04 Mar 1969
2nd SPS cutoff.	022:13:54.36	14:13:54	04 Mar 1969
3rd SPS ignition.	025:17:39.27	17:17:39	04 Mar 1969
3rd SPS cutoff.	025:22:19.15	17:22:19	04 Mar 1969
4th SPS ignition.	028:24:41.37	20:24:41	04 Mar 1969
4th SPS cutoff.	028:25:09.24	20:25:09	04 Mar 1969
Pressure suits donned.	041:00	09:00	05 Mar 1969
LMP entered LM.	043:15	11:15	05 Mar 1969
LM transferred to internal power.	043:40	11:40	05 Mar 1969
LM systems activated.	043:45	11:45	05 Mar 1969
CDR entered LM.	044:04	12:04	05 Mar 1969
Landing gear deployed.	045:00	13:00	05 Mar 1969
Portable life support systems prepared.	045:05	13:05	05 Mar 1969
CDR requested private communication regarding LMP illness.	045:39:05	13:39:05	05 Mar 1969
CAPCOM replies that he is ready to receive CDR's private communication.	045:51:56	13:51:56	05 Mar 1969
TV transmission.	046:28	14:28	05 Mar 1969
Self test of landing radar and rendezvous radar.	048:15	16:15	05 Mar 1969
DPS docked ignition.	049:41:34.46	17:41:34	05 Mar 1969
DPS docked cutoff.	049:47:45.97	17:47:46	05 Mar 1969
Landing radar self-test.	050:00	18:00	05 Mar 1969

Event	GET (hhh:mm:ss)	GMT Time	GMT Date
Transfer to CM started.	050:15	18:15	05 Mar 1969
LM deactivated.	051:00	19:00	05 Mar 1969
5th SPS ignition.	054:26:12.27	22:26:12	05 Mar 1969
5th SPS cutoff.	054:26:55.53	22:26:55	05 Mar 1969
Pressure suits removed.	055:00	23:00	05 Mar 1969
Pressure suits donned.	068:15	12:15	06 Mar 1969
Transfer to LM started.	069:45	13:45	06 Mar 1969
LM systems activated.	070:00	14:00	06 Mar 1969
CDR assessed LMP condition as excellent.	071:53	15:53	06 Mar 1969
LM depressurized.	072:45	16:45	06 Mar 1969
LM forward hatch open.	072:46	16:46	06 Mar 1969
CM depressurized.	072:59	16:59:00	06 Mar 1969
LMP started egress.	072:59:02	16:59:02	06 Mar 1969
CM side hatch open.	073:02:00	17:02:00	06 Mar 1969
CDR reported LMP's foot extending through LM forward hatch. LMP lowered EVA visor.	073:04	17:04	06 Mar 1969
LMP egress completed. Entered foot restraints. CDR photographed LMP activities.	073:07	17:07	06 Mar 1969
CDR passed 70 mm camera to LMP. LMP started photography.	073:10	17:10	06 Mar 1969
LMP ended 70 mm photography and handed camera to CDR. CMP photographed LM with 16 mm camera.	073:20	17:20	06 Mar 1969
CDR passed 16 mm camera to LMP. CMP activities photographed by LMP.	073:23	17:23	06 Mar 1969
CMP retrieved SM thermal samples.	073:26	17:26	06 Mar 1969
LMP passed 16 mm camera to CDR.	073:34	17:34	06 Mar 1969
LMP 16 mm camera failed. LMP evaluated handrail, retrieved LM thermal sample, and passed to CDR.	073:39	17:39	06 Mar 1969
LMP started handrail evaluation.	073:42	17:42	06 Mar 1969
LMP ingress started.	073:45	17:45	06 Mar 1969
LMP ingress completed.	073:46:03	17:46:03	06 Mar 1969
LM hatch closed.	073:48	17:48	06 Mar 1969
CM side hatch reported closed and locked.	073:49:23	17:49:23	06 Mar 1969
LM hatch reported locked.	073:49:56	17:49:56	06 Mar 1969
LM repressurized at 3.0 psi.	073:53	17:53	06 Mar 1969
CM repressurization started.	073:55	17:55	06 Mar 1969
CM repressurized at 2.7 psi.	074:02:00	18:02:00	06 Mar 1969
TV transmission started.	074:58	18:58	06 Mar 1969
TV transmission ended.	075:13	19:13	06 Mar 1969
CDR entered CM.	075:15	19:15	06 Mar 1969
LMP entered CM.	076:55	20:55	06 Mar 1969
Pressure suits removed.	077:15	21:15	06 Mar 1969
LMP entered LM to open translunar bus tie circuit breakers.	078:09	22:09	06 Mar 1969
Pressure suits donned.	086:00	06:00	07 Mar 1969
LMP entered LM.	088:05	08:05	07 Mar 1969
LM systems activated.	088:15	08:15	07 Mar 1969
CDR entered LM.	088:55	08:55	07 Mar 1969
Check LM systems.	089:05	09:05	07 Mar 1969
Rendezvous radar transponder test.	091:00	11:00	07 Mar 1969
Landing radar self-test.	091:55	11:55	07 Mar 1969
Rendezvous radar transponder test.	092:05	12:05	07 Mar 1969
Maneuver to undocking attitude.	092:22	12:22	07 Mar 1969
Unsuccessful undocking attempt. Capture latches failed to release.	092:38	12:38	07 Mar 1969
CSM/LM reported undocked.	092:39:36	12:39:36	07 Mar 1969
Formation flying and photography.	092:45	12:45	07 Mar 1969

Apollo 9 Timeline

Event	GET (hhh:mm:ss)	GMT Time	GMT Date
CSM/LM separation maneuver ignition.	093:02:54	13:02:54	07 Mar 1969
CSM/LM separation maneuver cutoff.	093:03:03.5	13:03:03	07 Mar 1969
LM descent propulsion phasing maneuver ignition.	093:47:35.4	13:47:35	07 Mar 1969
LM descent propulsion phasing maneuver cutoff.	093:47:54.4	13:47:54	07 Mar 1969
Landing radar self-test.	094:15	14:15	07 Mar 1969
Terminal phase initiation for abort.	094:57:53	14:57:53	07 Mar 1969
Rendezvous radar on.	095:10	15:10	07 Mar 1969
LM descent propulsion insertion maneuver ignition.	095:39:08.06	15:39:08	07 Mar 1969
LM descent propulsion insertion maneuver cutoff.	095:39:30.43	15:39:30	07 Mar 1969
CAPCOM reported, "Everything looks good for staging."	095:58:15	15:58:15	07 Mar 1969
LM coelliptic sequence initiation maneuver ignition. Approximate time of LM descent stage jettison.	096:16:06.54	16:16:06	07 Mar 1969
LM coelliptic sequence initiation maneuver cutoff.	096:16:38.25	16:16:38	07 Mar 1969
CDR reports that LM "staging went okay."	096:33:11	16:33:11	07 Mar 1969
LM constant differential height ignition.	096:58:15.0	16:58:15	07 Mar 1969
LM constant differential height cutoff.	096:58:17.9	16:58:17	07 Mar 1969
LM terminal phase initiation ignition.	097:57:59	17:57:59	07 Mar 1969
LM terminal phase initiation cutoff.	097:58:36.6	17:58:36	07 Mar 1969
1st RCS midcourse correction ignition.	098:25:19.66	18:25:19	07 Mar 1969
1st RCS midcourse correction cutoff.	098:25:23.57	18:25:23	07 Mar 1969
Terminal phase braking.	098:30:03	18:30:03	07 Mar 1969
Stationkeeping.	098:30:51.2	18:30:51	07 Mar 1969
Formation flying and photography.	098:40	18:40	07 Mar 1969
CSM/LM docked.	099:02:26	19:02:26	07 Mar 1969
CDR entered CM.	100:35	20:35	07 Mar 1969
LM prepared for jettison.	100:40	20:40	07 Mar 1969
LMP entered CM.	101:00	21:00	07 Mar 1969
LM ascent stage jettisoned.	101:22:45.0	21:22:45	07 Mar 1969
Post-jettison CSM separation maneuver.	101:32:44	21:32:44	07 Mar 1969
LM ascent engine depletion ignition.	101:53:15.4	21:53:15	07 Mar 1969
LM ascent engine depletion.	101:59:17.7	21:59:17	07 Mar 1969
Pressure suits removed.	102:00	22:00	07 Mar 1969
6th SPS ignition.	123:25:06.97	19:25:07	08 Mar 1969
6th SPS cutoff.	123:25:08.40	19:25:08	08 Mar 1969
Experiment S065 photography.	124:10	20:10	08 Mar 1969
CSM landmark tracking.	125:30	21:30	08 Mar 1969
CSM landmark tracking.	143:00	15:00	09 Mar 1969
Experiment S065 photography.	146:00	18:00	09 Mar 1969
Experiment S065 photography.	147:30	19:30	09 Mar 1969
Target of opportunity photography.	149:00	21:00	09 Mar 1969
Target of opportunity photography.	150:10	22:10	09 Mar 1969
7th SPS ignition.	169:39:00.36	17:39:00	10 Mar 1969
7th SPS cutoff.	169:39:25.26	17:39:25	10 Mar 1969
16 mm photography.	171:10	19:10	10 Mar 1969
Experiment S065 photography.	171:20	19:20	10 Mar 1969
Experiment S065 photography.	171:50	19:50	10 Mar 1969
Target of opportunity photography.	173:10	21:10	10 Mar 1969
Experiment S065 photography.	190:40	14:40	11 Mar 1969
Experiment S065 photography.	192:10	16:10	11 Mar 1969
Tracking of Pegasus II satellite started.	192:43	16:43	11 Mar 1969
Tracking of Pegasus II satellite ended.	192:44	16:44	11 Mar 1969

Event	GET (hhh:mm:ss)	GMT Time	GMT Date
High-gain antenna test.	193:10	17:10	11 Mar 1969
High-gain antenna test.	193:40	17:40	11 Mar 1969
Target of opportunity photography.	193:50	17:50	11 Mar 1969
Tracking of Pegasus II satellite started.	194:13	18:13	11 Mar 1969
Tracking of Pegasus II satellite ended.	194:15	18:15	11 Mar 1969
CSM landmark tracking.	195:10	19:10	11 Mar 1969
Target of opportunity photography.	195:30	19:30	11 Mar 1969
Target of opportunity photography.	213:25	13:25	12 Mar 1969
Observation of descent stage attempted.	213:50	13:50	12 Mar 1969
Target of opportunity photography.	215:00	15:00	12 Mar 1969
Experiment S065 photography.	215:10	15:10	12 Mar 1969
Target of opportunity photography.	215:30	15:30	12 Mar 1969
Experiment S065 photography.	216:10	16:10	12 Mar 1969
Target of opportunity photography.	216:20	16:20	12 Mar 1969
Experiment S065 photography.	216:40	16:40	12 Mar 1969
Target of opportunity photography.	217:00	17:00	12 Mar 1969
CSM landmark tracking.	217:50	17:50	12 Mar 1969
Passive thermal control evaluated.	218:30	18:30	12 Mar 1969
Passive thermal control evaluated.	222:00	22:00	12 Mar 1969
Tracking of ascent stage with optics started.	222:38:40	22:38:40	12 Mar 1969
Tracking of ascent stage with optics ended.	222:45:40	22:45:40	12 Mar 1969
8th SPS ignition (deorbit).	240:31:14.84	16:31:14	13 Mar 1969
8th SPS cutoff.	240:31:26.58	16:31:26	13 Mar 1969
CM/SM separation.	240:36:03.8	16:36:03	13 Mar 1969
Entry.	240:44:10.2	16:44:10	13 Mar 1969
Communication blackout started.	240:47:01	16:47:01	13 Mar 1969
Communication blackout ended.	240:50:43	16:50:43	13 Mar 1969
Radar contact with CM established by recovery aircraft.	240:51	16:51	13 Mar 1969
Drogue parachute deployed.	240:55:07.8	16:55:07	13 Mar 1969
Main parachute deployed.	240:55:59.0	16:55:59	13 Mar 1969
Recovery beacon contact with CM established by recovery aircraft. VHF voice contact with CM established by recovery helicopter.	240:57	16:57	13 Mar 1969
Visual contact with CM established by recovery helicopter.	240:58	16:58	13 Mar 1969
Splashdown (went to apex-up).	241:00:54	17:00:54	13 Mar 1969
Swimmers and flotation collar deployed.	241:07	17:07	13 Mar 1969
Flotation collar inflated.	241:14	17:14	13 Mar 1969
CM hatch opened.	241:27	17:27	13 Mar 1969
Crew aboard recovery helicopter.	241:45	17:45	13 Mar 1969
Crew aboard recovery ship.	241:49:33	17:49:33	13 Mar 1969
CM aboard recovery ship.	243:13	19:13	13 Mar 1969
Flight crew departed recovery ship.	263:00	15:00	14 Mar 1969
Flight crew arrived in Eleuthera, Bahamas.	264:30	16:30	14 Mar 1969
Deactivation of CM started at Norfolk Naval Air Station.	312:00	16:00	16 Mar 1969
Deactivation of CM completed.			
CM arrived at contractor's facility in Downey, CA.			21 Mar 1969
LM descent stage entry.	443:45	03:45	22 Mar 1969
LM ascent stage entry.			23 Oct 1981

APOLLO 10

The Fourth Mission:
Testing the LM in Lunar Orbit

Apollo 10 Summary

(18 May–26 May 1969)

Apollo 10 crew (l. to. r.): Gene Cernan, John Young, Tom Stafford (NASA S69-34385).

Background

Apollo 10 was a Type F mission, a lunar module piloted flight demonstration in lunar orbit, the dress rehearsal for the first piloted landing on the Moon. It was also the first time all members of a three-person crew had previously flown in space.

The primary objectives were:

• to demonstrate crew, space vehicle, and mission support facilities performance during a piloted lunar mission with command and service modules and lunar module; and

• to evaluate lunar module performance in the cislunar and lunar environment.

The mission events simulated those for a lunar landing mission. In addition, visual observations and stereoscopic strip photography of Apollo Landing Site 2 (first planned lunar landing site) would be attempted.

[1] Copyright United Features Syndicate.

The crew members were Colonel Thomas Patten Stafford (USAF), commander; Commander John Watts Young (USN), command module pilot; and Commander Eugene Andrew "Gene" Cernan (USN), lunar module pilot.

Selected as an astronaut in 1962, Stafford was making his third spaceflight. He had been pilot of Gemini 6-A and command pilot of Gemini 9-A. Born 17 September 1930 in Weatherford, Oklahoma, Stafford was 38 years old at the time of the Apollo 10 mission. He received a B.S. from the U.S. Naval Academy in 1952. His backup was Colonel Leroy Gordon Cooper, Jr. (USAF).

Young was also making his third spaceflight, having been pilot on Gemini 3 and command pilot of Gemini 10. Born 24 September 1930 in San Francisco, California, Young was 38 years old at the time of the Apollo 10 mission. Young received a B.S. in aeronautical engineering from the Georgia Institute of Technology in 1952, and was selected as an astronaut in 1962. His backup was Lt. Colonel Donn Fulton Eisele (USAF).

Cernan had been pilot of Gemini 9-A. Born 14 March 1934 in Chicago, Illinois, he was 35 years old at the time of the Apollo 10 mission. Cernan received a B.S. in electrical engineering from Purdue University in 1956 and an M.S. in aeronautical engineering from the U.S. Naval Postgraduate School in 1963, and was selected as an astronaut in 1963. His backup was Commander Edgar Dean Mitchell (USN).

The capsule communicators (CAPCOMs) were Major Charles Moss Duke, Jr. (USAF), Major Joe Henry Engle (USAF), Major Jack Robert Lousma (USMC), and Lt. Commander Bruce McCandless, II (USN). The support crew consisted of Engle, Lt. Col. James Benson Irwin (USAF), and Duke. The flight directors were Glynn S. Lunney and Gerald D. Griffin (first shift), Milton L. Windler (second shift), and M. P. "Pete" Frank (third shift).

The Apollo 10 launch vehicle was a Saturn V, designated SA-505. The mission also carried the designation Eastern Test Range #920. The CSM was designated CSM-106, and had the call-sign "Charlie Brown." The lunar module was designated LM-4, and had the call-sign "Snoopy." The call-signs were taken from the popular comic strip Peanuts[1] by Charles L. Schultz. For this mission, Snoopy the Beagle exchanged the goggles and scarf of the World War I flying ace for a space helmet. At the Manned Spacecraft Center, Snoopy was the symbol of qualify performance.

Launch Preparations

The terminal countdown was picked up at 01:00:00 GMT on 17 May 1969 and proceeded with no unscheduled holds. The primary LOX replenish pump failed to start at T-8 hours due to a blown fuse in the pump motor starter circuit. Troubleshooting and fuse replacement delayed LOX loading by 50 minutes but it was completed by T-4 hours 22 minutes. The lost time was made up during the scheduled 1-hour hold at T-3 hours 30 minutes.

A high pressure cell in the Atlantic Ocean off the New England coast caused southeasterly surface winds and brought moisture into the Cape Canaveral area, which contributed to overcast conditions. At launch time, cumulus clouds covered 40 percent of the sky (base 2,200 feet), altocumulus covered 20 percent (base 11,000 feet), and cirrus covered 100 percent (base not recorded); the temperature was 80.1° F; the relative humidity was 75 percent; and the barometric pressure was 14.779 lb/in^2. The winds, as measured by the anemometer on the light pole 60.0 feet above ground at the launch site, measured 19.0 knots at 142° from true north.

Ascent Phase

Apollo 10 was launched from Kennedy Space Center Launch Complex 39, Pad B, at a Range Zero time of 16:49:00 GMT (11:49:00 p.m. EDT) on 18 May 1969, and was the first piloted launch from this pad. The launch window extended to 21:09 GMT to take advantage of a sun elevation angle on the lunar surface of 11°.

Between 000:00:13.05 and 000:00:32.3, the vehicle rolled from a launch pad azimuth of 90° to a flight azimuth of 72.028°. The S-IC engine shut down at 000:02:41.63, followed by S-IC/S-II separation, and S-II engine ignition. The S-II engine shut down at 000:09:12.64 followed by separation from the S-IVB, which ignited at 000:09:16.9. The first S-IVB engine cutoff occurred at 000:11:43.76, with deviations from the planned trajectory of only -0.23 ft/sec in velocity and only -0.08 n mi in altitude.

The S-IC stage impacted the Atlantic Ocean at 000:08:59.12 at latitude 30.188° north and longitude 74.207° west, 348.80 n mi from the launch site. The S-II stage impacted the Atlantic Ocean at 000:20:17.89 at latitude 31.522° north and longitude 34.512° west, 2,389.29 n mi from the launch site.

Apollo 10 becomes the first piloted mission to lift off from Kennedy Space Center Pad 39B (NASA S69-34145).

The maximum wind conditions encountered during ascent were 82.6 knots at 270° from true north at 46,520 feet, and a maximum wind shear of 0.0203 sec^{-1} at 50,200 feet.

Parking orbit conditions at insertion, 000:11:53.76 (S-IVB cutoff plus 10 seconds to account for engine tailoff and other transient effects), showed an apogee and perigee of 100.32 by 99.71 n mi, an inclination of 32.546°, a period of 88.20 minutes, and a velocity of 25,567.88 ft/sec. The apogee and perigee were based upon a spherical Earth with a radius of 3,443.934 n mi.

The international designation for the CSM upon achieving orbit was 1969-043A and the S-IVB was designated 1969-043B. After undocking at the Moon, the LM would be designated 1969-018C.

Earth Orbit Phase

After inflight systems checks, the 343.08-second translunar injection maneuver (second S-IVB firing) was performed at 002:33:27.5. The S-IVB engine shut down at 2:39:10.58 and translunar injection occurred ten seconds later, after 1.5 Earth orbits lasting 2 hours 27 minutes 16.82 seconds, at a velocity of 35,585.83 ft/sec.

Translunar Phase

At 003:02:42.4, the CSM was separated from the S-IVB stage. It was transposed and then docked with the LM at 003:17:36.0. The docked spacecraft were ejected at 003:56:25.7 and a separation maneuver was performed at 004:39:09.8. The sequence was televised to Earth starting at 003:06:00 for 22 minutes and from 003:56:00 for 13 minutes 25 seconds. Additional television broadcasts during translunar coast included:

Television Transmissions—Translunar Coast

Start	Duration (mm:ss)	Subject
005:06:34	13:15	View of Earth and spacecraft interior
007:11:27	24:00	View of Earth and spacecraft interior
027:00:48	27:43	View of Earth and spacecraft interior
048:00:51	14:39	View of Earth and spacecraft interior (recorded)
048:24:00	03:51	View of Earth and spacecraft interior (recorded)
049:54:00	04:49	View of Earth
053:35:30	25:00	View of Earth and spacecraft interior
072:37:26	17:16	View of Earth and spacecraft interior

A ground command for propulsive venting of residual propellants targeted the S-IVB to go past the Moon. The closest approach of the S-IVB to the Moon was 1,680 n mi, at 078:51:03.6 on May 21 at 23:40 GMT. The trajectory after passing from the lunar sphere of influence resulted in a solar orbit with an aphelion and perihelion of 82.160 million by 73.330 million n mi, an inclination of 23.46, and a period of 344.88 days.

A preplanned, 7.1-second, midcourse correction of 49.2 ft/sec was executed at 026:32:56.8 and adjusted the trajectory to coincide with a July lunar landing trajectory. The maneuver was so accurate that two additional planned midcourse corrections were canceled. The passive thermal control technique was employed to maintain desired spacecraft temperatures throughout the translunar coast except when a specific attitude was required.

Young displays drawing of cartoon character for which LM was named (NASA S69-34076, Snoopy© United Features Syndicate).

Earth as seen from a distance of 100,000 n mi (NASA AS10-34-5026).

At 075:55:54.0, at an altitude of 95.1 n mi above the Moon, the service propulsion engine was fired for 356.1 seconds to insert the spacecraft into a lunar orbit of 170.0 by 60.2 n mi. The translunar coast had lasted 73 hours 22 minutes 29.5 seconds.

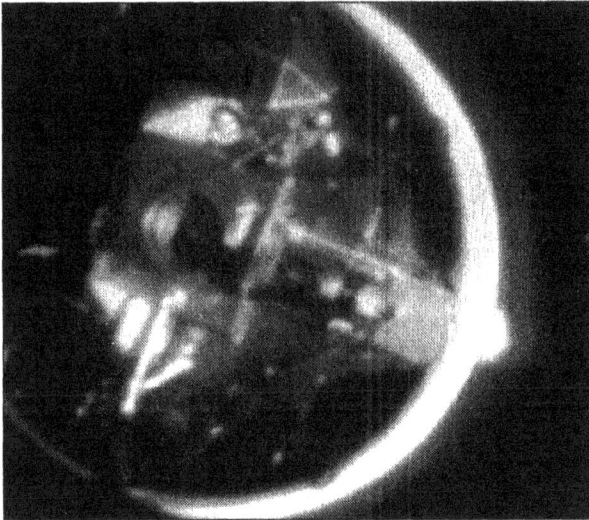

LM inside S-IB following separation from CSM (NASA S69-33994).

Lunar Orbit Phase

After two revolutions of tracking and ground updates, a 13.9-second maneuver was performed at 080:25:08.1 to circularize the orbit at 61.0 by 59.2 n mi.

Earthrise as seen from Apollo 10 (NASA AS10-27-3890).

A 29-minute 9-second scheduled color television transmission of the lunar surface was conducted at 080:45:00, with the crew describing the lunar features below them. The picture quality of lunar scenes was excellent.

The lunar module pilot entered the LM at 081:55 for two hours of "housekeeping" activities and some LM communications tests. The tests were terminated following the LM relay communications tests due to time limitations. Results were excellent, and the remaining tests were conducted later in the mission.

At 095:02, the commander and lunar module pilot entered to activate LM systems and discovered that the LM had moved 3.5 degrees out of line with the CM. The crew feared that separating the two spacecraft might shear off some of the latching pins, possibly preventing redocking. But mission control reported that as long as the alignment was less than six degrees, there would be no problem. Undocking occurred at 098:11:57 and was televised for 20 minutes 10 seconds starting at 098:13:00. During this period, the LM landing gear were deployed and all LM systems checked out.

A 8.3-second CSM reaction control system maneuver at 098:47:17.4 separated the CSM to about 30 feet from the LM. The CSM was in an orbit of 62.9 by 57.7 n mi at the time. Stationkeeping was initiated at this point while the command module pilot visually inspected the LM. The CSM reaction control system was then used to perform the separation maneuver directed radially downward toward the Moon's center. This maneuver provided a separation at descent orbit insertion of about 2 n mi from the LM.

CM after separation from LM (NASA AS10-27-3873).

Following stationkeeping, a 27.4-second LM descent propulsion system burn at 099:46:01.6 inserted the LM into a descent orbit of 60.9 by 8.5 n mi so that the resulting lowest point in the orbit occurred about 15° prior to lunar landing site 2.

Northwest view of Triesnecher crater with associated rille at bottom (NASA AS10-32-4819).

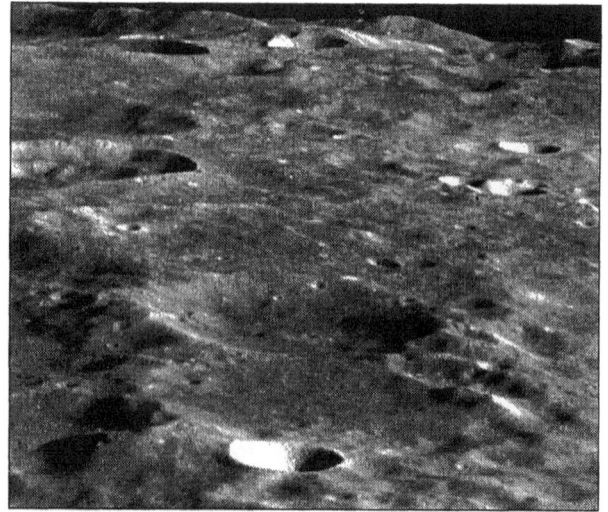

Lunar farside area near International Astronomical Union Crater 300, seen from the CM (NASA AS10-34-5173).

Numerous photographs of the lunar surface were taken. Some camera malfunctions were reported and although some communications difficulties were experienced, the crew provided a continuous commentary of their observations. An hour later, the LM made a low-level pass over Apollo landing site 2. The pass was highlighted by a test of the landing radar, visual observation of lunar lighting, stereoscopic strip photography, and execution of the phasing maneuver using the descent engine. The lowest measured point in the trajectory was 47,400 feet (7.8 n mi) above the lunar surface at 100:41:43.

The second LM maneuver, a 39.9-second descent propulsion system phasing burn at 100:58:25.93, established a lead angle equivalent to that which would occur at powered ascent cutoff during a lunar landing, and put the LM into an orbit of 190.1 by 12.1 n mi.

At 102:44:49, during preparations for rendezvous with the CSM, the LM started to wallow off slowly in yaw, and then stopped. At 102:45:12, it started a rapid roll accompanied by small pitch and yaw rates. The ascent stage was then separated from the descent stage at 102:45:17 at an altitude of 31.4 n mi and the motion was stopped eight seconds later. A 15.55-second firing of the ascent engine at 102:55:02.13 placed the ascent stage into an orbit of 46.5 by 11.0 n mi. The descent stage went into solar orbit.

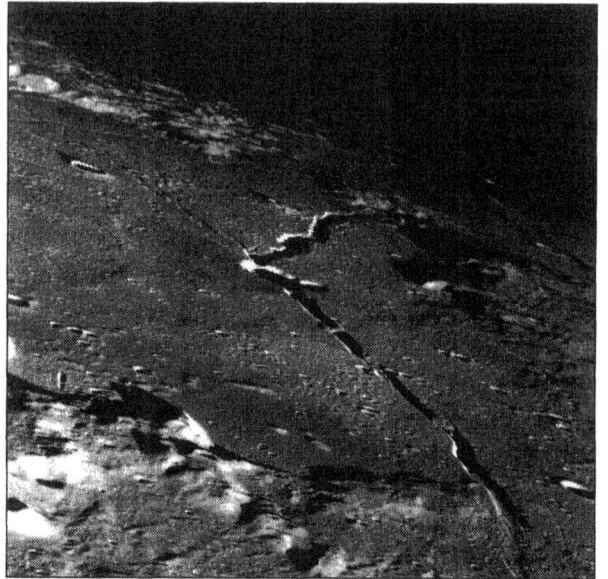

Photograph of the lunar nearside; crater Hyginus, near Central Bay, seen from the CM (NASA AS10-31-4650).

Analysis revealed that the cause of the anomalous motion was human error. Inadvertently, the control mode of the LM abort guidance system was left in AUTO rather than the required ATTITUDE HOLD mode for the staging maneuver. In AUTO, the abort guidance system drove the LM to acquire the CSM which was not in accordance with the planned attitude timeline. The commander took over manual control to reestablish the proper attitude.

At the orbital low point, the insertion maneuver was performed on time using the LM ascent propulsion system. This burn established the equivalent of the standard LM insertion orbit of a lunar landing mission (45 x 11.2 n mi). The LM coasted in that orbit for about one hour. The terminal maneuver occurred at about the midpoint of darkness, and braking during the terminal phase finalization was performed manually as planned.

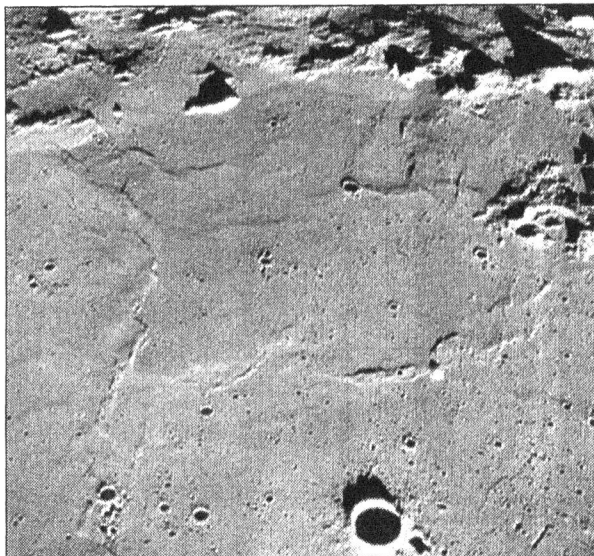

Apollo landing site #3. Crater Bruce is seen at the bottom right (NASA AS10-27-3907).

The rendezvous simulated one that would follow a normal ascent from the lunar surface. It started with a 27.3-second LM coelliptic sequence initiation maneuver at 103:45:55.3, which placed the spacecraft into an orbit of 48.7 by 40.7 n mi. This was followed by a 1.65-second constant differential height maneuver at 104:43:53.29 which raised the perigee to 42.1 n mi. The 16.50-second terminal phase initiation maneuver at 105:22:55.28 then raised the orbit to 58.3 by 46.8 n mi. Docking was complete at 106:22:02 at an altitude of 54.7 n mi after 8 hours 10 minutes 5 seconds of lunar flight.

Once docked, the LM crew members transferred the exposed film packets to the CM. The LM ascent stage was jettisoned at 108:24:36. A 6.3-second separation maneuver at 108:43:23.3 raised the orbit to 64.0 by 56.3 n mi. This

was followed by a 249.0-second remote control firing of the ascent engine to depletion at 108:52:05.5.

About one revolution after docking, the LM ascent propulsion system burn to depletion was commanded as planned, utilizing the LM ascent engine arming assembly. This burn was targeted to place the LM in a solar orbit. Communications were maintained until LM ascent stage battery depletion at about 120:00. The ascent batteries lasted about 12 hours after LM jettison.

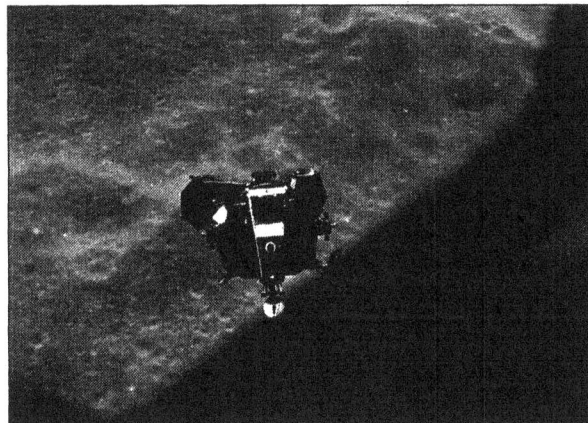

LM ascent stage prior to docking with the CM (NASA AS10-34-5112).

Prior to transearth injection, views of the lunar surface and spacecraft interior were transmitted to Earth for 24 minutes 12 seconds starting at 132:07:12. After a rest period, the crew conducted landmark tracking and photography exercises. During the remaining lunar orbital period or operation, 18 landmark sightings, and extensive stereo and oblique photographs were taken. Two scheduled TV periods were deleted because of crew fatigue.

Transearth injection was achieved at 137:39:13.7 at a velocity of 8,987.2 ft/sec, following a 164.8-second engine firing at 56.0 n mi altitude. The spacecraft had been in lunar orbit for 31 lunar orbits lasting 61 hours 37 minutes 23.6 seconds.

Transearth Phase

Transearth activities included a number of star-Earth horizon navigation sightings and the CSM S-band high gain reflectivity test which was conducted at 168:00. The passive thermal control technique and the navigation procedures used on the translunar portion of the mission were also used during the return trip. The only midcourse correction required was a 6.7-second, 2.2 ft/sec, maneuver at 188:49:58.0, three hours before CM/SM separation.

Six television transmissions were made on the return trip and were broadcast to Earth. The duration of the transmissions and the subjects were as follows:

Television Transmissions—Return Trip

Start	Duration (mm:ss)	Subject
137:50:51	43:03	View of Moon after transearth injection
139:30:16	06:55	View of Moon after transearth injection
147:23:00	11:25	View of receding Moon and spacecraft interior
152:29:19	29:05	View of Earth, Moon, and spacecraft interior
173:27:17	10:22	View of Earth and spacecraft interior
186:51:49	11:53	View of Earth and spacecraft interior

The service module was jettisoned at 191:33:26, and the CM entry followed a normal profile. The command module reentered Earth's atmosphere (400,000 feet altitude) at 191:48:54.5 at a velocity of 36,314 ft/sec, following a transearth coast of 54 hours 3 minutes 40.9 seconds[2]. The service module impacted the Pacific Ocean at a point estimated to be latitude 19.4° south and longitude 173.37° west.

Recovery

Apollo 10 CM on parachutes prior to splashdown (NASA S69-36594).

The parachute system effected a soft splashdown of the CM in the Pacific Ocean at 16:52:23 GMT (11:52:23 p.m. EDT) on 26 May. Mission duration was 192:023:23. The impact point was about 1.3 n mi from the target point and 2.9 n mi from the recovery ship U.S.S. *Princeton*. The splashdown site was estimated to be latitude 15.07° south and longitude 164.65° west.

After splashdown, the CM assumed an apex-up flotation attitude. The crew was retrieved by helicopter and was aboard the recovery ship 39 minutes after splashdown. The CM was recovered 57 minutes later. The estimated CM weight at splashdown was 10,901 pounds, and the estimated distance traveled for the mission was 721,250 n mi.

Helicopter lifts Apollo 10 CM from ocean following splashdown (NASA S69-21037).

At CM retrieval, the weather recorded onboard the *Princeton* showed 10 percent cloud cover at 2,000 feet and 20 percent at 7000 feet; visibility 10 n mi; wind speed five knots from 100° true north; air temperature unknown; water temperature 85° F; with waves to three feet.

The CM was offloaded from the *Princeton* on 31 May at Ford Island, Hawaii, and the Landing Safing Team began the evaluation and deactivation procedures at 18:00 GMT. Deactivation was completed at 05:56 GMT on 3 June. The CM was flown to Long Beach, California, where it arrived at 10:15 GMT on 4 June. It was trucked the same day to the North American Rockwell Space Division facility in Downey, California for postflight analysis.

All systems in the CSM and the LM were managed very well. Although some problems occurred, most were minor and none caused a constraint to completion of mission objectives. Valuable data concerning lunar gravitation were obtained during the 61 hours in lunar orbit.

---

[2] The *Guinness Book Of World Records* states that Apollo 10 holds the record for the fastest a human has ever traveled: 24,791 st mi per hour at 400,000 feet altitude (entry) on 26 May 1969. However, the Apollo 10 mission report states the maximum speed at entry was 36,397 feet per second, or 24,816 st mi per hour.

Apollo 10 crew receives "red carpet" reception aboard recovery ship U.S.S. *Princeton* (NASA S69-20544).

Spacecraft systems performance was satisfactory, and all mission objectives were accomplished. All detailed test objectives were satisfied with the exception of the LM steerable antenna and relay modes for voice and telemetry communications.

Conclusions

The Apollo 10 mission provided the concluding data and final environmental evaluation to proceed with a lunar landing. The following conclusions were made from an analysis of post-mission data:

1. The systems in the command and service modules and lunar module were operational for piloted lunar landing.

2. The crew activity timeline, in those areas consistent with the lunar landing profile, demonstrated that critical crew tasks associated with lunar module checkout, initial descent, and rendezvous were both feasible and practical without unreasonable crew workload.

3. The lunar module S-band communications capability using either the steerable or the omni-directional antenna was satisfactory at lunar distances.

4. The operating capability of the landing radar in the lunar environment during a descent propulsion firing was satisfactorily demonstrated for the altitudes experienced.

5. The range capability of the lunar module rendezvous radar was demonstrated in the lunar environment with excellent results. Used for the first time, VHF ranging information from the CM provided consistent correlation with radar range and range-rate data.

6. The lunar module abort guidance system capability to control an ascent propulsion system maneuver and to guide the spacecraft during rendezvous was demonstrated.

7. The capability of the Mission Control Center and the Manned Space Flight Network to control and monitor two vehicles at lunar distance during both descent and rendezvous operations was proven adequate for a lunar landing.

8. The lunar potential model was significantly improved over that of Apollo 8, and the orbit determination and prediction procedures proved remarkably more precise for both spacecraft in lunar orbit. After a combined analysis of Apollo 8 and 10 trajectory reconstructions, the lunar potential model was expected to be entirely adequate for support of lunar descent and ascent.

Apollos 10 and 11 crews during post-mission debriefing (NASA S69-35504).

Apollo 10 Objectives

Spacecraft Primary Objectives

1. To demonstrate crew/space vehicle/mission support facilities performance during a piloted lunar mission with a command and service module and lunar module. *Achieved.*

2. To evaluate lunar module performance in the cislunar and lunar environment. *Achieved.*

Spacecraft Primary Detailed Objectives

1. P11.15: To perform primary guidance and navigation control system/descent propulsion system undocked descent orbit insertion and a high thrust maneuver. *Achieved.*

2. P16.10: To perform manual and automatic acquisition, tracking, and communications with the Manned Space Flight Network using the steerable S-band antenna at lunar distance. *Achieved, despite some problems during the 13th lunar revolution.*

3. P16.14: To operate the landing radar at the closest approach to the Moon and during descent propulsion system burns. *Achieved.*

4. P20.66: To obtain data on the command module and lunar module crew procedures and timeline for the lunar orbit phase of a lunar landing mission. *Achieved.*

5. P20.78: To perform a lunar module active simulated lunar landing mission rendezvous. *Achieved.*

6. P20.91: To perform lunar landmark tracking in lunar orbit from the command and service module with the lunar module attached. *Achieved.*

7. P20.121: To perform lunar landmark tracking from the command and service module while in lunar orbit. *Achieved.*

Spacecraft Secondary Detailed Objectives

1. S1.39: To perform star-lunar landmark sightings during the transearth phase. *Achieved.*

2. S6.9: To perform a reflectivity test using the command and service module S-band high-gain antenna while docked. *Not achieved, canceled while docked. S-band communications lost because steerable antenna track mode not switched properly; however, operation of steerable antenna during abnormal staging excursions demonstrated ability of antenna to track under very high rates.*

3. S7.26: To obtain data on the passive thermal control system during a lunar orbit mission. *Achieved.*

4. S11.17: To obtain data to verify inertial measurement unit performance in the flight environment. *Achieved.*

5. S12.6: To obtain abort guidance system performance data in the flight environment. *Achieved.*

6. S12.8: To demonstrate reaction control system translation and attitude control of the staged lunar module using automatic and manual abort guidance system/control electronics system control. *Achieved.*

7. S12.9: To perform an unpiloted abort guidance system controlled ascent propulsion system burn. *Achieved.*

8. S12.10: To evaluate the ability of the abort guidance system to perform a lunar module active rendezvous. *Achieved.*

9. S13.13: To perform a long duration unpiloted ascent propulsion system burn. *Achieved.*

10. S13.14: To obtain supercritical helium system pressure data while in standby conditions and during all descent propulsion system engine firings. *Achieved.*

11. S16.12: To communicate with the Manned Space Flight Center using the lunar module S-band omni antennas at lunar distance. *Achieved, despite some problems during the 13th lunar revolution.*

12. S16.15: To obtain data on the rendezvous radar performance and capability near maximum range. *Achieved.*

13. S16.17: To demonstrate lunar module, command and service module/Manned Space Flight Center communications at lunar distance. *Achieved, despite some problems due to procedural errors.*

14. S20.46: To perform command and service module transposition, docking, and command and service module/lunar module ejection after the S-IVB translunar injection burn. *Achieved.*

15. S20.77: To obtain data on the operational capability of VHF ranging during a lunar module active rendezvous. *Achieved.*

16. S20.79: To demonstrate command and service module/lunar module passive thermal control modes during a lunar orbit mission. *Achieved.*

17. S20.80: To demonstrate operational support for a command and service module/lunar module orbit mission. *Achieved despite some communication problems.*

18. S20.82: To monitor primary guidance and navigation control system/abort guidance system performance during lunar orbit operations. *Achieved.*

19. S20.83: To obtain data on lunar module consumables for a simulated lunar landing mission, in lunar orbit, to determine lunar landing mission consumables. *Achieved.*

20. S20.86: To obtain data on the effects of lunar illumination and contrast conditions on crew visual perception while in lunar orbit. *Achieved.*

21. S20.95: To perform translunar midcourse corrections. *Achieved. Only one of four possible midcourse corrections was required.*

22. S20.117: To perform lunar orbit insertion using service propulsion system/guidance and navigation control system controlled burns with a docked command and service module/lunar module. *Achieved.*

Launch Vehicle Objectives

1. To demonstrate launch vehicle capability to inject the spacecraft into the specified translunar trajectory. *Achieved.*

2. To demonstrate launch vehicle capability to maintain a specified attitude for transposition, docking, and spacecraft ejection maneuver. *Achieved.*

3. To demonstrate S-IVB propellant dump and safing. *Achieved.*

4. To verify J-2 engine modifications. *Achieved.*

5. To confirm J-2 engine environment in S-II and S-IVB stages. *Achieved.*

6. To confirm launch vehicle longitudinal oscillations environment during S-IC stage burn period. *Achieved.*

7. To verify that modifications incorporated in the S-IC stage suppress low frequency longitudinal oscillations. *Achieved.*

8. To confirm launch vehicle longitudinal oscillation environment during S-II stage burn period. *Achieved.*

9. To demonstrate that early center engine cutoff for S-II stage suppresses low frequency longitudinal oscillations. *Achieved.*

Apollo 10 Spacecraft History

EVENT	DATE
LM #4 integrated test at factory.	25 May 1968
Individual and combined CM and SM systems test completed at factory.	08 Sep 1968
LM #4 final engineering evaluation acceptance test at factory.	02 Oct 1968
LM descent stage #4 ready to ship from factory to KSC.	09 Oct 1968
LM descent stage #4 delivered to KSC.	11 Oct 1968
LM ascent stage #4 ready to ship from factory to KSC.	12 Oct 1968
LM ascent stage #4 delivered to KSC.	16 Oct 1968
Integrated CM and SM systems test completed at factory.	19 Oct 1968
LM ascent stage #4 and descent stage #4 mated.	02 Nov 1968
LM #4 combined systems test completed.	06 Nov 1968
CM #106 and SM #106 delivered to KSC.	23 Nov 1968
CM #106 and SM #106 ready to ship from factory to KSC.	24 Nov 1968
CM #106 and SM #106 mated.	26 Nov 1968
Saturn S-IC stage #5 delivered to KSC.	27 Nov 1968
Saturn S-II stage #5 delivered to KSC.	03 Dec 1968
Saturn S-IVB stage #505 delivered to KSC.	03 Dec 1968
LM #4 altitude tests completed.	06 Dec 1968
Saturn V instrument unit #505 delivered to KSC.	15 Dec 1968
CSM #106 combined systems test completed.	16 Dec 1968
Launch vehicle erected.	30 Dec 1968
CSM #106 altitude tests completed.	17 Jan 1969
Launch vehicle propellant dispersion/malfunction overall test completed.	03 Feb 1969
CSM #106 moved to VAB.	06 Feb 1969
Spacecraft erected.	06 Feb 1969
LM #4 combined systems test completed.	10 Feb 1969
CSM #106 integrated systems test completed.	13 Feb 1969
CSM #106 electrically mated to launch vehicle.	27 Feb 1969
Space vehicle overall test completed.	03 Mar 1969
Space vehicle overall test #1 (plugs in) completed.	05 Mar 1969
Space vehicle and MLP #3 transferred to launch complex 39B.	11 Mar 1969
LM #4 flight readiness test completed.	27 Mar 1969
Emergency egress test completed.	28 Mar 1969
Space vehicle flight readiness test completed.	19 Apr 1969
Space vehicle hypergolic fuel loading completed.	25 Apr 1969
Saturn S-IC stage #5 RP-1 fuel loading completed.	02 May 1969
Space vehicle countdown demonstration test (wet) completed.	05 May 1969
Space vehicle countdown demonstration test (dry) completed.	06 May 1969

Apollo 10 Ascent Phase

Event	GET (hhh:mm:ss)	Altitude (n mi)	Range (n mi)	Earth Fixed Velocity (ft/sec)	Space Fixed Velocity (ft/sec)	Event Duration (sec)	Geocentric Latitude (deg N)	Longitude (deg E)	Space Fixed Flight Path Angle (deg)	Space Fixed Heading Angle (E of N)
Liftoff	000:00:00.58	0..035	0.000	1.3	1,340.4		28.4658	-80.6209	0.06	90.00
Mach 1 achieved	000:01:06.8	4.244	1.037	1,057.9	2,028.6		28.4714	-80.6023	27.82	85.03
Maximum dynamic pressure	000:01:22.6	7.137	2.893	1,623.4	2,645.8		28.4813	-80.5690	28.83	82.23
S-IC center engine cutoff[3]	000:02:15.16	23.430	25.009	5,299.0	6,473.20	141.56	28.5967	-80.1577	22.807	76.461
S-IC outboard engine cutoff	000:02:41.63	35.247	50.419	7,810.2	9,028.58	168.03	28.7182	-79.7090	18.946	75.538
S-IC/S-II separation[3]	000:02:42.31	35.580	51.223	7,833.4	9,052.79		28.7222	-79.6943	18.848	75.538
S-II center engine cutoff	000:07:40.61	96.710	599.079	17,310.1	18,630.15	296.56	30.9579	-69.4941	1.029	79.585
S-II outboard engine cutoff	000:09:12.64	101.204	883.670	21,309.9	22,632.02	388.59	31.7505	-64.0222	0.741	82.458
S-II/S-IVB separation[3]	000:09:13.50	101.247	886.634	21,317.81	22,639.93		31.7574	-63.9647	0.730	82.490
S-IVB 1st burn cutoff[3]	000:11:43.76	103.385	1,430.977	24,238.8	25,562.40	146.95	32.5150	-53.2920	-0.0064	88.497
Earth orbit insertion	000:11:53.76	103.334	1,469.790	24,244.3	25,567.88		32.5303	-52.5360	-0.0049	89.933

Apollo 10 Earth Orbit Phase

Event	GET (hhh:mm:ss)	Space Fixed Velocity (ft/sec)	Event Duration (sec)	Apogee (n mi)	Perigee (n mi)	Perigee (mins)	Inclination (deg)
Earth orbit insertion	000:11:53.76	25,567.88		100.32	99.71	88.20	32.546
S-IVB 2nd burn ignition	002:33:27.52	25,561.4					
S-IVB 2nd burn cutoff	002:39:10.58	35,585.83	343.06				31.701

Apollo 10 Translunar Phase

Event	GET (hhh:mm:ss)	Altitude (n mi)	Space Fixed Velocity (ft/sec)	Event Duration (sec)	Velocity Change (ft/sec)	Space Fixed Flight Path Angle (deg)	Space Fixed Heading Angle (E of N)
Translunar injection	002:39:20.58	179.920	35,562.96			7.379	61.065
CSM separated from S-IVB (ignition)	003:02:42.4	3,502.626	25,548.72			43.928	67.467
CSM SPS evasive maneuver ignition	004:39:09.8	17,938.5	14,220.2			65.150	91.21
CSM SPS evasive maneuver cutoff	004:39:12.7	17,944.7	14,203.7	2.9	18.8	65.100	91.22
Midcourse correction ignition	026:32:56.8	110,150.2	5,094.4			77.300	108.36
Midcourse correction cutoff	026:33:03.9	110,155.9	5,110.0	7.1	49.2	77.800	108.92

[3] Only the commanded time is available for this event.

Apollo 10 Lunar Orbit Phase

Event	GET (hhh:mm:ss)	Altitude (n mi)	Space Fixed Velocity (ft/sec)	Event Duration (sec)	Velocity Change (ft/sec)	Apogee (n mi)	Perigee (n mi)
Lunar orbit insertion ignition	075:55:54.0	95.1	8,232.3				
Lunar orbit insertion cutoff	076:01:50.1	61.2	5,471.9	356.1	2,982.4	170.0	60.2
Lunar orbit circularization ignition	080:25:08.1	60.4	5,484.7				
Lunar orbit circularization cutoff	080:25:22.0	59.3	5,348.9	13.9	139.0	61.0	59.2
CSM/LM undocked	098:11:57	58.1	5,357.8				
CSM/LM separation ignition	098:47:17.4	59.2	5,352.2				
CSM/LM separation cutoff	098:47:25.7	59.2	5,352.1	8.3	2.5	62.9	57.7
LM descent orbit insertion ignition	099:46:01.6	61.6	5,339.6				
LM descent orbit insertion cutoff	099:46:29.0	61.2	5,271.2	27.4	71.3	60.9	8.5
LM closest approach to lunar surface	100:41:43	7.8					
LM phasing ignition	100:58:25.93	17.7	5,212.4				
LM phasing cutoff	100:59:05.88	19.0	5,672.9	39.95	176.0	190.1	12.1
LM ascent stage/descent stage separated	102:45:16.9	31.4	5,605.6				
LM ascent orbit insertion ignition	102:55:02.13	11.6	5,705.2				
LM ascent orbit insertion cutoff	102:55:17.68	11.7	5,520.6	15.55	220.9	46.5	11.0
LM coelliptic sequence initiation ignition		44.7	5,335.5				
LM coelliptic sequence initiation cutoff	103:46:22.6	44.6	5,381.7	27.3	45.3	48.7	40.7
LM constant differential height ignition	104:43:53.28	44.3	5,394.7				
LM constant differential height cutoff	104:43:54.93	43.8	5,394.9	1.65	3.0	48.8	42.1
LM terminal phase initiation ignition	105:22:55.58	48.4	5,369.2				
LM terminal phase initiation cutoff	105:23:12.08	47.0	5,396.7	16.50	24.1	58.3	46.8
LM 1st midcourse correction	105:37:56				1.27		
LM 2nd midcourse correction	105:52:56				1.84		
LM braking	106:05:49				31.6	63.3	56.4
CSM/LM docked	106:22:02	54.7	5,365.9				
LM separation ignition	108:43:23.3	57.3	5,352.3				
LM separation cutoff	108:43:29.8	57.6	5,352.1	6.5	2.1	64.0	56.3
LM ascent propulsion system ignition	108:52:05.5	59.1	5,343.0				
LM ascent propulsion system depletion	108:56:14.5	89.7	9,056.4	249.0	4,600.0	2,211.6	56.2

Apollo 10 Transearth Phase

Event	GET (hhh:mm:ss)	Altitude (n mi)	Space Fixed Velocity (ft/sec)	Event Duration (sec)	Velocity Change (ft/sec)	Space Fixed Flight Path Angle (deg)	Space Fixed Heading Angle (E of N)
Transearth injection ignition	137:36:28.9	56.0	5,362.7			-0.44	-73.60
Transearth injection cutoff	137:39:13.7	56.5	8,987.2	164.8	3,680.3	2.53	-76.68
Midcourse correction ignition	188:49:58.0	25,570.4	12,540.0			-69.65	119.34
Midcourse correction cutoff	188:50:04.7	25,557.4	12,543.5	6.7	2.2	-69.64	119.34

Apollo 10 Timeline

Event	GET (hhh:mm:ss)	GMT Time	GMT Date
Terminal countdown started.	-028:00:00	01:00:00	17 May 1969
Scheduled 1-hour hold at T-3 hours 30 minutes.	-003:30:00	12:19:00	18 May 1969
Countdown resumed at T-3 hours 30 minutes.	-003:30:00	13:19:00	18 May 1969
Guidance reference release.	-000:00:16.978	16:48:43	18 May 1969
S-IC engine start command.	-000:00:08.9	16:48:51	18 May 1969
S-IC engine ignition (#5).	-000:00:06.4	16:48:53	18 May 1969
All S-IC engines thrust OK.	-000:00:01.6	16:48:58	18 May 1969
Range zero.	000:00:00.00	16:49:00	18 May 1969
All holddown arms released (1st motion) (1.06 g).	000:00:00.25	16:49:00	18 May 1969
Liftoff (umbilical disconnected).	000:00:00.58	16:49:00	18 May 1969
Tower clearance yaw maneuver started.	000:00:01.6	16:49:01	18 May 1969
Yaw maneuver ended.	000:00:10.0	16:49:10	18 May 1969
Pitch and roll maneuver started.	000:00:13.05	16:49:13	18 May 1969
Roll maneuver ended.	000:00:32.3	16:49:32	18 May 1969
Mach 1 achieved.	000:01:06.8	16:50:06	18 May 1969
Maximum dynamic pressure (694.232 lb/ft^2).	000:01:22.6	16:50:22	18 May 1969
Maximum bending moment (88,000,000 lbf-in).	000:01:24.6	16:50:24	18 May 1969
S-IC center engine cutoff command.	000:02:15.16	16:51:15	18 May 1969
Pitch maneuver ended.	000:02:38.7	16:51:38	18 May 1969
S-IC outboard engine cutoff.	000:02:41.63	16:51:41	18 May 1969
S-IC maximum total inertial acceleration (3.92 g).	000:02:41.71	16:51:41	18 May 1969
S-IC maximum Earth-fixed velocity.	000:02:41.96	16:51:41	18 May 1969
S-IC/S-II separation command.	000:02:42.31	16:51:42	18 May 1969
S-II engine start command.	000:02:43.05	16:51:43	18 May 1969
S-II ignition.	000:02:44.05	16:51:44	18 May 1969
S-II aft interstage jettisoned.	000:03:12.3	16:52:12	18 May 1969
Launch escape tower jettisoned.	000:03:17.8	16:52:17	18 May 1969
Iterative guidance mode initiated.	000:03:22.9	16:52:22	18 May 1969
S-IC apex.	000:04:26.87	16:53:26	18 May 1969
S-II center engine cutoff.	000:07:40.61	16:56:40	18 May 1969
S-II maximum total inertial acceleration (1.82 g).	000:07:40.69	16:56:40	18 May 1969
S-IC impact (theoretical).	000:08:59.12	16:57:59	18 May 1969
S-II outboard engine cutoff.	000:09:12.64	16:58:12	18 May 1969
S-II maximum Earth-fixed velocity. S-II/S-IVB separation command.	000:09:13.50	16:58:13	18 May 1969
S-IVB 1st burn start command.	000:09:13.60	16:58:13	18 May 1969
S-IVB 1st burn ignition.	000:09:16.81	16:58:16	18 May 1969
S-IVB ullage case jettisoned.	000:09:25.4	16:58:25	18 May 1969
S-II apex.	000:09:57.21	16:58:57	18 May 1969
S-IVB 1st burn cutoff command.	000:11:43.76	17:00:43	18 May 1969
S-IVB 1st burn maximum Earth-fixed velocity and maximum total inertial acceleration (0.70 g).	000:11:43.84	17:00:43	18 May 1969
Earth orbit insertion.	000:11:53.76	17:00:53	18 May 1969
Maneuver to local horizontal attitude and orbital navigation started.	000:12:04.1	17:01:04	18 May 1969
S-II impact (theoretical).	000:20:17.89	17:09:17	18 May 1969
S-IVB 2nd burn restart preparation.	002:23:49.26	19:12:49	18 May 1969
S-IVB 2nd burn restart command.	002:33:19.20	19:22:19	18 May 1969
S-IVB 2nd burn ignition.	002:33:27.52	19:22:27	18 May 1969
S-IVB 2nd burn cutoff.	002:39:10.58	19:28:10	18 May 1969
S-IVB 2nd burn maximum total inertial acceleration (1.49 g).	002:39:10.66	19:28:10	18 May 1969
S-IVB 2nd burn maximum Earth-fixed velocity. S-IVB safing procedures started.	002:39:11.30	19:28:11	18 May 1969
Translunar injection.	002:39:20.58	19:28:20	18 May 1969
Orbital navigation started.	002:39:29.6	19:28:29	18 May 1969

Apollo 10 Timeline

Event	GET (hhh:mm:ss)	GMT Time	GMT Date
CSM separated from S-IVB (ignition).	003:02:42.4	19:51:42	18 May 1969
CSM separated from S-IVB (cutoff).	003:02:45.7	19:51:45	18 May 1969
TV transmission started.	003:06:00	19:55:00	18 May 1969
CSM docked with LM/S-IVB.	003:17:36.0	20:06:36	18 May 1969
TV transmission ended.	003:28:00	20:17:00	18 May 1969
TV transmission started.	003:56:00	20:45:00	18 May 1969
CSM/LM ejected from S-IVB.	003:56:25.7	20:45:25	18 May 1969
TV transmission ended.	004:09:25	20:58:25	18 May 1969
CSM SPS evasive maneuver ignition.	004:39:09.8	21:28:09	18 May 1969
CSM SPS evasive maneuver cutoff.	004:39:12.7	21:28:12	18 May 1969
Maneuver to S-IVB slingshot attitude initiated.	004:42:15.8	21:31:15	18 May 1969
S-IVB lunar slingshot maneuver—APS ignition.	004:45:36.4	21:34:36	18 May 1969
S-IVB lead experiment—LOX lead started.	004:48:21.3	21:37:21	18 May 1969
S-IVB lead experiment—LOX lead ended.	004:48:30.3	21:37:30	18 May 1969
S-IVB lead experiment—LH$_2$ lead started.	004:50:09.9	21:39:09	18 May 1969
S-IVB lunar slingshot maneuver—APS cutoff.	004:50:17.0	21:39:17	18 May 1969
S-IVB lead experiment —LH$_2$ lead ended.	004:50:58.8	21:39:58	18 May 1969
S-IVB safing—LH$_2$ tank CVS open.	004:51:36.1	21:40:36	18 May 1969
S-IVB lunar slingshot maneuver—LOX dump started.	004:54:15.79	21:43:15	18 May 1969
S-IVB lunar slingshot maneuver—LOX dump ended.	004:59:16.00	21:48:16	18 May 1969
TV transmission started.	005:06:34	21:55:34	18 May 1969
S-IVB safing—LH$_2$ tank NPV valve open.	005:16:09.8	22:05:09	18 May 1969
TV transmission ended.	005:19:49	22:08:49	18 May 1969
S-IVB lunar impact maneuver—APS ignition.	005:28:55.8	22:17:55	18 May 1969
S-IVB lunar impact maneuver—APS cutoff.	005:29:04.9	22:18:04	18 May 1969
TV transmission started.	007:11:27	00:00:27	19 May 1969
TV transmission ended.	007:35:27	00:24:27	19 May 1969
Midcourse correction ignition.	026:32:56.8	19:21:56	19 May 1969
Midcourse correction cutoff.	026:33:03.9	19:22:03	19 May 1969
TV transmission started.	027:00:48	19:49:48	19 May 1969
TV transmission ended.	027:28:31	20:17:31	19 May 1969
High-gain antenna reacquisition test.	028:50	21:39	19 May 1969
TV transmission started (recorded).	048:00:51	16:49:51	20 May 1969
TV transmission ended.	048:15:30	17:04:30	20 May 1969
TV transmission started (recorded).	048:24:00	17:13:00	20 May 1969
TV transmission ended.	048:27:51	17:16:51	20 May 1969
TV transmission started.	049:54:00	18:43:00	20 May 1969
TV transmission ended.	049:58:49	18:47:49	20 May 1969
TV transmission started.	053:35:30	22:24:30	20 May 1969
TV transmission ended.	054:00:30	22:49:30	20 May 1969
Equigravisphere.	061:50:50	06:39:50	21 May 1969
TV transmission started.	072:37:26	17:26:26	21 May 1969
TV transmission ended.	072:54:42	17:43:42	21 May 1969
Lunar orbit insertion ignition.	075:55:54.0	20:44:54	21 May 1969
Lunar orbit insertion cutoff.	076:01:50.1	20:50:50	21 May 1969
Lunar surface observations.	076:30	21:19	21 May 1969
S-IVB closest approach to lunar surface.	078:51:03.6	23:40:03	21 May 1969
Lunar orbit circularization ignition.	080:25:08.1	01:14:08	22 May 1969
Lunar orbit circularization cutoff.	080:25:22.0	01:14:22	22 May 1969
TV transmission started.	080:44:40	01:33:40	22 May 1969
Lunar surface observations.	080:50 01:39		22 May 1969

Apollo 10 Timeline

Event	GET (hhh:mm:ss)	GMT Time	GMT Date
TV transmission ended.	081:13:49	02:02:49	22 May 1969
LM cabin pressurized.	081:30 02:19		22 May 1969
Transfer to LM power and systems checked.	081:55 02:44		22 May 1969
Transfer to LM power. Systems tested.	082:40 03:29		22 May 1969
Transfer to CM, hatch and tunnel closed.	084:30 05:19		22 May 1969
CDR and LMP entered LM to activate systems.	095:02 15:51		22 May 1969
Landing gear deployed.	098:00 18:49		22 May 1969
CSM/LM undocked.	098:11:57	19:00:57	22 May 1969
TV transmission started.	098:29:20	19:18:20	22 May 1969
CSM/LM separation maneuver ignition.	098:47:17.4	19:36:17	22 May 1969
CSM/LM separation maneuver cutoff.	098:47:25.7	19:36:25	22 May 1969
TV transmission ended.	098:49:30	19:38:30	22 May 1969
CSM rendezvous radar transponder anomaly.	098:51:54	19:40:54	22 May 1969
LM system checks.	099:00	19:49	22 May 1969
Descent orbit insertion ignition (SPS).	099:46:01.6	20:35:01	22 May 1969
Descent orbit insertion cutoff.	099:46:29.0	20:35:29	22 May 1969
LM near-lunar-surface activity.	100:40	21:29	22 May 1969
LM oriented for radar overpass test.	100:32:00	21:21:00	22 May 1969
LM acquisition of radar beam.	100:32:22	21:21:22	22 May 1969
LM closest approach to lunar surface.	100:41:43	21:30:43	22 May 1969
Phasing maneuver ignition.	100:58:25.93	21:47:25	22 May 1969
Phasing maneuver cutoff.	100:59:05.88	21:48:05	22 May 1969
LM ascent stage/descent stage separated.	102:45:16.9	23:34:16	22 May 1969
Ascent orbit insertion ignition.	102:55:02.13	23:44:02	22 May 1969
Ascent orbit insertion cutoff.	102:55:17.68	23:44:17	22 May 1969
Coelliptic sequence initiation ignition.	103:45:55.3	00:34:55	23 May 1969
Coelliptic sequence initiation cutoff.	103:46:22.6	00:35:22	23 May 1969
Constant differential height maneuver ignition.	104:43:53.28	01:32:53	23 May 1969
Constant differential height maneuver cutoff.	104:43:54.93	01:32:54	23 May 1969
Terminal phase initiation ignition.	105:22:55.58	02:11:55	23 May 1969
Terminal phase initiation cutoff.	105:23:12.08	02:12:12	23 May 1969
Midcourse correction (lunar orbit).	105:37:56	02:26:56	23 May 1969
Midcourse correction (lunar orbit).	105:52:56	02:41:56	23 May 1969
Braking maneuver.	106:05:49	02:54:49	23 May 1969
CSM/LM docked.	106:22:02	03:11:02	23 May 1969
CDR and LMP entered CM.	106:42	03:31	23 May 1969
LM closeout activities started.	107:20	04:09	23 May 1969
LM ascent stage jettisoned.	108:24:36	05:13:36	23 May 1969
LM separation maneuver ignition.	108:43:23.3	05:32:23	23 May 1969
LM separation maneuver cutoff.	108:43:29.8	05:32:29	23 May 1969
LM ascent propulsion system ignition.	108:52:05.5	05:41:05	23 May 1969
LM ascent propulsion system depletion.	108:56:14.5	05:45:14	23 May 1969
LM descent orbit insertion. Terminator-to-terminator strip photographs.	119:20	16:09	23 May 1969
Orbital navigation and landmark tracking.	124:30	21:19	23 May 1969
Orbital navigation and landmark tracking.	128:00	00:49	24 May 1969
TV transmission started.	132:07:12	04:56:12	24 May 1969
TV transmission ended.	132:31:24	05:20:24	24 May 1969
Target of opportunity photography.	133:00	05:49	24 May 1969
Target of opportunity and strip photography.	134:40	07:29	24 May 1969
Transearth injection ignition (SPS).	137:36:28.9	10:25:28	24 May 1969
Transearth injection cutoff.	137:39:13.7	10:28:13	24 May 1969

Apollo 10 Timeline

Event	GET (hhh:mm:ss)	GMT Time	GMT Date
TV transmission started.	137:50:51	10:39:51	24 May 1969
TV transmission ended.	138:33:54	11:22:54	24 May 1969
TV transmission started.	139:30:16	12:19:16	24 May 1969
TV transmission ended.	139:37:11	12:26:11	24 May 1969
TV transmission started.	147:23:00	20:12:00	24 May 1969
TV transmission ended.	147:34:25	20:23:25	24 May 1969
TV transmission started.	152:29:19	01:18:19	25 May 1969
TV transmission ended.	152:58:24	01:47:24	25 May 1969
CSM S-band high-gain reflectivity test.	168:00	16:49	25 May 1969
TV transmission started.	173:27:17	22:16:17	25 May 1969
TV transmission ended.	173:37:39	22:26:39	25 May 1969
TV transmission started.	186:51:49	11:40:49	26 May 1969
TV transmission ended.	187:03:42	11:52:42	26 May 1969
Midcourse correction ignition.	188:49:58.0	13:38:58	26 May 1969
Midcourse correction cutoff.	188:50:04.7	13:39:04	26 May 1969
Maneuver to entry attitude.	189:40	14:29	26 May 1969
CM/SM separation.	191:33:26	16:22:26	26 May 1969
Entry.	191:48:54.5	16:37:54	26 May 1969
Communication blackout started.	191:49:12	16:38:12	26 May 1969
Maximum entry g force (6.78 g).	191:50:14	16:39:14	26 May 1969
Visual contact with CM established by recovery forces.	191:51	16:40	26 May 1969
Radar contact with CM established by recovery ship.	191:52	16:41	26 May 1969
Communication blackout ended.	191:53:40	16:42:40	26 May 1969
Drogue parachute deployed	191:57:18.0	16:46:18	26 May 1969
Main parachute deployed.	191:58:05	16:47:05	26 May 1969
Splashdown (went to apex-up).	192:03:23	16:52:23	26 May 1969
Flotation collar inflated.	192:21	17:10	26 May 1969
CM hatch opened.	192:28	17:17	26 May 1969
Crew in life raft.	192:31	17:20	26 May 1969
Crew aboard recovery helicopter.	192:37	17:26	26 May 1969
Crew aboard recovery ship.	192:42	17:31	26 May 1969
CM aboard recovery ship.	193:39	18:28	26 May 1969

APOLLO 11

The Fifth Mission:
The First Lunar Landing

Apollo 11 Summary

(16 July–24 July 1969)

Apollo 11 crew (l. to r.): Neil Armstrong, Mike Collins, Buzz Aldrin (NASA S69-31740).

Background

Apollo 11 was a Type G mission, a piloted lunar landing demonstration. The primary objective of the Apollo program was to perform a piloted lunar landing and return safely to Earth.

It was only the second time an all-experienced crew had flown an American mission, and it would be the last until Space Shuttle mission STS-26 nearly two decades later.

The crew members for this historic mission were Neil Alden Armstrong, commander; Lt. Colonel Michael Collins (USAF), command module pilot; and Colonel Edwin Eugene "Buzz" Aldrin, Jr. (USAF), lunar module pilot.

Selected as an astronaut in 1962, Armstrong had been the first civilian ever to command an American space mission when he was command pilot of Gemini 8, which featured the first-ever docking of two vehicles in space. Apollo 11 made him the first civilian to command two missions. Armstrong was born 5 August 1930 in Wapakoneta, Ohio, and was 38 years old at the time of the Apollo 11 mission. He received a B.S. in aeronautical engineering from Purdue University in 1955 and an M.S. in aerospace engineering

from the University of Southern California in 1970, following the Apollo mission. His backup was Captain James Arthur Lovell, Jr. (USN).

Collins had been pilot of Gemini 10. He was born 31 October 1930 in Rome, Italy, and was 38 years old at the time of the Apollo 11 mission. Collins received a B.S. from the U.S. Military Academy in 1952 and was selected as an astronaut in 1963. His backup was Lt. Colonel William Alison Anders (USAF).

Aldrin had been pilot of Gemini 12. He was born 20 January 1930 in Montclair, New Jersey, and was 39 years old at the time of the Apollo 11 mission. Aldrin received a B.S. in mechanical engineering from the U.S. Military Academy in 1951 and an Sc.D. in astronautics from the Massachusetts Institute of Technology in 1963. Also in 1963, he was selected as an astronaut. Aldrin has the distinction of being the first astronaut with a doctorate to fly in space. His backup was Fred Wallace Haise, Jr.

The capsule communicators (CAPCOMs) for the mission were Major Charles Moss Duke, Jr. (USAF), Lt. Commander Ronald Ellwin Evans (USN), Lt. Commander Bruce McCandless II (USN), Lovell, Anders, Lt. Commander Thomas Kenneth "Ken" Mattingly II (USAF), Haise, Don Leslie Lind, Ph. D., Owen Kay Garriott, Jr., Ph. D., and Harrison Hagan "Jack" Schmitt, Ph. D. The support crew were Mattingly, Evans, Major William Reid Pogue (USAF), and John Leonard "Jack" Swigert, Jr. The flight directors were Clifford E. Charlesworth and Gerald D. Griffin (first shift), Eugene F. Kranz (second shift), and Glynn S. Lunney (third shift).

The Apollo 11 launch vehicle was a Saturn V, designated SA-506. The mission also carried the designation Eastern Test Range #5307. The CSM was designated CSM-107 and had the call-sign "Columbia." The lunar module was designated LM-5 and had the call-sign "Eagle."

Possible landing sites for Apollo 11 were under study by NASA's Apollo Site Selection Board for more than two years. Thirty sites were originally considered, but the list was shortened to three for the first lunar landing. Selection of the final sites was based on high-resolution photographs taken by the Lunar Orbiter satellite, plus close-up photographs and surface data provided by the Surveyor spacecraft, which landed on the Moon.

Apollo 11 lifts off from Kennedy Space Center Pad 39A (NASA S69-39526).

The original sites were located on the visible side of the Moon, within 45° east and west of the Moon's center and 5° north and south of its equator. The final site choices were based on the following factors:

Smoothness: Relatively few craters and boulders.

Approach: No large hills, high cliffs, or deep craters that could cause incorrect altitude signals to the lunar module landing radar.

Propellant Requirements: Least potential expenditure of spacecraft propellants.

Recycling: Effective launch preparation recycling if the countdown were delayed.

Free Return: Sites within reach of the spacecraft launched on a free return translunar trajectory.

Slope: Less than 2° slope in the approach path and landing area.

There were a number of considerations which determined the launch windows for a lunar landing mission. These considerations included illumination conditions at launch, launch pad azimuth, translunar injection geometry, sun elevation angle at the lunar landing site, illumination conditions at Earth splashdown, and the number and location of the lunar landing sites.

The time of a lunar landing was determined by the location of the lunar landing site and by the acceptable range of sun elevation angles. The range of these angles was from 5° to 14° and in a direction from east to west. Under these conditions, visible shadows of craters would aid the crew in recognizing topographical features. When the sun angle approached the descent angle, the mean value of which was 16°, visual resolution would be degraded by a "washout" phenomenon where backward reflectance was high enough to eliminate contrast. Sun angles above the flight path were not as desirable because shadows would not be readily visible unless the sun was significantly outside the descent plane. In addition, higher sun angles (greater than 18°) could be eliminated from consideration by planning the landing one day earlier where the lighting is at least 5°. Because lunar sunlight incidence changed about 0.5° per hour, the sun elevation angle restriction established a 16-hour period, which occurred every 29.5 days, when landing at a given site could be attempted. The number of Earth-launch opportunities for a given lunar month was equal to the number of candidate landing sites.

The time of launch was primarily determined by the allowable variation in launch pad azimuth and by the location of the Moon at spacecraft arrival. The spacecraft had to be launched into an orbital plane that contained the position of the Moon and its antipode at spacecraft arrival. A 34° launch pad azimuth variation afforded a launch period of 4 hours 30 minutes. This period was called the "daily launch window," the time when the direction of launch was within the required range to intercept the Moon.

Two launch windows occurred each day. One was available for a translunar injection out of Earth orbit in the vicinity of the Pacific Ocean, and the other was in the vicinity of the Atlantic Ocean. The injection opportunity over the Pacific Ocean was preferred because it usually permitted a daytime launch.

Launch Preparations

The terminal countdown started at T-28 hours, 21:00:00 GMT on 14 July. The scheduled holds of 11 hours at T-9 hours and 1 hour 32 minutes at T-3 hours 30 minutes were the only holds initiated. The start of the S-II stage LH2 loading was delayed 25 minutes due to a communications problem in the Pad Terminal Connection Room. However, the delay was recovered during the scheduled countdown hold at T-3 hours 30 minutes.

A high-pressure cell in the Atlantic Ocean off the North Carolina coast, along with a weak trough of low pressure located in the northeastern Gulf of Mexico, caused light southerly surface winds and brought moisture into the Cape Kennedy area. These circumstances contributed to the cloudy conditions and distant thunderstorms observed at launch time. Cumulus clouds covered 10 percent of the sky (base 2,400 feet), altocumulus covered 20 percent (base 15,000 feet), and cirrostratus covered 90 percent (base not recorded); the temperature was 84.9° F; the relative humidity was 73 percent; and the barometric pressure was 14.798 lb/in². The winds, as measured by the anemometer on the light pole 60.0 feet above ground at the launch site, measured 6.4 knots at 175° from true north.

Ascent Phase

Apollo 11 was launched from Kennedy Space Center Launch Complex 39, Pad A, at a Range Zero time of 13:32:00 GMT (09:32:00 a.m. EDT) on 16 July 1969. The planned launch window for Apollo 11 extended to 17:54:00 GMT to take advantage of a sun elevation angle on the lunar surface of 10.8°.

Lunar landing site 2 in the Sea of Tranquility, compared to the size of Washington, DC (NASA S69-38667).

Between 000:00:13.2 and 000:00:31.1, the vehicle rolled from a launch pad azimuth of 90° to a flight azimuth of 72.058°. The S-IC engine shut down at 000:02:41.63, followed by S-IC/S-II separation and S-II engine ignition. The S-II engine shut down at 000:09:08.22, followed by separation from the S-IVB, which ignited at 000:09:12.2. The first S-IVB engine cutoff occurred at 000:11:39.33, with deviations from the planned trajectory of only -0.6 ft/sec in velocity and only -0.1 n mi in altitude.

The S-IC stage impacted the Atlantic Ocean at 000:09:03.70 at latitude 30.212° north and longitude 74.038° west, 357.1 n mi from the launch site. The S-II stage impacted the Atlantic Ocean at 000:20:13.7 at latitude 31.535° north and longitude 34.844° west, 2,371.8 n mi from the launch site.

The maximum wind conditions encountered during ascent were 18.7 knots at 297° from true north at 37,400 feet, and a maximum wind shear of 0.0077 sec⁻¹ at 48,490 feet.

Parking orbit conditions at insertion, 000:11:49.34 (S-IVB cutoff plus 10 seconds to account for engine tailoff and other transient effects), showed an apogee and perigee of 100.4 by 98.9 n mi, and an inclination of 32.521°, a period of 88.18 minutes, and a velocity of 25,567.9 ft/sec. The apogee and perigee were based upon a spherical Earth with a radius of 3,443.934 n mi.

The international designation for the CSM upon achieving orbit was 1969-059A, and the S-IVB was designated 1969-059B. After undocking at the Moon, the LM ascent stage would be designated 1969-059C and the descent stage 1969-059D.

Earth Orbit Phase

After in-flight systems checks, the 346.87-second translunar injection maneuver (second S-IVB firing) was performed at 002:44:16.20. The S-IVB engine shut down at 002:50:03.03, and translunar injection occurred ten seconds later, after 1.5 Earth orbits lasting 2 hours 38 minutes 23.73 seconds, at a velocity of 35,567.3 ft/sec.

Translunar Phase

At 003:15:23.0, the CSM was separated from the S-IVB stage and transposed and docked with the LM at 003:24:03.7. The docked spacecraft were ejected from the S-IVB at 004:17:03.0, and a 2.93-second separation maneuver was performed at 004:40:01.72. A ground command for propulsive venting of residual propellants targeted the S-IVB to go past the Moon and into solar orbit. The closest approach of the S-IVB to the Moon was 1,825 n mi at 20:14:00 GMT on 19 July at 078:42:00. The trajectory after passing from the lunar sphere of influence resulted in a solar orbit with an aphelion and perihelion of 82.000 million by 72.520 million n mi, an inclination of 0.3836°, and a period of 342.00 days.

An unscheduled 16-minute television transmission was recorded at the Goldstone Tracking Station starting at 010:32. The tape was played back at Goldstone and transmitted to Houston starting at 011:26.

View of Earth at 98,000 n mi altitude following translunar injection (NASA AS11-36-5355).

Trajectory parameters after the translunar injection firing were nearly perfect. A 3.13-second midcourse correction of 20.9 ft/sec was made at 026:44:58.64 during the translunar phase. During the remaining periods of free-attitude flight, passive thermal control, a rotating "barbecue"-like maneuver, was used to maintain spacecraft temperatures within desired limits.

An unscheduled 50-minute television transmission was accomplished at 030:28, and a 36-minute scheduled transmission began at 033:59. The crew initiated a 96-minute color television transmission at 055:08. The picture resolution and general quality were exceptional. The coverage included the interior of the CM and LM and views of the exterior of the CM and Earth. Excellent views of the crew accomplishing probe and drogue removal, spacecraft tunnel hatch opening, LM housekeeping, and equipment testing were broadcast.

During the latter transmission, the commander and lunar module pilot transferred to the LM at 055:30 to make the initial inspection and preparations for the systems checks that would be made shortly after lunar orbit insertion. They returned to the CM at 058:00.

At 075:49:50.37, at an altitude of 86.7 n mi above the Moon, the service propulsion engine was fired for 357.53 seconds to insert the spacecraft into a lunar orbit of 169.7 by 60.0 n mi. The translunar coast had lasted 73 hours 5 minutes 34.83 seconds.

Earthrise over lunar surface following lunar orbit insertion (NASA AS11-44-6552).

Lunar Orbit/Lunar Surface Phase

During the second lunar orbit, at 078:20, a scheduled live color television transmission was accomplished, providing spectacular views of the lunar surface and the approach path to landing site 2.

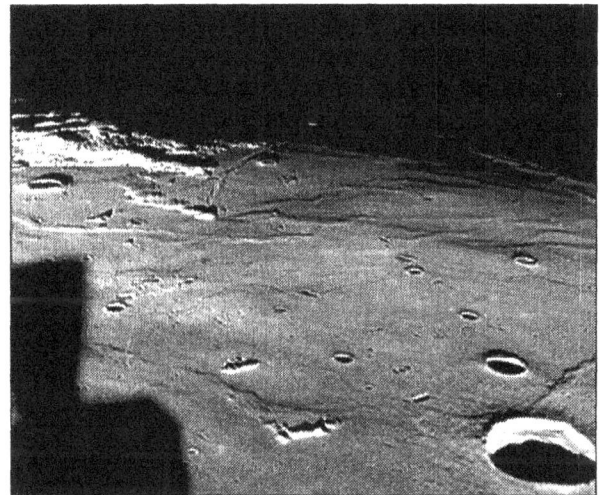

Approach to lunar landing site #3 in southwest Sea of Tranquility seen from LM while still docked to the CSM (NASA AS11-37-5437).

After two revolutions and a navigation update, a second service propulsion retrograde burn was made. The 16.88-second maneuver occurred at 080:11:36.75 and circularized the orbit at 66.1 by 54.5 n mi. The commander and lunar module pilot then transferred to the LM and, for about two hours, performed various housekeeping functions, a voice and telemetry test, and an oxygen purge system check. LM functions and consumables checked out well. Additionally, both cameras were checked and verified operational. The pair then returned to the CSM. At 095:20, they returned to the LM to perform a thorough check of all LM systems in preparation for descent.

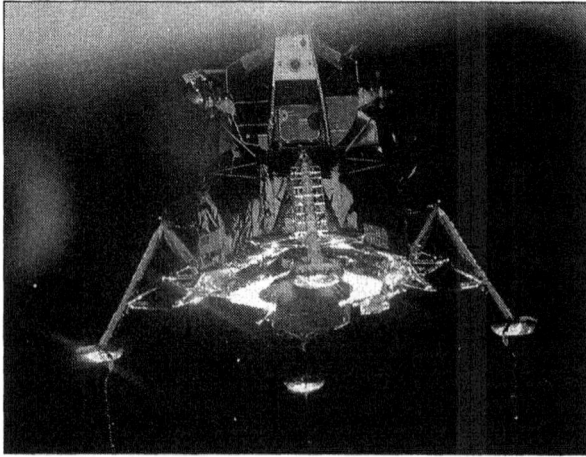

LM Eagle seen from CM Columbia following undocking (NASA AS11-44-6574).

Undocking occurred at 100:12:00.0 at an altitude of 62.9 n mi. This was followed by a CSM reaction control system 9.0-second separation maneuver at 100:39:52.9 directed radially downward toward the center of the Moon as planned. The LM descent orbit insertion maneuver was performed with a 30-second firing of the descent propulsion system at 101:36:14.0, which put the LM into an orbit of 58.5 by 7.8 n mi.

CM seen from LM following separation (NASA AS11-37-5445).

The 756.19-second powered descent engine burn was initiated at 102:33:05.01. The time was as planned, but the position at which powered descent initiation occurred was about 4 n mi farther downrange than expected. This resulted in the landing point being shifted downrange about 4 n mi.

The first of five alarms occurred at 102:38:22 because of a computer overload, but it was determined that it was safe to continue the landing. The crew checked the handling qualities of the LM at 102:41:53 and switched to automatic guidance ten seconds later. The landing radar switched to "low-scale" at 102:42:19 as the LM descended below 2,500 feet altitude. The LM was maneuvered manually 1,100 feet down range from the preplanned landing point during the final 2.5 minutes of descent. The final alarm occurred at 102:42:58, followed by the red-line low-level fuel quantity light at 102:44:28, just 72 seconds before landing.

During the final approach, the commander noted that the landing point toward which the spacecraft was headed was in the center of a large crater that appeared extremely rugged, with boulders of five to ten feet in diameter and larger. Consequently, he switched to manual attitude control to translate beyond the rough terrain area.

The LM landed on the Moon at 20:17:39 GMT (16:17:39 EDT) on 20 July 1969 at 102:45:39.9. Engine shutdown occurred 1.5 seconds later. The LM landed in Mare Tranquilitatis (Sea of Tranquility) at latitude 0.67408° north and longitude 23.47297° east at an angle to the surface of 4.5°, and about 3.75 n mi southwest of the planned point. Approximately 45 seconds of firing time remained at landing.[1]

For the first two hours on the lunar surface, the crew performed a checkout of all systems, configured the controls for lunar stay, and ate their first post-landing meal. A rest period had been planned to precede the extravehicular activity of exploring the lunar surface but was not needed.

After donning the back-mounted portable life support and oxygen purge systems, the commander prepared to exit the LM. The forward hatch was opened at 109:07:35 and the commander exited at 109:19:16. While descending the LM ladder, he deployed the Modular Equipment Stowage Assembly from the descent stage. A camera in the module provided live television coverage as he descended. The commander's left foot made first contact with the lunar surface at 02:56:15 GMT on 21 July (22:56:15 EDT on 20 July) at 109:24:15. His first words on the lunar surface were, "That's one small step for man, one giant leap for mankind."

[1] According to the *Apollo 11 Mission Report* (MSC-00171), postflight analysis revealed that there was 45 seconds of fuel remaining at lunar touchdown, not as little as 7 seconds as indicated by other sources.

The commander made a brief check of the LM exterior, indicating that penetration of the footpads was only about three to four inches and collapse of the LM footpad strut was minimal. He reported sinking about one-eighth inch into the fine, powdery surface material, which adhered readily to his lunar boots in a thin layer. There was no crater from the effects of the descent engine, and about one foot of clearance was observed between the engine bell and the lunar surface. He also reported that it was quite dark in the shadows of the LM, which made it difficult for him to see his footing.

Aldrin inside LM during first LM inspection (NASA AS11-36-5390).

He then collected a contingency sample of lunar soil from the vicinity of the LM ladder. He reported that although loose material created a soft surface, as he dug down six or eight inches he encountered very hard, cohesive material.

The commander then photographed the lunar module pilot as he exited at 109:39:00 and descended to the lunar surface at 109:43:15.

Following the LMP's descent to the surface, the crew unveiled a plaque mounted on the strut behind the ladder, and read its inscription to their worldwide television audience. The plaque read:

HERE MEN FROM THE PLANET EARTH
FIRST SET FOOT UPON THE MOON
JULY 1969, A.D.
WE CAME IN PEACE FOR ALL MANKIND.

The plaque featured the signatures of the three Apollo crew members and President Richard M. Nixon. Next, the commander removed the television camera from the descent stage, obtained a panorama, and placed the camera on its tripod in position to view the subsequent surface extravehicular operations.

Footprint in soft soil on lunar surface (NASA AS11-40-5877).

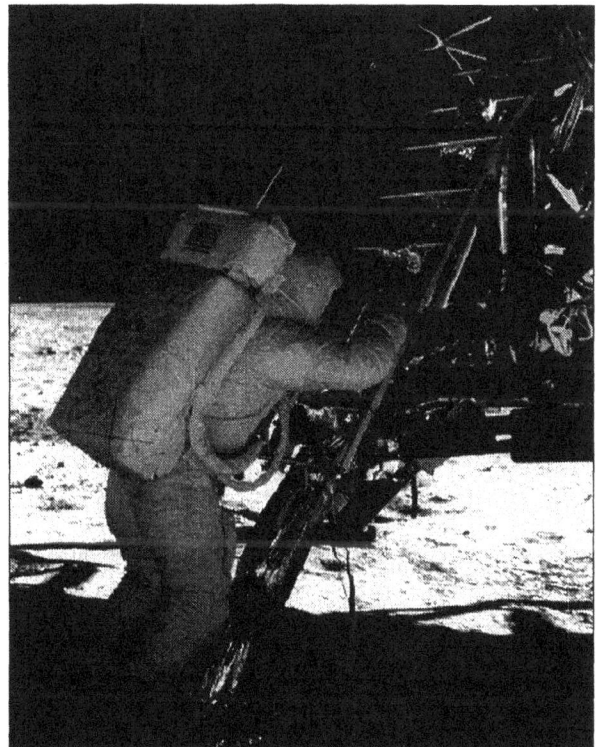

Aldrin steps from LM ladder onto lunar surface (NASA AS11-40-5869).

The lunar module pilot deployed the solar wind composition experiment on the lunar surface in direct sunlight and to the north of the LM as planned.

Photo of plaque on LM leg (NASA AS11-40-5899).

Aldrin deploys solar wind composition experiment (NASA AS11-40-5964).

At 110:09:43, the pair erected a three-by-five-foot United States flag on an eight-foot aluminum staff. A conversation between President Richard M. Nixon and the LM crew was held at 110:16:30. The conversation originated from the White House and included congratulations and good wishes.

During the environmental evaluation, the lunar module pilot indicated that he had to be careful of his center of mass in maintaining balance. He noted that the LM shadow had no significant effect on his backpack temperature. He also noted that his agility was better than expected and that

he was able to move about with great ease. Both crew members indicated that their mobility throughout this period exceeded all expectations. Also, indications were that metabolic rates were much lower than pre-mission estimates.

Armstrong and Aldrin set up U.S. flag on lunar surface (NASA S69-40308).

Aldrin prepares to deploy experiments from LM (NASA AS11-40-5927).

The commander collected a bulk sample, consisting of assorted surface material and rock chunks, and placed them in a sample return container. The crew then inspected the LM, finding the quads, struts, skirts, and antennas in satisfactory condition.

The passive seismic experiment package and laser ranging retroreflector were deployed south of the LM. Excellent PSEP data were obtained, including detection of the crew walking on the surface and, later, in the LM. The crew then collected more lunar samples, including two core samples and about 20 pounds of discretely selected material. The LMP had to exert considerable force to drive the core tubes six to eight inches into the lunar surface.

Aldrin deploys PSEP and LR3 (NASA AS11-40-5942).

This photo (Hasselblad image) of Armstrong on the lunar surface shows him at the MESA packing the bulk sample. U.S. flag is to the left and solar wind experiment is in the center (NASA AS11-40-5886).

The solar wind experiment was recovered after 1 hour 17 minutes exposure. The transfer of lunar sample containers to the LM began at 111:23. The crew entered the LM and closed the hatch at 111:39:13, thus ending humankind's first exploration of the Moon. The total time spent outside the LM was 2 hours 31 minutes 40 seconds; the total dis-

tance traveled was 3,300 feet (1 km); and the collected samples totaled 47.52 pounds (21.55 kg).[2] The farthest point traveled from the LM was 200 feet (60 m), when the commander visited a crater 108 feet in diameter (33 m) near the end of the extravehicular period.

Cropped close-up of Armstrong at the LM (NASA AS11-40-5886).

Armstrong inside LM following EVA (NASA AS11-37-5528).

Ignition of the ascent stage engine for liftoff occurred at 17:54:00 GMT on 21 July at 124:22:00.79. The LM had been on the lunar surface for 21 hours 36 minutes 20.9 seconds. An orbit of 48.0 by 9.4 n mi was achieved at 124:29:15.67, 434.88 seconds after liftoff.

Several rendezvous sequence maneuvers were required before docking could occur 3.5 hours later. A 47.3-second coelliptic orbit maneuver at 125:19:34.70 raised the orbit to 49.3 by 45.7 n mi. A 17.8-second constant delta height maneuver at

---

[2] Official total in kilograms as determined by the Lunar Receiving Laboratory in Houston.

126:17:49.6 lowered the orbit to 47.4 by 42.1 n mi. A 22.7-second terminal phase initiate maneuver at 127:03:51.8 brought the ascent stage to an orbit of 61.7 by 43.7 n mi. The 28.4-second terminal phase maneuver at 127:46:09.8 finalized the orbit at 63.0 by 56.5 for docking of the ascent stage and the CSM at 128:03:00.0. The two craft had been undocked for exactly 27 hours 51 minutes.

View of ascent stage prior to docking with CM. CMP Collins refers to this photo as "There They Are," meaning that his crewmates are in the ascent stage and all the rest of the humans we know about are on that spot in the background (AS11-44-6642).

In the process of maneuvering the LM to docking attitude, while avoiding direct sunlight in the forward window, the platform inadvertently reached gimbal lock, causing a brief and unexpected tumbling motion of the LM. A quick recovery was made and the docking was completed using the abort guidance system for attitude control.

After transfer of the crew and samples to the CSM, the ascent stage was jettisoned at 130:09:31.2 at an altitude of 61.6 n mi, and the CSM was prepared for transearth injection. A 7.2-second maneuver was made at 130:30:01.0 to separate the CM from the ascent stage; it resulted in an orbit of 62.7 by 54.0 n mi. The ascent stage would remain in lunar orbit for an indefinite period.

The 151.41-second transearth injection maneuver was performed at 135:23:42.28 at an altitude of 52.4 n mi. A nominal injection was achieved at 135:26:13.69 after 30 lunar orbits lasting 59 hours 30 minutes 25.79 seconds, at a velocity of 8,589.0 ft/sec.

Transearth Phase

As in translunar flight, only one midcourse correction was required, a 10.0-second, 4.8-ft/sec maneuver, at 150:29:57.4.

Passive thermal control was exercised for most of the transearth coast.

View of Moon after departure from an altitude of 10,000 n mi. (NASA AS11-44-6667).

An 18-minute television transmission was initiated at 155:36; it featured a demonstration of the effect of weightlessness on food and water, as well as brief scenes of the Moon and Earth. The final color television broadcast was made at 177:32. The 12.5-minute transmission featured a message of appreciation by each crew member to all the people who helped make the mission possible.

Recovery

Because of inclement weather in the planned recovery area, the splashdown point was moved 215 n mi down range. The weather in the new area was excellent: visibility 12 miles, waves to 3 feet, and wind 16 knots.

The service module was jettisoned at 194:49:12.7, and the CM entry followed an automatic entry profile. The command module reentered Earth's atmosphere (400,000 feet altitude) at 195:03:05.7 at a velocity of 36,194.4 ft/sec, following a transearth coast of 59 hours 36 minutes 52.0 seconds.

The parachute system effected splashdown of the CM in the Pacific Ocean at 16:50:35 GMT (12:50:35 EDT) on 24 July. Mission duration was 195:18:35. The impact point was 1.69 n mi from the target point and 13 n mi from the recovery ship U.S.S. *Hornet*. The splashdown site was estimated to be latitude 13.30° north and longitude 169.15° west.

Apollo 11 crew in raft while waiting for helicopter retrieval (NASA S69-21698).

After splashdown, the CM assumed an apex-down flotation attitude but was successfully returned to the normal flotation position in 7 minutes 40 seconds by the inflatable bag uprighting system. After splashdown, the crew donned biological isolation garments and exited the CM into a rubber boat, where they were scrubbed down with an iodine solution to protect against "lunar germs." They were then retrieved by helicopter and taken to the primary recovery ship, where they arrived 63 minutes after splashdown. The CM was recovered 125 minutes later. The estimated CM weight at splashdown was 10,873 pounds, and the estimated distance traveled for the mission was 828,743 n mi.

The crew, the recovery physician, and a recovery technician, along with lunar samples, entered the Mobile Quarantine Facility aboard the recovery ship for transport to the Lunar Receiving Laboratory in Houston.

The CM and Mobile Quarantine Facility were offloaded from the *Hornet* in Hawaii 00:15 GMT on 27 July. The Mobile Quarantine Facility was loaded aboard a C-141 aircraft and flown to Houston, where it arrived at 06:00 GMT on 28 July.

The crew arrived at the Lunar Receiving Laboratory four hours later. The safing of the CM pyrotechnics was completed at 02:05 GMT on 27 July. The CM was taken to Ford Island for deactivation, after which it was transferred to Hickam Air Force Base, Hawaii, and flown on a C-133 aircraft to Houston, where it arrived at 23:17 GMT on 30 July.

The crew and spacecraft were released from quarantine on 10 August. On 14 August the spacecraft was delivered to the North American Rockwell Space Division facility in Downey, California, for postflight analysis.

Wearing biological isolation garments, crew enters the Mobile Quarantine Facility aboard recovery ship U.S.S. *Hornet* (NASA S69-40753).

President Richard M. Nixon welcomes Apollo 11 crew home (NASA S69-21365).

All spacecraft systems performed satisfactorily. With the completion of the Apollo 11 mission, the national objective of landing humans on the Moon and returning them safely to Earth before the end of the decade was accomplished.

Mission Director Chris Kraft (holding flag) celebrates success of Apollo 11 mission with other NASA officials (NASA S69-40302).

Conclusions

The Apollo 11 mission, including a piloted lunar landing and surface exploration, was conducted with skill, precision, and relative ease. The excellent performance of the spacecraft in the preceding four missions and the thorough planning in all aspects of the program permitted the safe and efficient execution of this mission. The following conclusions were made from an analysis of post-mission data:

1. The effectiveness of pre-mission training was reflected in the skill and precision with which the crew executed the lunar landing. Manual control while maneuvering to the desired landing point was satisfactorily exercised.

2. The planned techniques involved in the guidance, navigation, and control of the descent trajectory were good. Performance of the landing radar met all expectations in providing the information required for descent.

3. The extravehicular mobility units were adequately designed to enable the crew to conduct the planned activities. Adaptation to 1/6 g was relatively quick, and mobility on the lunar surface was easy.

4. The two-person pre-launch checkout and countdown for ascent from the lunar surface were well planned and executed.

5. The timeline activities for all phases of the lunar landing mission were well within the crew's capability to perform the required tasks.

6. The quarantine operation from spacecraft landing until release of the crew, spacecraft, and lunar samples from the Lunar

Receiving Laboratory was accomplished successfully and without any violation of the quarantine.

7. No microorganisms of extraterrestrial origin were recovered from either the crew or the spacecraft.

8. Hardware problems, as experienced on previous piloted missions, did not unduly hamper the crew or compromise crew safety or mission objectives.

9. The Mission Control Center and the Manned Space Flight Network proved to be adequate for controlling and monitoring all phases of flight, including the descent, surface activities, and ascent phases of the mission.

Apollo 11 Objectives

Spacecraft Primary Objective

To perform a piloted lunar landing and return. *Achieved.*

Spacecraft Secondary Objectives

1. To perform selenological inspection and sampling.

   a. Contingency sample collection. *Achieved.*

   b. Lunar surface characteristics. *Achieved.*

   c. Bulk sample collection. *Achieved.*

   d. Lunar environment visibility. *Achieved.*

2. To obtain data to assess the capability and limitations of the astronaut and his equipment in the lunar surface environment.

   a. Lunar surface extravehicular operations. *Achieved.*

   b. Lunar surface operations with extravehicular mobility unit. *Achieved.*

   c. Landing effects on lunar module. *Achieved.*

   d. Location of landed lunar module. *Partially achieved. The LM crew was unable to make observations of lunar features during descent. The command module pilot was therefore unable to locate the lunar module through the sextant. Toward the end of the lunar surface stay, the location of the lunar module was determined from the lunar module rendezvous radar tracking data, which was confirmed post-mission using descent photographic data.*

e. Assessment of contamination by lunar material. *Achieved.*

f. Television coverage. *Achieved.*

g. Photographic coverage. *Achieved.*

1) Long-distance photographic coverage from the command module.

2) Lunar mapping photography from orbit.

3) Landed lunar module location.

4) Sequence photography during descent, lunar stay, and ascent.

5) Still photographs through the lunar module window.

6) Still photographs on the lunar surface.

7) Close-up stereo photography.

Apollo 11 bulk rock samples collected during the mission (NASA S69-45519).

Experiments

1. Passive seismic experiment. *Achieved.*

2. Lunar field geology. *Partially achieved. Although 2 core tube samples and 15 pounds of additional lunar samples were obtained, time constraints precluded collection of these samples with the degree of documentation originally planned. In addition, time did not permit the collection of a lunar environmental sample or a gas analysis sample in the two special containers provided. It was, however, possible to obtain the desired results using other samples contained in the regular sample return containers.*

3. Laser ranging retroreflector experiment. *Achieved.*

4. Solar wind composition. *Achieved.*

5. Cosmic ray detection. *Achieved.*

6. Pilot describing function. *Achieved.*

Core tube sample #10004 (NASA S69-40945).

Launch Vehicle Objectives

1. To launch on a variable 72° to 108° flight azimuth and insert the S-IVB, instrument unit, and spacecraft into a circular Earth parking orbit. *Achieved.*

2. To restart the S-IVB during either the second or third revolution, and inject the S-IVB, instrument unit, and spacecraft into the planned translunar trajectory. *Achieved.*

3. To provide the required attitude control for the S-IVB, instrument unit, and spacecraft during the transposition, docking, and ejection maneuver. *Achieved.*

4. To use residual S-IVB propellants and auxiliary propulsion system after final launch vehicle/spacecraft separation, to safe the S-IVB, and to minimize the possibility of the following, in order of priority:

a. S-IVB/instrument unit recontact with the spacecraft. *Achieved.*

b. S-IVB/instrument unit Earth impact. *Achieved.*

c. S-IVB/instrument unit lunar impact. *Achieved.*

Apollo 11 Spacecraft History

EVENT	DATE
Individual and combined CM and SM systems test completed at factory.	12 Oct 1968
LM #5 integrated test at factory.	21 Oct 1968
Integrated CM and SM systems test completed at factory.	06 Dec 1968
LM #5 final engineering evaluation acceptance test at factory.	13 Dec 1968
LM ascent stage #5 ready to ship from factory to KSC.	07 Jan 1969
LM ascent stage #5 delivered to KSC.	08 Jan 1969
Spacecraft/LM adapter #14 delivered to KSC.	10 Jan 1969
LM descent stage #5 ready to ship from factory to KSC.	11 Jan 1969
LM descent stage #5 delivered to KSC.	12 Jan 1969
CSM #107 quads delivered to KSC.	15 Jan 1969
Saturn S-IVB stage #506 delivered to KSC.	18 Jan 1969
Saturn S-IVB stage #506 delivered to KSC.	19 Jan 1969
CM #107 and SM #107 ready to ship from factory to KSC.	22 Jan 1969
CM #107 and SM #107 delivered to KSC.	23 Jan 1969
CM #107 and SM #107 mated.	29 Jan 1969
Saturn S-II stage #6 delivered to KSC.	06 Feb 1969
LM ascent stage #5 and descent stage #5 mated.	14 Feb 1969
CSM #107 combined systems test completed.	17 Feb 1969
LM #5 combined systems test completed.	17 Feb 1969
Saturn S-IC stage #6 delivered to KSC.	20 Feb 1969
Saturn S-IC stage #6 erected.	21 Feb 1969
Saturn V instrument unit #506 delivered to KSC.	27 Feb 1969
Saturn S-II stage #6 erected.	04 Mar 1969
Saturn S-IVB stage #506 erected.	05 Mar 1969
Saturn V instrument unit #506 erected.	05 Mar 1969
CSM #107 altitude test with prime crew completed.	18 Mar 1969
LM #5 altitude test with prime crew completed.	21 Mar 1969
CSM #107 altitude tests completed.	24 Mar 1969
LM #5 altitude tests completed.	25 Mar 1969
Launch vehicle propellant dispersion/malfunction overall test completed.	27 Mar 1969
CSM #107 moved to VAB.	14 Apr 1969
Spacecraft erected.	14 Apr 1969
LM #5 combined systems test completed.	18 Apr 1969
CSM #107 integrated systems test completed.	22 Apr 1969
CSM #107 electrically mated to launch vehicle.	05 May 1969
Space vehicle overall test completed.	06 May 1969
Space vehicle overall test #1 (plugs in) completed.	14 May 1969
Space vehicle and MLP #1 transferred to launch complex 39A.	20 May 1969
Mobile service structure transferred to launch complex 39A.	22 May 1969
LM #4 flight readiness test completed.	02 Jun 1969
Space vehicle flight readiness test completed.	06 Jun 1969
Saturn S-IC stage #6 RP-1 fuel loading completed.	25 Jun 1969
Space vehicle countdown demonstration test (wet) completed.	02 Jul 1969
Space vehicle countdown demonstration test (dry) completed.	03 Jul 1969

Apollo II Ascent Phase

Event	GET (hhh:mm:ss)	Altitude (n mi)	Range (n mi)	Earth Fixed Velocity (ft/sec)	Space Fixed Velocity (ft/sec)	Event Duration (deg E)	Geocentric Latitude (deg N)	Longitude (deg E)	Space Fixed Flight Path Angle (deg)	Space Fixed Heading Angle (E of N)
Liftoff	000:00:00.63	0.032	0.000	1.5	1,340.7		28.4470	-80.6041	0.06	90.00
Mach 1 achieved	000:01:06.30	4.236	1.044	1,054.1	2,023.9		28.4523	-80.5853	27.88	85.32
Maximum dynamic pressure	000:01:23.00	7.326	3.012	1,653.4	2,671.9		28.4624	-80.5499	29.23	82.41
S-IC center engine cutoff[3]	000:02:15.20	23.761	25.067	5,320.8	6,492.8	141.6	28.5739	-81.1517	22.957	76.315
S-IC outboard engine cutoff	000:02:41.63	35.701	50.529	7,851.9	9,068.6	168.03	28.7007	-79.6908	19.114	75.439
S-IC/S-II separation[3]	000:02:42.30	36.029	51.323	7,882.9	9,100.6		28.7046	-79.6764	19.020	75.436
S-II center engine cutoff	000:07:40.62	97.280	601.678	17,404.8	18,725.5	296.62	30.9513	-69.4309	0.897	79.646
S-II outboard engine cutoff	000:09:08.22	101.142	873.886	21,368.2	22,690.8	384.22	31.7089	-64.1983	0.619	82.396
S-II/S-IVB separation[3]	000:09:09.00	101.175	876.550	21,377.0	22,699.6		31.7152	-64.1467	0.611	82.426
S-IVB 1st burn cutoff	000:11:39.33	103.202	1,421.959	24,237.6	25,561.6	147.13	32.4865	-53.4588	0.011	88.414
Earth orbit insertion	000:11:49.33	103.176	1,460.697	24,243.9	25,567.8					

Apollo II Earth Orbit Phase

Event	GET (hhh:mm:ss)	Space Fixed Velocity (ft/sec)	Event Duration (sec)	Velocity Change (ft/sec)	Apogee (n mi)	Perigee (n mi)	Period (mins)	Inclination (deg)
Earth orbit insertion	000:11:49.33	25,567.8			100.4	98.9	88.18	32.521
S-IVB 2nd burn ignition	002:44:16.20	25,560.2						
S-IVB 2nd burn cutoff	002:50:03.03	35,568.3	346.83	10,008.1				31.386

Apollo II Translunar Phase

Event	GET (hhh:mm:ss)	Altitude (n mi)	Space Fixed Velocity (ft/sec)	Event Duration (sec)	Velocity Change (ft/sec)	Space Fixed Flight Path Angle (deg)	Space Fixed Heading Angle (E of N)
Translunar injection	002:50:13.03	180.581	35,545.6			7.367	60.073
CSM separated from S-IVB	003:15:23.00	3,815.190	24,962.5			45.148	93.758
CSM docked with LM/S-IVB	003:24:03.70	5,317.6	22,662.5			44.94	99.57
CSM/LM evasive maneuver ignition	004:40:01.72	16,620.8	14,680.0			64.30	113.73
CSM/LM evasive maneuver cutoff	004:40:04.65	16,627.3	14,663.0	2.93	19.7	64.25	113.74
Midcourse correction ignition	026:44:58.64	109,475.3	5,025.0			77.05	120.88
Midcourse correction cutoff	026:45:01.77	109,477.2	5,010.0	3.13	20.9	76.88	120.87

[3] Only the commanded time is available for this event.

Apollo 11 Lunar Orbit Phase

Event	GET (hhh:mm:ss)	Altitude (n mi)	Space Fixed Velocity (ft/sec)	Event Duration (sec)	Velocity Change (ft/sec)	Apogee (n mi)	Perigee (n mi)
Lunar orbit insertion ignition	075:49:50.37	86.7	8,250.0				
Lunar orbit insertion cutoff	075:55:47.90	60.1	5,479.0	357.53	2917.5	169.7	60.0
Lunar orbit circularization ignition	080:11:36.75	61.8	5,477.3				
Lunar orbit circularization cutoff	080:11:53.63	61.6	5,338.3	16.88	158.8	66.1	54.5
CSM/LM undocked	100:12:00.00	62.9	5,333.8				
CSM/LM separation ignition	100:39:52.90	62.7	5,332.7				
CSM/LM separation cutoff	100:40:01.90	62.5	5,332.2	9.0	2.7	63.7	56.0
LM descent orbit insertion ignition	101:36:14.00	56.4	5,364.9				
LM descent orbit insertion cutoff	101:36:44.00	57.8	5,284.9	30.0	76.4	64.3	55.6
LM powered descent initiation	102:33:05.01	6.4	5,564.9			58.5	7.8
LM powered descent cutoff	102:45:41.40			756.39			
LM lunar liftoff ignition	124:22:00.79						
LM orbit insertion cutoff	124:29:15.67	10.0	5,537.9	434.88	6,070.1	48.0	9.4
LM coelliptic sequence initiation ignition	125:19:35.00	47.4	5,328.1				
LM coelliptic sequence initiation cutoff	125:20:22.00	48.4	5,376.6	47.0	51.5	49.3	45.7
LM constant differential height ignition	126:17:49.60						
LM constant differential height cutoff	126:18:07.40			17.8	19.9	47.4	42.1
LM terminal phase initiation ignition	127:03:51.80	44.1	5,391.5				
LM terminal phase initiation cutoff	127:04:14.50	44.0	5,413.2	22.7	25.3	61.7	43.7
LM 1st midcourse correction	127:18:30.80				1.0		
LM 2nd midcourse correction	127:33:30.80				1.5		
LM terminal phase finalize ignition	127:46:09.80	7.6	5,339.7				
LM terminal phase finalize cutoff	127:46:38.20			28.4	31.4	63.0	56.5
LM begin braking	127:36:57.30						
LM begin stationkeeping	127:52:05.30						
CSM/LM docked	128:03:00.00	60.6	5,341.5				
LM ascent stage jettisoned	130:09:31.20	61.6	5,335.9				
CSM/LM final separation ignition	130:30:01.00	62.7	5,330.1				
CSM/LM final separation cutoff	130:30:08.10	62.7	5,326.9	7.2	2.2	62.7	54.0

Apollo 11 Transearth Phase

Event	GET (hhh:mm:ss)	Altitude (n mi)	Space Fixed Velocity (ft/sec)	Event Duration (sec)	Velocity Change (ft/sec)	Space Fixed Flight Path Angle (deg)	Space Fixed Heading Angle (E of N)
Transearth injection ignition	135:23:42.28	52.4	5,376.0			-0.03	-62.77
Transearth injection cutoff	135:26:13.69	58.1	8,589.0	151.41	3,279.0	5.13	-62.60
Midcourse correction ignition	150:29:57.40	169,087.2	4,075.0			-80.34	129.30
Midcourse correction cutoff	150:30:07.40	169,080.6	4,074.0	10.0	4.8	-80.41	129.30
CM/SM separation	194:49:12.70	1,778.3	29,615.5			-35.26	69.27

Apollo 11 Timeline

Event	GET (hhh:mm:ss)	GMT Time	GMT Date
Terminal countdown started.	-028:00:00.00	21:00:00	14 Jul 1969
Scheduled 11-hour hold at T-9 hours.	-009:00:00.00	16:00:00	15 Jul 1969
Countdown resumed at T-9 hours.	-009:00:00.00	03:00:00	16 Jul 1969
Scheduled 1-hour 32-minute hold at T-3 hours 30 minutes.	-003:30:00.00	08:30:00	16 Jul 1969
Countdown resumed at T-3 hours 30 minutes.	-003:30:00.00	10:02:00	16 Jul 1969
Guidance reference release.	-000:00:16.968	13:31:43	16 Jul 1969
S-IC engine start command.	-000:00:08.90	13:31:51	16 Jul 1969
S-IC engine ignition (#5).	-000:00:06.40	13:31:53	16 Jul 1969
All S-IC engines thrust OK.	-000:00:01.60	13:31:58	16 Jul 1969
Range zero.	000:00:00.00	13:32:00	16 Jul 1969
All holddown arms released (1st motion).	000:00:00.30	13:32:00	16 Jul 1969
Liftoff (umbilical disconnected) (1.07 g).	000:00:00.63	13:32:00	16 Jul 1969
Tower clearance yaw maneuver started.	000:00:01.70	13:32:01	16 Jul 1969
Yaw maneuver ended.	000:00:09.70	13:32:09	16 Jul 1969
Pitch and roll maneuver started.	000:00:13.20	13:32:13	16 Jul 1969
Roll maneuver ended.	000:00:31.10	13:32:31	16 Jul 1969
Mach 1 achieved.	000:01:06.30	13:33:06	16 Jul 1969
Maximum dynamic pressure (735.17 lb/ft^2).	000:01:23.00	13:33:23	16 Jul 1969
Maximum bending moment (33,200,000 lbf-in).	000:01:31.50	13:33:31	16 Jul 1969
S-IC center engine cutoff command.	000:02:15.20	13:34:15	16 Jul 1969
Pitch maneuver ended.	000:02:40.00	13:34:40	16 Jul 1969
S-IC outboard engine cutoff.	000:02:41.63	13:34:41	16 Jul 1969
S-IC maximum total inertial acceleration (3.94 g).	000:02:41.71	13:34:41	16 Jul 1969
S-IC maximum Earth-fixed velocity. S-IC/S-II separation command.	000:02:42.30	13:34:42	16 Jul 1969
S-II engine start command.	000:02:43.04	13:34:43	16 Jul 1969
S-II ignition.	000:02:44.00	13:34:44	16 Jul 1969
S-II aft interstage jettisoned.	000:03:12.30	13:35:12	16 Jul 1969
Launch escape tower jettisoned.	000:03:17.90	13:35:17	16 Jul 1969
Iterative guidance mode initiated.	000:03:24.10	13:35:24	16 Jul 1969
S-IC apex.	000:04:59.10	13:36:59	16 Jul 1969
S-II center engine cutoff.	000:07:40.62	13:39:40	16 Jul 1969
S-II maximum total inertial acceleration (1.82 g).	000:07:40.70	13:39:40	16 Jul 1969
S-IC impact (theoretical).	000:09:03.70	13:41:03	16 Jul 1969
S-II outboard engine cutoff.	000:09:08.22	13:41:08	16 Jul 1969
S-II maximum Earth-fixed velocity. S-II/S-IVB separation command.	000:09:09.00	13:41:09	16 Jul 1969
S-IVB 1st burn start command.	000:09:09.20	13:41:09	16 Jul 1969
S-IVB 1st burn ignition.	000:09:12.20	13:41:12	16 Jul 1969
S-IVB ullage case jettisoned.	000:09:21.00	13:41:21	16 Jul 1969
S-II apex.	000:09:47.00	13:41:47	16 Jul 1969
S-IVB 1st burn cutoff.	000:11:39.33	13:43:39	16 Jul 1969
S-IVB 1st burn maximum total inertial acceleration (0.69 g).	000:11:39.41	13:43:39	16 Jul 1969
Earth orbit insertion. S-IVB 1st burn maximum Earth-fixed velocity.	000:11:49.33	13:43:49	16 Jul 1969
Maneuver to local horizontal attitude started.	000:11:59.30	13:43:59	16 Jul 1969
Orbital navigation started.	000:13:21.10	13:45:21	16 Jul 1969
S-II impact (theoretical).	000:20:13.70	13:52:13	16 Jul 1969
S-IVB 2nd burn restart preparation.	002:34:38.20	16:06:38	16 Jul 1969
S-IVB 2nd burn restart command.	002:44:08.20	16:16:08	16 Jul 1969
S-IVB 2nd burn ignition (STDV open).	002:44:16.20	16:16:16	16 Jul 1969
S-IVB 2nd burn cutoff.	002:50:03.03	16:22:03	16 Jul 1969

Apollo 11 Timeline

Event	GET (hhh:mm:ss)	GMT Time	GMT Date
S-IVB 2nd burn maximum total inertial acceleration (1.45 g).	002:50:03.11	16:22:03	16 Jul 1969
S-IVB 2nd burn maximum Earth-fixed velocity.	002:50:03.50	16:22:03	16 Jul 1969
S-IVB safing procedures started.	002:50:03.80	16:22:03	16 Jul 1969
Translunar injection.	002:50:13.03	16:22:13	16 Jul 1969
Maneuver to local horizontal attitude started.	002:50:23.00	16:22:23	16 Jul 1969
Orbital navigation started.	002:50:23.90	16:22:23	16 Jul 1969
Maneuver to transposition and docking attitude started.	003:05:03.90	16:37:03	16 Jul 1969
CSM separated from S-IVB.	003:15:23.00	16:47:23	16 Jul 1969
CSM separation maneuver ignition.	003:17:04.60	16:49:04	16 Jul 1969
CSM separation maneuver cutoff.	003:17:11.70	16:49:11	16 Jul 1969
CSM docked with LM/S-IVB.	003:24:03.70	16:56:03	16 Jul 1969
CSM/LM ejected from S-IVB.	004:17:03.00	17:49:03	16 Jul 1969
CSM/LM evasive maneuver from S-IVB ignition.	004:40:01.72	18:12:01	16 Jul 1969
CSM/LM evasive maneuver from S-IVB cutoff.	004:40:04.65	18:12:04	16 Jul 1969
S-IVB maneuver to lunar slingshot attitude initiated.	004:41:07.60	18:13:07	16 Jul 1969
S-IVB lunar slingshot maneuver—LH$_2$ tank CVS opened.	004:51:07.70	18:23:07	16 Jul 1969
S-IVB lunar slingshot maneuver—LOX dump started.	005:03:07.60	18:35:07	16 Jul 1969
S-IVB lunar slingshot maneuver—LOX dump ended.	005:04:55.80	18:36:55	16 Jul 1969
S-IVB lunar slingshot maneuver—APS ignition.	005:37:47.60	19:09:47	16 Jul 1969
S-IVB lunar slingshot maneuver—APS cutoff.	005:42:27.80	19:14:27	16 Jul 1969
S-IVB maneuver to communications attitude initiated.	005:42:48.80	19:14:48	16 Jul 1969
TV transmission started (recorded at Goldstone and transmitted to Houston at 011:26).	010:32:00.00	00:04:00	17 Jul 1969
TV transmission ended.	010:48:00.00	00:20:00	17 Jul 1969
Midcourse correction ignition.	026:44:58.64	16:16:58	17 Jul 1969
Midcourse correction cutoff.	026:45:01.77	16:17:01	17 Jul 1969
TV transmission started.	030:28:00.00	20:00:00	17 Jul 1969
TV transmission ended.	031:18:00.00	20:50:00	17 Jul 1969
TV transmission started.	033:59:00.00	23:31:00	17 Jul 1969
TV transmission ended.	034:35:00.00	00:07:00	18 Jul 1969
TV transmission started.	055:08:00.00	20:40:00	18 Jul 1969
CDR and LMP entered LM for initial inspection.	055:30:00.00	21:02:00	18 Jul 1969
TV transmission ended.	056:44:00.00	22:16:00	18 Jul 1969
CDR and LMP entered CM.	057:55:00.00	23:27:00	18 Jul 1969
Equigravisphere.	061:39:55.00	03:11:55	19 Jul 1969
Lunar orbit insertion ignition.	075:49:50.37	17:21:50	19 Jul 1969
Lunar orbit insertion cutoff.	075:55:47.90	17:27:47	19 Jul 1969
Sighting of an illumination in the Aristarchus region. 1st time a lunar transient event sighted by an observer in space.	077:13:00.00	18:45:00	19 Jul 1969
TV transmission started.	078:20:00.00	19:52:00	19 Jul 1969
S-IVB closest approach to lunar surface.	078:42:00.00	20:14:00	19 Jul 1969
TV transmission ended.	079:00:00.00	20:32:00	19 Jul 1969
Lunar orbit circularization ignition.	080:11:36.75	21:43:36	19 Jul 1969
Lunar orbit circularization cutoff.	080:11:53.63	21:43:53	19 Jul 1969
LMP entered CM for initial power-up and system checks.	081:10:00.00	22:42:00	19 Jul 1969
LMP entered CM.	083:35:00.00	01:07:00	20 Jul 1969
CDR and LMP entered LM for final preparations for descent.	095:20:00.00	12:52:00	20 Jul 1969
LMP entered CM.	097:00:00.00	14:32:00	20 Jul 1969
LMP entered LM.	097:30:00.00	15:02:00	20 Jul 1969
LM system checks started.	097:45:00.00	15:17:00	20 Jul 1969

Apollo 11 Timeline

Event	GET (hhh:mm:ss)	GMT Time	GMT Date
LM system checks ended.	100:00:00.00	17:32:00	20 Jul 1969
CSM/LM undocked.	100:12:00.00	17:44:00	20 Jul 1969
CSM/LM separation maneuver ignition.	100:39:52.90	18:11:52	20 Jul 1969
CSM/LM separation maneuver cutoff.	100:40:01.90	18:12:01	20 Jul 1969
LM descent orbit insertion ignition (LM SPS).	101:36:14.00	19:08:14	20 Jul 1969
LM descent orbit insertion cutoff.	101:36:44.00	19:08:44	20 Jul 1969
LM acquisition of data.	102:17:17.00	19:49:17	20 Jul 1969
LM landing radar on.	102:20:53.00	19:52:53	20 Jul 1969
LM abort guidance aligned to primary guidance.	102:24:40.00	19:56:40	20 Jul 1969
LM yaw maneuver to obtain improved communications.	102:27:32.00	19:59:32	20 Jul 1969
LM altitude 50,000 feet.	102:32:55.00	20:04:55	20 Jul 1969
LM propellant settling firing started.	102:32:58.00	20:04:58	20 Jul 1969
LM powered descent engine ignition.	102:33:05.01	20:05:05	20 Jul 1969
LM fixed throttle position.	102:33:31.00	20:05:31	20 Jul 1969
LM face-up maneuver completed.	102:37:59.00	20:09:59	20 Jul 1969
LM 1202 alarm.	102:38:22.00	20:10:22	20 Jul 1969
LM radar updates enabled.	102:38:45.00	20:10:45	20 Jul 1969
LM altitude less than 30,000 feet and velocity less than 2,000 feet per second (landing radar velocity update started).	102:38:50.00	20:10:50	20 Jul 1969
LM 1202 alarm.	102:39:02.00	20:11:02	20 Jul 1969
LM throttle recovery.	102:39:31.00	20:11:31	20 Jul 1969
LM approach phase entered.	102:41:32.00	20:13:32	20 Jul 1969
LM landing radar antenna to position 2.	102:41:37.00	20:13:37	20 Jul 1969
LM attitude hold mode selected (check of LM handling qualities).	102:41:53.00	20:13:53	20 Jul 1969
LM automatic guidance enabled.	102:42:03.00	20:14:03	20 Jul 1969
LM 1201 alarm.	102:42:18.00	20:14:18	20 Jul 1969
LM landing radar switched to low scale.	102:42:19.00	20:14:19	20 Jul 1969
LM 1202 alarm.	102:42:43.00	20:14:43	20 Jul 1969
LM 1202 alarm.	102:42:58.00	20:14:58	20 Jul 1969
LM landing point redesignation.	102:43:09.00	20:15:09	20 Jul 1969
LM altitude hold.	102:43:13.00	20:15:13	20 Jul 1969
LM abort guidance attitude updated.	102:43:20.00	20:15:20	20 Jul 1969
LM rate of descent landing phase entered.	102:43:22.00	20:15:22	20 Jul 1969
LM landing radar data not good.	102:44:11.00	20:16:11	20 Jul 1969
LM landing data good.	102:44:21.00	20:16:21	20 Jul 1969
LM fuel low-level quantity light.	102:44:28.00	20:16:28	20 Jul 1969
LM landing radar data not good.	102:44:59.00	20:16:59	20 Jul 1969
LM landing radar data good.	102:45:03.00	20:17:03	20 Jul 1969
1st evidence of surface dust disturbed by descent engine.	102:44:35.00	20:16:35	20 Jul 1969
LM lunar landing.	102:45:39.90	20:17:39	20 Jul 1969
LM powered descent engine cutoff.	102:45:41.40	20:17:41	20 Jul 1969
Decision made to proceed with EVA prior to first rest period.	104:40:00.00	22:12:00	20 Jul 1969
Preparation for EVA started.	106:11:00.00	23:43:00	20 Jul 1969
EVA started (hatch open).	109:07:33.00	02:39:33	21 Jul 1969
CDR completely outside LM on porch.	109:19:16.00	02:51:16	21 Jul 1969
Modular equipment stowage assembly deployed (CDR).	109:21:18.00	02:53:18	21 Jul 1969
First clear TV picture received.	109:22:00.00	02:54:00	21 Jul 1969
CDR at foot of ladder (starts to report, then pauses to listen).	109:23:28.00	02:55:28	21 Jul 1969
CDR at foot of ladder and described surface as "almost like a powder."	109:23:38.00	02:55:38	21 Jul 1969

Apollo II Timeline

Event	GET (hhh:mm:ss)	GMT Time	GMT Date
1st step taken on lunar surface (CDR). "That's one small step for a man...one giant leap for mankind."	109:24:15.00	02:56:15	21 Jul 1969
CDR started surface examination and description, assessed mobility and described effects of LM descent engine.	109:24:48.00	02:56:48	21 Jul 1969
CDR ended surface examination. LMP started to send down camera.	109:26:54.00	02:58:54	21 Jul 1969
Camera installed on RCU bracket, LEC stored on secondary strut of LM landing gear.	109:30:23.00	03:02:23	21 Jul 1969
Surface photography (CDR).	109:30:53.00	03:02:53	21 Jul 1969
Contingency sample collection started (CDR).	109:33:58.00	03:05:58	21 Jul 1969
Contingency sample collection ended (CDR).	109:37:08.00	03:09:08	21 Jul 1969
LMP started egress from LM.	109:39:57.00	03:11:57	21 Jul 1969
LMP at top of ladder. Descent photographed by CDR.	109:41:56.00	03:13:56	21 Jul 1969
LMP on lunar surface.	109:43:16.00	03:15:16	21 Jul 1969
Surface examination and examination of landing effects on surface and on LM started (CDR, LMP).	109:43:47.00	03:15:47	21 Jul 1969
Insulation removed from modular equipment stowage assembly (CDR).	109:49:06.00	03:21:06	21 Jul 1969
TV camera focal distance adjusted (CDR).	109:51:35.00	03:23:35	21 Jul 1969
Plaque unveiled (CDR).	109:52:19.00	03:24:19	21 Jul 1969
Plaque read (CDR).	109:52:40.00	03:24:40	21 Jul 1969
TV camera redeployed. Panoramic TV view started (CDR).	109:59:28.00	03:31:28	21 Jul 1969
TV camera placed in final deployment position (CDR).	110:02:53.00	03:34:53	21 Jul 1969
Solar wind composition experiment deployed (LMP).	110:03:20.00	03:35:20	21 Jul 1969
United States flag deployed (CDR, LMP).	110:09:43.00	03:41:43	21 Jul 1969
Evaluation of surface mobility started (LMP).	110:13:15.00	03:45:15	21 Jul 1969
Evaluation of surface mobility end (LMP).	110:16:02.00	03:48:02	21 Jul 1969
Presidential message from White House and response from CDR.	110:16:30.00	03:48:30	21 Jul 1969
Presidential message and CDR response ended.	110:18:21.00	03:50:21	21 Jul 1969
Evaluation of trajectory of lunar soil when kicked (LMP) and bulk sample collection started (CDR).	110:20:06.00	03:52:06	21 Jul 1969
Evaluation of visibility in lunar sunlight (LMP).	110:10:24.00	03:42:24	21 Jul 1969
Evaluation of thermal effects of sun and shadow inside the suit (LMP).	110:25:09.00	03:57:09	21 Jul 1969
Evaluation of surface shadows and colors (LMP).	110:28:22.00	04:00:22	21 Jul 1969
LM landing gear inspection and photography (LMP).	110:34:13.00	04:06:13	21 Jul 1969
Bulk sample completed (CDR).	110:35:36.00	04:07:36	21 Jul 1969
LM landing gear inspection and photography (CDR, LMP).	110:46:36.00	04:18:36	21 Jul 1969
Scientific equipment bay doors opened.	110:53:38.00	04:25:38	21 Jul 1969
Passive seismometer deployed.	110:55:42.00	04:27:42	21 Jul 1969
Lunar ranging retroreflector deployed (CDR).	111:03:57.00	04:35:57	21 Jul 1969
1st passive seismic experiment data received on Earth.	111:08:39.00	04:40:39	21 Jul 1969
Collection of documented samples started (CDR/LMP).	111:11:00.00	04:43:00	21 Jul 1969
Solar wind composition experiment retrieved (LMP) .	111:20:00.00	04:52:00	21 Jul 1969
LMP inside LM.	111:29:39.00	05:01:39	21 Jul 1969
Sample containers transferred (LMP).	111:30:00.00	05:02:00	21 Jul 1969
EVA ended. CDR inside LM, assisted and monitored by LMP.	111:37:00.00	05:09:00	21 Jul 1969
EVA ended (hatch closed).	111:39:13.00	05:11:13	21 Jul 1969
LM equipment jettisoned.	114:05:00.00	07:37:00	21 Jul 1969
LM lunar liftoff ignition (LM APS).	124:22:00.79	17:54:00	21 Jul 1969
LM orbit insertion cutoff.	124:29:15.67	18:01:15	21 Jul 1969
Coelliptic sequence initiation ignition.	125:19:35.00	18:51:35	21 Jul 1969
Coelliptic sequence initiation cutoff.	125:20:22.00	18:52:22	21 Jul 1969

Apollo 11 Timeline

Event	GET (hhh:mm:ss)	GMT Time	GMT Date
Constant differential height maneuver ignition.	126:17:49.60	19:49:49	21 Jul 1969
Constant differential height maneuver cutoff.	126:18:29.20	19:50:29	21 Jul 1969
Terminal phase initiation ignition.	127:03:51.80	20:35:51	21 Jul 1969
Terminal phase initiation cutoff.	127:04:14.50	20:36:14	21 Jul 1969
LM 1st midcourse correction.	127:18:30.80	20:50:30	21 Jul 1969
LM 2nd midcourse correction.	127:33:30.80	21:05:30	21 Jul 1969
Braking started.	127:36:57.30	21:08:57	21 Jul 1969
Terminal phase finalize ignition.	127:46:09.80	21:18:09	21 Jul 1969
Terminal phase finalize cutoff.	127:46:38.20	21:18:38	21 Jul 1969
Stationkeeping started.	127:52:05.30	21:24:05	21 Jul 1969
CSM/LM docked.	128:03:00.00	21:35:00	21 Jul 1969
CDR entered CM.	129:20:00.00	22:52:00	21 Jul 1969
LMP entered CM.	129:45:00.00	23:17:00	21 Jul 1969
LM ascent stage jettisoned.	130:09:31.20	23:41:31	21 Jul 1969
CSM/LM final separation ignition.	130:30:01.00	00:02:01	22 Jul 1969
CSM/LM final separation cutoff.	130:30:08.20	00:02:08	22 Jul 1969
Transearth injection ignition (SPS).	135:23:42.28	04:55:42	22 Jul 1969
Transearth injection cutoff.	135:26:13.69	04:58:13	22 Jul 1969
Midcourse correction ignition.	150:29:57.40	20:01:57	22 Jul 1969
Midcourse correction cutoff.	150:30:07.40	20:02:07	22 Jul 1969
TV transmission started.	155:36:00.00	01:08:00	23 Jul 1969
TV transmission ended.	155:54:00.00	01:26:00	23 Jul 1969
TV transmission started.	177:10:00.00	22:42:00	23 Jul 1969
TV transmission ended.	177:13:00.00	22:45:00	23 Jul 1969
TV transmission started.	177:32:00.00	23:04:00	23 Jul 1969
TV transmission ended.	177:44:00.00	23:16:00	23 Jul 1969
CM/SM separation.	194:49:12.70	16:21:12	24 Jul 1969
Entry.	195:03:05.70	16:35:05	24 Jul 1969
Drogue parachute deployed.	195:12:06.90	16:44:06	24 Jul 1969
Visual contact with CM established by aircraft.	195:07:00.00	16:39	24 Jul 1969
Radar contact with CM established by recovery ship.	195:08:00.00	16:40	24 Jul 1969
VHF voice contact and recovery beacon contact established.	195:14:00.00	16:46	24 Jul 1969
Splashdown (went to apex-down).	195:18:35.00	16:50:35	24 Jul 1969
CM returned to apex-up position.	195:26:15.00	16:58:15	24 Jul 1969
Flotation collar inflated.	195:32:00.00	17:04	24 Jul 1969
Hatch opened for crew egress.	195:49:00.00	17:21	24 Jul 1969
Crew egress.	195:57:00.00	17:29	24 Jul 1969
Crew aboard recovery ship.	196:21:00.00	17:53	24 Jul 1969
Crew entered mobile quarantine facility.	196:26:00.00	17:58	24 Jul 1969
CM lifted from water.	198:18:00.00	19:50	24 Jul 1969
CM secured to quarantine facility.	198:26:00.00	19:58	24 Jul 1969
CM hatch reopened.	198:33:00.00	20:05	24 Jul 1969
Sample return containers 1 and 2 removed from CM.	200:28:00.00	22:00	24 Jul 1969
Container 1 removed from mobile quarantine facility.	202:00:00.00	23:32	24 Jul 1969
Container 2 removed from mobile quarantine facility.	202:33:00.00	00:05	25 Jul 1969
Container 2 and film flown to Johnston Island.	207:43:00.00	05:15	25 Jul 1969
Container 1 flown to Hickam Air Force Base, HI.	214:13:00.00	11:45	25 Jul 1969
Container 2 and film arrived in Houston.	218:43:00.00	16:15	25 Jul 1969
Container 1, film, and biological samples arrived in Houston.	225:41:00.00	23:13	25 Jul 1969

Apollo 11 Timeline

Event	GET (hhh:mm:ss)	GMT Time	GMT Date
CM decontaminated and hatch secured.	229:28:00.00	03:00	26 Jul 1969
Mobile quarantine facility secured.	231:03:00.00	04:35	26 Jul 1969
Mobile quarantine facility and CM offloaded.	250:43:00.00	00:15	27 Jul 1969
Safing of CM pyrotechnics completed.	252:33:00.00	02:05	27 Jul 1969
Mobile quarantine facility arrived in Houston.	280:28:00.00	06:00	28 Jul 1969
Flight crew in Lunar Receiving Laboratory.	284:28:00.00	10:00	28 Jul 1969
CM delivered to Lunar Receiving Laboratory.	345:45:00.00	23:17	30 Jul 1969
Passive seismic experiment turned off.	430:26:46.00	11:58:46	03 Aug 1969
Crew released from quarantine.			10 Aug 1969

APOLLO 12

The Sixth Mission:
The Second Lunar Landing

Apollo 12 Summary

(14 November–24 November 1969)

Apollo 12 crew (l. to r.): Pete Conrad, Dick Gordon, Al Bean (NASA S69-38852).

Background

Apollo 12 was a Type H mission, a precision piloted lunar landing demonstration and systematic lunar exploration. It was the second successful human landing on the Moon.

The primary objectives were:

• to perform selenological inspection, survey, and sampling in a mare area;

• to deploy the Apollo Lunar Surface Experiments Package (ALSEP);

• to develop techniques for a point landing capability;

• to further develop human capability to work in the lunar environment; and

• to obtain photographs of candidate exploration sites.

The all-Navy crew included Commander Charles "Pete" Conrad, Jr. (USN), commander; Commander Richard Francis "Dick" Gordon, Jr. (USN), command module pilot; and Commander Alan LaVern Bean (USN), lunar module pilot.

Selected as an astronaut in 1962, Conrad was making his third spaceflight. He had been pilot of Gemini 5 and command pilot of Gemini 11. Born 2 June 1930 in Philadelphia, Pennsylvania, Conrad was 39 years old at the time of the Apollo 12 mission. He received a B.S. in aeronautical engineering from Princeton University in 1953. His backup was Colonel David Randolph Scott (USAF).[1]

Gordon had been pilot of Gemini 11. Born 5 October 1929 in Seattle, Washington, he was 40 years old at the time of the Apollo 12 mission. Gordon received a B.S. in chemistry from the University of Washington in 1951, and was selected as an astronaut in 1963. His backup was Major Alfred Merrill Worden (USAF).

Bean was making his first spaceflight. Born 15 March 1932 in Wheeler, Texas, he was 37 years old at the time of the Apollo 12 mission. Bean received a B.S. in aeronautical engineering from the University of Texas in 1955, and was selected as an astronaut in 1963. His backup was Lt. Colonel James Benson Irwin (USAF).

The capsule communicators (CAPCOMs) for the mission were Lt. Colonel Gerald Paul Carr (USMC), Edward George Gibson, Ph.D., Commander Paul Joseph Weitz (USN), Don Leslie Lind, Ph. D., Scott, Worden, and Irwin. For this mission, there were also four civilian backup CAPCOMs: Dickie K. Warren, James O. Rippey, James L. Lewis, and Michael R. Wash. The support crew members were Carr, Weitz, and Gibson. The flight directors were Gerald D. Griffin (first shift), M. P. "Pete" Frank (second shift), Clifford E. Charlesworth (third shift), and Milton L. Windler (fourth shift).

The Apollo 12 launch vehicle was a Saturn V, designated SA-507. The mission also carried the designation Eastern Test Range #2793. The CSM was designated CSM-108, and had the call-sign "Yankee Clipper." The lunar module was designated LM-6, and had the call-sign "Intrepid."

Launch Preparations

The terminal countdown started at T-28 hours at 02:00:00 GMT on 12 November. Scheduled holds occurred at T-9 hours for 9 hours 22 minutes and at T-3 hours 30 minutes for one hour. During spacecraft preparations on November

---

[1] Conrad died 8 July 1999 in Ojai, CA, as a result of injuries sustained in a motorcycle accident.

12, a leak developed in the CSM LH$_2$ tank No. 2 during cryogenic loading. The tank was drained and replaced using a tank from the Apollo 13 CSM. An unscheduled hold was initiated on 13 November at T-17 hours (12:00:00 GMT) for retanking cryogenics in the CSM. Loading was completed in six hours and the count resumed at 19:00:00 GMT. The scheduled hold at T-9 hours was reduced by six hours, thereby averting a launch delay.

A cold front was moving slowly southward through the central section of Florida. This front produced the rain showers and overcast conditions that existed over the pad at launch time. Stratocumulus clouds covered 100 percent of the sky (base 2,100 feet), the temperature was 68.0° F, the relative humidity was 92 percent, and the barometric pressure was 14.621 lb/in^2. Winds, as measured by the anemometer on the light pole 60.0 feet above ground at the launch site, measured 13.2 knots at 280° from true north.

Ascent Phase

Apollo 12 was launched from Kennedy Space Center Launch Complex 39, Pad A, at a Range Zero time of 16:22:00 GMT (11:22:00 a.m. EST) on 14 November 1969. The planned launch window extended to 19:26:00 GMT to take advantage of a sun elevation angle on the lunar surface of 5.1°.

Apollo 12 was the first Saturn vehicle launched during a rainstorm, following the decision to waive Manned Space Flight Center Launch Mission Rule 1-404, which stated:

"The vehicle will not be launched when its flight path will carry it through a cumulonimbus (thunderstorm) cloud formation."

The reason for the rule was that the Saturn V was not designed to withstand thunderstorm weather conditions during launch.

Between 000:00:12.8 and 000:00:32.3, the vehicle rolled from a launch pad azimuth of 90° to a flight azimuth of 72.029°.

At 000:00:36.5 there were numerous space vehicle indications of a massive electrical disturbance, followed by a second disturbance at 000:00:52. The crew reported that, in their opinion, the vehicle had been struck by lightning, and that the fuel cells in the service module were disconnected and that all A/C power in the spacecraft was lost. Numerous indicator lamps were illuminated at this time.

Apollo 12 lifts off from Kennedy Space Center Pad 39A (NASA S69-58883).

Ground camera data, telemetered data, and launch computers later showed that the vehicle had indeed been struck by lightning. Virtually no discernible effects were noted on the launch vehicle during the second disturbance.

Atmospheric electrical factors and the fact that the vehicle did not have the capacitance to store sufficient energy to produce the effects noted indicated that the first discharge was triggered by the vehicle. The second disturbance may have been due to a lesser lightning discharge. The launch vehicle hardware and software suffered no significant effects, and the mission proceeded as scheduled. Because the lightning was self-induced, and because the vehicle did not fly through cumulonimbus clouds, it was determined that Rule 1-404 had not been violated.

The S-IC engine shut down at 000:02:41.74, followed by S-IC/S-II separation, and S-II engine ignition. The S-II engine shut down at 000:09:12.34 followed by separation from the S-IVB, which ignited at 000:09:16.16. The first S-IVB engine cutoff occurred at 000:11:33.91, with deviations from the planned trajectory of only -1.9 ft/sec in velocity and only 0.2 n mi in altitude.

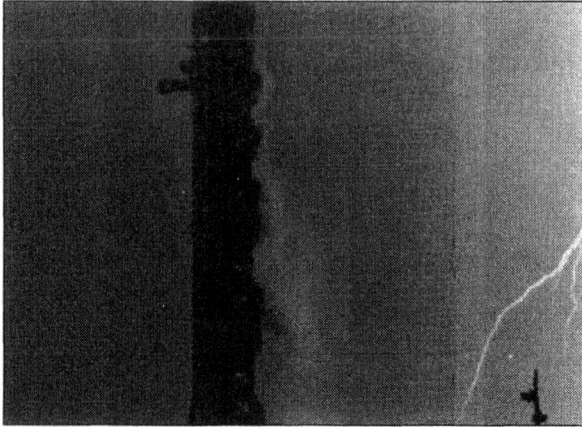

Electrical discharge between clouds and ground 36.5 seconds after liftoff, when vehicle was at an altitude of 6,000 feet (NASA S69-60068).

The S-IC stage impacted the Atlantic Ocean at 000:09:14.5 at latitude 30.273° north and longitude 73.895° west, 365.2 n mi from the launch site. The S-II stage impacted the Atlantic Ocean at 000:20:21.6 at latitude 31.465° north and longitude 34.214° west, 2,404.4 n mi from the launch site.

The maximum wind conditions encountered during ascent were 92.5 knots at 245° from true north at 46,670 feet, and a maximum wind shear of 0.0183 sec^{-1} at 46,750 feet.

Parking orbit conditions at insertion, 000:11:43.91 (S-IVB cutoff plus 10 seconds to account for engine tailoff and other transient effects), showed an apogee and perigee of 100.1 by 97.8 n mi, an inclination of 32.540°, a period of 88.16 minutes, and a velocity of 25,565.9 ft/sec. The apogee and perigee were based upon a spherical Earth with a radius of 3,443.934 n mi.

The international designation for the CSM upon achieving orbit was 1969-099A and the S-IVB was designated 1969-099B. After undocking at the Moon, the LM ascent stage would be designated 1969-099C and the descent stage 1969-099D.

Earth Orbit Phase

After inflight systems checks, made with extra care because of the two lightning strikes, the 341.24-second translunar injection maneuver (second S-IVB firing) was performed at 002:47:22.7. The S-IVB engine shut down at 002:53:03.94 and translunar injection occurred ten seconds later, at a velocity of 35,419.3 ft/sec after 1.5 Earth orbits lasting 2 hours 41 minutes 30.03 seconds.

Translunar Phase

For the first time, an Apollo vehicle was targeted for a high-pericynthion free-return translunar profile, a trajectory that would achieve satisfactory Earth entry within the reaction control velocity correction capability.

The major advantage of the new profile, termed a "hybrid" non-free-return trajectory, was the greater mission planning flexibility. This profile permitted a daylight launch to the planned landing site and a greater performance margin for the service propulsion system. The hybrid profile was constrained so that a safe return using the descent propulsion system could be made following a failure to enter lunar orbit.

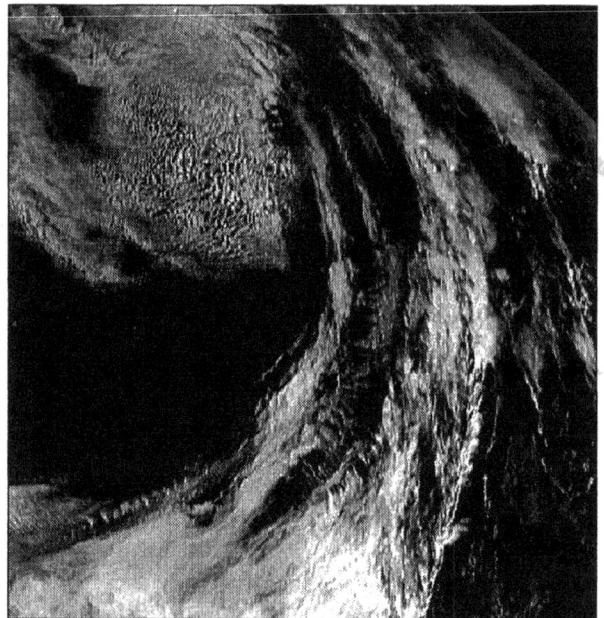

Earth view at three and one half hours into the mission (NASA AS12-50-7325).

At 003:18:04.9, the CSM was separated from the S-IVB stage, transposed, and docked with the LM at 003:26:53.3. Onboard television, transmitted from 003:25 to 004:28, clearly showed the docking. The docked spacecraft were ejected from the S-IVB at 004:13:00.9. An S-IVB auxiliary propulsion system evasive maneuver was performed at 004:27 and was also observed on television.

A ground command for propulsive venting of residual propellants targeted the S-IVB to go past the Moon and into solar orbit. However, due to an excessively long ullage engine burn, the distance of closest approach to the Moon did not provide sufficient energy to allow the S-IVB to escape the Earth-Moon system, and it was placed into an elliptical orbit around Earth and the Moon. However, the objectives of not striking the spacecraft, Earth, or the Moon were achieved. The closest approach of the S-IVB to the Moon was 3,082 n mi at 085:48.

LM inside S-IVB following separation (NASA AS12-50-7328).

To insure that the electrical transients experienced during launch had not affected the LM systems, the commander and lunar module pilot entered the LM earlier than planned, at 007:20, to perform some of the housekeeping and systems checks. The checks indicated that the LM systems were satisfactory.

One midcourse correction was required during translunar coast, a 9.19-second, 61.8 ft/sec maneuver at 030:52:44.36. It placed the spacecraft on the desired hybrid, non-free-return circumlunar trajectory. Good quality television coverage of the preparations for this burn was received for 47 minutes, starting at 030:18.

A 56-minute television transmission began at 062:52. It provided excellent color pictures of the CM, intravehicular transfer, the LM interior, and brief shots of Earth and the Moon.

At 083:25:23.36, at an altitude of 82.5 n mi above the Moon, the service propulsion engine was fired for 352.25 seconds to insert the spacecraft into a lunar orbit of 168.8

by 62.6 n mi. The translunar coast had lasted 80 hours 38 minutes 1.67 seconds.

Crescent view of Earth on the way to the Moon (NASA AS12-50-7362).

Earthrise over lunar surface following lunar orbit insertion (NASA AS12-47-6891).

Lunar Orbit/Lunar Surface Phase

During the first lunar orbit, good quality television coverage of the surface was received for about 33 minutes, beginning at 084:00. The crew provided excellent descriptions of the lunar features while transmitting sharp pictures back to Earth.

Two revolutions later, at 087:48:48.08, a 16.91-second maneuver was performed to circularize the orbit at 66.1 by

54.3 n mi. On the next revolution, the LM crew transferred to the LM to perform various housekeeping chores and communication checks.

At 104:20, the commander entered the LM, followed by the lunar module pilot at 105:00 to prepare for descent to the lunar surface. The two spacecraft were undocked at 107:54:02.3 at an altitude of 63.0 n mi, followed by a 14.4-second separation maneuver at 108:24:36.9. At 109:23:39.9, a 29.0-second descent orbit insertion maneuver placed the LM into an orbit of 60.6 by 8.1 n mi.

The 717.0-second powered descent initiation ignition occurred at 8.0 n. mi. at 110:20:38.1, and landing occurred at 06:54:36 GMT (01:54:36 a.m. EST) on 19 November at 110:32:36.2 (the engine was shut down 1.1 seconds before landing). The spacecraft landed in the Oceanus Procellarum region (Ocean of Storms) at latitude 3.01239° south and longitude 23.42157° west, at an angle of 4° to 5° to the surface. Approximately 103 seconds of engine firing time remained at landing.

One objective of the mission was to achieve a precision landing near the Surveyor III spacecraft, which had landed on 20 April 1967.[2] The LM landed just 535 feet from Surveyor.

LM following separation. Large crater in foreground is Ptolemaeus (NASA AS12-51-7507).

During the next CSM revolution, the commander reported a visual sighting of the CSM orbiting overhead. On the following revolution, the command module pilot reported sighting the Surveyor III spacecraft as well as the LM northwest of Surveyor III.

Three hours after landing, the crew members began preparations for egress. The commander exited the hatch and deployed the modularized equipment stowage assembly

and automatically activated a color television camera which permitted his actions to be televised to Earth.

Conrad descends the LM ladder as seen by Bean inside the LM cabin (NASA AS12-46-6716).

Before reaching the surface, the commander reported seeing Surveyor III about 600 feet away and also stated that the LM had landed about 25 feet from the lip of a crater. He was on the lunar surface at 115:22:22. His description indicated that the lunar surface was quite soft and loosely packed, causing his boots to dig in as he walked.

The lunar module pilot descended to the lunar surface at 115:51:50.

Shortly after the television camera was removed from its bracket on the LM, transmission was lost when the camera was pointed at the Sun. Lithium hydroxide canisters and the contingency sample were transferred to the LM cabin as planned. The S-band erectable antenna, and solar wind composition experiment were deployed, and the United States flag was erected at 116:19:31.

Except for minor difficulty removing the radioisotope thermoelectric generator fuel element from the cask, the removal of the Apollo Lunar Surface Experiments Package (ALSEP), transport, and deployment were nominal.

The ALSEP deployment site was estimated to be 600 to 700 feet from the LM. Shortly after deployment, the passive seismometer transmitted to Earth the crew members' footsteps as they returned to the LM.

On the return traverse, the crew collected a core tube sample and additional surface samples. They entered the LM and the closed the hatch at 119:06:36. The first extravehic-

[2] The COSPAR designation for Surveyor III was 1967-035A. The NORAD designation was 02756.

ular activity period lasted 3 hours 56 minutes 3 seconds. The crew walked 3,300 feet (1 km) and collected an estimated 36.8 pounds (16.7 kg) of samples.

Bean steps from LM ladder to the lunar surface (NASA AS12-46-6729).

Conrad erects U.S. flag at landing site (NASA AS12-47-6897). There are no still images of Bean by the flag.

At 119:47:13.23, the CSM performed a plane change maneuver of 18.23 seconds which changed the orbit to 62.5 by 57.6 n mi.

The second extravehicular activity period began at 131:33, after a seven-hour rest period. The crew first cut the cable and stored the inoperative LM TV camera in the equipment transfer bag for return to Earth and failure analysis. The commander then went to the ALSEP site to check the leveling of the lunar atmosphere detector. As he approached the instrument, it recorded a higher atmosphere, which was attributed to the outgassing of his suit.

Bean removes the RTG fuel element from its cask (NASA AS12-46-6790).

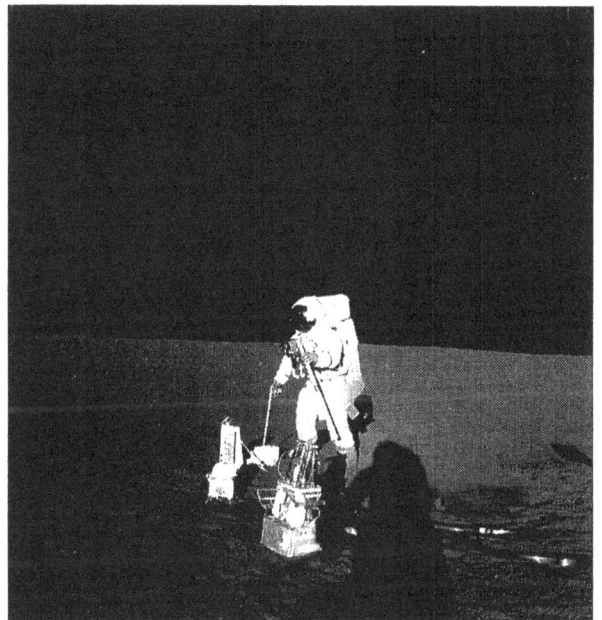

Bean deploys the ALSEP during the first EVA (NASA AS12-47-6919).

Astronaut movement on the lunar surface was recorded on the passive seismometer and on the lunar surface magnetometer. In addition, the commander rolled a grapefruit-sized rock down the wall of Head Crater, about 300 to 400 feet from the passive seismometer. No significant response was detected on any of the four axes.

During the geological traverse towards Surveyor III, the crew members obtained the desired photographic panoramas, stereo photographs, core samples (two single and one double), an eight-inch-deep trench sample, lunar environment samples, and assorted rock, dirt, bedrock, and "molten" samples. They reported seeing fine dust buildup on all sides of larger rocks and that soil color seemed to become lighter as they dug deeper.

Conrad at Modular Equipment Stowage Assembly (MESA). S-band antenna is at right (NASA AS12-47-6988).

Bean drives core tube sample into the lunar surface (NASA AS12-49-7286).

The crew photographed the Surveyor III and removed parts of it including the soil scoop. They reported that the Surveyor footpad marks were still visible and that the

entire spacecraft had a brown appearance. The glass parts were not broken, only warped slightly on their mountings, and therefore were not retrieved.

Bean holds a special environmental sample container filled with lunar soil. Conrad's reflection can be seen in Bean's visor (NASA AS12-49-7278).

After the return traverse, the crew retrieved the solar wind composition experiment after 18 hours 42 minutes exposure. The Apollo lunar surface close-up camera was used to take stereo pictures in the vicinity of the LM during the last few minutes of surface activity. Before reentering the LM, the crew members dusted each other off. The lunar module pilot entered the LM at 135:08, received samples, parts, and equipment from the commander, who then reentered at 135:20. Expendable equipment was jettisoned at 136:55, and the cabin was repressurized.

The second extravehicular activity period lasted 3 hours 49 minutes 15 seconds. The distance traveled was 4,300 feet (1.3 km); an estimated 38.8 pounds (17.6 kg) of samples were collected. The crew entered the LM at 135:22:00, thus ending the second human exploration of the Moon.

Mobility and portable life support system operation, as for Apollo 11, were excellent throughout both extravehicular periods. For the mission, the total time spent outside the LM was 7 hours 45 minutes 18 seconds, the total distance traveled was 7,600 feet (2.3 km), and the collected samples totaled 75.73 pounds (34.35 kg, official total in kilograms as determined by the Lunar Receiving Laboratory in Houston). The farthest point traveled from the LM was 1,362 feet.

Conrad uses tongs to pick up rock from lunar surface (NASA AS12-47-6932).

Conrad stands by Surveyor III. Note the lunar module on the horizon (NASA AS12-48-7133).

During the LM lunar surface stay, the S-158 lunar multi-spectral photography experiment was completed by the command module pilot in the CSM. In addition, photography of three desirable targets of opportunity was obtained. The areas were the Wall of Theophilus and two future Apollo landing sites, Fra Mauro and Descartes.

Ignition of the ascent stage engine for lunar liftoff occurred at 142:03:47.78. The LM had been on the lunar surface for 31 hours 31 minutes 12.0 seconds.

The 423.2-second burn was 1.2 seconds longer than planned and placed the spacecraft into an orbit of 46.3 by 8.8 n mi at 142:10:59.9. Several rendezvous sequence maneuvers were required before docking could occur three and a half hours later. A 41.1-second coelliptic orbit maneuver at 143:01:51.0 raised the orbit to 51.0 by 41.5 n mi. A 13.0-second constant

differential height maneuver at 144:00:02.6 lowered the orbit to 44.4 by 40.4 n mi. A 26-second terminal phase initiate maneuver occurred at 144:36:26 and brought the ascent stage to an orbit of 60.2 by 43.8 n mi. Finally, the ascent stage made a 38.0-second maneuver at 145:19:29.3 to finalize the orbit at 62.3 by 58.3 n mi for docking with the CM at 145:36:20.2 at an altitude of 58.1 n mi. The two craft had been undocked for 37 hours 42 minutes 17.9 seconds. Good quality television was transmitted from the CSM for 24 minutes during the final portions of the rendezvous sequence.

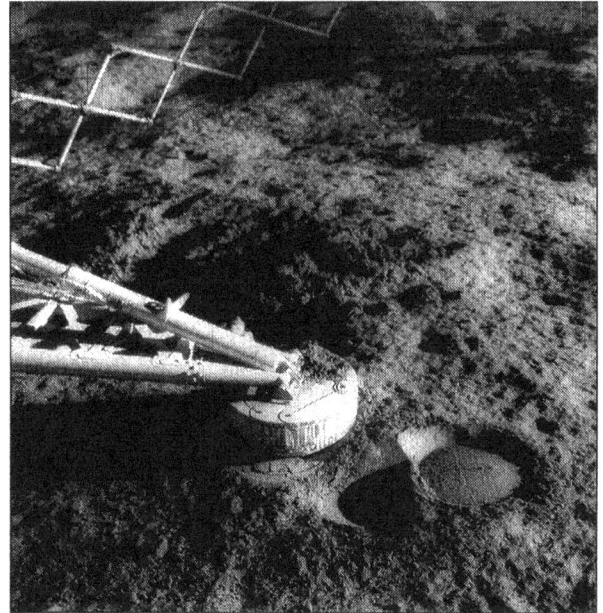

Evidence that Surveyor III bounced when it landed is the footpad imprint seen to the right (AS12-48-7110).

After the transfer of the crew and samples to the CSM, the ascent stage was jettisoned at 147:59:31.6, and the CSM was prepared for transearth injection. The ascent stage was then maneuvered by remote control to impact the lunar surface. A 5.4-second maneuver was made at 148:04:30.9 to separate the CSM from the ascent stage, and resulted in an orbit of 62.0 by 57.5 n mi. An 82.1-second ascent stage deorbit firing was made at 149:28:14.8 at 57.6 n mi altitude. The firing depleted the ascent stage propellants, and impact occurred at 149:55:16.4, at a point estimated to be latitude 3.94° south and longitude 21.20° west, 40 n mi east southeast of the Apollo 12 landing site and 5 n mi from the target.

During the final lunar orbits, extensive landmark tracking and photography from lunar orbit were conducted. A 500 mm long-range lens was used to obtain mapping and training data for future missions.

Prior to transearth injection, a 19.25-second plane change maneuver at 159:04:45.47 altered the CSM orbit to 64.7 by 56.8 n mi. Following a 130.32-second maneuver at 63.3 n mi altitude at 172:27:16.81, transearth injection was achieved at 172:29:27.13 at a velocity of 8,351 ft/sec after 45 lunar orbits lasting 88 hours 58 minutes 11.52 seconds. Good quality television of the receding Moon and the spacecraft interior was received for about 38 minutes, beginning about 20 minutes after transearth injection.

Transearth Phase

A small midcourse correction was made at 188:27:15.8. It was a 4.4-second, 2.0-ft/sec maneuver, delayed one hour to allow additional crew rest. The final television transmission included the spacecraft interior and a question and answer period with scientists and members of the press. It began at 224:07 and lasted for approximately 37 minutes.

The final midcourse correction, a 5.7-second, 2.4-ft/sec maneuver, was made at 241:21:59.7.

Recovery

The service module was jettisoned at 244:07:20.1, and command module entry (400,000 feet altitude) occurred at 244:22:19.09 at a velocity of 36,116 ft/sec, following a transearth coast of 71 hours, 52 minutes and 52.0 seconds. Following separation from the CM, the service module reaction control system was fired to depletion. However, no radar acquisition nor visual sightings by the crew or recovery personnel were made, and it was believed that the service module became unstable during the depletion firing and did not execute the velocity change required to skip out of Earth's atmosphere into the planned high-apogee orbit. Instead, it probably entered the atmosphere and impacted before detection.

Eclipse of the Sun by Earth as seen during transearth flight (NASA S80-37-37406).

Apollo 12 about to impact the surface of the Pacific Ocean (NASA S69-22728).

Sea-state conditions were fairly rough, and the parachute system effected an extremely hard splashdown of the CM in the Pacific Ocean at 20:58:24 GMT (03:58:24 p.m. EST) on 24 November 1969. The force of the impact, about 15 g, not only knocked loose portions of the heat shield, but caused the 16 mm sequence camera to separate from its bracket and strike the LMP above the right eye. Mission duration was 244:36:25. The impact point was about 2.0 n mi from the target point and 3.91 n mi from the recovery ship U.S.S. *Hornet*. The splashdown site was estimated to be latitude 15.78° south and longitude 165.15° west.

After splashdown, the CM assumed an apex-down attitude, but was successfully returned to the normal flotation position in 4 minutes 26 seconds by the inflatable bag uprighting system.

Apollo 12 crew in raft following egress from CM (l. to r.): Conrad, Bean and Gordon (NASA S69-22271).

Biological isolation precautions similar to those of Apollo 11 were taken. The crew was retrieved by helicopter and was aboard the recovery ship 60 minutes after splashdown. The crew immediately entered the mobile quarantine facility. The CM was recovered 48 minutes later. The estimated CM weight at splashdown was 11,050 pounds, and the estimated distance traveled for the mission was 828,134 n mi.

The mobile quarantine facility was offloaded from the *Hornet* in Hawaii at 02:18 GMT on 29 November, followed shortly by the CM. The mobile quarantine facility was loaded aboard a C-141 aircraft and flown to Ellington Air Force Base, Houston, Texas, where it arrived at 11:50 GMT. The crew entered the Lunar Receiving Laboratory two hours later.

The CM was taken to Hickam Air Force Base, Hawaii, for deactivation. Upon completion of deactivation, at 14:15 GMT on 1 December, the CM was flown to Ellington Air Force Base on a C-133 aircraft, and delivered to the Lunar Receiving Laboratory at 19:30 GMT on 2 December.

The crew was released from quarantine on 10 December. The CM was released soon after, and on 11 January was delivered to the North American Rockwell Space Division facility in Downey, California, for postflight analysis.

Apollo 12 crew aboard recovery ship U.S.S. *Hornet* enter the mobile quarantine facility (NASA S69-22849).

Wives of Apollo 12 crew greet them when the mobile quarantine facility arrives at Ellington AFB, Texas (NASA S69-60760).

Conclusions

The Apollo 12 mission demonstrated the capability for performing a precision lunar landing, which was a requirement for the success of future lunar surface explorations. The excellent performance of the spacecraft, the crew, and the supporting ground elements resulted in a wealth of scientific information. The following conclusions were made from an analysis of post-mission data:

1. The effectiveness of crew training, mission planning, and real-time navigation from the ground resulted in a precision landing near a previously landed Surveyor spacecraft and well within the desired landing footprint.

2. A hybrid non-free-return translunar profile was flown to demonstrate a capability for additional maneuvering which would be required for future landings at greater latitudes.

3. The timeline activities and metabolic loads associated with the extended lunar surface scientific exploration were within the capability of the crew and the portable life support systems.

4. An ALSEP was deployed for the first time and, despite some operating anomalies, returned valuable scientific data in a variety of study areas.

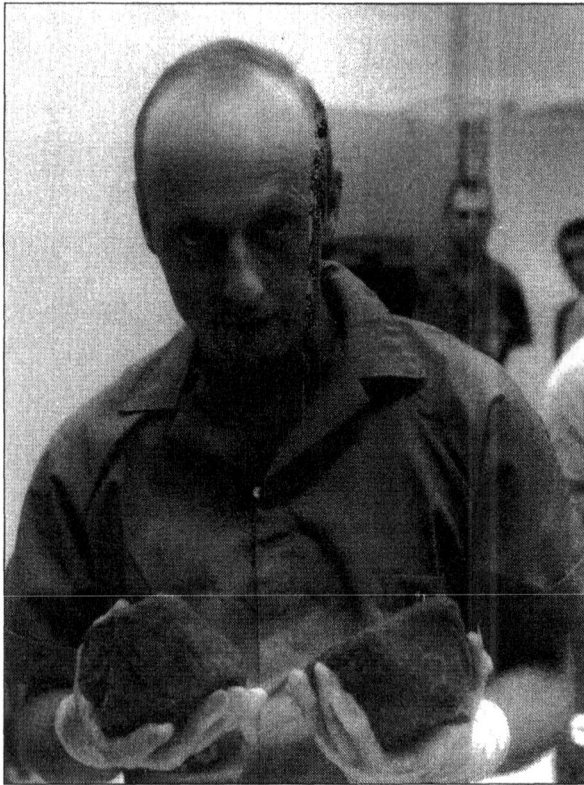

Conrad holds two lunar rock samples while in the Lunar Receiving Laboratory in Houston, Texas (NASA S69-60424).

Apollo 12 Objectives

Spacecraft Primary Objectives

1. To perform selenological inspection, survey, and sampling in a mare area. *Achieved.*

2. To deploy the ALSEP consistent with a seismic event. *Achieved.*

   a. S-031: Passive seismic experiment.

   b. S-034: Lunar surface magnetometer experiment.

   c. S-035: Solar wind spectrometer experiment.

   d. S-036: Suprathermal ion detector experiment.

   e. S-058: Cold cathode ion gauge experiment.

3. To develop techniques for a point landing capability. *Achieved.*

4. To further develop human capability to work in the lunar environment. *Achieved.*

5. To obtain photographs of candidate exploration sites. *Achieved.*

Detailed Objectives

1. Contingency sample collection. *Achieved.*

2. Lunar surface extravehicular operations. *Achieved.*

3. Portable life support system recharge. *Achieved.*

4. Selected sample collection. *Achieved.*

5. Photographs of candidate exploration sites.

   a. 70 mm stereoscopic photography of the ground track from terminator to terminator during two passes over the three sites, with concurrent 16 mm sextant sequence photography during the first pass. *Partially achieved. The first 70 mm pass and the concurrent 16 mm sextant sequence were accomplished. However, the necessity to repeat high resolution photography did not provide sufficient time to complete the second stereoscopic pass.*

   b. Landmark tracking of a series of four landmarks bracketing the three sites included in the stereoscopic photography, and performed during two subsequent, successive orbits. *Partially achieved. First series accomplished. However, necessity to repeat high resolution photography did not provide sufficient time to complete second series. Real-time decision assigning higher priority to landmark tracking allowed tracking of two landmarks associated with Fra Mauro and Descartes and completion of about one-fourth of the second stereoscopic pass.*

   c. High resolution photographs using a 500 mm lens, and additional high resolution oblique photography. *Partially achieved. The first photographs were of Herschel instead of Lalande due to crew error. A first attempt to obtain high resolution photographs of Fra Mauro and Descartes was unsuccessful because of a camera malfunction. On a second attempt, photographs were obtained of Fra Mauro and an area slightly east of Descartes.*

6. Lunar surface characteristics. *Achieved.*

7. Lunar environment visibility. *Achieved.*

8. Landed lunar module location. *Achieved.*

9. Photographic coverage. *Achieved.*

10. Television coverage.

a. A crew member descending to the lunar surface. *Achieved.*

b. An external view of the landed lunar module. *Not achieved. The camera was damaged immediately after it was removed from its stowage compartment.*

c. The lunar surface in the general vicinity of the lunar module. *Not achieved. The camera was damaged immediately after it was unstowed.*

d. Panoramic coverage of distant terrain features. *Not achieved. The camera was damaged immediately after it was removed from its stowage compartment.*

e. A crew member during extravehicular activity. *Not achieved. The camera was damaged immediately after it was removed from its stowage compartment.*

11. Surveyor III investigation. *Achieved.*

12. Selenodetic reference point update. *Achieved.*

Experiments

1. Lunar field geology. *Achieved.*

2. Solar wind composition. *Achieved.*

3. Lunar multispectral photography. *Achieved.*

4. Pilot describing function. *Achieved.*

5. Lunar dust detector. *Achieved.*

Launch Vehicle Objectives

1. To launch on a flight azimuth between 72° and 96° and insertion of the S-IVB/instrument unit/spacecraft into a circular Earth parking orbit. *Achieved.*

2. To restart the S-IVB during either the second or third revolution and injection of the S-IVB/instrument unit/spacecraft into the planned translunar trajectory. *Achieved.*

3. To provide the required attitude control for the S-IVB/instrument unit/spacecraft during the transposition, docking, and ejection maneuver. *Achieved.*

4. To use the S-IVB auxiliary propulsion system burn to execute a launch vehicle evasive maneuver after ejection of the command and service module/lunar module from the S-IVB/instrument unit. *Achieved.*

5. To use residual S-IVB propellants and the auxiliary propulsion system to maneuver to a trajectory that utilizes lunar gravity to insert the expended S-IVB/instrument unit into a solar orbit ("slingshot"). *Not achieved. The S-IVB/instrument unit failed to achieve solar orbit.*

6. To vent and dump all remaining gases and liquids to safe the S-IVB/instrument unit. *Achieved.*

Apollo 12 Spacecraft History

EVENT	DATE
LM #6 integrated test at factory.	31 Dec 1968
Individual and combined CM and SM systems test completed at factory.	20 Jan 1969
Integrated CM and SM systems test completed at factory.	03 Feb 1969
LM #6 final engineering evaluation acceptance test at factory.	18 Feb 1969
Saturn S-IVB stage #507 delivered to KSC.	10 Mar 1969
LM descent stage #6 ready to ship from factory to KSC.	22 Mar 1969
LM ascent stage #6 ready to ship from factory to KSC.	23 Mar 1969
LM ascent stage #6 and LM descent stage #6 delivered to KSC.	24 Mar 1969
CM #108 and SM #108 ready to ship from factory to KSC.	27 Mar 1969
CM #108 and SM #108 delivered to KSC.	28 Mar 1969
CM #108 and SM #108 mated.	02 Apr 1969
CSM #108 combined systems test completed.	21 Apr 1969
Saturn S-II stage #7 delivered to KSC.	21 Apr 1969
LM ascent stage #6 and descent stage #6 mated.	28 Apr 1969
LM #6 combined systems test completed.	01 May 1969
Saturn S-IC stage #7 delivered to KSC.	03 May 1969
Spacecraft/LM adapter #15 delivered to KSC.	06 May 1969
Saturn S-IC stage #7 erected on MLP #2.	07 May 1969
Saturn V instrument unit #507 delivered to KSC.	08 May 1969
Saturn S-II stage #7 erected.	05 May 1969
Saturn S-IVB stage #507 erected.	05 May 1969
Saturn V instrument unit #507 erected.	05 May 1969
CSM #108 altitude test with prime crew completed.	07 Jun 1969
CSM #108 altitude tests completed.	09 Jun 1969
CSM #108 altitude test with backup crew completed.	10 Jun 1969
Launch vehicle propellant dispersion/malfunction overall test completed.	12 Jun 1969
LM #6 altitude test with backup crew completed.	13 Jun 1969
LM #6 altitude test with prime crew completed.	16 Jun 1969
Spacecraft moved to VAB.	20 Jun 1969
LM #6 landing gear installed.	22 Jun 1969
LM #6 mated to spacecraft/LM adapter #15.	23 Jun 1969
CSM #108 mated to spacecraft/LM adapter #15.	27 Jun 1969
CSM #108 moved to VAB.	30 Jun 1969
Spacecraft erected.	01 Jul 1969
LM #6 combined systems test completed.	05 Jul 1969
CSM #108 integrated systems test completed.	07 Jul 1969
CSM #108 electrically mated to launch vehicle.	16 Jul 1969
Space vehicle overall test completed.	17 Jul 1969
Space vehicle electrically mated.	17 Aug 1969
Space vehicle overall test #1 (plugs in) completed.	21 Aug 1969
Space vehicle and MLP #2 transferred to launch complex 39A.	08 Sep 1969
Mobile service structure transferred to launch complex 39A.	10 Sep 1969
LM #5 flight readiness test completed.	18 Sep 1969
Space vehicle flight readiness test completed.	30 Sep 1969
Saturn S-IC stage #7 RP-1 fuel loading completed.	20 Oct 1969
Space vehicle countdown demonstration test (wet) completed.	28 Oct 1969
Space vehicle countdown demonstration test (dry) completed.	29 Oct 1969

Apollo 12 Ascent Phase

Event	GET (hhh:mm:ss)	Altitude (n mi)	Range (n mi)	Earth Fixed Velocity (ft/sec)	Space Fixed Velocity (ft/sec)	Event Duration (sec)	Geocentric Latitude (deg N)	Longitude (deg E)	Space Fixed Flight Path Angle (deg)	Space Fixed Heading Angle (E of N)
Liftoff	000:00:00.68	0.032	0.000	0.0	1,340.7		28.4470	-80.6041	0.07	90.00
1st lightning strike[3]	000:00:36.5	1.053	0.062	387.9	1,445.7		28.4469	-80.6030	15.40	89.29
2nd lightning strike	000:00:52	2.374	0.399	692.1	1,690.4		28.4487	-80.5968	22.74	87.32
Mach 1 achieved	000:01:06.1	4.215	1.228	1,067.6	2,057.7		28.4532	-80.5820	27.13	84.84
Maximum dynamic pressure	000:01:21.1	6.934	3.019	1,601.4	2,617.3		28.4627	-80.5498	29.02	82.10
S-IC center engine cutoff	000:02:15.24	24.158	25.441	5,334.5	6,494.4	141.7	28.5794	-80.1463	23.944	76.115
S-IC outboard engine cutoff	000:02:41.74	36.773	50.616	7,821.4	9,024.5	168.2	28.7069	-79.6913	20.513	75.231
S-IC/S-II separation	000:02:42.4	37.118	51.338	7,850.3	9,054.2		28.7107	-79.6773	20.430	75.228
S-II center engine cutoff	000:07:40.75	100.463	599.172	17,453.5	18,775.3	297.55	30.9599	-69.4827	0.502	79.632
S-II outboard engine cutoff	000:09:12.34	102.801	884.711	21,508.8	22,831.7	389.14	31.7508	-63.9914	0.442	82.501
S-II/S-IVB separation	000:09:13.20	102.827	887.667	21,517.8	22,840.7		31.7576	-63.9341	0.432	82.533
S-IVB 1st burn cutoff	000:11:33.91	103.093	1,399.874	24,236.6	25,560.2	137.31	32.4933	-53.8956	-0.015	88.146
Earth orbit insertion	000:11:43.91	103.086	1,438.608	24,242.3	25,565.9		32.5128	-53.1311	-0.014	88.580

Apollo 12 Earth Orbit Phase

Event	GET (hhh:mm:ss)	Space Fixed Velocity (ft/sec)	Event Duration (sec)	Velocity Change (ft/sec)	Apogee (n mi)	Perigee (n mi)	Period (mins)	Inclination (deg)
Earth orbit insertion	000:11:43.91	25,565.9			100.1	97.8	88.16	32.540
S-IVB 2nd burn ignition	002:47:22.80	25,555.4						
S-IVB 2nd burn cutoff	002:53:03.94	35,419.3	341.14	10515				30.360

Apollo 12 Translunar Phase

Event	GET (hhh:mm:ss)	Altitude (n mi)	Space Fixed Velocity (ft/sec)	Event Duration (sec)	Velocity Change (ft/sec)	Space Fixed Flight Path Angle (deg)	Space Fixed Heading Angle (E of N)
Translunar injection	002:53:13.94	199.023	35,389.9			8.584	63.902
CSM separated from S-IVB	003:18:04.9	3,819.258	24,865.5			45.092	100.194
CSM docked with LM/S-IVB	003:26:53.3	5,337.7	22,534			49.896	105.29
CSM/LM ejected from S-IVB	004:13:00.9	12,506.3	16,451.1			60.941	114.52
S-IVB APS evasive maneuver	004:29:21.4			80.0	9.5		
Midcourse correction ignition	030:52:44.36	116,929.1	4,317.4			75.833	120.80
Midcourse correction cutoff	030:52:53.55	116,935.4	4,297.5	9.19	61.8	76.597	120.05

[3] Data for this event reflects postflight trajectory reconstruction for 36 seconds Ground Elapsed Time.

Apollo 12 Lunar Orbit Phase

Event	GET (hhh:mm:ss)	Altitude (n mi)	Space Fixed Velocity (ft/sec)	Event Duration (sec)	Velocity Change (ft/sec)	Apogee (n mi)	Perigee (n mi)
Lunar orbit insertion ignition	083:25:23.36	83.91	8,173.6			NA	64.94
Lunar orbit insertion cutoff	083:31:15.61	62.91	5,470.1	352.25	2,889.5	170.20	61.66
Lunar orbit circularization ignition	087:48:48.08	62.79	5,470.6			170.37	61.42
Lunar orbit circularization cutoff	087:49:04.99	62.74	5,331.4	16.91	165.2	66.10	54.59
CSM/LM undocked	107:54:02.3	63.02	5,329.0			63.08	56.91
CSM/LM separation ignition	108:24:36.8	59.22	5,350.0			63.91	56.99
CSM/LM separation cutoff	108:24:51.2	59.15	5,350.5	14.4	2.4	64.06	56.58
LM descent orbit insertion ignition	109:23:39.9	60.52	5,343.0			63.27	57.25
LM descent orbit insertion cutoff	109:24:08.9	61.52	5,268.0	29.0	72.4	61.53	8.70
LM powered descent initiation	110:20:38.1	7.96	5,566.4			62.30	7.96
LM powered descent cutoff	110:32:35.1			717.0			
CSM plane change ignition	119:47:13.23	62.20	5,333.5			62.50	57.61
CSM plane change cutoff	119:47:31.46	62.20	5,683.4	18.23	349.9	62.50	57.60
LM lunar liftoff ignition	142:03:48						
LM orbit insertion	142:10:59.9	9.97	5,542.5		6,057	51.93	9.21
LM ascent stage cutoff	142:11:01.78			474.0			
LM coelliptic sequence initiation ignition	143:01:51.0	51.46	5,310.3			52.51	9.94
LM coelliptic sequence initiation cutoff	143:02:32.1	51.48	5,354.9	41.1	45	51.49	41.76
LM constant differential height ignition	144:00:02.6						
LM constant differential height cutoff	144:00:15.6			13.0	13.8	44.40	40.40
LM terminal phase initiation ignition	144:36:26	44.50	5,382.5			44.73	40.91
LM terminal phase initiation cutoff	144:36:52			26.0	29	60.20	43.80
LM 1st midcourse correction	144:51:29						
LM 2nd midcourse correction	145:06:29						
LM terminal phase finalize ignition	145:19:29.3						
LM terminal phase finalize cutoff	145:20:07.3			38.0	40	62.30	58.30
CSM/LM docked	145:36:20.2	58.14	5,357.1			63.43	58.04
CSM/LM separation ignition	148:04:30.9	59.94	5,347.4			64.66	59.08
CSM/LM separation cutoff	148:04:36.3			5.4	1.0	62.00	57.50
LM ascent stage deorbit ignition	149:28:14.8	57.62	5,361.8			63.52	57.94
LM ascent stage deorbit cutoff	149:29:36.9	57.42	5,176.8	82.1	196.2	57.59	-63.15
CSM orbit plane change ignition	159:04:45.47	58.70	5,353.2			64.23	56.58
CSM orbit plane change cutoff	159:05:04.72	58.90	5,353.0	19.25	381.8	64.66	56.81

Apollo 12 Transearth Phase

Event	GET (hhh:mm:ss)	Altitude (n mi)	Space Fixed Velocity (ft/sec)	Event Duration (sec)	Velocity Change (ft/sec)	Space Fixed Flight Path Angle (deg)	Space Fixed Heading Angle (E of N)
Transearth injection ignition	172:27:16.81	63.60	5,322.9			-0.202	-115.73
Transearth injection cutoff	172:29:27.13	66.00	8,350.4	130.32	3,042.0	2.718	-116.45
Midcourse correction ignition	188:27:15.8	180,031.2	3,035.6			-78.444	91.35
Midcourse correction cutoff	188:27:20.2	180,029.0	3,036.0	4.4	2.0	-78.404	91.36
Midcourse correction ignition	241:21:59.7	25,059.0	12,082.9			-68.547	96.00
Midcourse correction cutoff	241:22:05.4	25,048.3	12,084.7	5.7	2.4	-68.547	96.01
CM/SM separation	244:07:20.1	1,949.5	29,029.1			-36.454	98.17

Apollo 12 Timeline

Event	GET (hhh:mm:ss)	GMT Time	GMT Date
Terminal countdown started.	-028:00:00	02:00:00	13 Nov 1969
Unscheduled 6-hour hold at T-17 hours to replace CSM LH2 tank #2 due to leak.	-017:00:00	13:00:00	13 Nov 1969
Countdown resumed at T-17 hours.	-017:00:00	19:00:00	13 Nov 1969
Scheduled 9-hour 22-minute hold at T-9 hours (shortened by 6 hours to avert launch delay).	-009:00:00	03:00:00	14 Nov 1969
Countdown resumed at T-9 hours.	-009:00:00	06:22:00	14 Nov 1969
Scheduled 1-hour hold at T-3 hours 30 minutes.	-003:30:00	11:52:00	14 Nov 1969
Countdown resumed at T-3 hours 30 minutes.	-003:30:00	12:52:00	14 Nov 1969
Guidance reference release.	-000:00:16.968	16:21:43	14 Nov 1969
S-IC engine start command.	-000:00:08.9	16:21:51	14 Nov 1969
S-IC engine ignition (#5).	-000:00:06.5	16:21:53	14 Nov 1969
All S-IC engines thrust OK.	-000:00:01.4	16:21:58	14 Nov 1969
Range zero.	000:00:00.00	16:22:00	14 Nov 1969
All holddown arms released (1st motion) (1.09 g).	000:00:00.25	16:22:00	14 Nov 1969
Liftoff (umbilical disconnected).	000:00:00.68	16:22:00	14 Nov 1969
Tower clearance yaw maneuver started.	000:00:02.4	16:22:02	14 Nov 1969
Yaw maneuver ended.	000:00:10.2	16:22:10	14 Nov 1969
Pitch and roll maneuver started.	000:00:12.8	16:22:12	14 Nov 1969
Roll maneuver ended.	000:00:32.3	16:22:32	14 Nov 1969
1st electrical disturbance (lightning).	000:00:36.5	16:22:36	14 Nov 1969
2nd electrical disturbance (lightning).	000:00:52	16:22:52	14 Nov 1969
Mach 1 achieved.	000:01:06.1	16:23:06	14 Nov 1969
Maximum bending moment achieved (37,000,000 lbf-in).	000:01:17.5	16:23:17	14 Nov 1969
Maximum dynamic pressure (682.95 lb/ft^2).	000:01:21.1	16:23:21	14 Nov 1969
S-IC center engine cutoff command.	000:02:15.24	16:24:15	14 Nov 1969
Fuel cell power restored to buses.	000:02:22	16:24:22	14 Nov 1969
Pitch maneuver ended.	000:02:38.1	16:24:38	14 Nov 1969
S-IC outboard engine cutoff.	000:02:41.74	16:24:41	14 Nov 1969
S-IC maximum total inertial acceleration (3.91 g).	000:02:41.82	16:24:41	14 Nov 1969
S-IC maximum Earth-fixed velocity.	000:02:42.18	16:24:42	14 Nov 1969
S-IC/S-II separation command.	000:02:42.4	16:24:42	14 Nov 1969
S-II engine start command.	000:02:43.17	16:24:43	14 Nov 1969
S-II ignition.	000:02:43.2	16:24:43	14 Nov 1969
S-II aft interstage jettisoned.	000:03:12.4	16:25:12	14 Nov 1969
Launch escape tower jettisoned.	000:03:17.9	16:25:17	14 Nov 1969
Iterative guidance mode initiated.	000:03:22.5	16:25:22	14 Nov 1969
S-IC apex.	000:04:35.6	16:26:35	14 Nov 1969
S-II center engine cutoff.	000:07:40.75	16:29:40	14 Nov 1969
S-II maximum total inertial acceleration (1.83 g).	000:07:40.83	16:29:40	14 Nov 1969
S-II outboard engine cutoff.	000:09:12.34	16:31:12	14 Nov 1969
S-II maximum Earth-fixed velocity, S-II/S-IVB separation command.	000:09:13.20	16:31:13	14 Nov 1969
S-IVB 1st burn start command.	000:09:13.30	16:31:13	14 Nov 1969
S-IC impact (theoretical).	000:09:14.5	16:31:14	14 Nov 1969
S-IVB 1st burn ignition.	000:09:16.60	16:31:16	14 Nov 1969
S-IVB ullage case jettisoned.	000:09:25.1	16:31:25	14 Nov 1969
S-II apex.	000:09:41.7	16:31:41	14 Nov 1969
S-IVB 1st burn cutoff.	000:11:33.91	16:33:33	14 Nov 1969
S-IVB 1st burn maximum total inertial acceleration (0.69 g).	000:11:33.99	16:33:34	14 Nov 1969
Earth orbit insertion; S-IVB 1st burn maximum Earth-fixed velocity.	000:11:43.91	16:33:43	14 Nov 1969
Maneuver to local horizontal attitude started.	000:11:54.2	16:33:54	14 Nov 1969
Orbital navigation started.	000:13:15.1	16:35:15	14 Nov 1969
S-II impact (theoretical).	000:20:21.6	16:42:21	14 Nov 1969

Apollo 12 Timeline

Event	GET (hhh:mm:ss)	GMT Time	GMT Date
S-IVB 2nd burn restart preparation.	002:37:44.50	18:59:44	14 Nov 1969
S-IVB 2nd burn restart command.	002:47:15.10	19:09:15	14 Nov 1969
S-IVB 2nd burn ignition.	002:47:22.80	19:09:22	14 Nov 1969
S-IVB 2nd burn cutoff.	002:53:03.94	19:15:03	14 Nov 1969
S-IVB 2nd burn maximum total inertial acceleration (1.48 g).	002:53:04.02	19:15:04	14 Nov 1969
S-IVB 2nd burn maximum Earth-fixed velocity.	002:53:04.32	19:15:04	14 Nov 1969
1st LOX tank NPV valve open.	002:53:04.6	19:15:04	14 Nov 1969
1st LH_2 tank latching valve open.	002:53:04.6	19:15:04	14 Nov 1969
S-IVB safing procedures started.	002:53:04.6	19:15:04	14 Nov 1969
Translunar injection.	002:53:13.94	19:15:13	14 Nov 1969
Maneuver to local horizontal attitude and orbital navigation started.	002:53:24.4	19:15:24	14 Nov 1969
1st LH_2 tank CVS open.	002:54:20	19:16:00	14 Nov 1969
Cold helium dump start.	002:54:55	19:16:00	14 Nov 1969
1st LOX tank NPV valve closed.	002:57:05	19:19:00	14 Nov 1969
2nd LH_2 tank latching valve open.	003:04:00	19:26:00	14 Nov 1969
Cold helium dump stop.	003:08:30	19:30:00	14 Nov 1969
1st LH_2 tank latching valve closed.	003:08:03.9	19:30:03	14 Nov 1969
Maneuver to transposition and docking attitude started.	003:08:05.0	19:30:05	14 Nov 1969
1st LH_2 tank CVS closed.	003:09:05	19:31:00	14 Nov 1969
CSM separated from S-IVB.	003:18:04.9	19:40:04	14 Nov 1969
TV transmission started.	003:25	19:47	14 Nov 1969
CSM docked with LM/S-IVB.	003:26:53.3	19:48:53	14 Nov 1969
Ambient repressurization helium dump start.	003:53:05	20:15:00	14 Nov 1969
Engine start-tank dump start.	003:53:04.9	20:15:04	14 Nov 1969
Cold helium dump start.	003:54:00	20:16:00	14 Nov 1969
Ambient repressurization helium dump stop.	003:54:07	20:16:00	14 Nov 1969
Engine start-tank dump stop.	003:55:34.9	20:17:34	14 Nov 1969
2nd LH_2 tank latching valve closed.	003:56:35	20:18:00	14 Nov 1969
Cold helium dump stop.	004:08:35	20:30:00	14 Nov 1969
CSM/LM ejected from S-IVB.	004:13:00.9	20:35:00	14 Nov 1969
Observation and photography of two ventings from the S-IVB burner area started.	004:19:20	20:41:20	14 Nov 1969
Maneuver to evasive attitude start.	004:20:00	20:42:00	14 Nov 1969
Maneuver to evasive attitude stop.	004:23:20	20:45:00	14 Nov 1969
Stage control helium dump start.	004:26:40	20:48:00	14 Nov 1969
1st APS evasive maneuver ignition.	004:26:40	20:48:00	14 Nov 1969
Observation and photography of S-IVB ventings ended.	004:26:40	20:48:40	14 Nov 1969
Cold helium dump start.	004:26:41.2	20:48:41	14 Nov 1969
TV transmission ended.	004:28	20:50	14 Nov 1969
1st APS evasive maneuver cutoff.	004:28:00	20:50:00	14 Nov 1969
S-IVB APS ullage evasive maneuver started.	004:28:01.4	20:50:01	14 Nov 1969
S-IVB APS ullage evasive maneuver ended.	004:29:21.4	20:51:21	14 Nov 1969
2nd LH_2 tank CVS open.	004:36:20.4	20:58:20	14 Nov 1969
S-IVB slingshot maneuver—Propulsive LH_2 vent (CVS).	004:36:20.4	20:58:20	14 Nov 1969
Maneuver to slingshot attitude.	004:36:21.0	20:58:21	14 Nov 1969
S-IVB maneuver to lunar slingshot attitude for solar orbit initiated.	004:36:21.0	20:58:21	14 Nov 1969
LOX dump start.	004:48:00.2	21:10:00	14 Nov 1969
S-IVB slingshot maneuver—LOX dump started.	004:48:00.2	21:10:00	14 Nov 1969
LOX dump stop.	004:48:58.2	21:10:58	14 Nov 1969
S-IVB slingshot maneuver—LOX dump ended.	004:48:58.2	21:10:58	14 Nov 1969
2nd LOX tank NPV valve open.	004:49:00	21:11:00	14 Nov 1969
Cold helium dump stop.	004:50:50	21:12:00	14 Nov 1969

Apollo 12 Timeline

Event	GET (hhh:mm:ss)	GMT Time	GMT Date
3rd LH$_2$ tank latching valve open.	004:50:07.2	21:12:07	14 Nov 1969
Engine control helium dump start.	004:58:00	21:20:00	14 Nov 1969
Engine control helium dump stop.	005:05:30	21:27:00	14 Nov 1969
Programmed APS ignition.	005:23:20.4	21:45:20	14 Nov 1969
S-IVB slingshot maneuver—APS ullage ignition (planned).	005:23:20.4	21:45:20	14 Nov 1969
Stage control helium dump stop.	005:26:40	21:48:00	14 Nov 1969
Programmed APS cutoff.	005:28:20.4	21:50:20	14 Nov 1969
S-IVB slingshot maneuver—APS ullage cutoff.	005:28:20.4	21:50:20	14 Nov 1969
Ground commanded APS ignition.	005:29:10	21:51:00	14 Nov 1969
S-IVB slingshot maneuver—APS ullage ignition (unplanned).	005:29:13.2	21:51:13	14 Nov 1969
Ground commanded APS cutoff.	005:33:40	21:55:00	14 Nov 1969
LMP entered LM.	010:40	03:02	15 Nov 1969
LMP entered CM.	010:50	03:12	15 Nov 1969
TV transmission started.	030:18	22:40	15 Nov 1969
Midcourse correction ignition.	030:52:44.36	23:14:44	15 Nov 1969
Midcourse correction cutoff.	030:52:53.55	23:14:53	15 Nov 1969
S-IVB slingshot maneuver—APS ullage cutoff.	005:33:43.2	21:55:43	14 Nov 1969
S-IVB maneuver to communications attitude initiated.	005:36:37.0	21:58:37	14 Nov 1969
LMP entered LM.	007:20	23:42	14 Nov 1969
LM inspection.	007:30	23:52	14 Nov 1969
LMP entered CM.	008:10	00:32	15 Nov 1969
TV transmission ended.	031:05	23:27	15 Nov 1969
TV transmission started.	062:52	07:14	17 Nov 1969
LMP entered LM.	063:10	07:32	17 Nov 1969
LMP entered CM.	063:45	08:07	17 Nov 1969
TV transmission ended.	063:48	08:10	17 Nov 1969
Equigravisphere.	068:30:00	12:52:00	17 Nov 1969
Rendezvous transponder activation and self-test.	079:35	23:57	17 Nov 1969
System checks for lunar orbit insertion maneuver.	082:00	02:22	18 Nov 1969
Lunar orbit insertion ignition.	083:25:23.36	03:47:23	18 Nov 1969
Lunar orbit insertion cutoff.	083:31:15.61	03:53:15	18 Nov 1969
TV transmission started.	084:00	04:22	18 Nov 1969
TV transmission ended.	084:33	04:55	18 Nov 1969
S-IVB closest approach to lunar surface.	085:48	06:10	18 Nov 1969
System checks for lunar orbit circularization maneuver.	086:30	06:52	18 Nov 1969
Lunar orbit circularization ignition.	087:48:48.08	08:10:48	18 Nov 1969
Lunar orbit circularization cutoff.	087:49:04.99	08:11:05	18 Nov 1969
LMP entered LM.	089:20	09:42	18 Nov 1969
LM activation and checkout.	089:45	10:07	18 Nov 1969
LM deactivation and LMP transferred back to CM.	090:30	10:52	18 Nov 1969
LM landing radar altitude lock.	100:24:00	20:46:00	18 Nov 1969
LM landing radar velocity lock.	100:24:04	20:46:04	18 Nov 1969
LMP entered LM.	103:45	00:07	19 Nov 1969
LM system checks.	104:04	00:26	19 Nov 1969
CDR entered LM.	104:20	00:42	19 Nov 1969
LM system checks.	104:30	00:52	19 Nov 1969
LMP entered CM.	104:40	01:02	19 Nov 1969
LMP entered LM. System checks.	105:00	01:22	19 Nov 1969
TV transmission started.	107:50	04:12	19 Nov 1969
CSM/LM undocked.	107:54:02.3	04:16:02	19 Nov 1969
CSM/LM separation maneuver ignition.	108:24:36.8	04:46:36	19 Nov 1969

Apollo 12 Timeline

Event	GET (hhh:mm:ss)	GMT Time	GMT Date
CSM/LM separation maneuver cutoff.	108:24:51.2	04:46:51	19 Nov 1969
TV transmission ended.	108:30	04:52	19 Nov 1969
LM descent orbit insertion ignition (SPS).	109:23:39.9	05:45:39	19 Nov 1969
LM descent orbit insertion cutoff.	109:24:08.9	05:46:08	19 Nov 1969
LM powered descent engine ignition.	110:20:38.1	06:42:38	19 Nov 1969
LM throttle up.	110:21:05	06:43:05	19 Nov 1969
LM landing site correction initiated.	110:22:03	06:44:03	19 Nov 1969
LM landing site correction entered.	110:22:27	06:44:27	19 Nov 1969
LM "permit landing radar updates" entered.	110:24:09	06:46:09	19 Nov 1969
LM state-vector update allowed.	110:24:25	06:46:25	19 Nov 1969
LM "permit landing radar updates" exited.	110:24:31	06:46:31	19 Nov 1969
LM abort guidance system altitude updated.	110:26:08	06:48:08	19 Nov 1969
LM velocity update initiated.	110:26:24	06:48:24	19 Nov 1969
LM X-axis override inhibited.	110:26:39	06:48:39	19 Nov 1969
LM throttle recovery.	110:27:01	06:49:01	19 Nov 1969
LM abort guidance system altitude updated.	110:27:26	06:49:26	19 Nov 1969
LM approach phase entered.	110:29:11	06:51:11	19 Nov 1969
LM landing point designator enabled.	110:29:14	06:51:14	19 Nov 1969
LM landing radar antenna to position 2.	110:29:18	06:51:18	19 Nov 1969
LM abort guidance system altitude updated.	110:29:20	06:51:20	19 Nov 1969
LM redesignation right.	110:29:44	06:51:44	19 Nov 1969
LM landing radar switched to low scale.	110:29:47	06:51:47	19 Nov 1969
LM redesignation long.	110:30:02	06:52:02	19 Nov 1969
LM redesignation long.	110:30:06	06:52:06	19 Nov 1969
LM redesignation right.	110:30:12	06:52:12	19 Nov 1969
LM redesignation short.	100:30:30	20:52:30	18 Nov 1969
LM landing radar data recovery.	110:31:37	06:53:37	19 Nov 1969
LM redesignation right.	110:30:42	06:52:42	19 Nov 1969
LM landing radar data recovery.	110:31:24	06:53:24	19 Nov 1969
LM landing radar data dropout.	110:31:27	06:53:27	19 Nov 1969
LM attitude hold mode selected.	110:30:46	06:52:46	19 Nov 1969
LM rate of descent landing phase entered.	110:30:50	06:52:50	19 Nov 1969
LM landing radar data dropout.	110:31:18	06:53:18	19 Nov 1969
1st photographic evidence of surface dust disturbed by descent engine.	110:31:44	06:53:44	19 Nov 1969
LM premature low level fuel light on tank #2.	110:31:59.6	06:53:59	19 Nov 1969
LM landing radar data dropout.	110:32:00	06:54:00	19 Nov 1969
LM landing radar data recovery.	110:32:04	06:54:04	19 Nov 1969
Lunar dust completely obscured landing site.	110:32:11	06:54:11	19 Nov 1969
LM powered descent engine cutoff.	110:32:35.1	06:54:35	19 Nov 1969
LM lunar landing.	110:32:36.2	06:54:36	19 Nov 1969
1st EVA started (egress).	115:10:35	11:32:35	19 Nov 1969
CDR on lunar surface. Environmental familiarization.	115:22:22	11:44:22	19 Nov 1969
Contingency sample collected (CDR). CDR activities photographed (LMP).	115:25:41	11:47:41	19 Nov 1969
Equipment bag transferred (LMP to CDR).	115:38:53	12:00:53	19 Nov 1969
Contingency sample site photographed (CDR).	115:46:57	12:08:57	19 Nov 1969
LMP egress.	115:49:41	12:11:41	19 Nov 1969
LMP on lunar surface.	115:51:50	12:13:50	19 Nov 1969
S-band antenna deployed (CDR).	116:09:38	12:31:38	19 Nov 1969
Solar wind composition experiment deployed (LMP).	116:13:17	12:35:17	19 Nov 1969
United States flag deployed (CDR).	116:19:31	12:41:31	19 Nov 1969
LM inspection complete (LMP).	116:31:46	12:53:46	19 Nov 1969

Apollo 12 Timeline

Event	GET (hhh:mm:ss)	GMT Time	GMT Date
Panoramic photography complete (CDR).	116:25:51	12:47:51	19 Nov 1969
Experiment package unloaded (CDR, LMP).	116:32	12:54	19 Nov 1969
Experiment package transferred (CDR, LMP).	116:52	13:14	19 Nov 1969
Experiment package deployed (CDR) and activated (LMP).	117:01	13:23	19 Nov 1969
Return traverse started (CDR, LMP).	118:00	14:22	19 Nov 1969
Sample container packing started (CDR).	118:27	14:49	19 Nov 1969
Core tube sample gathered (LMP).	118:35	14:57	19 Nov 1969
LMP on ladder for ingress.	118:50:46	15:12:46	19 Nov 1969
LMP inside LM.	118:52:18	15:14:18	19 Nov 1969
Equipment transfer bag in LM (CDR to LMP).	118:56:19	15:18:19	19 Nov 1969
Sample return container in LM (CDR to LMP).	118:58:30	15:20:30	19 Nov 1969
CDR on LM footpad.	119:02:11	15:24:11	19 Nov 1969
CDR inside LM.	119:05:17	15:27:17	19 Nov 1969
1st EVA ended (hatch closed).	119:06:36	15:28:36	19 Nov 1969
CSM plane change ignition (SPS).	119:47:13.23	16:09:13	19 Nov 1969
CSM plane change cutoff.	119:47:31.46	16:09:31	19 Nov 1969
Debriefing for 1st EVA.	120:45	17:07	19 Nov 1969
CDR set foot on lunar surface.	131:37	03:59	20 Nov 1969
2nd EVA started (egress).	131:32:45	03:54:45	20 Nov 1969
Safety monitoring of CDR descent to surface by LMP.	131:35	03:57	20 Nov 1969
CDR transferred equipment bag.	131:39	04:01	20 Nov 1969
CDR prepared for traverse. LMP began egress.	131:44	04:06	20 Nov 1969
Contrast chart photographs taken by LMP.	131:49	04:11	20 Nov 1969
Initial geological traverse started (CDR).	132:00	04:22	20 Nov 1969
Initial geological traverse started (LMP).	132:11	04:33	20 Nov 1969
Core tube sample gathered (CDR).	133:23	05:45	20 Nov 1969
Final geological traverse started (CDR).	133:36	05:58	20 Nov 1969
Surveyor spacecraft inspected (CDR, LMP).	133:53	06:15	20 Nov 1969
Sample container packing and close-up photographs (LMP).	134:46	07:08	20 Nov 1969
Solar wind composition experiment retrieved.	134:55	07:17	20 Nov 1969
Ingress (LMP).	135:08	07:30	20 Nov 1969
Ingress (CDR) started.	135:20	07:42	20 Nov 1969
Equipment transferred (CDR to LMP).	135:11	07:33	20 Nov 1969
2nd EVA ended (ingress completed).	135:22:00	07:44:00	20 Nov 1969
LM equipment jettisoned.	136:55	90:17	20 Nov 1969
Debriefing for 2nd EVA.	138:20	10:42	20 Nov 1969
LM coelliptic sequence initiation cutoff.	143:02:32.1	15:24:32	20 Nov 1969
LM constant differential height maneuver ignition.	144:00:02.6	16:22:02	20 Nov 1969
LM constant differential height maneuver cutoff.	144:00:15.6	16:22:15	20 Nov 1969
LM terminal phase initiation ignition.	144:36:26	16:58:26	20 Nov 1969
LM terminal phase initiation cutoff.	144:36:52	16:58:52	20 Nov 1969
LM lunar liftoff ignition (LM APS).	142:03:47.78	14:25:47	20 Nov 1969
LM ascent stage orbit insertion.	142:10:59.9	14:32:59	20 Nov 1969
LM ascent stage cutoff.	142:11:01.78	14:33:01	20 Nov 1969
LM RCS trim burn (due to overburn on ascent) cutoff.	142:11:51.78	14:33:51	20 Nov 1969
LM coelliptic sequence initiation ignition.	143:01:51.0	15:23:51	20 Nov 1969
LM 1st midcourse correction.	144:51:29	17:13:29	20 Nov 1969
LM 2nd midcourse correction.	145:06:29	17:28:29	20 Nov 1969
LM terminal phase finalize ignition.	145:19:29.3	17:41:29	20 Nov 1969
LM terminal phase finalize cutoff.	145:20:07.3	17:42:07	20 Nov 1969
CSM/LM docked.	145:36:20.2	17:58:20	20 Nov 1969

Apollo 12 Timeline

Event	GET (hhh:mm:ss)	GMT Time	GMT Date
CDR entered CM.	147:05	19:27	20 Nov 1969
LMP entered CM.	147:20	19:42	20 Nov 1969
LM ascent stage jettisoned.	147:59:31.6	20:21:31	20 Nov 1969
LM ascent stage separation maneuver ignition.	148:04:30.9	20:26:30	20 Nov 1969
LM ascent stage separation maneuver cutoff.	148:04:36.3	20:26:36	20 Nov 1969
LM ascent stage deorbit ignition.	149:28:14.8	21:50:14	20 Nov 1969
LM ascent stage deorbit cutoff.	149:29:36.9	21:51:36	20 Nov 1969
LM ascent stage impact on lunar surface.	149:55:16.4	22:17:16	20 Nov 1969
CSM lunar orbit plane change ignition.	159:04:45.47	7:26:45	21 Nov 1969
CSM lunar orbit plane change cutoff.	159:05:04.72	07:27:04	21 Nov 1969
CSM landmark tracking and photography.	160:15	08:37	21 Nov 1969
CSM landmark tracking and photography.	165:05	13:27	21 Nov 1969
CSM landmark tracking and photography.	166:50	15:12	21 Nov 1969
CSM landmark tracking and photography.	171:20	19:42	21 Nov 1969
Transearth injection ignition (SPS).	172:27:16.81	20:49:16	21 Nov 1969
Transearth injection cutoff.	172:29:27.13	20:51:27	21 Nov 1969
TV transmission started.	172:45	21:07	21 Nov 1969
TV transmission ended.	173:23	21:45	21 Nov 1969
Midcourse correction ignition.	188:27:15.8	12:49:15	22 Nov 1969
Midcourse correction cutoff.	188:27:20.2	12:49:20	22 Nov 1969
High-gain antenna test started.	191:15	15:37	22 Nov 1969
High-gain antenna test ended.	194:00	18:22	22 Nov 1969
High-gain antenna test started.	214:00	14:22	23 Nov 1969
High-gain antenna test ended.	216:40	17:02	23 Nov 1969
TV transmission started.	224:07	0:29	24 Nov 1969
TV transmission ended.	224:44	1:06	24 Nov 1969
Midcourse correction ignition.	241:21:59.7	17:43:59	24 Nov 1969
Midcourse correction cutoff.	241:22:05.4	17:44:05	24 Nov 1969
CM/SM separation.	244:07:20.1	20:29:20	24 Nov 1969
Entry.	244:22:19.09	20:44:19	24 Nov 1969
Drogue parachute deployed	244:30:39.7	20:52:39	24 Nov 1969
Radar contact with CM established by recovery ship.	244:24	20:46	24 Nov 1969
Main parachute deployed.	244:31:30.2	20:53:30	24 Nov 1969
S-band contact with CM established by rescue aircraft.	244:28	20:50	24 Nov 1969
VHF recovery beacon contact established with CM by recovery forces.	244:31	20:53	24 Nov 1969
VHF voice contact with CM established by aircraft and recovery ship.	244:32	20:54	24 Nov 1969
Splashdown (went to apex-down).	244:36:25	20:58:25	24 Nov 1969
CM returned to apex-up position.	244:40:51	21:02:51	24 Nov 1969
Swimmers deployed to CM.	244:46	21:08	24 Nov 1969
Flotation collar inflated.	244:53	21:15	24 Nov 1969
Hatch opened for respirator transfer.	245:14	21:36	24 Nov 1969
Hatch opened for crew egress.	245:18	21:40	24 Nov 1969
Crew aboard recovery ship.	245:36	21:58	24 Nov 1969
Crew entered mobile quarantine facility.	245:44	22:06	24 Nov 1969
CM lifted from water.	246:24	22:46	24 Nov 1969
CM secured to quarantine facility.	247:53	00:15	25 Nov 1969
CM hatch opened.	248:18	00:40	25 Nov 1969
Sample containers 1 and 2 removed from CM.	249:30	01:52	25 Nov 1969
Container 1 removed from mobile quarantine facility.	250:52	03:14	25 Nov 1969
Container 1, controlled temperature shipping container 1, and film flown to Samoa.	254:18	06:40	25 Nov 1969
Container 2 removed from mobile quarantine facility.	255:49	08:11	25 Nov 1969

Apollo 12 Timeline

Event	GET (hhh:mm:ss)	GMT Time	GMT Date
Container 2, remainder of biological samples and film flown to Samoa.	259:08	11:30	25 Nov 1969
Container 1, controlled temperature shipping container 1, and film arrived in Houston.	268:23	20:45	25 Nov 1969
CM hatch secured and decontaminated.	270:01	22:23	25 Nov 1969
Mobile quarantine facility secured after removal of transfer tunnel.	271:08	23:30	25 Nov 1969
Container 2, remainder of biological samples, and film arrived in Houston.	276:26	04:48	26 Nov 1969
Mobile quarantine facility and CM offloaded in Hawaii.	345:56	02:18	29 Nov 1969
Safing of CM pyrotechnics completed.	352:18	08:40	29 Nov 1969
Mobile quarantine facility arrived at Ellington Air Force Base.	355:28	11:50	29 Nov 1969
Flight crew in Lunar Receiving Laboratory.	357:28	13:50	29 Nov 1969
Deactivation of CM fuel and oxidizer completed.	405:53	14:15	1 Dec 1969
CM delivered to Lunar Receiving Laboratory.	435:08	19:30	2 Dec 1969

APOLLO 13

The Seventh Mission:
The Third Lunar Landing Attempt

Apollo 13 Summary

(11 April–17 April 1970)

Original Apollo 13 crew (l. to r.): Jim Lovell, Ken Mattingly, Fred Haise (NASA S69-62224).

Preflight portrait of Apollo 13 flight crew (l. to r.) Haise, Jack Swigert, Lovell (NASA 70-H-724).

Portrait of crew taken after mission (l. to r.): Lovell, Swigert, Haise (NASA S70-36485).

[1] Major Charles Moss Duke, Jr. (USAF).

Background

Apollo 13 was planned as a Type H mission, a precision piloted lunar landing demonstration and systematic lunar exploration. It was, however, aborted during translunar flight because of the loss of all the oxygen stored in two tanks in the service module.

The primary objectives were:

- to perform selenological inspection, survey, and sampling of materials in a preselected region of the Fra Mauro formation;

- to deploy and activate an Apollo lunar surface experiments package;

- to further develop human capability to work in the lunar environment; and

- to obtain photographs of candidate exploration sites.

The crew members were Captain James Arthur Lovell, Jr. (USN), commander; John Leonard "Jack" Swigert, Jr. [SWY-girt], command module pilot; and Fred Wallace Haise, Jr., lunar module pilot. Swigert was backup command module pilot, but Lt. Commander Thomas Kenneth "Ken" Mattingly, II (USN), the prime command module pilot, had been exposed to rubella (German measles) by a member of the backup crew[1] eight days before the scheduled launch date, and results of his pre-mission physical examination revealed he had no immunity to the disease. Consequently, on the day prior to launch, and after several days of intense training with the prime crew, Swigert was named to replace Mattingly.

Selected as an astronaut in 1962, Lovell was making his fourth spaceflight and second trip to the Moon, the first person ever to achieve those milestones. He had been pilot of Gemini 7, command pilot of Gemini 12, and command module pilot of Apollo 8, the first piloted mission to the Moon. Lovell was born 25 March 1928 in Cleveland, Ohio, and was 42 years old at the time of the Apollo 13 mission. He received a B.S. from the U.S. Naval Academy in 1952. His backup for the mission was Commander John Watts Young (USN).

The original command module pilot, Mattingly would have been making his first spaceflight. Born 17 March 1936 in Chicago, Illinois, he was 34 years old at the time of the Apollo 13 mission. He received a B.S. in aeronautical engineering from Auburn University in 1958, and was selected as an astronaut in 1966.

Swigert was making his first spaceflight. Born 30 August 1931 in Denver, Colorado, he was 38 years old at the time of the Apollo 13 mission. Swigert received a B.S. in mechanical engineering from the University of Colorado in 1953, an M.S. in aerospace science from Rensselaer Polytechnic Institute in 1965, and an M.B.A. from the University of Hartford in 1967. He was selected as an astronaut in 1966.[2]

Haise was also making his first spaceflight. Born in Biloxi, Mississippi, on 14 November 1993, he was 36 years old at the time of the Apollo 13 mission. Haise received a B.S. in aeronautical engineering from the University of Oklahoma in 1959, and was selected as an astronaut in 1966. His back-up was Major Charles Moss Duke, Jr. (USAF).

The capsule communicators (CAPCOMs) for the mission were Joseph Peter Kerwin, M.D., Vance DeVoe Brand, Major Jack Robert Lousma (USMC), Young, and Mattingly. The support crew were Lousma, Brand, and Major William Reid Pogue (USAF). The flight directors were Milton L. Windler (first shift), Gerald D. Griffin (second shift), Eugene F. Kranz (third shift), and Glynn S. Lunney (fourth shift).

The Apollo 13 launch vehicle was a Saturn V, designated SA-508. The mission also carried the designation Eastern Test Range #3381. The CSM was designated CSM-109, and had the call-sign "Odyssey." The lunar module was designated LM-7, and had the call-sign "Aquarius."

Launch Preparations

The terminal countdown was picked up at T-28 hours at 05:00:00 GMT on 10 April. Scheduled holds were 9 hours 13 minutes at T-9 hours and one hour duration at T-3 hours 30 minutes.

At launch time, a cold front extended from a low pressure cell in the North Atlantic, becoming stationary through northern Florida and along the Gulf Coast to a low pressure area located in southern Louisiana. The frontal intensity was weak in northern Florida but became stronger in the northwestern Gulf of Mexico/Louisiana area. Surface winds in the Kennedy Space Center area were light and variable. Generally, winds in the lower part of the troposphere were light, permitting the sea breeze to switch the surface wind to the east southeast by early afternoon. Altocumulus clouds covered 40 percent of the sky (base 19,000 feet) and cirrostratus 100 percent (base 26,000 feet), the temperature was 75.9° F, the relative humidity was 57 percent; and the barometric pressure was 14.676 lb/in². The winds, as measured by the anemometer on the light pole

60.0 feet above ground at the launch site measured 12.2 knots at 105° from true north.

Ascent Phase

Apollo 13 was launched from Kennedy Space Center Launch Complex 39, Pad A, at a Range Zero time of 19:13:00 GMT (02:13:00 p.m. EST) on 11 April 1970. The planned launch window extended to 22:36:00 GMT to take advantage of a sun elevation angle on the lunar surface of 10.0°.

Apollo 13 lifts off from Kennedy Space Center Pad 39A (NASA S70-34853).

Between 000:00:12.6 and 000:00:32.1, the vehicle rolled from a launch pad azimuth of 90° to a flight azimuth of 72.043°. The S-IC engine shut down at 000:02:43.60, followed by S-IC/S-II separation, and S-II engine ignition. Due to high amplitude oscillations in the propulsion/structural system, the S-II center engine shut down at 000:05:30.64, 2 minutes 12 seconds earlier than planned. The early shutdown caused considerable deviations from the planned trajectory. The altitude at shutdown was 10.7 n mi lower and the velocity was 5,685.3 ft/sec slower than expected.

[2] Swigert died of complications from bone marrow cancer treatments on 27 December 1982 in Washington, DC.

With CAPCOM Joe Kerwin (r.), original Apollo 13 CMP Ken Mattingly monitors communications during liftoff (NASA S70-34628).

The S-II engine shut down at 000:09:52.64 followed by separation from the S-IVB, which ignited at 000:09:56.90, both 34 seconds late. The first S-IVB engine cutoff occurred 44 seconds late, at 000:12:29.83, with deviations from the planned trajectory of only -1.9 ft/sec in velocity and only 0.2 n mi in altitude.

The S-IC stage impacted the Atlantic Ocean at 000:09:06.9 at latitude 30.177° north and longitude 74.065° west, 355.3 n mi from the launch site. The S-II stage impacted the Atlantic Ocean at 000:20:58.1 at latitude 32.320° north and longitude 33.289° west, 2,452.6 n mi from the launch site.

The maximum wind conditions encountered during ascent were 108.13 knots at 252° from true north at 44,540 feet, and a maximum wind shear of 0.0166 sec⁻¹ at 50,610 feet.

Despite the early shutdown of the S-II center engine, parking orbit conditions at insertion, 000:12:39.83 (S-IVB cutoff plus 10 seconds to account for engine tailoff and other transient effects), showed a nearly nominal apogee and perigee of 100.3 by 99.3 n mi, a period of 88.19 minutes, an inclination of 32.547°, and a velocity of 25,565.9 ft/sec. The apogee and perigee were based upon a spherical Earth with a radius of 3,443.934 n mi.

The international designation for the CSM upon achieving orbit was 1970-029A and the S-IVB was designated 1970-029B. After undocking prior to Earth entry, the LM would be designated 1970-029C.

After orbital insertion, all launch vehicle and spacecraft systems were verified and preparations were made for translu-

nar injection. Onboard television was initiated at 001:35 for about five-and-a-half minutes.

The 350.75-second translunar injection maneuver (second S-IVB firing) was performed at 002:35:46.30. The S-IVB engine shut down at 002:41:37.15 and translunar injection occurred ten seconds later, after 1,5 Earth orbits lasting 2 hours 29 minutes 7.3 seconds, at a velocity of 35,562.7 ft/sec.

Translunar Phase

At 003:06:38.9, the CSM was separated from the S-IVB stage and onboard television was initiated at 003:09 for about 72 minutes to show the docking, ejection, and interior and exterior views of the CM. Transposition and docking with the LM occurred at 003:19:08.8. The docked spacecraft were ejected from the S-IVB at 004:01:00.8, and an 80.2-second separation maneuver was initiated by the S-IVB auxiliary propulsion system at 004:18:00.6.

On previous lunar missions, the S-IVB stage had been maneuvered by ground command into a trajectory such that it would pass by the Moon and go into a solar orbit. For Apollo 13, the S-IVB was targeted to hit the Moon so that the vibrations resulting from the impact could be sensed by the Apollo 12 seismic station and telemetered to Earth for study.

A 217.2-second lunar impact maneuver was made at 005:59:59.5. The S-IVB impacted the lunar surface at 077:56:39.7. The seismic signals lasted three hours 20 minutes, and were so strong that the Apollo 12 seismometer gain had to be reduced to keep the recording on the scale. The suprathermal ion detector recorded a jump in the number of ions from zero at impact to 2,500 and then back to zero. It was theorized that the impact drove particles from the lunar surface up to 200,000 feet above the moon, where they were ionized by sunlight. The impact point was latitude 2.5° south and longitude 27.9° west, 35.4 n mi from the target point and 75 n mi from the Apollo 12 seismometer. At impact, the S-IVB weighed 29,599 pounds and was traveling 2,579 ft/sec.

Good quality television coverage of the preparations and performance of the second midcourse correction burn was received for 49 minutes beginning at 030:13.

Photographs of Earth were taken during the early part of translunar coast to support an analysis of atmospheric winds. At 030:40:49.65, a 3.49-second midcourse correction lowered the closest point of spacecraft approach to the

Moon to an altitude of 60 miles. Before this maneuver, the spacecraft had been on a free-return trajectory, in which the spacecraft would have looped around the Moon and returned to Earth without requiring a major maneuver.

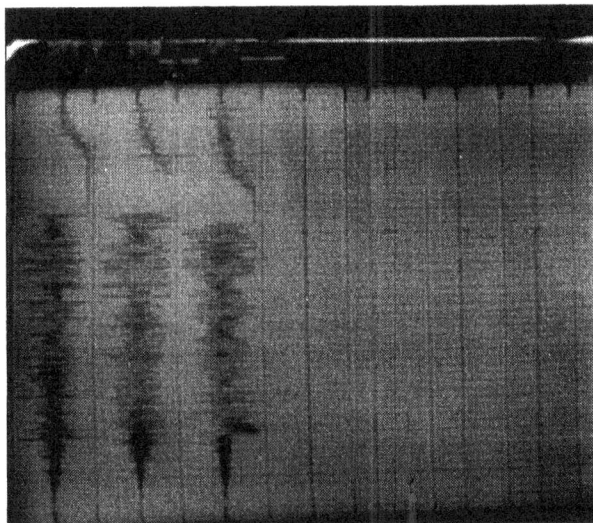

Seismic recording of the S-IVB stage impacting the lunar surface as planned (NASA S70-34985).

Through the first 46 hours of the mission, telemetered data and crew observations indicated that the performance of oxygen tank 2 was normal. At 046:40:02, the crew routinely turned on the fans in oxygen tank 2. Within three seconds, the oxygen tank 2 quantity indication changed from a normal reading of about 82 percent full to an obviously incorrect "off-scale high" reading of over 100 percent. Analysis of the electrical wiring of the quantity gauge shows that this erroneous reading could have been caused by either a short circuit or an open circuit in the gauge wiring or a short circuit between the gauge plates. Subsequent events indicated that a short was the more likely failure mode.

At 047:54:50 and at 051:07:44, the oxygen tank 2 fans were turned on again, with no apparent adverse effects. The quantity gauge continued to read off-scale high.

Following a rest period, the Apollo 13 crew began preparations for activating and powering up the LM for checkout. At 053:27, the commander and lunar module pilot were cleared to enter the LM to commence inflight inspection of the LM. A television transmission of the spacecraft interior started at 055:14 and ended at 055:46. The crew moved back into the CM and the LM hatch was closed at 055:50.

At 055:52:31, a master alarm on the CM caution and warning system alerted the crew to a low pressure indica-

tion in the cryogenic hydrogen tank 1. This tank had reached the low end of its normal operating pressure range several times previously during the flight. At 055:52:58, flight controllers requested the crew to turn on the cryogenic system fans and heaters.

The command module pilot acknowledged the fan cycle request at 55:53:06, and data indicate that current was applied to the oxygen tank 2 fan motors at 055:53:20, followed by a power transient in the stabilization control system.

About 90 seconds later, at 055:54:53.555, telemetry from the spacecraft was lost almost totally for 1.8 seconds. During the period of data loss, the caution and warning system alerted the crew to a low voltage condition on DC main bus B. At about the same time, the crew heard a loud "bang" and realized that a problem existed in the spacecraft.

When the crew heard the bang and got the master alarm for low DC main bus B voltage, the commander was in the lower equipment bay of the command module, stowing the television camera which had just been in use. The lunar module pilot was in the tunnel between the CSM and the LM, returning to the CSM. The command module pilot was in the left-hand couch, monitoring spacecraft performance. Because of the master alarm indicating low voltage, the command module pilot moved across to the right-hand couch where CSM voltages can be observed. He reported that voltages were "looking good" at 055:56:10 and also reported hearing "…a pretty good bang…" a few seconds before. At this time, DC main bus B had recovered and fuel cell 3 did not fail for another 90 seconds. He also reported fluctuations in the oxygen tank 2 quantity, followed by a return to the off-scale high position.

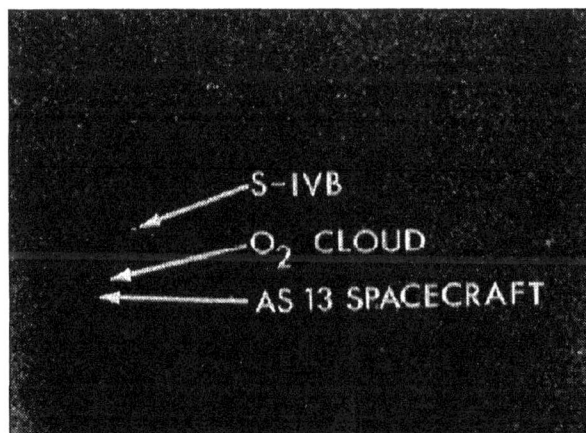

Telescopic photograph showing the Apollo 13 spacecraft, S-IVB stage and oxygen cloud formed following the SM explosion (NASA S70-34857).

The commander reported, "...We're venting something...into space..." at 056:09:07, followed at 056:09:58 by the lunar module pilot's report that fuel cell 1 was off-line. Less than half an hour later, he reported that fuel cell 3 was also off-line.

When fuel cells' 1 and 3 electrical output readings went to zero, the ground controllers could not be certain that the cells had not somehow been disconnected from their respective busses and were not otherwise alright. Attention continued to be focused on electrical problems.

Astronaut Al Shepard, scheduled to command Apollo 14, monitors communications between crew and ground regarding oxygen cell failure (NASA S70-34904).

Five minutes after the accident, controllers asked the crew to connect fuel cell 3 to DC main bus B in order to be sure that the configuration was known. When it was realized that fuel cells 1 and 3 were not functioning, the crew was directed to perform an emergency powerdown to lower the load on the remaining fuel cell. Fuel cell 2 was shut down at 058:00, followed 10 minutes later by powerdown of the CM computer and platform.

Observing the rapid decay in oxygen tank 1 pressure, controllers asked the crew to switch power to the oxygen tank 2 instrumentation. When this was done, and it was realized that oxygen tank 2 had failed, the extreme seriousness of the situation became clear.

Several attempts were then made to save the remaining oxygen in oxygen tank 1, but the pressure continued to decrease. It was obvious by about 90 minutes after the accident that the oxygen tank 1 leak could not be stopped and that shortly it would be necessary to use the LM as a "lifeboat" for the remainder of the mission. The resultant loss of oxygen made the three fuel cells inoperative. This

left the CM batteries, normally used only during reentry, as the sole power source. The only oxygen left was contained in a surge tank and repressurization packages used to repressurize the CM after cabin venting. The LM became the only source of sufficient electrical power and oxygen to permit a safe return to Earth, and led to the decision to abort the Apollo 13 mission. By 058:40, the LM had been activated, the inertial guidance reference transferred from the CSM guidance system to the LM guidance system, and the CSM systems were turned off.

From l. to r., Director of Flight Crew Operations Donald "Deke" Slayton and astronauts Ken Mattingly, Vance Brand, Jack Lousma, and John Young evaluate Apollo 13's situation (NASA S70-34902).

The remainder of the mission was characterized by two main activities: planning and conducting the necessary propulsion maneuvers to return the spacecraft to Earth, and managing the use of consumables in such a way that the LM, which is designed for a basic mission with two crew members for a relatively short duration, could support three crew members and serve as the actual control vehicle for the time required.

A number of propulsion options were developed and considered. It was necessary to return the spacecraft to a free-return trajectory and to make any required midcourse corrections. Normally, the SM service propulsion system would be used for such maneuvers. However, because of the high electrical power requirements for that engine, and in view of its uncertain condition and the uncertain nature of the structure of the SM after the accident, it was decided to use the LM descent engine if possible.

The spacecraft was then maneuvered back into a free-return trajectory at 061:29:43.49 by firing the LM descent engine for 34.23 seconds. It then looped behind the Moon and was out of contact with the Earth tracking stations between 077:08:35 and 077:33:10, a total of 24 minutes 35 seconds.[3]

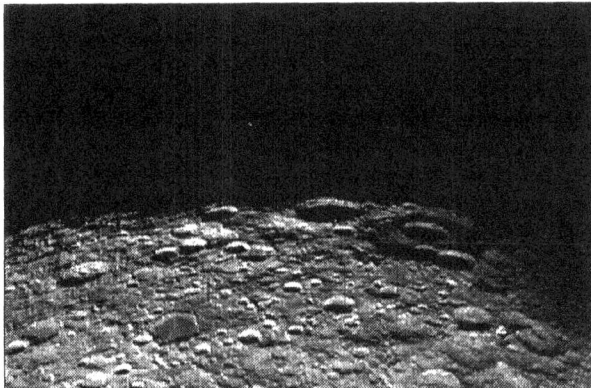

Crater IAU 221 on lunar farside (center of photo on horizon) (NASA AS13-62-8918).

Flight controllers calculated that the minimum practical return time for Apollo 13 was 133 hours total mission time to the Atlantic Ocean, and the maximum was 152 hours to the Indian Ocean. Since recovery forces were deployed in the Pacific, a return path was selected for splashdown there at 142:40.

A 263.82-second transearth injection maneuver using the LM descent propulsion system was executed at 079:27:38.95 to speed up the return to Earth by 860.5 ft/sec after the docked spacecraft had swung around the far side of the Moon.

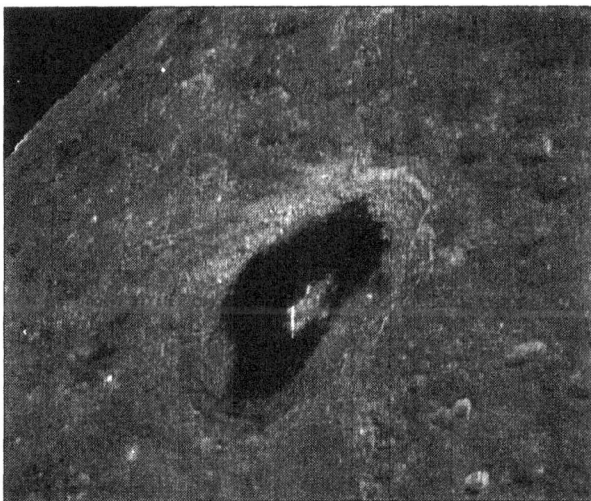

Lunar farside, showing crater Tsiolkovsky (NASA AS13-60-8659).

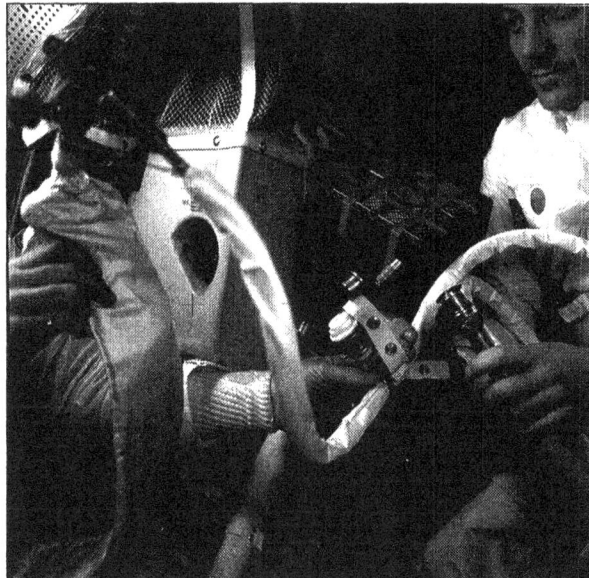

Swigert with temporary hose connections and apparatus required when the crew moved from the CM to the LM (NASA AS13-62-9004).

Guidance errors during the transearth injection maneuver necessitated a 14-second transearth midcourse correction of 7.8 ft/sec, using the descent propulsion system at 105:18:42.0 to bring the projected entry flight-path angle within the specified limits. During the transearth coast period, the docked spacecraft were maneuvered into a passive thermal control mode.

The most critical consumables were water, used to cool the CSM and LM systems during use; CSM and LM battery power, the CSM batteries being for use during reentry and the LM batteries being needed for the rest of the mission; LM oxygen for breathing; and lithium hydroxide (LiOH) filter canisters used to remove carbon dioxide from the spacecraft cabin atmosphere.

These consumables, and in particular the water and LiOH canisters, appeared to be extremely marginal in quantity shortly after the accident, but once the LM was powered down to conserve electric power and to generate less heat and thus use less water, the situation improved greatly. Engineers in Houston developed a method that allowed the crew to use materials on board to fashion a device allowing use of the CM LiOH canisters in the LM cabin atmosphere cleaning system. At splashdown, many hours of each consumable remained available.

The unprecedented powered-down state of the CM required several new procedures for entry. The CM was

[3] Source of lunar occultation times unknown, but appear to be more accurate expressions of times in Apollo 13 Mission Operations Report, p. III-26. 1992 Guinness Book of World Records, page 118, states that Apollo 13 holds the record for farthest distance traveled from Earth: 248,655 st mi at 1:21 a.m. British Daylight Time 15 April 1970 at 158 miles above the Moon, the equivalent of 216,075 n mi 00:21 GMT 15 April 1970 (08:21 p.m. EST, 14 April) at an apolune of 137 n mi.

briefly powered up to assess the operational capability of critical systems. Also, the CM entry batteries were charged through the umbilical connectors that had supplied power from the LM while the CM was powered down.

Box used to house CM lithium hydroxide canister used to purge carbon dioxide from the LM "lifeboat" (NASA AS13-62-8929).

Approximately six hours before entry, the passive thermal control mode was discontinued, and a final midcourse correction was made using the LM reaction control system to refine the flight-path angle slightly. The 21.50-second maneuver of 3.0 ft/sec was made at 137:40:13.00.

View of Mission Operations Control Room (MOCR) during transearth flight (NASA S70-34986).

Less than half an hour later, at 138:01:48.0, the service module was jettisoned, which afforded the crew an opportunity to observe and photograph the damage caused by the failed oxygen tank.

The crew viewed the SM and reported that an entire panel was missing near the S-band high-gain antenna, the fuel cells on the shelf above the oxygen shelf were tilted, the high-gain antenna was damaged, and a great deal of debris was exposed.

View of damaged service module taken from 16 mm film (NASA S70-35703).

The LM was retained until 141:30:00.2, about 70 minutes before entry, to minimize usage of CM electrical power. At undocking, normal tunnel pressure provided the necessary force to separate the two spacecraft. All other events were the same as a normal mission.

View of LM following jettison, about an hour prior to splashdown (NASA AS13-59-8562).

View of Earth during transearth flight. Visible are parts of southwestern U.S. and northwestern Mexico. Baja California is clearly seen (NASA AS13-60-8588).

Recovery

The command module reentered the Earth's atmosphere (400,000 feet altitude) at 142:40:45.7 at a velocity of 36,210.6 ft/sec, following a transearth coast of 63 hours 8 minutes 42.9 seconds. Some pieces of the LM survived entry and projected trajectory data indicated that they struck the open sea between Samoa and New Zealand.

Flight controllers gather around the console of Shift 4 Flight Director Glynn Lunney to review weather maps of the proposed splashdown site in the south Pacific Ocean (NASA S70-35014).

After a harrowing mission, the Apollo 13 CM finally splashes down in the Pacific (NASA S70-35638).

The parachute system effected splashdown of the CM in the Pacific Ocean at 18:07:41 GMT (01:07:41 p.m. EST) on 17 April. Mission duration was 142:54:41.

Swigert (l.) and Haise (center) in life raft as Navy team assists crew from the CM. Lovell is exiting through hatch (NASA S70-35610).

The impact point was about 1.0 n mi from the target point and 3.5 n mi from the recovery ship U.S.S. *Iwo Jima*. The splashdown site was at latitude 21.63° south and longitude 165.37° west. After splashdown, the CM assumed an apex-up flotation attitude. The crew was retrieved by helicopter and aboard the recovery 45 minutes after splashdown.

Crew exits recovery helicopter aboard U.S.S. *Iwo Jima* (NASA KSC-70PC-0130).

The CM was recovered 43 minutes later. The estimated CM weight at splashdown was 11,133 pounds, and the estimated distance traveled for the mission was 541,103 n mi.

CM is loaded aboard the recovery ship (NASA S70-35632).

The crew departed the *Iwo Jima* by aircraft at 18:20 GMT on 18 April and arrived in Houston 03:30 GMT on 20 April. The *Iwo Jima* arrived with the CM at Hawaii at 19:30 GMT on 24 April. Deactivation was completed on 26 April.

The CM was delivered to the North American Rockwell Space Division facility in Downey, California, for postflight analysis, arriving at 14:00 GMT on 27 April.

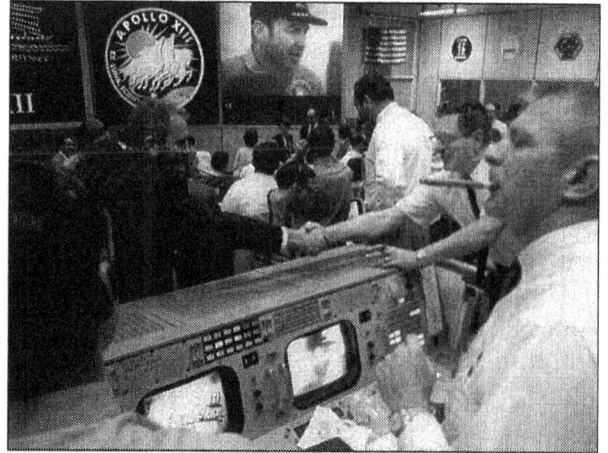

Flight Director Gene Kranz relaxes after the safe return of the Apollo 13 crew (NASA S70-35145).

Former Apollo Program Director Lt. Gen. Sam Phillips (left) NASA Administrator Dr. Thomas Paine (center), and Dr. George Low celebrate safe return of Apollo 13 crew (NASA S70-35148).

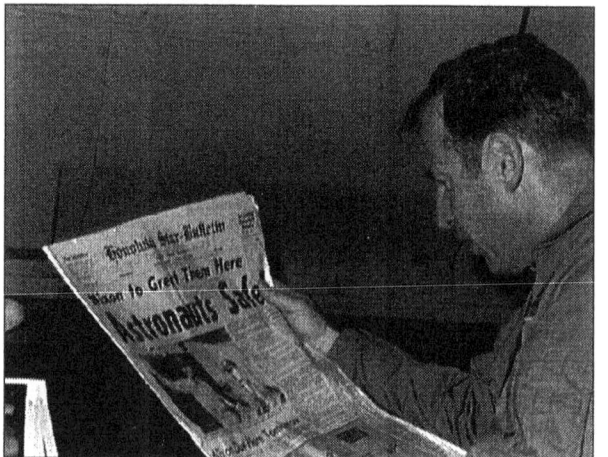

Lovell reads newspaper account of Apollo 13 recovery (NASA S70-15501).

Conclusions

The Apollo 13 accident was nearly catastrophic. Only outstanding performances on the part of the crew, ground support personnel, and the excellent performance of the LM systems made the safe return of the crew possible.

President Richard M. Nixon awards the Presidential Medal of Freedom to the Apollo 13 crew at Hickam AFB, Hawaii (NASA S70-15511).

The following conclusions were made from an analysis of post-mission data:

1. The mission was aborted because of the total loss of primary oxygen in the service module. This loss resulted from an incompatibility between switch design and pre-mission procedures, a condition which, when combined with an abnormal pre-mission detanking procedure, caused an inflight shorting and a rapid oxidation within one of two redundant storage tanks. The oxidation then resulted in a loss of pressure integrity in the related tank and eventually in the remaining tank.

2. The concept of a backup crew was proven for the first time when, three days prior to launch, the backup command module pilot was substituted for his prime crew counterpart, who was exposed and found susceptible to rubella (German measles).

3. The performance of lunar module systems demonstrated an emergency operational capability. Lunar module systems supported the crew for a period twice their intended design lifetime.

4. The effectiveness of pre-mission crew training, especially in conjunction with ground personnel, was reflected in the skill and precision with which the crew responded to the emergency.

5. Although the mission was not a complete success, a lunar flyby mission, including three planned experiments (lightning phe-

nomena, Earth photography, and S-IVB lunar impact), were completed and data were derived with respect to the capabilities of the lunar module.

Report of the Apollo 13 Review Board

On 17 April 1970, NASA Administrator Thomas O. Paine established the Apollo 13 Review Board, naming Edgar M. Cortright, director of the NASA Langley Research Center, as chairman. Cortright's eight-member panel met for nearly two months, and submitted their final report on 15 June. Neil Armstrong, commander of the recent Apollo 11 mission, was the only astronaut on the board. William Anders, lunar module pilot of Apollo 8, and executive secretary of the National Aeronautics and Space Council, was one of three observers.

The evidence pointed strongly to an electrical short circuit with arcing as the initiating event. About 2.7 seconds after the fans were turned on in the SM oxygen tanks, an 11.1-ampere current spike and simultaneously a voltage-drop spike were recorded in the spacecraft electrical system. Immediately thereafter, current drawn from the fuel cells decreased by an amount consistent with the loss of power to one fan. No other changes in spacecraft power were being made at the time. No power was on the heaters in the tanks (the quantity gauge and temperature sensor were very low power devices). The next anomalous event recorded was the beginning of a pressure rise in oxygen tank 2 thirteen seconds later. Such a time lag was possible with low-level combustion at the time. These facts pointed to the likelihood that an electrical short circuit with arcing occurred in the fan motor or its wires to initiate the accident sequence. The energy available from the short circuit was probably 10 to 20 joules. Tests conducted during the investigation showed that this energy is more than adequate to ignite Teflon of the type contained within the tank. This likelihood of electrical initiation is enhanced by the high probability that the electrical wires within the tank were damaged during abnormal tanking operations at KSC prior to launch.

Data were not adequate to determine precisely the way in which the oxygen tank 2 system lost its integrity. However, available information, analyses, and tests performed during this investigation indicate that most probably combustion within the pressure vessel ultimately led to localized heating and failure at the pressure vessel closure. It is at this point, the upper end of the quantity probe, that the Inconel conduit is located, through which the Teflon-insulated wires enter the pressure vessel. It is likely that the combustion progressed along the wire insulation and reached this location where all of the wires come together. This, possibly augmented by ignition of the metal in the upper end of the probe, led to weakening and failure of the closure or the conduit, or both.

Failure at this point would lead immediately to pressurization of the tank dome, which is equipped with a rupture disc rated at about 75 psi. Rupture of this disc or of the entire dome would then release oxygen, accompanied by combustion products, into bay 4. Spacecraft accelerations recorded at this time were probably caused by this release.

Release of the oxygen then began to pressurize the oxygen shelf space of bay 4. If the holes formed in the pressure vessel were large enough and formed rapidly enough, the escaping oxygen alone would be adequate to blow off the bay 4 panel. However, it is also quite possible that the escape of oxygen was accompanied by combustion of Mylar and Kapton (used extensively as thermal insulation in the oxygen shelf compartment and in the tank dome), which would augment the pressure caused by the oxygen itself. The slight temperature increases recorded at various SM locations indicate that combustion external to the tank probably took place. The ejected panel then struck the high-gain antenna, disrupting communications from the spacecraft for the 1.8 seconds.

How the Problem Occurred

Following is a list of factors that led to the accident:

• After assembly and acceptance testing, oxygen tank 2 that flew on Apollo 13 was shipped from Beech Aircraft Corporation to North American Rockwell (NR) in apparently satisfactory condition.

• It is now known, however, that the tank contained two protective thermostatic switches on the heater assembly, which were inadequate and would subsequently fail during ground test operations at Kennedy Space Center (KSC).

• In addition, it is probable that the tank contained a loosely fitting fill tube assembly. This assembly was probably displaced during subsequent handling, which included an incident at the prime contractor's arc plant in which the tank was jarred.

• In itself, the displaced fill tube assembly was not particularly serious, but it led to the use of improvised detanking procedures at KSC which almost certainly set the stage for the accident.

• Although Beech did not encounter any problem in detanking during acceptance tests, it was not possible to detank oxygen tank 2 using normal procedures at KSC. Tests and analyses indi-

cated that this was due to gas leakage through the displaced fill tube assembly.

• The special detanking procedures at KSC subjected the tank to an extended period of heater operation and pressure cycling. These procedures had not been used before, and the tank had not been qualified by test for the conditions experienced. However, the procedures did not violate the specifications that governed the operation of the heaters at KSC.

• In reviewing these procedures before the flight, officials of NASA, NR, and Beech did not recognize the possibility of damage due to overheating. Many of these officials were not aware of the extended heater operation. In any event, adequate thermostatic switches might have been expected to protect the tank.

• A number of factors contributed to the presence of inadequate thermostatic switches in the heater assembly. The original 1962 specifications from NR to Beech Aircraft Corporation for the tank and heater assembly specified the use of 28 V DC power, which is used in the spacecraft. In 1965, NR issued a revised specification which stated that the heaters should use a 65 V DC power supply for tank pressurization; this was the power supply used at KSC to reduce pressurization time. Beech ordered switches for the Block II tanks but did not change the switch specifications to be compatible with 65 V DC.

• The thermostatic switch discrepancy was not detected by NASA, NR, or Beech in their review of documentation, nor did tests identify the incompatibility of the switches with the ground support equipment at KSC, since neither qualification nor acceptance testing required switch cycling under load as should have been done. It was a serious oversight in which all parties shared.

• The thermostatic switches could accommodate the 65 V DC during tank pressurization because they normally remained cool and closed. However, they could not open without damage with 65 V DC power applied. They were never required to do so until the special detanking. During this procedure, as the switches started to open when they reached their upper temperature limit, they were welded permanently closed by the resulting arc and were rendered inoperative as protective thermostats.

• Failure of the thermostatic switches to open could have been detected at KSC if switch operation had been checked by observing heater current readings on the oxygen tank heater control panel. Although it was not recognized at that time, the tank temperature readings indicated that the heaters had reached their temperature limit and switch opening should have been expected.

• As shown by subsequent tests, failure of the thermostatic switches probably permitted the temperature of the heater tube assembly to reach about 1,000° F in spots during the continuous eight-hour period of heater operation. Such heating has been shown by tests to severely damage the Teflon insulation on the fan motor wires in the vicinity of the heater assembly. From that time on, including pad occupancy, the oxygen tank 2 was in a hazardous condition when filled with oxygen and electrically powered.

• It was not until nearly 56 hours into the mission, however, that the fan motor wiring, possibly moved by the fan stirring, short circuited and ignited its insulation by means of an electric arc. The resulting combustion in the oxygen tank probably overheated and failed the wiring conduit where it enters the tank, and possibly a portion of the tank itself.

• The rapid expulsion of high-pressure oxygen which followed, possibly augmented by combustion of insulation in the space surrounding the tank, blew off the outer panel to bay 4 of the SM, caused a leak in the high-pressure system of oxygen tank 1, damaged the high-gain antenna, caused other miscellaneous damage, and aborted the mission.

Apollo 13 Objectives

Spacecraft Primary Objectives

1. To perform selenological inspection, survey, and sampling of materials in a preselected region of the Fra Mauro formation. *Not attempted.*

2. To deploy and activate an Apollo lunar surface experiments package. *Not attempted.*

3. To further develop human capability to work in the lunar environment. *Not attempted.*

4. To obtain photographs of candidate exploration sites. *Not attempted.*

Detailed Objectives

1. Television coverage. *Not attempted.*

2. Contingency sample collection. *Not attempted.*

3. Selected sample collection. *Not attempted.*

4. Landing accuracy improvement techniques. *Not attempted.*

5. Photographs of candidate exploration sites. *Not attempted.*

6. Extravehicular communication system performance. *Not attempted.*

7. Lunar soil mechanics. *Not attempted.*

8. Dim light photography. *Not attempted.*

9. Selenodetic reference point update. *Not attempted.*

10. CSM orbital science photography. *Not attempted.*

11. Transearth lunar photography. *Not attempted.*

12. EMU water consumption measurement. *Not attempted.*

13. Thermal coating degradation. *Not attempted.*

Experiments

1. ALSEP III: Apollo Lunar Surface Experiments Package. *Not attempted.*

    a. Passive seismic experiment.

    b. Heat flow experiment.

    c. Charged particle lunar environment experiment.

    d. Cold cathode gauge experiment.

    e. Lunar dust detection.

2. S-059: Lunar field geology. *Not attempted.*

3. S-080: Solar wind composition. *Not attempted.*

4. S-164: S-band transponder exercise. *Not attempted.*

5. S-170: Downlink bistatic radar observations of the Moon. *Not attempted.*

6. S-178: Gegenschein from lunar orbit. *Not attempted.*

7. S-184: Lunar surface close-up photography. *Not attempted.*

8. T-029: Pilot describing function. *Achieved.*

Launch Vehicle Objectives

1. To launch on a flight azimuth between 72° and 96° and insert the S-IVB/instrument unit/spacecraft into the planned circular Earth parking orbit. *Achieved.*

2. To restart the S-IVB during either the second or third revolution and inject the S-IVB/instrument unit/spacecraft into the planned translunar trajectory. *Achieved.*

3. To provide the required attitude control for the S-IVB/instrument unit/spacecraft during transposition, docking, and ejection. *Achieved.*

4. To perform an evasive maneuver after ejection of the command and service module/lunar module from the S-IVB/instrument unit. *Achieved.*

5. To attempt to impact the S-IVB/instrument unit on the lunar surface within 350 kilometers (189 nautical miles) of latitude 3° south, longitude 30° west. *Achieved.*

6. To determine actual impact point within 5.0 kilometers (2.7 nautical miles) and time of impact within one second. *Achieved.*

7. To vent and dump the remaining gases and propellants to safe the S-IVB/instrument unit. *Achieved.*

Apollo 13 Spacecraft History

EVENT	DATE
Individual and combined CM and SM systems test completed at factory.	16 Mar 1969
Integrated CM and SM systems test completed at factory.	08 Apr 1969
LM 7 final engineering evaluation acceptance test at factory.	18 May 1969
LM 7 integrated test at factory.	18 May 1969
Saturn S-IVB stage 508 delivered to KSC.	13 Jun 1969
Saturn S-IC stage 8 delivered to KSC.	16 Jun 1969
Saturn S-IC stage 8 erected on MLP 3.	18 Jun 1969
LM ascent stage 7 ready to ship from factory to KSC.	24 Jun 1969
CM 109 and SM 109 ready to ship from factory to KSC.	25 Jun 1969
LM descent stage 7 ready to ship from factory to KSC.	25 Jun 1969
CM 109 and SM 109 delivered to KSC.	26 Jun 1969
LM ascent stage 7 delivered to KSC.	27 Jun 1969
LM descent stage 7 delivered to KSC.	28 Jun 1969
Saturn S-II stage 8 delivered to KSC.	29 Jun 1969
CM 109 and SM 109 mated.	30 Jun 1969
CSM 109 combined systems test completed.	07 Jul 1969
Saturn V instrument unit 508 delivered to KSC.	07 Jul 1969
LM ascent stage 7 and descent stage 7 mated.	15 Jul 1969
Saturn S-II stage 8 erected.	17 Jul 1969
Spacecraft/LM adapter 16 delivered to KSC.	18 Jul 1969
LM 7 combined systems test completed.	22 Jul 1969
Saturn S-IVB stage 508 erected.	31 Jul 1969
Saturn V instrument unit 508 erected.	01 Aug 1969
Launch vehicle electrical systems test completed.	29 Aug 1969
CSM 109 altitude tests completed.	12 Sep 1969
LM 7 altitude tests completed.	20 Sep 1969
Launch vehicle propellant dispersion/malfunction overall test completed.	21 Oct 1969
Launch vehicle service arm overall test completed.	04 Dec 1969
CSM 109 moved to VAB.	09 Dec 1969
Spacecraft erected.	10 Dec 1969
Space vehicle and MLP 3 transferred to launch complex 39A.	15 Dec 1969
CSM 109 integrated systems test completed.	05 Jan 1970
LM 7 combined systems test completed.	05 Jan 1970
CSM 109 electrically mated to launch vehicle.	18 Jan 1970
Space vehicle overall test 1 (plugs in) completed.	20 Jan 1970
LM 6 flight readiness test completed.	24 Jan 1970
Space vehicle flight readiness test completed.	26 Feb 1970
Saturn S-IC stage 8 RP-1 fuel loading completed.	16 Mar 1970
Space vehicle countdown demonstration test (wet) completed.	25 Mar 1970
Space vehicle countdown demonstration test (dry) completed.	26 Mar 1970

Apollo 13 Ascent Phase

Event	GET (hhh:mm:ss)	Altitude (n mi)	Range (n mi)	Earth Fixed Velocity (ft/sec)	Space Fixed Velocity (ft/sec)	Event Duration (deg E)	Geocentric Latitude (deg N)	Longitude (deg E)	Space Fixed Flight Path Angle (deg)	Space Fixed Heading Angle (E of N)
Liftoff	000:00:00.61	0.032	0.000	0.9	1,340.7		28.4470	-80.6041	0.04	90.00
Mach 1 achieved	000:01:08.4	4.394	1.310	1,095.2	2,087.5		28.4533	-80.5804	27.34	85.14
Maximum dynamic pressure	000:01:21.3	6.727	2.829	1,550.6	2,566.2		28.4608	-80.5529	28.98	82.96
S-IC center engine cutoff[4]	000:02:15.18	23.464	24.266	5,162.8	6,328.2	141.9	28.5677	-80.1654	23.612	76.609
S-IC outboard engine cutoff	000:02:43.60	36.392	50.991	7,787.3	9,002.5	170.3	28.6989	-79.6810	19.480	75.696
S-IC/S-II separation[4]	000:02:44.3	36.739	51.815	7,820.8	9,036.3		28.7029	-79.6660	19.383	75.693
S-II center engine 5 cutoff	000:05:30.64	86.183	298.100	11,566.6	12,859.6	164.64	29.8167	-75.1433	4.158	76.956
S-II to complete CECO[4]	000:07:42.6	97.450	580.109	15,583.8	16,904.3	132.00	30.8785	-69.8409	0.77	79.40
S-II outboard engine cutoff	000:09:52.64	102.112	964.578	21,288.0	22,610.8	426.64	31.9133	-62.4374	0.657	83.348
S-II/S-IVB separation[4]	000:09:53.50	102.150	967.505	21,301.6	22,624.5		31.9193	-62.3805	0.650	83.380
S-IVB 1st burn cutoff	000:12:29.83	103.469	1,533.571	24,236.4	25,560.4	152.93	32.5241	-51.2552	0.004	89.713
Earth orbit insertion	000:12:39.83	103.472	1,572.300	24,242.1	25,566.1		32.5249	-50.4902	0.005	90.148

Apollo 13 Earth Orbit Phase

Event	GET (hhh:mm:ss)	Space Fixed Velocity (ft/sec)	Event Duration (sec)	Velocity Change (ft/sec)	Apogee (n mi)	Perigee (n mi)	Period (mins)	Inclination (deg)
Earth orbit insertion	000:12:39.83	25,566.1		35,538.4	100.3	99.3	88.19	32.547
S-IVB 2nd burn ignition	002:35:46.30	25,573.2						
S-IVB 2nd burn cutoff	002:41:37.15	35,562.6	350.85	10,039.0				31.818

Apollo 13 Translunar Phase

Event	GET (hhh:mm:ss)	Altitude (n mi)	Space Fixed Velocity (ft/sec)	Event Duration (sec)	Velocity Change (ft/sec)	Space Fixed Flight Path Angle (deg)	Space Fixed Heading Angle (E of N)
Translunar injection	002:41:47.15	182.445	35,538.4	7.635	59.318		
CSM separated from S-IVB	003:06:38.9	3,778.582	25,029.2	45.030	72.315		
CSM docked with LM/S-IVB	003:19:08.8	5,934.90	21,881.4	51.507	79.351		
CSM/LM ejected from S-IVB	004:01:00.8	12,455.83	16,619.0	61.092	91.491		
Midcourse correction ignition (CM SPS)	030:40:49.65	121,381.93	4,682.5	77.464	112.843		
Midcourse correction cutoff	030:40:53.14	121,385.43	4,685.6	3.49	23.2	77.743	112.751
Midcourse correction ignition (LM DPS)	061:29:43.49	188,371.38	3,065.8	79.364	115.464		
Midcourse correction cutoff	061:30:17.72	188,393.19	3,093.2	34.23	37.8	79.934	116.54

[4] Only the commanded time is available for this event.

Apollo 13 Transearth Phase

Event	GET (hhh:mm:ss)	Altitude (n mi)	Space Fixed Velocity (ft/sec)	Event Duration (sec)	Velocity Change (ft/sec)	Space Fixed Flight Path Angle (deg)	Space Fixed Heading Angle (E of N)
Transearth injection ignition (LM DPS)	079:27:38.95	5,465.26	4,547.7			72.645	-116.308
Transearth injection cutoff	079:32:02.77	5,658.68	5,020.2	263.82	860.5	64.784	-117.886
Midcourse correction ignition (LM DPS)	105:18:28.0	152,224.32	4,457.8			-79.673	114.134
Midcourse correction cutoff	105:18:42.0	152,215.52	4,456.6	14.00	7.8	-79.765	114.242
Midcourse correction ignition (LM RCS)	137:39:51.5	37,808.58	10,109.1			-72.369	118.663
Midcourse correction cutoff	137:40:13.00	37,776.05	10,114.6	21.50	3.2	-72.373	118.660
SM separation	138:01:48.0	35,694.93	10,405.9			-71.941	118.824
LM jettisoned	141:30:00.2	11,257.48	17,465.9			-60.548	120.621

Apollo 13 Timeline

Event	GET (hhh:mm:ss)	GMT Time	GMT Date
Terminal countdown started at T-28 hours.	-028:00:00	05:00:00	10 Apr 1970
Scheduled 9-hour 13 minute hold at T-9 hours.	-009:00:00	00:00:00	11 Apr 1970
Countdown resumed at T-9 hours.	-009:00:00	09:13:00	11 Apr 1970
Scheduled 1-hour hold at T-3 hours 30 minutes.	-003:30:00	14:43:00	11 Apr 1970
Countdown resumed at T-3 hours 30 minutes.	-003:30:00	15:43:00	11 Apr 1970
Guidance reference release.	-000:00:16.961	19:12:43	11 Apr 1970
S-IC engine start command.	-000:00:08.9	19:12:51	11 Apr 1970
S-IC engine ignition (5).	-000:00:06.7	19:12:53	11 Apr 1970
All S-IC engines thrust OK.	-000:00:01.4	19:12:58	11 Apr 1970
Range zero.	000:00:00.00	19:13:00	11 Apr 1970
All holddown arms released (1st motion) (1.06 g).	000:00:00.3	19:13:00	11 Apr 1970
Liftoff (umbilical disconnected).	000:00:00.61	19:13:00	11 Apr 1970
Tower clearance yaw maneuver started.	000:00:02.3	19:13:02	11 Apr 1970
Yaw maneuver ended.	000:00:10.0	19:13:10	11 Apr 1970
Pitch and roll maneuver started.	000:00:12.6	19:13:12	11 Apr 1970
Roll maneuver ended.	000:00:32.1	19:13:32	11 Apr 1970
Mach 1 achieved.	000:01:08.4	19:14:08	11 Apr 1970
Maximum bending moment achieved (69,000,000 lbf-in).	000:01:16	19:14:16	11 Apr 1970
Maximum dynamic pressure (651.63 lb/ft^2).	000:01:21.3	19:14:21	11 Apr 1970
S-IC center engine cutoff command.	000:02:15.18	19:15:15	11 Apr 1970
Pitch maneuver ended.	000:02:43.3	19:15:43	11 Apr 1970
S-IC outboard engine cutoff.	000:02:43.60	19:15:43	11 Apr 1970
S-IC maximum total inertial acceleration (3.83 g).	000:02:43.70	19:15:43	11 Apr 1970
S-IC maximum Earth-fixed velocity.	000:02:44.10	19:15:44	11 Apr 1970
S-IC/S-II separation command.	000:02:44.3	19:15:44	11 Apr 1970
S-II engine start command.	000:02:45.0	19:15:45	11 Apr 1970
S-II ignition.	000:02:46.0	19:15:46	11 Apr 1970
S-II aft interstage jettisoned.	000:03:14.3	19:16:14	11 Apr 1970
Launch escape tower jettisoned.	000:03:21.0	19:16:21	11 Apr 1970
Iterative guidance mode initiated.	000:03:24.5	19:16:24	11 Apr 1970
S-IC apex.	000:04:31.7	19:17:31	11 Apr 1970
S-II center engine cutoff (S-II engine 5 cutoff 132.36 seconds early).	000:05:30.64	19:18:30	11 Apr 1970
S-II command to complete CECO.	000:07:42.6	19:20:42	11 Apr 1970
S-II maximum total inertial acceleration (1.66 g).	000:08:57.00	19:21:57	11 Apr 1970
S-IC impact (theoretical).	000:09:06.9	19:22:06	11 Apr 1970
S-II outboard engine cutoff (34.53 seconds later than planned).	000:09:52.64	19:22:52	11 Apr 1970
S-II maximum Earth-fixed velocity; S-II/S-IVB separation command.	000:09:53.50	19:22:53	11 Apr 1970
S-IVB 1st burn start command.	000:09:53.60	19:22:53	11 Apr 1970
S-IVB 1st burn ignition.	000:09:56.90	19:22:56	11 Apr 1970
S-IVB ullage case jettisoned.	000:10:05.4	19:23:05	11 Apr 1970
S-II apex.	000:10:32.2	19:23:32	11 Apr 1970
S-IVB 1st burn cutoff (9 seconds later than planned).	000:12:29.83	19:25:29	11 Apr 1970
S-IVB 1st burn maximum total inertial acceleration (0.58 g).	000:12:30.00	19:25:30	11 Apr 1970
S-IVB 1st burn maximum Earth-fixed velocity.	000:12:30.50	19:25:30	11 Apr 1970
Earth orbit insertion.	000:12:39.83	19:25:39	11 Apr 1970
Maneuver to local horizontal attitude started.	000:12:50.1	19:25:50	11 Apr 1970
Orbital navigation started.	000:14:10.4	19:27:10	11 Apr 1970
S-II impact (theoretical).	000:20:58.1	19:33:58	11 Apr 1970
TV transmission started.	001:37 20:50		11 Apr 1970
TV transmission ended.	001:43 20:56		11 Apr 1970
S-IVB 2nd burn restart preparation.	002:26:08.10	21:39:08	11 Apr 1970

Apollo 13 Timeline

Event	GET (hhh:mm:ss)	GMT Time	GMT Date
S-IVB 2nd burn restart command.	002:35:38.10	21:48:38	11 Apr 1970
S-IVB 2nd burn ignition.	002:35:46.30	21:48:46	11 Apr 1970
S-IVB 2nd burn cutoff.	002:41:37.15	21:54:37	11 Apr 1970
S-IVB 2nd burn maximum total inertial acceleration (1.43 g).	002:41:37.23	21:54:37	11 Apr 1970
S-IVB safing procedures started.	002:41:37.9	21:54:37	11 Apr 1970
Translunar injection.	002:41:47.15	21:54:47	11 Apr 1970
Maneuver to local horizontal attitude and orbital navigation started.	002:44:08	21:57:08	11 Apr 1970
S-IVB 2nd burn maximum Earth-fixed velocity.	002:53:53.6	22:06:53	11 Apr 1970
Maneuver to transposition and docking attitude started.	002:56:38.3	22:09:38	11 Apr 1970
CSM separated from S-IVB.	003:06:38.9	22:19:38	11 Apr 1970
TV transmission started.	003:09	22:22	11 Apr 1970
CSM docked with LM/S-IVB.	003:19:08.8	22:32:08	11 Apr 1970
CSM/LM ejected from S-IVB.	004:01:00.8	23:14:00	11 Apr 1970
S-IVB maneuver to evasive APS burn attitude.	004:09:00	23:22:00	11 Apr 1970
S-IVB APS evasive maneuver ignition.	004:18:00.6	23:31:00	11 Apr 1970
S-IVB APS evasive maneuver cutoff.	004:19:20.8	23:32:20	11 Apr 1970
TV transmission ended.	004:20	23:33	11 Apr 1970
S-IVB maneuver to LOX dump attitude initiated.	004:27:40.0	23:40:40	11 Apr 1970
S-IVB lunar impact maneuver—CVS vent opened.	004:34:39.4	23:47:39	11 Apr 1970
S-IVB lunar impact maneuver—LOX dump started.	004:39:19.4	23:52:19	11 Apr 1970
S-IVB lunar impact maneuver—CVS vent opened.	004:39:39.4	23:52:39	11 Apr 1970
S-IVB lunar impact maneuver—LOX dump ended.	004:40:07.4	23:53:07	11 Apr 1970
Maneuver to attitude for final S-IVB APS burn initiated.	005:48:07.8	01:01:07	12 Apr 1970
S-IVB lunar impact maneuver—APS ignition.	005:59:59.5	01:12:59	12 Apr 1970
S-IVB lunar impact maneuver—APS cutoff.	006:03:36.7	01:16:36	12 Apr 1970
Earth weather photography started.	007:17:14	02:30:14	12 Apr 1970
Unsuccessful passive thermal control attempt.	007:43:02	02:56:02	12 Apr 1970
Earth weather photography ended.	011:17:19	06:30:19	12 Apr 1970
2nd S-IVB transposition maneuver (unplanned) initiated by launch vehicle digital computer.	013:42:33	08:55:33	12 Apr 1970
Unplanned S-IVB velocity increase of 5 feet per second which altered lunar impact trajectory closer to target point.	019:29:10	14:42:10	12 Apr 1970
TV transmission started.	030:13	01:26	13 Apr 1970
Midcourse correction ignition (SPS)—transfer to hybrid non-free return trajectory.	030:40:49.65	01:53:49	13 Apr 1970
Midcourse correction cutoff.	030:40:53.14	01:53:53	13 Apr 1970
TV transmission ended.	031:02	02:15	13 Apr 1970
Photograph Comet Bennett.	031:50	03:03	13 Apr 1970
Unsuccessful passive thermal control attempt.	032:21:49	03:34:49	13 Apr 1970
Crew turned on fans in oxygen tank 2 (routine procedure).	046:40:02	17:15	13 Apr 1970
Cryogenic oxygen tank 2 quantity gauge indicated "off-scale high," of over 100 percent (probably due to short circuit). first indication of a problem.	046:40:05	17:18:05	13 Apr 1970
Cryogenic oxygen tank 2 quantity probe short circuited.	046:40:08	17:21:08	13 Apr 1970
Oxygen tank 2 fans turned on again with no apparent adverse affects. Quantity gauge continued to read "off-scale high."	047:54:50	19:03:50	13 Apr 1970
Oxygen tank 2 fans turned on again with no apparent adverse affects. Quantity gauge continued to read "off-scale high."	051:07:44	22:57:44	13 Apr 1970
CDR and LMP cleared to enter the LM to commence inflight inspection.	053:27	00:40	14 Apr 1970
LMP entered LM.	054:20	01:33	14 Apr 1970
CDR entered LM.	054:35	01:48	14 Apr 1970
LM system checks.	054:40	01:53	14 Apr 1970
TV transmission started.	055:14	02:27	14 Apr 1970
CDR and LMP returned to CM.	055:30	02:43	14 Apr 1970

Apollo 13 Timeline

Event	GET (hhh:mm:ss)	GMT Time	GMT Date
TV transmission ended.	055:46	02:59	14 Apr 1970
Tunnel hatch closed.	055:50	03:03	14 Apr 1970
Master caution and warning triggered by low hydrogen pressure in tank 1. Alarm turned off after 4 seconds.	055:52:31	03:05:31	14 Apr 1970
CAPCOM: "13, we've got one more item for you, when you get a chance. We'd like you to stir up your cryo tanks. In addition, I have shaft and trunnion..."	055:52:58	03:05:58	14 Apr 1970
LMP (Swigert): "Okay."	055:53:06	03:06:06	14 Apr 1970
CAPCOM: "...for looking at the Comet Bennett, if you need it."	055:53:07	03:06:07	14 Apr 1970
LMP: "Okay. Stand by."	055:53:12	03:06:12	14 Apr 1970
Oxygen tank 1 fans on.	055:53:18	03:06:18	14 Apr 1970
Oxygen tank 1 pressure decreased 8 psi due to normal destratification. Spacecraft current increased by 1 ampere.	055:53:19	03:06:19	14 Apr 1970
Oxygen tank 2 fans on. Stabilization control system electrical disturbance indicated a power transient.	055:53:20	03:06:20	14 Apr 1970
Oxygen tank 2 pressured decreased 4 psi.	055:53:21	03:06:21	14 Apr 1970
Electrical short in tank 2 (stabilization control system electrical disturbance indicated a power transient).	055:53:22.718	03:06:22	14 Apr 1970
1.2-volt decrease in AC bus 2 voltage.	055:53:22.757	03:06:22	14 Apr 1970
11.1 ampere "spike" recorded in fuel cell 3 current followed by drop in current and rise in voltage typical of removal of power from one fan motor, indicating opening of motor circuit.	055:53:22.772	03:06:22	14 Apr 1970
Oxygen tank 2 pressure started to rise for 24 seconds.	055:53:36	03:06:36	14 Apr 1970
11 volt decrease in AC bus 2 voltage for one sample.	055:53:38.057	03:06:38	14 Apr 1970
Stabilization control system electrical disturbance indicated a power transient.	055:53:38.085	03:06:38	14 Apr 1970
22.9-ampere "spike" recorded in fuel cell 3 current, followed by drop in current and rising voltage typical of one fan motor, indicating opening of another motor circuit.	055:53:41.172	03:06:41	14 Apr 1970
Stabilization control system electrical disturbance indicated a power transient.	055:53:41.192	03:06:41	14 Apr 1970
Oxygen tank 2 pressure rise ended at a pressure of 953.8 psia.	055:54:00	03:07:00	14 Apr 1970
Oxygen tank 2 pressure started to rise again.	055:54:15	03:07:15	14 Apr 1970
Oxygen tank 2 quantity dropped from full scale (to which it had failed at 046:40) for two seconds and then read 75.3 percent full. This indicated the gauge short circuit may have corrected itself.	055:54:30	03:07:30	14 Apr 1970
Oxygen tank 2 temperature started to rise rapidly.	055:54:31	03:07:31	14 Apr 1970
Flow rate of oxygen to all three fuel cells started to decrease.	055:54:43	03:07:43	14 Apr 1970
Oxygen tank 2 pressure reached maximum value of 1,008.3 psia.	055:54:45	03:07:45	14 Apr 1970
Oxygen tank 2 temperature rises 40 F° for one sample (invalid reading).	055:54:48	03:07:48	14 Apr 1970
Oxygen tank 2 quantity jumped to off-scale high and then started to drop until the time of telemetry loss, indicating a failed sensor.	055:54:51	03:07:51	14 Apr 1970
Oxygen tank 2 temperature read -151.3 F. Last valid indication.	055:54:52	03:07:52	14 Apr 1970
Oxygen tank 2 temperature suddenly went off-scale low, indicating a failed sensor.	055:54:52.703	03:07:52	14 Apr 1970
Last telemetered pressure from oxygen tank 2 before telemetry loss was 995.7 psia.	055:54:52.763	03:07:52	14 Apr 1970
Sudden accelerometer activity on X, Y, and Z axes.	055:54:53.182	03:07:53	14 Apr 1970
Body-mounted roll, pitch, and yaw rate gyros showed low-level activity for 1/4 second.	055:54:53.220	03:07:53	14 Apr 1970
Oxygen tank 1 pressure dropped 4.2 psi.	055:54:53.323	03:07:53	14 Apr 1970
2.8-amp rise in total fuel cell current.	055:54:53.5	03:07:53	14 Apr 1970
X, Y, and Z accelerations in CM indicate 1.17 g, 0.65 g, and 0.65 g.	055:54:53.542	03:07:53	14 Apr 1970
Telemetry loss for 1.8 seconds. Master caution and warning triggered by DC main bus B undervoltage. Alarm turned off in 6 seconds. Indications were that the cryogenic oxygen tank 2 lost pressure during this time period and the panel separated. It was at this time that the crew heard a loud bang.	055:54:53.555	03:07:53	14 Apr 1970
Nitrogen pressure in fuel cell 1 went off-scale low indicating a failed sensor.	055:54:54.741	03:07:54	14 Apr 1970

Apollo 13 Timeline

Event	GET (hhh:mm:ss)	GMT Time	GMT Date
Telemetry recovered.	055:54:55.35	03:07:55	14 Apr 1970
Service propulsion system engine valve body temperature started a rise of 1.65 F° in 7 seconds. DC main bus A decreased 0.9 volt to 28.5 volts and DC main bus B decreased 0.9 volt to 29.0 volts. Total fuel cell current was 15 amps higher than the final value before telemetry loss. High current continued for 19 seconds. Oxygen tank 2 temperature read off-scale high after telemetry recovery, probably indicating failed sensors. Oxygen tank 2 pressure read off-scale low following telemetry recovery, indicating a broken supply line, a tank pressure below 19 psia, or a failed sensor. Oxygen tank 1 pressure read 781.9 psia and started to drop steadily. Pressure drops over a period of 130 minutes to the point at which it was insufficient to sustain operation of fuel cell 2.	055:54:56	03:07:56	14 Apr 1970
Oxygen tank 2 quantity read off-scale high following telemetry recovery indicating a failed sensor.	055:54:57	03:07:57	14 Apr 1970
The reaction control system helium tank C temperature began a 1.66 F° increase in 36 seconds.	055:54:59	03:07:59	14 Apr 1970
Oxygen flow rates to fuel cells 1 and 3 approached zero after decreasing for 7 seconds.	055:55:01	03:08:01	14 Apr 1970
Surface temperature of SM oxidizer tank in bay 3 started a 3.8 F° increase in a 15 second period. Service propulsion system helium tank temperature started a 3.8 F° increase in a 32-second period.	055:55:02	03:08:02	14 Apr 1970
DC main bus A voltage recovered to 29.0 volts. DC main bus B recovered to 28.8.	055:55:09	03:08:09	14 Apr 1970
LMP: "Okay, Houston, we've had a problem here."	055:55:20	03:08:20	14 Apr 1970
CAPCOM: "This is Houston. Say again, please."	055:55:28	03:08:28	14 Apr 1970
CDR (Lovell): "Houston, we've had a problem. We've had a main B bus undervolt."	055:55:35	03:08:35	14 Apr 1970
CAPCOM: "Roger. Main B bus undervolt." 055:55:42 03:08:42 14 Apr. 1970 Oxygen tank 2 temperature started steady drop lasting 59 seconds, indicating a failed sensor.	055:55:49	03:08:49	14 Apr 1970
CMP (Haise): "Okay. Right now, Houston, the voltage is—is looking good. And we had a pretty large bang associated with the caution and warning there. And as I recall, main B was the one that had an amp spike on it once before."	055:56:10	03:09:10	14 Apr 1970
CAPCOM: "Roger, Fred."	055:56:30	03:09:30	14 Apr 1970
Oxygen tank 2 quantity became erratic for 69 seconds before assuming an off-scale-low state, indicating a failed sensor.	055:56:38	03:09:38	14 Apr 1970
CMP: "In the interim here, we're starting to go ahead and button up the tunnel again."	055:56:54	03:09:54	14 Apr 1970
CMP: "That jolt must have rocked the sensor on—see now—-oxygen quantity 2. It was oscillating down around 20 to 60 percent. Now it's full-scale high."	055:57:04	03:10:04	14 Apr 1970
Master caution and warning triggered by DC main bus B undervoltage. Alarm was turned off in six seconds.	055:57:39	03:10:39	14 Apr 1970
DC main bus B dropped below 26.25 volts and continued to fall rapidly.	055:57:40	03:10:40	14 Apr 1970
CDR: "Okay. And we're looking at our service module RCS helium 1. We have—B is barber poled and D is barber poled, helium 2, D is barber pole, and secondary propellants, I have A and C barber pole." AC bus fails within two seconds.	055:57:44	03:10:44	14 Apr 1970
Fuel cell 3 failed.	055:57:45	03:10:45	14 Apr 1970
Fuel cell current started to decrease.	055:57:59	03:10:59	14 Apr 1970
Master caution and warning caused by AC bus 2 being reset.	055:58:02	03:11:02	14 Apr 1970
Master caution and warning triggered by DC main bus A undervoltage.	055:58:06	03:11:06	14 Apr 1970
DC main bus A dropped below 26.25 volts and in the next few seconds leveled off at 25.5 volts.	055:58:07	03:11:07	14 Apr 1970
CMP: "AC 2 is showing zip."	055:58:07	03:11:07	14 Apr 1970
CMP: "Yes, we got a main bus A undervolt now, too, showing. It's reading about 25 and a half. Main B is reading zip right now."	055:58:25	03:11:25	14 Apr 1970
Master caution and warning triggered by high hydrogen flow rate to fuel cell 2.	056:00:06	03:13:06	14 Apr 1970
CDR: "...It looks to me, looking out the hatch, that we are venting something. We are venting something out into the—into space."	056:09:07	03:22:07	14 Apr 1970
LMP reported fuel cell 1 off line.	056:09:58	03:22:58	14 Apr 1970
Emergency power-down.	056:33:49	03:46:49	14 Apr 1970

Apollo 13 Timeline

Event	GET (hhh:mm:ss)	GMT Time	GMT Date
LMP reported fuel cell 3 off line.	056:34:46	03:47:46	14 Apr 1970
CDR and LMP entered LM.	057:43	04:56	14 Apr 1970
Shutdown of fuel cell #2.	058:00	05:13	14 Apr 1970
CM computer and platform powered down.	058:10	05:23	14 Apr 1970
CSM systems powered down. LM systems powered up.	058:40	05:53:00	14 Apr 1970
Midcourse correction ignition to free-return trajectory (LM DPS).	061:29:43.49	08:42:43	14 Apr 1970
Midcourse correction cutoff.	061:30:17.72	08:43:17	14 Apr 1970
LM systems powered down.	062:50	10:03	14 Apr 1970
Lunar occultation entered.	077:08:35	00:21:35	15 Apr 1970
Lunar occultation exited.	077:33:10	00:46:10	15 Apr 1970
S-IVB impact on lunar surface.	077:56:39.7	01:09:39	15 Apr 1970
LM systems powered up.	078:00	01:13	15 Apr 1970
Abort guidance system to primary guidance system aligned.	078:10	01:23	15 Apr 1970
Transearth injection ignition (LM DPS).	079:27:38.95	02:40:39	15 Apr 1970
Transearth injection cutoff.	079:32:02.77	02:45:02	15 Apr 1970
LM systems powered down.	082:10	05:23	15 Apr 1970
Apparent short-circuit in LM electrical system, accompanied by a "thump" in vicinity of descent stage and observation of venting for several minutes in area of LM descent batteries 1 and 2.	097:13:53	20:26:53	15 Apr 1970
LM configured for midcourse correction.	100:00	23:13	15 Apr 1970
CSM power configuration for telemetry established.	101:20	00:33	16 Apr 1970
CM powered up.	101:53	01:06	16 Apr 1970
LM systems powered up.	104:50	04:03	16 Apr 1970
Midcourse correction ignition (LM DPS).	105:18:28.0	04:31:28	16 Apr 1970
Midcourse correction cutoff.	105:18:42.0	04:31:42	16 Apr 1970
Passive thermal control started.	105:20	04:33	16 Apr 1970
LM systems powered down.	105:50	05:03	16 Apr 1970
LM power transferred to CSM.	112:05	11:18	16 Apr 1970
Battery A charge initiated.	112:20	11:33	16 Apr 1970
Battery A charge terminated. Battery B charge initiated.	126:10	01:23	17 Apr 1970
Battery B charge terminated.	128:10	03:23	17 Apr 1970
LM systems powered up.	133:35	08:48	17 Apr 1970
Platform aligned.	134:40	09:53	17 Apr 1970
Preparation for midcourse correction.	136:30	11:43	17 Apr 1970
Midcourse correction ignition (LM RCS).	137:39:51.5	12:52:51	17 Apr 1970
Midcourse correction cutoff.	137:40:13.00	12:53:13	17 Apr 1970
SM separation.	138:01:48.0	13:14:48	17 Apr 1970
SM photographed.	138:15	13:28	17 Apr 1970
CM powered up.	140:10	15:23	17 Apr 1970
Platform aligned.	140:40	15:53	17 Apr 1970
LM maneuvered to undocking attitude.	140:50	16:03	17 Apr 1970
LM jettisoned.	141:30:00.2	16:43:00	17 Apr 1970
Entry.	142:40:45.7	17:53:45	17 Apr 1970
Drogue parachute deployed. S-band contact with CM established by recovery aircraft.	142:48	18:01	17 Apr 1970
Visual contact with CM established by recovery helicopters.	142:49	18:02	17 Apr 1970
Visual contact with CM established by recovery ship. Voice contact with CM established by recovery helicopters.	142:50	18:03	17 Apr 1970
Main parachute deployed. Splashdown (went to apex-up)	142:54:41	18:07:41	17 Apr 1970
Swimmers deployed to retrieve main parachutes.	142:56	18:09	17 Apr 1970
1st swimmer deployed to CM.	143:03	18:16	17 Apr 1970
Flotation collar inflated.	143:11	18:24	17 Apr 1970

Apollo 13 Timeline

Event	GET (hhh:mm:ss)	GMT Time	GMT Date
Life preserver unit delivered to lead swimmer.	143:18	18:31	17 Apr 1970
CM hatch opened for crew egress.	143:19	18:32	17 Apr 1970
Crew egress.	143:22	18:35	17 Apr 1970
Crew aboard recovery helicopter.	143:29	18:42	17 Apr 1970
Crew aboard recovery ship.	143:40	18:53	17 Apr 1970
CM aboard recovery ship.	144:23	19:36	17 Apr 1970
Flight crew departed recovery ship via Samoa and Hawaii.	167:07	18:20	18 Apr 1970
Flight crew arrived in Hawaii.	199:22	02:35	19 Apr 1970
Flight crew arrived in Houston.	224:17	03:30	20 Apr 1970
Recovery ship arrived in Hawaii.	312:17	19:30	24 Apr 1970
Safing of CM pyrotechnics completed.	343:22	02:35	26 Apr 1970
Deactivation of fuel and oxidizer completed.	360:15	19:28	26 Apr 1970
CM arrived at contractor's facility in Downey, CA.	378:47	14:00	27 Apr 1970

APOLLO 14

The Eighth Mission:
The Third Lunar Landing

Apollo 14 Summary

(31 January–09 February 1971)

Apollo 14 crew (l. to r.): Stu Roosa, Al Shepard, Ed Mitchell (NASA S70-55387).

Background

Apollo 14 was a Type H mission, a precision piloted lunar landing demonstration and systematic lunar exploration. It was the third successful lunar landing.

The primary objectives were:

• to perform selenological inspection, survey, and sampling of materials in a preselected region of the Fra Mauro formation;

• to deploy and activate the Apollo lunar surface experiments package;

• to develop human capability of working in the lunar environment;

• to obtain photographs of candidate exploration sites.

Although the primary mission objectives for Apollo 14 were the same as those of Apollo 13, provisions were made for returning a significantly greater quantity of lunar material and scientific data than had been possible previously. An innovation that allowed an increase in the range of lunar surface exploration and in the amount of material collected was the provision of a collapsible two-wheeled cart, the modular equipment transporter (MET), for carrying tools, cameras, a portable magnetometer, and lunar samples.

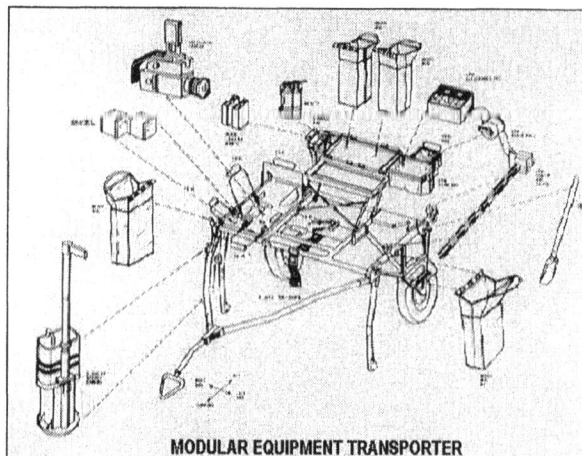

Line drawing of the Modular Equipment Transporter (MET) (NASA S70-50762).

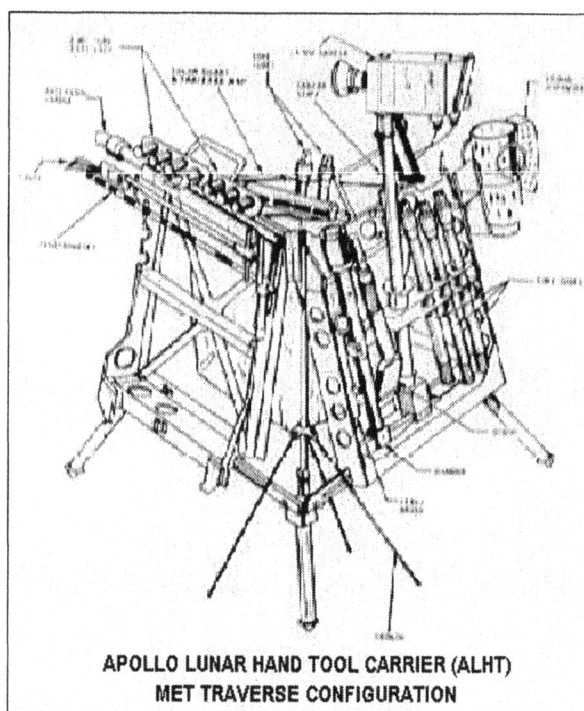

Line drawing of the Apollo Lunar Hand Tool Carrier in configuration for use with the MET (NASA S70-50763).

An investigation into the cause of the Apollo 13 cryogenic oxygen tank failure led to three significant changes in the CSM cryogenic oxygen storage and electrical power systems for Apollo 14 and future missions. The internal construction of the oxygen tanks was modified, a third oxygen tank was added, and an auxiliary battery was installed. These changes were also incorporated into all subsequent spacecraft.

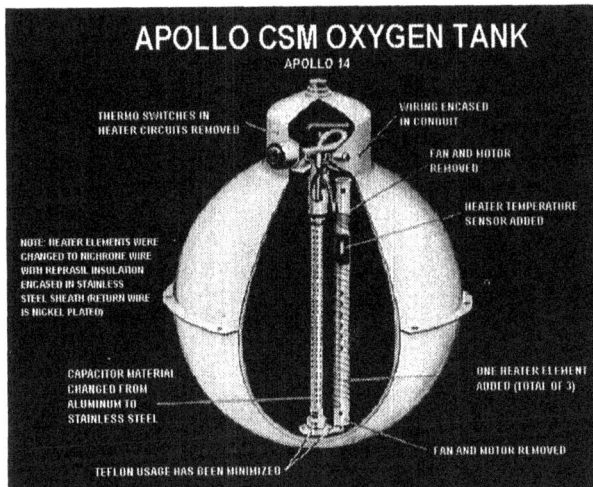

APOLLO CSM OXYGEN TANK
APOLLO 14

THERMO SWITCHES IN
HEATER CIRCUITS REMOVED

WIRING ENCASED
IN CONDUIT

FAN AND MOTOR
REMOVED

HEATER TEMPERATURE
SENSOR ADDED

NOTE: HEATER ELEMENTS WERE
CHANGED TO NICHROME WIRE
WITH REFRASIL INSULATION
ENCASED IN STAINLESS
STEEL SHEATH RETURN WIRE
IS NICKEL PLATED

CAPACITOR MATERIAL
CHANGED FROM
ALUMINUM TO
STAINLESS STEEL

ONE HEATER ELEMENT
ADDED (TOTAL OF 3)

FAN AND MOTOR REMOVED

TEFLON USAGE HAS BEEN MINIMIZED

Diagram of CSM oxygen tank modified following the Apollo 13 accident (NASA S71-16745).

The crew members were Captain Alan Bartlett Shepard, Jr. (USN), commander; Major Stuart Allen Roosa (USAF), command module pilot; and Commander Edgar Dean Mitchell (USN), lunar module pilot.

Selected as one of the original astronauts in 1959, Shepard became the first American in space when he piloted the initial Mercury suborbital mission (MR-3). Shepard subsequently developed an ear disorder, Meniere's syndrome, which caused the Navy to forbid him to fly solo in jet planes, and which forced NASA to ground him. He then became chief of the astronaut office. In 1969, however, Shepard underwent experimental surgery that corrected the problem. He was restored to full status in May and assigned to command Apollo 14 in August. Shepard was born 18 November 1923 in East Derry, New Hampshire, and at 47 years old, he was to become the oldest person to walk on the Moon. He received a B.S. from the U.S. Naval Academy in 1944.[1] His backup for the mission was Captain Eugene Andrew "Gene" Cernan (USN).

Roosa and Mitchell were making their first space flights. Roosa, born 16 August 1933 in Durango, Colorado, was 37 years old at the time of the Apollo 14 mission. He received a B.S. in aeronautical engineering from the University of Colorado in 1960 and was selected as an astronaut in 1966.[2] His backup was Commander Ronald Ellwin Evans (USN).

Mitchell, born 17 September 1930 in Hereford, Texas, was 40 years old. He received a B.S. in industrial management from the Carnegie Institute of Technology in 1952, a B.S. in aeronautical engineering from the U.S. Naval Postgraduate

School in 1961, and an Sc.D. in aeronautics and astronautics from the Massachusetts Institute of Technology in 1964. He was selected as an astronaut in 1966. His backup was Lt. Colonel Joe Henry Engle (USAF).

The capsule communicators (CAPCOMs) for the mission were Major Charles Gordon Fullerton (USAF), Lt. Commander Bruce McCandless II (USN), Fred Wallace Haise, Jr., and Evans. The support crew were McCandless, Lt. Colonel William Reid Pogue (USAF), Fullerton, and Phillip Kenyon Chapman, Sc.D. The flight directors were M.P. "Pete" Frank and Glynn S. Lunney (first shift), Milton L. Windler (second shift), Gerald D. Griffin (third shift), and Glynn S. Lunney (fourth shift).

The Apollo 14 launch vehicle was a Saturn V, designated SA-509. The mission also carried the designation Eastern Test Range #7194. The CSM was designated CSM-110 and had the call-sign "Kitty Hawk." The lunar module was designated LM-8 and had the call-sign "Antares."

Launch Preparations

The terminal countdown was picked up at T-28 hours at 06:00:00 GMT on 30 January 1971. Scheduled holds were initiated at T-9 hours for 9 hours 23 minutes and at T-3 hours 30 minutes for one hour.

At launch time, a cold front extended through northern Florida. Scattered rain shower activity existed to the south of this front throughout the morning of launch, but the showers did not reach the launch area until just before the scheduled launch time. A band of cumulus congestus clouds with showers developed about 30 minutes before scheduled launch time along a line extending from Orlando toward the northern Merritt Island Launch Area (MILA). This, and the threat of lightning, necessitated a 40-minute 2-second hold at T-8 minutes until the showers had moved a sufficient distance from the launch complex. Although it was raining prior to launch, there was no rain at the pad at the time of launch, but the vehicle did travel through the cloud decks.

Surface winds in the Cape Canaveral area were fairly light and westerly. Cumulus clouds covered 70 percent of the sky (base 4,000 feet) and altocumulus covered 20 percent (base 8,000 feet), the temperature was 71.1° F, the relative humidity was 86 percent, and the barometric pressure was 14.652 lb/in^2. The winds, as measured by the anemometer on the light pole 60.0 feet above ground at the launch site, measured 9.7 knots at 255° from true north. The winds, as

---

[1] Shepard died of leukemia 21 July 1998 in Community Hospital on the Monterey Peninsula, CA.
[2] Roosa died of complications from pancreatitis 12 December 1994 in Washington, DC. (NASA Headquarters Release No. 94-210).

measured at 530 feet above the launch site, measured 16.5 knots at 275° from true north.

The weather delay required the flight azimuth to be changed from 72.067° to 75.5579° east of north.

Ascent Phase

Apollo 14 was launched from Kennedy Space Center Launch Complex 39, Pad A, at a Range Zero time of 21:03:02 GMT (04:03:02 p.m. EST) on 31 January 1971. The planned launch window was from 20:23:00 GMT on 31 January to 00:12:00 GMT on 1 February to take advantage of a sun elevation angle on the lunar surface of 10.3°.

Between 000:00:12.814 and 000:00:28.000, the vehicle rolled from a launch pad azimuth of 90° to a flight azimuth of 75.558°. The S-IC engine shut down at 000:02:44.094, followed by S-IC/S-II separation and S-II engine ignition. The S-II engine shut down at 000:09:19.05, followed by separation from the S-IVB, which ignited at 000:09:23.4. The first S-IVB engine cutoff occurred at 000:11:40.56, with deviations from the planned trajectory of only -2.6 ft/sec in velocity; altitude was exactly as planned.

The maximum wind conditions encountered during ascent were 102.6 knots at 255° from true north at 43,270 feet, and a maximum wind shear of 0.0201 sec⁻¹ at 43,720 feet.

Parking orbit conditions at insertion, 000:11:50.56 (S-IVB cutoff plus 10 seconds to account for engine tailoff and other transient effects), showed an apogee and perigee of 100.1 by 98.9 n mi, an inclination of 31.120°, a period of 88.18 minutes, and a velocity of 25,565.9 ft/sec. The apogee and perigee were based upon a spherical Earth with a radius of 3,443.934 n mi.

The international designation for the CSM upon achieving orbit was 1971-008A, and the S-IVB was designated 1971-008B. After undocking at the moon, the LM ascent stage would be designated 1971-008C and the descent stage 1971-008D.

After inflight systems checks, the 350.84-second translunar injection maneuver (second S-IVB firing) was performed at 002:28:32.40. The S-IVB engine shut down at 002:34:23.24 and translunar injection occurred ten seconds later, at a velocity of 35,541.0 ft/sec after 1.5 Earth orbits lasting 2 hours 22 minutes 42.68 seconds.

Apollo 14 lifts off from Kennedy Space Center Pad 39A (NASA S71-18398).

Translunar Phase

At 003:02:29.4, the CSM was separated from the S-IVB stage. Transposition occurred normally; however, six docking attempts were required before the CSM was successfully docked with the LM at 004:56:56.0. The docked spacecraft were ejected from the S-IVB by a 6.9-second maneuver at 005:47:14.4, and an 80.2-second separation maneuver was performed at 006:04:01.7. Inflight examination of the docking probe revealed no problems. It was therefore assumed that the capture-latch assembly must not have been in the locked configuration during the first five attempts.

As on Apollo 13, the S-IVB stage was targeted to impact the Moon within a prescribed area to supply seismic data. A 252.2-second auxiliary propulsion system lunar impact maneuver was performed at 009:04:11.2 to accomplish that objective. The S-IVB impacted the lunar surface at 082:37:52.17. The impact point was latitude 8.07° south and longitude 26.04° west, 159 n mi from the target point, and 94 n mi southwest of the Apollo 12 seismometer. The seismometer recorded the impact 37 seconds later and responded to vibrations for more than three hours. At impact, the S-IVB weighed 30,836 pounds and was traveling 8,343 ft/sec.

Translunar activities included star and Earth horizon calibration sightings in preparation for a cislunar navigation exercise to be performed during transearth coast, and dim-light photography of the Earth. A 10.19-second midcourse correction was made at 030:36:07.91. At 060:30, the commander and lunar module pilot transferred to the LM for two hours of housekeeping and systems checks. While there, the crew photographed a wastewater dump from the CM to obtain data for a particle contamination study being conducted for the Skylab program. A second midcourse correction, a 0.65-second maneuver, was made at 076:58:11.98 to achieve the final trajectory desired for lunar-orbit insertion.

At 054:33:36, a clock update was performed to compensate for the weather hold during the launch countdown. This procedure, which added 40 minutes 20.9 seconds to the mission time clock, was an aid to the command module pilot while in lunar orbit because it eliminated the need for numerous updates to his flight log.

At 081:56:40.70, at an altitude of 87.4 n mi above the Moon, the service propulsion engine was fired for 370.84 seconds to insert the spacecraft into a lunar orbit of 169.0 by 58.1 n mi. The translunar coast had lasted 79 hours 28 minutes 18.30 seconds.

Lunar Orbit/Lunar Surface Phase

At 086:10:52.97, a 20.81-second service propulsion system maneuver was performed and established the descent orbit of 58.8 by 9.1 n mi in preparation for undocking of the LM. On previous missions, the descent orbit insertion maneuver had been performed with the LM descent propulsion system. A change was made on this mission to allow a greater margin of LM propellant for landing in a more rugged area.

The commander and lunar module pilot entered the LM at 101:20 to perform system checks and prepare for undocking. A 2.7-second firing of the service module reaction control system separated the CM from the LM at 103:47:41.6 and resulted in an orbit of 60.2 by 7.8 n mi. A 4.02-second maneuver at 105:11:46.11 circularized the CSM to 63.9 by 56.0 n mi.

Following vehicle separation and before powered descent, ground personnel detected the presence of an abort command at a computer input channel although the crew had not depressed the abort switch. The failure was isolated to the abort switch, and to prevent an unwanted abort, a workaround procedure was developed. The procedure was

followed and the 764.61-second powered descent was performed successfully at 108:02:26.52 at an altitude of 7.8 n mi.

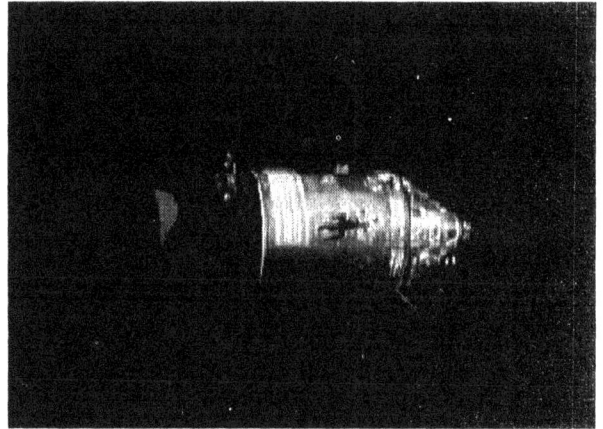

CSM photographed against black background following separation from LM (NASA AS14-66-9344).

Approximately six minutes after initial actuation of the landing radar, the system switched to the low-range scale, forcing the trackers into the narrow-band mode of operation. This ranging scale problem would have prevented acquisition of radar data until late in the descent—and prevented a lunar landing—but it was corrected by cycling the circuit breaker on and off manually.

Near-vertical view of the Apollo 14 Fra Mauro landing site (NASA S70-49764).

First contact with the lunar surface occurred at 09:18:11 GMT (04:18:11 a.m. EST) on 5 February at 108:15:09.30, with engine shutdown 1.83 seconds later. The spacecraft landed in the Fra Mauro highlands at latitude 3.64530° south and longitude 17.47136° west, the intended landing site for Apollo 13, on a slope of about 7°. Approximately 70 seconds of engine firing time remained at landing.

Preparations for the initial period of lunar surface exploration began two hours after landing, and cabin depressurization began at 113:39:11. The first extravehicular activity began 49 minutes late due to intermittent PLSS communications during the EVA preparations. Proper communications were established during a rerun of the checklist. The cause was believed to have been an LM configuration problem. A recycling of the audio circuit breaker cleared the problem.

The commander exited at 113:47. He was followed eight minutes later by the lunar module pilot, whose first task was to collect the contingency sample.

Shepard shades his eyes after descending to the lunar surface for the first time (NASA AS14-66-9230).

During the first extravehicular period, the crew deployed the television, S-band antenna, and solar wind experiment; deployed and loaded the modularized equipment; collected samples; and photographed activities, panoramas, and equipment.

At 115:46, the pair began their trip to the Apollo lunar surface experiments package deployment site, about 500 feet west of the LM. They also deployed the laser-ranging retro-reflector 100 feet west of the ALSEP. The first ALSEP data were received on Earth at 116:47:58.

Several problems were encountered during the deployment of the ALSEP package. They were as follows: difficulty in releasing the Boyd bolt on the suprathermal ion detector; stiffness in the cable between the suprathermal ion detector and the cold cathode ion gauge, which caused the cold

cathode ion gauge to fall over; low transmitter strength on the central station; noisy data from the suprathermal ion detector experiment; and failure of five of the active seismic experiment thumper initiators to fire.

Laser-Ranging Retro-Reflector, set up during EVA-1 (NASA AS14-67-9386).

Although communications were nominal during this period, gradual degradation of the television picture resolution was noted during the latter part of the EVA.

View of the Passive Seismic Experiment set up during EVA-1 (NASA AS14-67-9362).

The crew entered the LM and the cabin was repressurized at 118:27:01. The first extravehicular activity period lasted 4 hours 47 minutes 50 seconds. The distance traveled was 3,300 feet (1 km); an estimated 45.2 pounds (20.5 kg) of samples were collected.

Shepard holds U.S. flag during EVA-1 (NASA AS14-66-9232).

LM seen from a distance as the crew traverses the lunar surface during EVA-1. Note MET tracks (NASA AS14-67-9367).

View of the left rear quadrant of the LM on the lunar surface as seen during EVA-1 (NASA AS14-66-9277).

Components of the Apollo Lunar Surface Experiments Package (ALSEP) deployed during EVA-1 (NASA AS14-67-9376).

Shepard walks to the MET during EVA-1. Mitchell is in the background setting up various lunar surface experiments (NASA S71-19510).

During the lunar surface operations, the CSM made an 18.50-second plane change maneuver at 117:29:33.17, which adjusted the orbit to 67.1 by 57.7 n mi.

The second extravehicular activity began with cabin depressurization at 131:08:13, 27 minutes earlier than planned, and commander egress at 131:13, followed by the lunar module pilot seven minutes later.

In preparation for an excursion to the area of Cone Crater, 0.7 n mi (1.3 km) east-northeast of the landing site, the crew prepared and loaded the modular equipment transporter. They experienced difficulties in navigating the slopes and fell 30 minutes behind schedule. As a result, they only reached a point within 50 feet (15 m) from the rim of the crater. Nevertheless, the objectives associated with reaching the vicinity of this crater were achieved.

At Station C-Prime during EVA-1, in a field of small boulders, Shepard takes a series of panoramic photographs. He is about 250 feet from the southern rim of Cone Crater (NASA AS14-64-9099).

Mitchell looks at traverse map during EVA-2. Note lunar dust clinging to the left leg of his suit (NASA AS14-64-9089).

En route to Cone Crater, photographs, various samples, and terrain descriptions were obtained. Rock and soil samples were collected in a blocky field near the rim.

During EVA-2, Shepard stands near a boulder referred to later as "The Big Rock" (NASA AS14-68-9414).

On the return to the LM, the crew also obtained magnetometer measurements at two sites along the traverse. An estimated 1.5-foot trench was dug and samples were taken. An unsuccessful attempt to obtain a triple core tube sample was made, but other containerized samples were collected.

An alignment adjustment was made to the ALSEP Central Station's antenna just prior to crew ingress preparations in order to improve the signal strength being received at the Manned Space Flight Network ground stations. This improved signal strength approximately 1/2 db; however, data could still be received by the 30-foot antenna.

Before reentering the LM, the commander dropped two golf balls on the surface. Using a golf club face attached to the handle of the contingency sample collector, he hit the first ball into a crater and sent the next one into the lunar night.

The second extravehicular activity period lasted 4 hours 34 minutes 41 seconds. The distance traveled was 9,800 feet (3 km); an estimated 49.2 pounds (22.3 kg) of samples were collected. The crew reentered the LM and the cabin was repressurized at 135:42:54, thus ending the Apollo program's third piloted exploration of the Moon.

While the landing crew was on the lunar surface, the command module pilot performed tasks to obtain data for scientific analyses and future mission planning. These tasks included orbital science photography of the lunar surface, photography of the proposed Descartes landing site for site selection studies, photography of the lunar surface under high-sun-angle lighting conditions for operations planning, photography of low-brightness astronomical light sources, and photography of the Gegenschein and Moulton Point regions.

"Weird Rock," a large boulder approximately five feet wide, photographed by Shepard during EVA-2 (NASA AS14-64-9135).

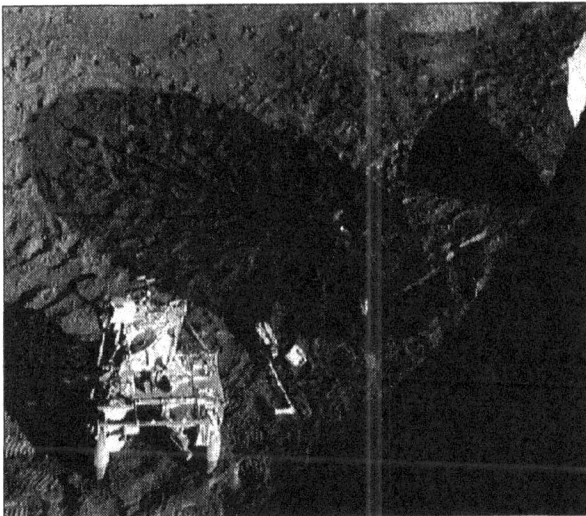

View of the MET as seen from inside the LM after EVA-2. Shadow of the erectable S-band antenna can be seen (NASA AS14-66-9340).

For the mission, the total time spent outside the LM was 9 hours 22 minutes 31 seconds, the total distance traveled was 13,100 feet (4 km), and the collected samples totaled 93.21 pounds (42.28 kg; official total in kilograms as determined by the Lunar Receiving Laboratory in Houston). The farthest point traveled from the LM was 4,770 feet.

Ignition of the ascent stage engine for lunar liftoff occurred at 18:48:42 GMT (13:48:42 EST) on 6 February at 141:45:40. The LM had been on the lunar surface for 33 hours 30 minutes 31 seconds.

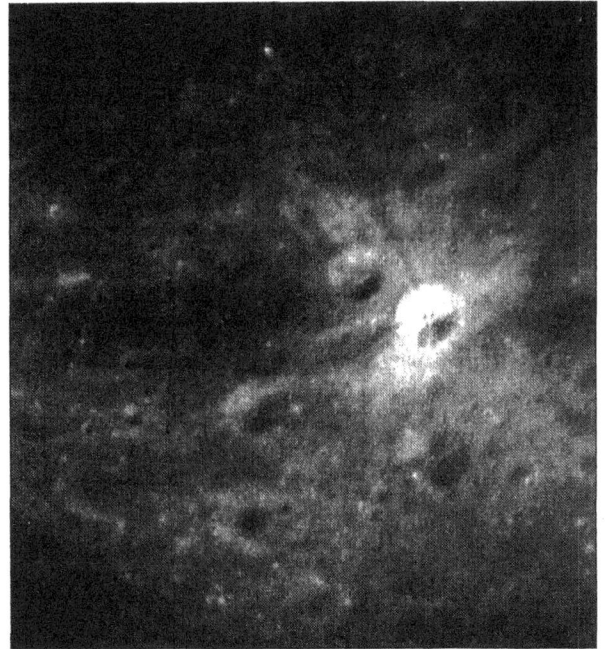

View of Descartes, proposed landing site for Apollo 16 (NASA AS14-69-9560).

The 432.1-second firing of the ascent engine placed the vehicle directly into a 51.7 by 8.5 n mi orbit, the first use of a direct lunar orbit rendezvous in the Apollo program. However, a 12.1-second vernier adjustment was required at 141:56:49.4 and altered the orbit to 51.2 by 8.4 n mi.

A 3.6-second terminal phase initiate maneuver at 142:30:51.1 and two small midcourse corrections brought the ascent stage to an orbit of 60.1 by 46.0 n mi. The ascent stage made a 26.7-second maneuver at 143:13:29.1 to finalize the orbit at 61.5 by 58.2 n mi for docking with the CSM at 143:32:50.5 at an altitude of 58.6 n mi. The two craft had been undocked for 39 hours 45 minutes 8.9 seconds. During the braking phase for docking, telemetry indicated that the abort guidance system had failed, but no caution and warning signals were on. A cycling of all circuit breakers and switches did not remedy this condition. During docking, no probe/drogue problems were experienced. The probe was returned for postflight analysis. Television during rendezvous and docking was excellent and clearly showed the docking maneuver.

After transfer of the crew and samples to the CM, the ascent stage was jettisoned at 145:44:58.0, and the CSM was prepared for transearth injection. The ascent stage was then maneuvered by remote control to impact the lunar surface. A 15.8-second maneuver was made at 145:49:42.5 to separate the CM from the ascent stage, and resulted in an orbit of 63.4 by 56.8 n mi. A 76.2-second deorbit firing at 57.2 n mi altitude depleted the ascent stage propellants, and impact occurred at 147:42:23.4. The impact point was latitude 3° 25' 12" south and longitude 19° 40' 1" west, 36 n mi west of the Apollo 14 landing site, 62 n mi from the Apollo 12 landing site, and 7 n mi from the target.

View of the CSM from the LM during rendezvous (NASA AS14-66-9348).

On Apollo 14, special dust control procedures were used to effectively decrease the amount of lunar surface dust in the cabins. On previous missions, dust adhering to equipment being returned to Earth had created a problem.

Following a 149.23-second maneuver at 148:36:02.30, transearth injection was achieved at 148:38:31.53 at a velocity of 8,505 ft/sec after 34 lunar orbits lasting 66 hours 35 minutes 39.99 seconds.

Transearth Phase

During transearth coast, a 3.0-second midcourse correction of 0.5 ft/sec was made at 165:34:56.69 using the service module reaction control system. In addition, a special oxygen flowrate test was performed to evaluate the system for planned extravehicular activities on subsequent missions, and a navigation exercise simulating a return to Earth without ground control was conducted using only the guidance and navigation system. Scientific investigations included televised demonstrations of electrophoretic sepa-

ration, liquid transfer, heat flow and convection, and composite casting under zero-gravity conditions.

Recovery

The service module was jettisoned at 215:32:42.2, and the CM followed a normal entry profile. The command module reentered the Earth's atmosphere (400,000 feet altitude) at 215:47:45.3 at a velocity of 36,170 ft/sec following a transearth coast of 67 hours 9 minutes 13.8 seconds. The service module should have entered Earth's atmosphere and its debris should have landed in the Pacific Ocean 650 n mi southwest of the CM splashdown; however, no radar data or sightings confirmed the entry or impact.

The parachute system effected splashdown of the CM in the Pacific Ocean at 21:05:00 GMT (16:05:00 EST) on 9 February.

CM about to splash down at the end of the mission (NASA S71-18753).

Mission duration was 216:01:58.1. The impact point was about 0.6 n mi from the target point and 3.8 n mi from the recovery ship U.S.S. New Orleans. The splashdown site was estimated to be latitude 27.02° south and longitude 172.67° west.

After splashdown, the CM assumed an apex-up flotation attitude. The crew was retrieved by helicopter and was aboard the recovery ship 48 minutes after splashdown. The CM was recovered 76 minutes later. The estimated CM weight at splashdown was 11,481.2 pounds, and the estimated distance traveled for the mission was 1,000,279 n mi.

Crew in raft following splashdown (l. to r.): Mitchell, Roosa, Shepard (NASA S71-19475).

Apollo 14 crew on prime recovery ship U.S.S. *New Orleans* (NASA S71-19473).

The crew remained aboard the *New Orleans* in the mobile quarantine facility until they departed by aircraft for Pago Pago, Samoa, at 17:46 GMT on 11 February. They were then transferred to a second mobile quarantine facility aboard a C-141 aircraft and flown to Ellington Air Force Base, Houston, where they arrived at 09:34 GMT on 12 February, following a refueling stop at Norton Air Force Base, California. The crew entered the lunar receiving laboratory at 11:35 GMT the same day.

After the *New Orleans* arrived at Hawaii, the CM and first mobile quarantine facility were offloaded at 21:30 GMT on 17 February. The first mobile quarantine facility was sent by aircraft to Houston, where it arrived at 07:40 GMT on 18 February. The CM was taken to Hickam Air Force Base, Hawaii, for deactivation. Upon completion of deactivation, at 23:00 GMT on 19 February, it was transferred to Ellington Air Force Base via C-133 aircraft, where it arrived at 21:45 GMT on 22 February.

The crew and medical support personnel were released from quarantine on 26 February, and the CM and lunar samples were released on 14 April. The tests showed no evidence of lunar microorganisms at the three sites explored, and this was considered to be sufficient justification for discontinuing the quarantine procedures on future missions.

On 8 April 1971, the CM was delivered to the North American Rockwell Space Division facility in Downey, California, for postflight analysis.

Conclusions

The Apollo 14 mission was the third successful lunar landing and demonstrated excellent performance of all contributing elements, thereby resulting in the collection of a wealth of scientific information. All of the objectives and experiment operations were accomplished satisfactorily except for some desired photography that could not be obtained.

The following conclusions were made from an analysis of post-mission data:

1. Cryogenic oxygen system hardware modifications and changes made as a result of the Apollo 13 failure satisfied, within safe limits, all system requirements for future missions, including extravehicular activity.

2. The advantages of piloted space flight were again clearly demonstrated on this mission by the crew's ability to diagnose and work around hardware problems and malfunctions which otherwise might have resulted in mission termination.

3. Navigation was the most difficult lunar surface task because of problems in finding and recognizing small features, reduced visibility in the up-sun and down-sun directions, and the inability to judge distances.

4. Rendezvous within one orbit of lunar ascent was demonstrated for the first time in the Apollo program. This type of rendezvous reduces the time between lunar liftoff and docking by approxi-

mately two hours from that required on previous missions. The timeline activities, however, are greatly compressed.

Mitchell and Shepard examine some of the rock samples they brought back from the Moon (NASA S71-20375).

5. On previous lunar missions, lunar surface dust adhering to equipment being returned to Earth had created a problem in both spacecraft. The special dust control procedures and equipment used on this mission were effective in lowering the overall level of dust.

6. Onboard navigation without air-to-ground communications was successfully demonstrated during the transearth phase of the mission to be sufficiently accurate for use as a contingency mode of operation during future missions.

7. Launching through cumulus clouds with tops up to 10,000 feet was demonstrated to be a safe launch restriction for the prevention of triggered lightning. The cloud conditions at liftoff were at the limit of this restriction and no triggered lightning was recorded during the launch phase.

Apollo 14 Objectives

Spacecraft Primary Objectives

1. To perform selenological inspection, survey, and sampling of materials in a preselected region of the Fra Mauro formation. *Achieved.*

2. To deploy and activate the Apollo lunar surface experiments package (ALSEP). *Achieved.*

3. To develop human capability to work in the lunar environment. *Achieved.*

4. To obtain photographs of candidate exploration sites. *Achieved.*

Detailed Objectives

Contingency sample collection. Achieved.

1. Photographs of a candidate exploration site. *Partially achieved. On the low-altitude pass (fourth revolution), the camera malfunctioned and no usable photography was obtained of Descartes. During the stereo strip photographic pass, the S-band high-gain antenna malfunctioned, and no usable high-bit-rate telemetry, and consequently, no camera shutter-open data, were obtained.*

2. Visibility at high-sun angles. *Partially achieved. The last of four sets of observations was deleted to provide another opportunity to photograph the Descartes area; however, sufficient data were collected to verify that the visibility analytical model could be used for Apollo planning purposes.*

3. Modular equipment transporter evaluation. *Achieved.*

4. Selenodetic reference point update. *Achieved.*

5. Command and service module orbital science photography. *Partially achieved. The lunar topographic camera malfunctioned, and the Hasselblad 70 mm camera with the 500 mm lens was substituted. The photography was excellent, but the resolution was considerably lower than possible with the lunar topographic camera.*

6. Assessment of extravehicular activity operation limits. *Achieved.*

7. Command and service module oxygen flow rate. *Achieved.*

8. Transearth lunar photography. *Partially achieved. Excellent photography of the lunar surface was obtained, but no lunar topographic photography was obtained because of a camera malfunction.*

9. Thermal coating degradation. *Achieved.*

10. Dim-light photography. *Achieved.*

Detailed Objectives Added During Mission

1. S-IVB photography. *Not achieved. The S-IVB could not be identified on the film during post-mission analysis.*

2. Command and service module water-dump photography. *Partially achieved. Although some water particles were seen on photographs of the water dump, there was no indication of the "snowstorm" described by the crew.*

Experiments

1. ALSEP IV: Apollo Lunar Surface Experiments Package.

    a. Lunar passive seismology. *Achieved.*

    b. Lunar active seismology. *Achieved.*

    c. Suprathermal ion detector. *Achieved.*

    d. Charged particle lunar environment. *Achieved.*

    e. Cold cathode gauge. *Achieved.*

    f. Lunar dust detector. *Achieved.*

2. Lunar geology investigation. *Achieved.*

3. Laser-ranging retro-reflector. *Achieved.*

4. Solar wind composition. *Achieved.*

5. S-band transponder. *Achieved.*

6. Downlink bistatic radar observation of the Moon. *Achieved.*

7. Apollo window meteoroid experiment. *Achieved.*

8. Gegenschein from lunar orbit. *Achieved.*

9. Portable magnetometer. *Achieved.*

10. Soil mechanics. *Achieved.*

11. Bone mineral measurement. *Achieved.*

Inflight Demonstrations

1. Electrophoretic separation (Marshall Space Flight Center). *Achieved.*

2. Heat flow and convection (Marshall Space Flight Center). *Achieved.*

3. Liquid transfer (Lewis Research Center). *Achieved.*

4. Composite casting (Marshall Space Flight Center). *Achieved.*

Operational Tests

1. For Manned Spacecraft Center

    a. Lunar gravity measurement (using the lunar module primary guidance system). *Achieved.*

    b. Hydrogen maser test (a network and unified S-band investigation sponsored by the Goddard Space Flight Center). *Achieved.*

2. For Department of Defense

    a. Chapel Bell (classified Department of Defense test). *Results classified.*

    b. Radar skin tracking. *Results classified.*

    c. Ionospheric disturbance from missiles. *Results classified.*

    d. Acoustic measurement of missile exhaust noise. *Results classified.*

    e. Army acoustic test. *Results classified.*

    f. Long-focal-length optical system. *Results classified.*

Launch Vehicle Objectives

1. To launch on a flight azimuth between 72° and 96° and insert the S-IVB/instrument unit/spacecraft into the planned circular Earth parking orbit. *Achieved.*

2. To restart the S-IVB during either the second or third revolution and inject the S-IVB/instrument unit/spacecraft into the planned translunar trajectory. *Achieved.*

3. To provide the required attitude control for the S-IVB/instrument unit/spacecraft during transposition, docking, and ejection. *Achieved.*

4. To perform an evasive maneuver after ejection of the command and service module/lunar module from the S-IVB/instrument unit. *Achieved.*

5. To attempt to impact the S-IVB/instrument unit on the lunar surface within 350 kilometers (189 nautical miles) of latitude 01° 35' 06" south, longitude 33° 15' west. *Achieved.*

6. To determine actual impact point within 5.0 kilometers (2.7 nautical miles) and time of impact within one second. *Achieved.*

7. To vent and dump the remaining gases and propellants to safe the S-IVB/instrument unit after final launch vehicle/spacecraft separation. *Achieved.*

8. To verify the operation of the liquid oxygen feedline accumulator systems installed on the S-II stage center engine. *Achieved.*

Apollo 14 Spacecraft History

EVENT	DATE
Individual and combined CM and SM systems test completed at factory.	02 Apr 1969
Integrated CM and SM systems test completed at factory.	07 May 1969
LM #8 final engineering evaluation acceptance test at factory.	25 Aug 1969
LM #8 integrated test at factory.	25 Aug 1969
LM ascent stage #8 ready to ship from factory to KSC.	08 Nov 1969
LM descent stage #8 ready to ship from factory to KSC.	13 Nov 1969
CM #110 and SM #110 ready to ship from factory to KSC.	17 Nov 1969
CM #110 and SM #110 delivered to KSC.	19 Nov 1969
CM #110 and SM #110 mated.	24 Nov 1969
LM ascent stage #8 delivered to KSC.	24 Nov 1969
LM descent stage #8 delivered to KSC.	24 Nov 1969
Saturn S-IC stage #9 delivered to KSC.	11 Jan 1970
Saturn S-IC stage #9 erected on MLP #2.	14 Jan 1970
LM ascent stage #8 and descent stage #8 mated.	20 Jan 1970
Saturn S-IVB stage #509 delivered to KSC.	20 Jan 1970
Saturn S-II stage #9 delivered to KSC.	21 Jan 1970
LM #8 combined systems test completed.	22 Jan 1970
CSM #110 combined systems test completed.	02 Feb 1970
Spacecraft/LM adapter #17 delivered to KSC.	31 Mar 1970
Saturn V instrument unit #509 delivered to KSC.	06 May 1970
Saturn S-II stage #9 erected.	12 May 1970
Saturn S-IVB stage #509 erected.	13 May 1970
Saturn V instrument unit #509 erected.	14 May 1970
Launch vehicle electrical systems test completed.	04 Jun 1970
LM #8 altitude tests completed.	22 Jun 1970
Launch vehicle propellant dispersion/malfunction overall test completed.	07 Jul 1970
CSM #110 altitude tests completed.	01 Aug 1970
Launch vehicle service arm overall test completed.	21 Oct 1970
CSM #110 moved to VAB.	04 Nov 1970
Spacecraft erected.	04 Nov 1970
Space vehicle and MLP #2 transferred to launch complex 39A.	09 Nov 1970
LM #8 combined systems test completed.	16 Nov 1970
CSM #110 integrated systems test completed.	18 Nov 1970
CSM #110 electrically mated to launch vehicle.	13 Dec 1970
LM #7 flight readiness test completed.	14 Dec 1970
Space vehicle overall test #1 (plugs in) completed.	14 Dec 1970
Space vehicle flight readiness test completed.	19 Dec 1970
Saturn S-IC stage #9 RP-1 fuel loading completed.	08 Jan 1971
Space vehicle countdown demonstration test (wet) completed.	18 Jan 1971
Space vehicle countdown demonstration test (dry) completed.	19 Jan 1971

Apollo 14 Ascent Phase

Event	GET (hhh:mm:ss)	Altitude (n mi)	Range (n mi)	Earth Fixed Velocity (ft/sec)	Space Fixed Velocity (ft/sec)	Event Duration (sec)	Geocentric Latitude (deg N)	Longitude (deg E)	Space Fixed Flight Path Angle (deg)	Space Fixed Heading Angle (E of N)
Liftoff	000:00:00.57	0.060	0.000	1.1	1,340.7		28.4470	-80.6041	0.05	90.00
Mach 1 achieved	000:01:08.0	4.337	1.379	1,077.0	2,082.4		28.4521	-80.5787	26.80	86.06
Maximum dynamic pressure	000:01:21.0	6.649	2.886	1,524.6	2,540.5		28.4580	-80.5509	28.77	84.61
S-IC center engine cutoff	000:02:15.14	23.202	24.169	5,103.0	6,283.6	141.6	28.5441	-80.1598	23.554	79.228
S-IC outboard engine cutoff	000:02:44.10	36.317	51.132	7,741.7	8,972.5	170.6	28.6516	-79.6634	19.584	78.468
S-IC/S-II separation	000:02:44.8	36.663	51.947	7,773.0	9,004.8		28.6548	-79.6484	19.489	78.468
S-II center engine cutoff	000:07:43.09	98.091	594.709	17,212.7	18,554.4	296.59	30.3347	-69.4425	0.829	82.809
S-II outboard engine cutoff	000:09:19.05	101.556	890.920	21,562.5	22,905.8	392.55	30.8611	-63.7444	0.621	85.784
S-II/S-IVB separation	000:09:20.00	101.596	894.194	21,573.8	22,917.2		30.8654	-63.6810	0.612	85.818
S-IVB 1st burn cutoff	000:11:40.56	103.091	1,406.287	24,215.6	25,559.9	137.16	31.0978	-53.7349	-0.004	91.245
Earth orbit insertion	000:11:50.56	103.086	1,444.989	24,221.6	25,565.8		31.0806	-52.9826	-0.003	91.656

Apollo 14 Earth Orbit Phase

Event	GET (hhh:mm:ss)	Space Fixed Velocity (ft/sec)	Event Duration (sec)	Velocity Change (ft/sec)	Apogee (n mi)	Perigee (n mi)	Period (mins)	Inclination (deg)
Earth orbit insertion	000:11:50.56	25,565.8			100.1	98.9	88.18	31.120
S-IVB 2nd burn ignition	002:28:32.40	25,579.0						
S-IVB 2nd burn cutoff	002:34:23.24	35,535.5	350.84	10,366.5				30.835

Apollo 14 Translunar Phase

Event	GET (hhh:mm:ss)	Altitude (n mi)	Space Fixed Velocity (ft/sec)	Event Duration (sec)	Velocity Change (ft/sec)	Space Fixed Flight Path Angle (deg)	Space Fixed Heading Angle (E of N)
Translunar injection	002:34:33.24	179.544	35,511.6			7.480	65.583
CSM separated from S-IVB	003:02:29.4	4,289.341	24,102.3			46.810	65.369
CSM docked with LM/S-IVB	004:56:56.7	20,603.4	13,204.1			66.31	84.77
CSM/LM ejection ignition	005:47:14.4	26,299.6	11,723.5			68.54	87.76
CSM/LM ejection cutoff	005:47:21.3			6.9	0.8		
Midcourse correction ignition	030:36:07.91	118,515	4,437.9			76.47	101.98
Midcourse correction cutoff	030:36:18.10	118,522.1	4,367.2	10.19	71.1	76.95	102.23
Midcourse correction ignition	076:58:11.98	11,900.3	3,711.4			-80.1	295.57
Midcourse correction cutoff	076:58:12.63	11,899.7	3,713.1	0.65	3.5	-80.1	295.65

Apollo 14 Lunar Orbit Phase

Event	GET (hhh:mm:ss)	Altitude (n mi)	Space Fixed Velocity (ft/sec)	Event Duration (sec)	Velocity Change (ft/sec)	Apogee (n mi)	Perigee (n mi)
Lunar orbit insertion ignition	081:56:40.70	87.4	8,061.4				
Lunar orbit insertion cutoff	082:02:51.54	64.2	5,458.5	370.84	3,022.4	169.0	58.1
Descent orbit insertion ignition	086:10:52.97	59.2	5,484.8				
Descent orbit insertion cutoff	086:11:13.78	59	5,279.5	20.81	205.7	58.8	9.1
CSM/LM undocking/separation ignition	103:47:41.6	30.5	5,435.8				
CSM/LM undocking/separation cutoff	103:47:44.3			2.7	0.8	60.2	7.8
CSM orbit circularization ignition	105:11:46.11	60.5	5,271.3				
CSM orbit circularization cutoff	105:11:50.13	60.3	5,342.1	4.02	77.2	63.9	56.0
LM powered descent initiation	108:02:26.52	7.8	5,565.6				
LM powered descent cutoff	108:15:11.13			764.61			
CSM plane change ignition	117:29:33.17	62.1	5,333.1				
CSM plane change cutoff	117:29:51.67	62.1	5,333.3	18.50	370.5	62.1	57.7
LM lunar liftoff ignition	141:45:40						
Lunar ascent orbit cutoff	141:52:52.1			432.1	6,066.1	51.7	8.5
LM vernier adjustment ignition	141:56:49.4	11.1	5,548.5				
LM vernier adjustment cutoff	141:57:01.5			12.1	10.3	51.2	8.4
LM terminal phase initiation ignition	142:30:51.1	44.8	5,396.6				
LM terminal phase initiation cutoff	142:30:54.7			3.6	88.5	60.1	46.0
LM terminal phase finalize ignition	143:13:29.1	58.8	5,365.5				
LM terminal phase finalize cutoff	143:13:55.8			26.7	32	61.5	58.2
CSM/LM docked	143:32:50.5	58.6	5,353.5				
LM ascent stage jettisoned	145:44:58.0	59.9	5,344.6				
CSM/LM final separation ignition	145:49:42.5	60.6	5,341.7				
CSM/LM final separation cutoff	145:49:58.3			15.8	3.4	63.4	56.8
LM ascent stage deorbit ignition	147:14:16.9	57.2	5,358.7				
LM ascent stage fuel depletion	147:15:33.1	57.2	5,177	76.2	186.1	56.7	-59.8

Apollo 14 Transearth Phase

Event	GET (hhh:mm:ss)	Altitude (n mi)	Space Fixed Velocity (ft/sec)	Event Duration (sec)	Velocity Change (ft/sec)	Space Fixed Flight Path Angle (deg)	Space Fixed Heading Angle (E of N)
Transearth injection ignition	148:36:02.30	60.9	5,340.6			-0.17	260.81
Transearth injection cutoff	148:38:31.53	66.5	8,505	149.23	3,460.6	5.29	266.89
Midcourse correction ignition	165:34:56.69	176,713.8	3,593.2			-79.61	124.88
Midcourse correction cutoff	165:34:59.69			3.00	0.5		
CM/SM separation	215:32:42.2	1,965	29,050.8			-36.62	117.11

Apollo 14 Timeline

Event	GET (hhh:mm:ss)	GMT Time	GMT Date
Terminal countdown started.	-028:00:00	06:00:00	30 Jan 1971
Scheduled 9-hour 23-minute hold at T-9 hours.	-009:00:00	01:00:00	31 Jan 1971
Countdown resumed at T-9 hours.	-009:00:00	10:23:00	31 Jan 1971
Scheduled 1-hour hold at T-3 hours 30 minutes.	-003:30:00	15:53:00	31 Jan 1971
Countdown resumed at T-3 hours 30 minutes.	-003:30:00	16:53:00	31 Jan 1971
Unscheduled 40-minute 2-second weather hold at T-8 minutes.	-000:08:02	20:15:00	31 Jan 1971
Countdown resumed at T-8 minutes.	-000:08:02	20:55:02	31 Jan 1971
Guidance reference release.	-000:00:16.960	21:02:45	31 Jan 1971
S-IC engine start command.	-000:00:08.9	21:02:53	31 Jan 1971
S-IC engine ignition (#5).	-000:00:06.5	21:02:55	31 Jan 1971
All S-IC engines thrust OK.	-000:00:01.6	21:03:00	31 Jan 1971
Range zero.	000:00:00.00	21:03:02	31 Jan 1971
All holddown arms released (1st motion) (1.05 g).	000:00:00.2	21:03:02	31 Jan 1971
Liftoff (umbilical disconnected).	000:00:00.57	21:03:02	31 Jan 1971
Tower clearance yaw maneuver started.	000:00:01.958	21:03:04	31 Jan 1971
Yaw maneuver ended.	000:00:09.896	21:03:11	31 Jan 1971
Pitch and roll maneuver started.	000:00:12.814	21:03:14	31 Jan 1971
Roll maneuver ended.	000:00:28.000	21:03:30	31 Jan 1971
Mach 1 achieved.	000:01:08.0	21:04:10	31 Jan 1971
Maximum bending moment (116,000,000 lbf-in).	000:01:16	21:04:18	31 Jan 1971
Maximum dynamic pressure (655.80 lb/ft²).	000:01:21.0	21:04:23	31 Jan 1971
S-IC center engine cutoff command.	000:02:15.14	21:05:17	31 Jan 1971
Pitch maneuver ended.	000:02:44.088	21:05:46	31 Jan 1971
S-IC outboard engine cutoff.	000:02:44.10	21:05:46	31 Jan 1971
S-IC maximum total inertial acceleration (3.82 g).	000:02:44.18	21:05:46	31 Jan 1971
S-IC maximum Earth-fixed velocity.	000:02:44.59	21:05:46	31 Jan 1971
S-IC/S-II separation command.	000:02:44.8	21:05:46	31 Jan 1971
S-II engine start command.	000:02:45.5	21:05:47	31 Jan 1971
S-II ignition.	000:02:46.5	21:05:48	31 Jan 1971
S-II aft interstage jettisoned.	000:03:14.8	21:06:16	31 Jan 1971
Launch escape tower jettisoned.	000:03:20.7	21:06:22	31 Jan 1971
Iterative guidance mode initiated.	000:03:25.912	21:06:27	31 Jan 1971
S-IC apex.	000:04:31.8	21:07:33	31 Jan 1971
S-II center engine cutoff.	000:07:43.09	21:10:45	31 Jan 1971
S-II maximum total inertial acceleration (1.81 g).	000:07:43.17	21:10:45	31 Jan 1971
S-IC impact (theoretical).	000:09:06.2	21:12:08	31 Jan 1971
S-II outboard engine cutoff.	000:09:19.05	21:12:21	31 Jan 1971
S-II/S-IVB separation command.	000:09:20.00	21:12:22	31 Jan 1971
S-II maximum Earth-fixed velocity.	000:09:20.07	21:12:22	31 Jan 1971
S-IVB 1st burn start command.	000:09:20.1	21:12:22	31 Jan 1971
S-IVB 1st burn ignition.	000:09:23.4	21:12:25	31 Jan 1971
S-IVB ullage case jettisoned.	000:09:31.8	21:12:33	31 Jan 1971
S-II apex.	000:10:00.2	21:13:02	31 Jan 1971
S-IVB 1st burn cutoff command.	000:11:40.56	21:14:42	31 Jan 1971
S-IVB 1st burn maximum total inertial acceleration (0.67 g).	000:11:40.66	21:14:42	31 Jan 1971
Earth orbit insertion. S-IVB 1st burn maximum Earth-fixed velocity.	000:11:50.56	21:14:52	31 Jan 1971
Maneuver to local horizontal attitude started.	000:12:02.092	21:15:04	31 Jan 1971
Orbital navigation started.	000:13:22.323	21:16:24	31 Jan 1971
S-II impact (theoretical).	000:20:46.3	21:23:48	31 Jan 1971
S-IVB 2nd burn restart preparation.	002:18:54.20	23:21:56	31 Jan 1971
S-IVB 2nd burn restart command.	002:28:24.10	23:31:26	31 Jan 1971

Apollo 14 Timeline

Event	GET (hhh:mm:ss)	GMT Time	GMT Date
S-IVB 2nd burn ignition.	002:28:32.40	23:31:34	31 Jan 1971
S-IVB 2nd burn cutoff.	002:34:23.24	23:37:25	31 Jan 1971
S-IVB 2nd burn maximum total inertial acceleration (1.45 g).	002:34:23.34	23:37:25	31 Jan 1971
S-IVB 2nd burn maximum Earth-fixed velocity.	002:34:23.67	23:37:25	31 Jan 1971
S-IVB safing procedures started.	002:34:23.9	23:37:25	31 Jan 1971
Translunar injection.	002:34:33.24	23:37:35	31 Jan 1971
Orbital navigation started.	002:36:54.841	23:39:56	31 Jan 1971
Maneuver to local horizontal attitude started.	002:36:55.064	23:39:57	31 Jan 1971
Maneuver to transposition and docking attitude started.	002:51:04.339	23:54:06	31 Jan 1971
Maneuver to transposition and docking attitude ended.	002:55:23.37	23:58:25	31 Jan 1971
1st docking attempt—2nd contact.	003:14:01.5	00:17:03	01 Feb 1971
TV transmission started.	003:05	00:08	01 Feb 1971
CSM separated from S-IVB.	003:02:29.4	00:05:31	01 Feb 1971
1st docking attempt—1st contact.	003:13:53.7	00:16:55	01 Feb 1971
1st docking attempt—3rd contact.	003:14:04.45	00:17:06	01 Feb 1971
1st docking attempt—4th contact.	003:14:09.0	00:17:11	01 Feb 1971
2nd docking attempt.	003:14:43.7	00:17:45	01 Feb 1971
3rd docking attempt.	003:16:43.4	00:19:45	01 Feb 1971
4th docking attempt.	003:23:41.7	00:26:43	01 Feb 1971
5th docking attempt.	004:32:29.3	01:35:31	01 Feb 1971
6th docking attempt.	004:56:44.9	01:59:46	01 Feb 1971
CSM docked with LM/S-IVB (initial docking latch triggered).	004:56:56.7	01:59:58	01 Feb 1971
TV transmission ended.	005:00	02:03	01 Feb 1971
CSM/LM ejected from S-IVB (RCS ignition).	005:47:14.4	02:50:16	01 Feb 1971
CSM/LM ejected from S-IVB (RCS cutoff).	005:47:21.3	02:50:23	01 Feb 1971
Maneuver to attitude for S-IVB APS evasive burn initiated.	005:55:30	02:58:32	01 Feb 1971
S-IVB APS evasive maneuver ignition.	006:04:01.7	03:07:03	01 Feb 1971
S-IVB APS evasive maneuver cutoff.	006:05:21.9	03:08:23	01 Feb 1971
Maneuver to S-IVB LOX dump attitude initiated.	006:13:43.0	03:16:45	01 Feb 1971
S-IVB lunar impact maneuver—CVS vent for lunar targeting velocity change started.	006:20:40.5	03:23:42	01 Feb 1971
S-IVB lunar impact maneuver—LOX dump started.	006:25:20.5	03:28:22	01 Feb 1971
S-IVB lunar impact maneuver—CVS vent for lunar targeting velocity change started.	006:25:40.5	03:28:42	01 Feb 1971
S-IVB lunar impact maneuver—LOX dump ended.	006:26:08.5	03:29:10	01 Feb 1971
Maneuver to attitude required for final S-IVB APS burn initiated.	008:43:41.0	05:46:43	01 Feb 1971
S-IVB lunar impact maneuver—APS ignition.	008:59:59.0	06:03:01	01 Feb 1971
S-IVB lunar impact maneuver—APS cutoff.	009:04:11.2	06:07:13	01 Feb 1971
TV transmission started.	011:00	08:03	01 Feb 1971
Hatch, probe, and drogue removed for inspection.	011:30	08:33	01 Feb 1971
TV transmission ended.	012:12	09:15:02	01 Feb 1971
Midcourse correction ignition (SPS).	030:36:07.91	03:39:09	02 Feb 1971
Midcourse correction cutoff.	030:36:18.10	03:39:20	02 Feb 1971
Earth darkside dim-light photography.	031:00	04:03	02 Feb 1971
S-IVB photography.	034:00	07:03	02 Feb 1971
Lunar topographic camera unstowed and checked out.	034:15	07:18	02 Feb 1971
Bistatic radar frequency check.	052:00	01:03	03 Feb 1971
Mission clock updated (000:040:02.9 added).	054:53:36	03:56:38	03 Feb 1971
LM pressurization started.	059:50	08:53	03 Feb 1971
TV transmission started.	060:05	09:08	03 Feb 1971
Preparation for LM ingress.	060:10	09:13	03 Feb 1971
CDR and LMP entered LM.	060:30	09:33	03 Feb 1971
TV transmission ended.	060:42	09:45	03 Feb 1971

Apollo 14 Timeline

Event	GET (hhh:mm:ss)	GMT Time	GMT Date
LM system checks.	061:40	10:43	03 Feb 1971
Water dump photography.	061:50	10:53	03 Feb 1971
CDR and LMP entered CM.	062:20	11:23	03 Feb 1971
Equigravisphere.	066:09:01	15:12:03	03 Feb 1971
LM cabin pressurized.	075:20	00:23	04 Feb 1971
Midcourse correction ignition (SPS).	076:58:11.98	02:01:14	04 Feb 1971
Midcourse correction cutoff.	076:58:12.63	02:01:14	04 Feb 1971
LM ascent battery test started.	078:20	03:23	04 Feb 1971
LM ascent battery test ended.	080:20	05:23	04 Feb 1971
Lunar orbit insertion ignition (SPS).	081:56:40.70	06:59:42	04 Feb 1971
Lunar orbit insertion cutoff.	082:02:51.54	07:05:53	04 Feb 1971
S-IVB impact on lunar surface.	082:37:52.17	07:40:54	04 Feb 1971
CSM landmark tracking.	085:10	10:13	04 Feb 1971
Descent orbit insertion ignition (SPS).	086:10:52.97	11:13:55	04 Feb 1971
Descent orbit insertion cutoff.	086:11:13.78	11:14:15	04 Feb 1971
CSM landmark tracking.	087:10	12:13	04 Feb 1971
Descartes photographed.	088:50	13:53	04 Feb 1971
LM pressurized.	101:05	02:08	05 Feb 1971
Docking tunnel opened. CDR and LMP entered LM.	101:20	02:23	05 Feb 1971
LM activation and system checks.	101:30	02:33	05 Feb 1971
CSM/LM undocking and separation ignition (SM RCS).	103:47:41.6	04:50:43	05 Feb 1971
CSM/LM undocking and separation cutoff.	103:47:44.3	04:50:46	05 Feb 1971
CSM landmark tracking.	104:20	05:23	05 Feb 1971
LM landing site observation.	104:30	05:33	05 Feb 1971
CSM orbit circularization ignition (SPS).	105:11:46.11	06:14:48	05 Feb 1971
CSM orbit circularization cutoff.	105:11:50.13	06:14:52	05 Feb 1971
Checkout of LM descent propulsion system and landing radar.	105:40	06:43	05 Feb 1971
CSM landmark tracking.	106:20	07:23	05 Feb 1971
CSM orbital science photography.	107:50	08:53	05 Feb 1971
LM landing radar on.	107:57:18.66	09:00:20	05 Feb 1971
LM false "data good" indications from landing radar.	107:52:46.66	08:55:48	05 Feb 1971
LM landing radar switched to low scale.	107:57:34.66	09:00:36	05 Feb 1971
LM loading abort bit work-around routine started.	107:58:13.80	09:01:15	05 Feb 1971
LM powered descent engine ignition (DPS).	108:02:26.52	09:05:28	05 Feb 1971
LM manual throttle-up to full throttle position.	108:02:53.80	09:05:55	05 Feb 1971
LM manual target (landing site) update.	108:04:49.80	09:07:51	05 Feb 1971
LM throttle down.	108:08:47.68	09:11:49	05 Feb 1971
LM landing radar switched to high scale.	108:08:50.66	09:11:52	05 Feb 1971
LM landing radar velocity data good.	108:09:10.66	09:12:12	05 Feb 1971
LM Landing radar range data good.	109:09:12.66	10:12:14	05 Feb 1971
LM altitude updates enabled.	109:09:35.80	10:12:37	05 Feb 1971
LM approach phase program selected.	109:11:09.80	10:14:11	05 Feb 1971
LM pitchover started.	108:11:10.42	09:14:12	05 Feb 1971
LM landing radar redesignation enabled.	108:11:51.60	09:14:53	05 Feb 1971
LM radar antenna to position 2.	108:11:52.66	09:14:54	05 Feb 1971
LM attitude hold mode selected.	108:13:07.38	09:16:09	05 Feb 1971
LM landing phase program selected.	108:13:09.80	09:16:11	05 Feb 1971
LM lunar landing (left pad touchdown).	108:15:09.30	09:18:11	05 Feb 1971
LM powered descent engine cutoff.	108:15:11.13	09:18:13	05 Feb 1971
LM right, forward, and aft pad touchdown.	108:15:11.40	09:18:13	05 Feb 1971
CSM landmark tracking.	109:30	10:33	05 Feb 1971

Apollo 14 Timeline

Event	GET (hhh:mm:ss)	GMT Time	GMT Date
LM lunar surface navigation.	110:00	11:03	05 Feb 1971
CSM Gegenschein photography.	110:40	11:43	05 Feb 1971
CSM backward-looking zero phase observations.	111:20	12:23	05 Feb 1971
CSM forward-looking zero phase observations.	112:20	13:23	05 Feb 1971
CSM zodiacal light photography.	112:50	13:53	05 Feb 1971
1st EVA started (LM cabin depressurization started).	113:39:11	14:42:13	05 Feb 1971
Egress started (CDR). Pre-egress operations started (LMP).	113:47	14:50	05 Feb 1971
1st EVA television transmission started.	113:50	14:53	05 Feb 1971
CDR on lunar surface. Environmental familiarization, modular equipment transporter unloading, and television deployment (CDR).	113:51	14:54	05 Feb 1971
LMP egress.	113:55	14:58	05 Feb 1971
Environmental familiarization, contingency sample collection (LMP).	113:57	15:00	05 Feb 1971
CSM tracking of landed LM.	114:10	15:13	05 Feb 1971
S-band antenna deployment started (CDR).	114:12	15:15	05 Feb 1971
Solar wind composition experiment deployed (LMP).	114:13	15:16	05 Feb 1971
Laser-ranging retroreflector unloading started (LMP).	114:14	15:17	05 Feb 1971
Expendables transferred (CDR).	114:22	15:25	05 Feb 1971
LMP ingress.	114:23	15:26	05 Feb 1971
S-band antenna switching (LMP).	114:25	15:28	05 Feb 1971
LMP egress.	114:37	15:40	05 Feb 1971
Camera setup (LMP).	114:39	15:42	05 Feb 1971
United States flag deployed and photographed.	114:41	15:44	05 Feb 1971
LM and site inspection (CDR). Traverse to television started (LMP).	114:47	15:50	05 Feb 1971
TV panorama (LMP).	114:50	15:53	05 Feb 1971
Modular equipment transporter deployment (LMP). CSM landmark tracking.	115:00	16:03	05 Feb 1971
TV transfer to scientific equipment bay (CDR).	115:05	16:08	05 Feb 1971
Experiment package offloading started (CDR and LMP).	115:08	16:11	05 Feb 1971
TV positioning (CDR).	115:22	16:25	05 Feb 1971
Modular equipment transporter loading (CDR).	115:25	16:28	05 Feb 1971
Traverse to experiment package deployment site (CDR, LMP).	115:46	16:49	05 Feb 1971
Experiment package system interconnect, thumper and geophone unloading started (LMP).	116:03	17:06	05 Feb 1971
Experiment package system interconnect, passive seismic experiment offloading, laser-ranging retroreflector deployment.	116:04	17:07	05 Feb 1971
Mortar offloaded (LMP).	116:26	17:29	05 Feb 1971
Charged particle lunar environment experiment deployment (CDR).	116:30	17:33	05 Feb 1971
Suprathermal ion detector experiment unloading and deployment (LMP).	116:34	17:37	05 Feb 1971
Deployment of experiment package antenna, passive seismic experiment, and laser-ranging retroreflector and sample collection (CDR).	116:35	17:38	05 Feb 1971
CSM galactic survey photography.	116:40	17:43	05 Feb 1971
Penetrometer activity (LMP).	116:45	17:48	05 Feb 1971
Geophone deployment started (LMP).	116:47	17:50	05 Feb 1971
1st ALSEP data received on Earth.	116:47:58	17:51:00	05 Feb 1971
Thumper activity (LMP).	117:02	18:05	05 Feb 1971
CSM plane change ignition (SPS).	117:29:33.17	18:32:35	05 Feb 1971
CSM plane change cutoff.	117:29:51.67	18:32:53	05 Feb 1971
Mortar pack arming started (LMP).	117:37	18:40	05 Feb 1971
Return traverse started (CDR).	117:38	18:41	05 Feb 1971
Return traverse started (LMP).	117:42	18:45	05 Feb 1971
EVA closeout (LMP).	117:54	18:57	05 Feb 1971
Sample collection (CDR).	118:00	19:03	05 Feb 1971
EVA closeout (CDR).	118:03	19:06	05 Feb 1971

Apollo 14 Timeline

Event	GET (hhh:mm:ss)	GMT Time	GMT Date
CSM Earthshine photography.	118:10	19:13	05 Feb 1971
LMP ingress.	118:15	19:18	05 Feb 1971
EVA ended (LMP).	118:18	19:21	05 Feb 1971
CDR ingress.	118:19	19:22	05 Feb 1971
1st EVA television transmission ended.	118:20	19:23	05 Feb 1971
1st EVA ended (cabin pressurization started).	118:27:01	19:30:03	05 Feb 1971
VHF bistatic radar test started.	119:10	20:13	05 Feb 1971
CSM orbital science photography.	129:30	06:33	06 Feb 1971
S-band bistatic radar test started.	129:45	06:48	06 Feb 1971
VHF and S-band bistatic radar tests terminated.	130:20	07:23	06 Feb 1971
2nd EVA started (cabin depressurization started).	131:08:13	08:11:15	06 Feb 1971
CDR egress.	131:13	08:16	06 Feb 1971
Familiarization and transferal of equipment transfer bag (CDR). LMP egress. CSM vertical and orbital science photography.	131:20	08:23	06 Feb 1971
Modular equipment transporter preparation (LMP).	131:21	08:24	06 Feb 1971
Modular equipment transporter loading (CDR).	131:28	08:31	06 Feb 1971
Lunar portable magnetometer offloading (CDR).	131:38	08:41	06 Feb 1971
Lunar portable magnetometer offloading (LMP).	131:39	08:42	06 Feb 1971
2nd EVA television transmission started.	131:40	08:43	06 Feb 1971
Evaluation of modular equipment transporter track (CDR).	131:43	08:46	06 Feb 1971
Lunar portable magnetometer operation (LMP).	131:44	08:47	06 Feb 1971
Departed LM for station A (CDR).	131:46	08:49	06 Feb 1971
Departed LM for station A (LMP).	131:48	08:51	06 Feb 1971
Station A activity (CDR/LMP).	131:54	08:57	06 Feb 1971
CSM galactic survey photography.	132:25	09:28	06 Feb 1971
Departed station A for station B (CDR/LMP).	132:26	09:29	06 Feb 1971
Station B activity (CDR/LMP).	132:34	09:37	06 Feb 1971
Departed station B for station Delta (CDR/LMP).	132:39	09:42	06 Feb 1971
CSM lunar libration photography.	132:35	09:38	06 Feb 1971
Station Delta activity (CDR/LMP).	132:42	09:45	06 Feb 1971
Departed station Delta for station B1 (LMP).	132:44	09:47	06 Feb 1971
Departed station Delta for station B1 (CDR/LMP).	132:45	09:48	06 Feb 1971
Station B1 activity (CDR/LMP).	132:48	09:51	06 Feb 1971
Departed station B1 for station B2 (CDR/LMP).	132:52	09:55	06 Feb 1971
Station B2 activity (CDR/LMP).	132:57	10:00	06 Feb 1971
Departed station B2 for station B3 (CDR/LMP).	133:00	10:03	06 Feb 1971
Station B3 activity (CDR/LMP).	133:14	10:17	06 Feb 1971
Departed station B2 for station C prime (CDR/LMP).	133:16	10:19	06 Feb 1971
Station C prime activity (CDR/LMP).	133:22	10:25	06 Feb 1971
Departed station C prime for station C1 (CDR/LMP).	133:38	10:41	06 Feb 1971
Station C1 activity (CDR/LMP).	133:40	10:43	06 Feb 1971
Departed station C1 for station C2 (CDR/LMP).	133:46	10:49	06 Feb 1971
Station C2 activity (CDR/LMP).	133:52	10:55	06 Feb 1971
Departed station C2 for station E (CDR/LMP).	133:54	10:57	06 Feb 1971
Station E activity (CDR/LMP).	134:00	11:03	06 Feb 1971
Departed station E for station F (CDR/LMP).	134:02	11:05	06 Feb 1971
Station F activity (CDR/LMP).	134:06	11:09	06 Feb 1971
Departed station F for station G (CDR/LMP).	134:09	11:12	06 Feb 1971
Station G activity (CDR/LMP).	134:11	11:14	06 Feb 1971
Departed station G for station G1 (CDR/LMP).	134:47	11:50	06 Feb 1971
Station G1 activity (CDR/LMP).	134:49	11:52	06 Feb 1971

Apollo 14 Timeline

Event	GET (hhh:mm:ss)	GMT Time	GMT Date
Departed station G1 for LM (CDR/LMP).	134:52	11:55	06 Feb 1971
EVA closeout (CDR).	134:55	11:58	06 Feb 1971
EVA closeout (LMP).	134:57	12:00	06 Feb 1971
Solar wind composition experiment retrieved.	135:13	12:16	06 Feb 1971
CSM contingency photography of Descartes.	135:20	12:23	06 Feb 1971
EVA ended (LMP).	135:25	12:28	06 Feb 1971
EVA ended (CDR). Post-EVA activity operations prior to LM cabin repressurization (LMP).	135:35	12:38	06 Feb 1971
Post-EVA activity operations prior to LM cabin repressurization (CDR). 2nd EVA television transmission ended.	135:41	12:44	06 Feb 1971
2nd EVA ended (cabin repressurization started).	135:42:54	12:45:56	06 Feb 1971
LM cabin depressurized, equipment jettisoned, cabin repressurized.	136:40	13:43	06 Feb 1971
CSM landmark tracking started.	137:10	14:13	06 Feb 1971
CSM landmark tracking ended.	137:55	14:58	06 Feb 1971
Rendezvous radar activation and self-test.	138:40	15:43	06 Feb 1971
CSM backward-looking zero phase observations and orbital science photography.	139:00	16:03	06 Feb 1971
CSM forward-looking zero phase observations.	139:55	16:58	06 Feb 1971
LM lunar liftoff ignition (LM APS).	141:45:40	18:48:42	06 Feb 1971
Lunar ascent orbit cutoff.	141:52:52.1	18:55:54	06 Feb 1971
Vernier adjustment ignition (LM RCS).	141:56:49.4	18:59:51	06 Feb 1971
Vernier adjustment cutoff.	141:57:01.5	19:00:03	06 Feb 1971
Terminal phase initiation ignition.	142:30:51.1	19:33:53	06 Feb 1971
Terminal phase initiation cutoff.	142:30:54.7	19:33:56	06 Feb 1971
LM 1st midcourse correction.	142:45	19:48	06 Feb 1971
LM 2nd midcourse correction.	143:00	20:03	06 Feb 1971
Terminal phase finalize ignition.	143:13:29.1	20:16:31	06 Feb 1971
Terminal phase finalize cutoff.	143:13:55.8	20:16:57	06 Feb 1971
TV transmission started.	143:15	20:18	06 Feb 1971
TV transmission ended.	143:20	20:23	06 Feb 1971
TV transmission started.	143:28	20:31	06 Feb 1971
CSM/LM docked.	143:32:50.5	20:35:52	06 Feb 1971
TV transmission ended.	143:35	20:38	06 Feb 1971
Equipment and samples transferred to CM.	144:00	21:03	06 Feb 1971
LM ascent stage jettisoned.	145:44:58.0	22:48:00	06 Feb 1971
CSM/LM final separation ignition (SM RCS).	145:49:42.5	22:52:44	06 Feb 1971
CSM/LM final separation cutoff.	145:49:58.3	22:53:00	06 Feb 1971
Contamination control.	146:20	23:23	06 Feb 1971
LM ascent stage deorbit ignition (LM RCS).	147:14:16.9	00:17:18	07 Feb 1971
LM ascent stage fuel depletion.	147:15:33.1	00:18:35	07 Feb 1971
LM ascent stage impact on lunar surface.	147:42:23.4	00:45:25	07 Feb 1971
Apollo 12 LM impact point and Apollo 13 and Apollo 14 S-IVB impact points photographed.	147:45	00:48	07 Feb 1971
Transearth injection ignition (SPS).	148:36:02.30	01:39:04	07 Feb 1971
Transearth injection cutoff.	148:38:31.53	01:41:33	07 Feb 1971
Lunar photography.	148:55	01:58	07 Feb 1971
Cislunar navigation started.	163:30	16:33	07 Feb 1971
Cislunar navigation ended.	164:20	17:23	07 Feb 1971
Midcourse correction ignition (SM RCS).	165:34:56.69	18:37:58	07 Feb 1971
Midcourse correction cutoff.	165:34:59.69	18:38:01	07 Feb 1971
Cislunar navigation started.	165:40	18:43	07 Feb 1971
Cislunar navigation ended.	166:50	19:53	07 Feb 1971
Oxygen flow rate test attitude started.	167:25	20:28	07 Feb 1971
Oxygen flow rate test started.	167:50	20:53	07 Feb 1971

Apollo 14 Timeline

Event	GET (hhh:mm:ss)	GMT Time	GMT Date
Oxygen flow rate test ended.	169:00	22:03	07 Feb 1971
Oxygen flow rate test attitude ended.	170:40	23:43	07 Feb 1971
Contamination control.	171:20	00:23	08 Feb 1971
TV transmission started.	171:30	00:33	08 Feb 1971
Inflight demonstrations started.	171:50	00:53	08 Feb 1971
Inflight demonstrations ended.	172:09	01:12	08 Feb 1971
TV transmission ended.	172:20	01:23	08 Feb 1971
Light flash experiment started.	190:50	19:53	08 Feb 1971
Light flash experiment ended.	191:50	20:53	08 Feb 1971
TV transmission started.	194:29	23:32	08 Feb 1971
TV transmission ended.	194:52	23:55	08 Feb 1971
Earth darkside dim-light photography.	197:44	02:47	09 Feb 1971
CM/SM separation.	215:32:42.2	20:35:44	09 Feb 1971
Entry.	215:47:45.3	20:50:47	09 Feb 1971
Communication blackout started.	215:48:02	20:51:04	09 Feb 1971
Communication blackout ended.	215:51:19	20:54:21	09 Feb 1971
S-band contact with CM established by recovery forces.	215:52	20:55	09 Feb 1971
Radar contact with CM established by recovery ship.	215:53	20:56	09 Feb 1971
Drogue parachute deployed.	215:56:08	20:59:10	09 Feb 1971
Visual contact with CM established by recovery helicopter.	215:57	21:00	09 Feb 1971
Voice contact with CM established by recovery ship.	215:58	21:01	09 Feb 1971
Main parachute deployed.			
Splashdown (went to apex-up).	216:01:58.1	21:05:00	09 Feb 1971
VHF beacon contact established with CM by recovery helicopter.	216:04	21:07	09 Feb 1971
Swimmers deployed to CM.	216:09	21:12	09 Feb 1971
Flotation collar inflated.	216:17	21:20	09 Feb 1971
Decontamination swimmer deployed.	216:24	21:27	09 Feb 1971
Hatch opened for crew egress.	216:37	21:40	09 Feb 1971
Crew in life raft.	216:38	21:41	09 Feb 1971
Crew aboard recovery helicopter.	216:45	21:48	09 Feb 1971
Crew aboard recovery ship.	216:50	21:53	09 Feb 1971
Crew entered mobile quarantine facility.	217:00	22:03	09 Feb 1971
CM aboard recovery ship.	218:06	23:09	09 Feb 1971
1st sample flight departed recovery ship.	246:52	03:55	11 Feb 1971
Flight crew departed recovery ship.	260:43	17:46	11 Feb 1971
1st sample flight arrived in Houston.	263:54	20:57	11 Feb 1971
Flight crew arrived in Houston.	276:31	09:34	12 Feb 1971
Flight crew in Lunar Receiving Laboratory.	278:32	11:35	12 Feb 1971
Mobile quarantine facility and CM offloaded in HI.	408:27	21:30	17 Feb 1971
Mobile quarantine facility arrived in Houston.	418:37	07:40	18 Feb 1971
RCS deactivation completed.	457:57	23:00	19 Feb 1971
CM arrived in Houston.	528:42	21:45	22 Feb 1971
CM delivered to Lunar Receiving Laboratory.	530:27	23:30	22 Feb 1971
Crew released from quarantine.			

APOLLO 15

The Ninth Mission:
The Fourth Lunar Landing

Apollo 15 Summary

Apollo 15 crew (l. to r.): Dave Scott, Al Worden, Jim Irwin (NASA S71-37963).

Background

Apollo 15 was the first of the three Type J missions, consisting of extensive scientific investigations of the Moon on the lunar surface and from lunar orbit. It was designed to conduct exploration of the Moon over longer periods, over greater ranges, and with more instruments for scientific data acquisition than on previous Apollo missions. Major modifications and augmentations to the basic Apollo hardware were made. The most significant was the installation of a scientific instrument module in one of the service module bays for scientific investigations from lunar orbit. Other hardware changes consisted of LM modifications to accommodate a greater payload and permit a longer stay on the lunar surface, the provision of a lunar rover vehicle (LRV), and a scientific subsatellite to be deployed into lunar orbit.

Planned to be used on this and the next two lunar missions, the LRV was a four-wheeled, lightweight vehicle designed to greatly extend the area that could be explored on the lunar surface. The LRV had five major systems: mobility, crew station, navigation, power, and thermal control. Auxiliary equipment included the lunar communications relay unit with high and low gain antennas, ground control television assembly, a motion picture camera, scientific equipment, astronaut tools, and sample stowage bags. It was 10 feet 2 inches long, and 44.8 inches high, with a 7.5-foot wheelbase. Two 36-volt batteries provided power, although one alone would provide enough power for all LRV systems. Earth weight was 462 pounds, with a payload capacity of 1,080 pounds, including two astronauts and their life support equipment (about 800 pounds), communication equipment (100 pounds), scientific equipment, photographic gear (120 pounds), and lunar samples (60 pounds). For the flight to the Moon, the LRV was folded and stowed in Quad 1 of the LM descent stage. After landing, the astronauts would manually deploy the vehicle and prepare it for cargo loading and operation. The LRV was designed to operate for 78 hours during the lunar day, and could travel a cumulative distance of 35 nautical miles, within a five-mile radius from the LM.

Lunar Rover Vehicle (LRV) to be used for the first time on Apollo 15 to significantly extend the area the astronauts could explore within the constraints of time and consumables (NASA S71-00166).

The chosen landing site was an area near the foot of the Montes Apenninus (Apennine Mountains) and adjacent to Hadley Rille.

The primary objectives for Apollo 15 were:

- to perform selenological inspection, survey, and sampling of materials and surface features in a preselected area of the Hadley-Apennine region;

- to emplace and activate surface experiments;

- to evaluate the capability of the Apollo equipment to provide extended lunar surface stay time, increased extravehicular operations, and surface mobility; and

- to conduct inflight experiments and photographic tasks from lunar orbit.

The all-Air Force crew included Colonel David Randolph Scott (USAF), commander; Major Alfred Merrill Worden [WARD-in] (USAF), command module pilot; and Lt. Colonel James Benson Irwin (USAF), lunar module pilot.

Selected as an astronaut in 1963, Scott had been pilot of Gemini 8, the first docking of two vehicles in space, and command module pilot of Apollo 9, the first flight test of the LM. Born 6 June 1932 in San Antonio, Texas, he was 39 years old at the time of the Apollo 15 mission. Scott received a B.S. from the U.S. Military Academy in 1954 and a M.S. in aeronautics and astronautics from the Massachusetts Institute of Technology in 1962. His backup for the mission was Captain Richard Francis "Dick" Gordon, Jr. (USN).

Worden and Irwin were making their first spaceflights. Worden was born 7 February 1932 in Jackson, Michigan, and was 39 years old at the time of the Apollo 15 mission. He received a B.S. in military science from the U.S. Military Academy in 1955, a M.S. in astronautical and aeronautical engineering and a M.S. in instrumentation engineering from the University of Michigan in 1963, and was selected as an astronaut in 1966. His backup was Vance DeVoe Brand.

Born 17 March 1930 in Pittsburgh, Pennsylvania, Irwin was 41 years old at the time of the Apollo 15 mission. He received a B.S. in naval science from the U.S. Naval Academy in 1951, and a M.S. in aeronautical engineering and an M.S. in instrumentation engineering from the University of Michigan in 1957, and was selected as an astronaut in 1966.[1] His backup was Harrison Hagan "Jack" Schmitt, Ph.D.

The capsule communicators (CAPCOMs) for the mission were Joseph Percival Allen IV, Ph.D., Major Charles Gordon Fullerton (USAF), Karl Gordon Henize, Ph.D., Commander Edgar Dean Mitchell (USN/Sc.D.), Robert Alan Ridley Parker, Ph.D., Schmitt, Captain Alan Bartlett Shepard, Jr. (USN), Gordon, and Brand. The support crew were Henize, Allen, and Parker. The flight directors were Gerald D. Griffin (first shift), Milton L. Windler (second shift), and Glynn S. Lunney and Eugene F. Kranz (third shift).

The Apollo 15 launch vehicle was a Saturn V, designated SA-510. The mission also carried the designation Eastern Test Range #7744. The CSM was designated CSM-112, and had the call-sign "Endeavour." The lunar module was designated LM-10 and had the call-sign "Falcon."

Launch Preparations

The terminal countdown was picked up at T-28 hours at 23:00:00 GMT on 24 July 1971. Scheduled holds were initiated at T-9 hours for 9 hours 34 minutes and at T-3 hours 30 minutes for 1 hour.

At launch time, the Cape Kennedy launch area was experiencing fair weather resulting from a ridge of high pressure extending westward, from the Bermuda High, through central Florida. Cirrus clouds covered 70 percent of the sky (base 25,000 feet), the temperature was 85.6° F, the relative humidity was 68 percent, and the barometric pressure was 14.788 lb/in^2. The winds, as measured by the anemometer on the light pole 60.0 feet above ground at the launch site, measured 9.9 knots at 156° from true north. The winds, as measured at 530 feet above the launch site, measured 10.5 knots at 158° from true north.

Ascent Phase

Apollo 15 liftoff from Kennedy Space Center Pad 39A (NASA S71-41356).

Apollo 15 was launched from Kennedy Space Center Launch Complex 39, Pad A, at a Range Zero time of 13:34:00 GMT (09:34:00 a.m. EDT) on 26 July 1971. The planned launch window for Apollo 15 extended to 16:11:00 GMT to take advantage of a sun elevation angle on the lunar surface of 12.0°.

---

[1] Irwin, who had a history of heart trouble, died of a heart attack on 08 August 1991 in Glenwood Springs, Colorado.

Between 000:00:12.21 and 000:00:23.02, the vehicle rolled from a launch pad azimuth of 90° to a flight azimuth of 80.088°. The S-IC engine shut down at 000:02:39.56, followed by S-IC/S-II separation, and S-II engine ignition. The S-II engine shut down at 000:09:09.06 followed by separation from the S-IVB, which ignited at 000:09:13.20. The first S-IVB engine cutoff occurred at 000:11:34.67, with deviations from the planned trajectory of only -2.0 ft/sec in velocity and only 0.4 n mi in altitude.

The maximum wind conditions encountered during ascent were 36.2 knots at 63° from true north at 45,110 feet and a maximum wind shear of 0.0110 sec^{-1} at 36,830 feet.

Parking orbit conditions at insertion, 000:11:44.67 (S-IVB cutoff plus 10 seconds to account for engine tailoff and other transient effects), showed an apogee and perigee of 91.5 by 89.6 n mi, an inclination of 29.679°, a period of 87.84 minutes, and a velocity of 25,602.7 ft/sec. The apogee and perigee were based upon a spherical Earth with a radius of 3,443.934 n mi.

The international designation for the CSM upon achieving orbit was 1971-063A and the S-IVB was designated 1971-063B. After undocking at the Moon, the LM ascent stage would be designated 1971-063C, the descent stage 1971-063E, and the subsatellite 1971-063D.

Translunar Phase

After inflight systems checks, the 350.79-second translunar injection maneuver (second S-IVB firing) was performed at 002:50:02.90. The S-IVB engine shut down at 002:55:53.61 and translunar injection occurred ten seconds later at a velocity of 35,606.5 ft/sec after 1.5 Earth orbits lasting 2 hours 44 minutes 19.02 seconds.

At 003:22:27.2, the CSM was separated from the S-IVB stage, transposed, and docked at 003:33:49.50. Onboard color television was initiated to cover the docking. The docked spacecraft were ejected from the S-IVB at 004:18:01.2, and an 80.2-second separation maneuver was performed at 004:40:01.8.

At 005:46:00.7, the S-IVB tanks were vented and the auxiliary propulsion system was fired for 241.2 seconds to target the S-IVB for a lunar impact. An additional 71-second maneuver was made at 010:00:01, about 30 minutes later than planned. The late burn provided additional tracking time to compensate for any trajectory perturbations introduced by liquid oxygen and liquid hydrogen tank venting. The S-IVB impacted the lunar surface at 079:24:41.55. The

impact point was latitude 0.99° south and longitude 11.89° west, 83 n mi from the target point, 191 n mi from the Apollo 12 seismometer, and 102 n mi from the Apollo 14 seismometer. At impact, the S-IVB weighed 30,880 pounds and was traveling 8,455 ft/sec.

View of Earth from approximately 30,000 n mi (NASA AS15-91-12343).

Two minor midcourse corrections were required during translunar flight to assure proper lunar orbit injection. The first was a 0.8-second maneuver at 028:40:22.00 that produced a change in velocity of 5.3 ft/sec.

The second midcourse correction was performed with the service propulsion system bank A in order to provide better analysis of an apparent intermittent short. Because power could still be applied to the valve with a downstream short, bank A could be operated satisfactorily in the manual mode for subsequent firings. The redundant bank B system was nominal and could be used for automatic starting and shutdown.

The LM crew entered the LM at 033:56 for checkout, approximately 50 minutes earlier than scheduled. LM communications checks were performed between 034:21 and 034:45. Good quality voice and data were received even though the Goldstone tracking station in California was not yet configured correctly during the initial portion of the down-voice backup checks. Approximately 15 minutes later, the downlink carrier lock was lost for a minute and a half; however, because other stations were tracking, data loss was reduced to just a few seconds.

A television transmission of the CSM and LM interiors was broadcast between 034:55 and 035:46. Camera operation was nominal, but the picture quality varied with the lighting of the scene observed. During the checkout of the LM, the crew discovered the range/range rate exterior cover glass was broken, removing the helium barrier. Subsequent ground testing qualified the unprotected meter for use during the remainder of the mission in the spacecraft ambient atmosphere.

Intravehicular transfer and LM housekeeping began at 056:26, about an hour and a half earlier than scheduled. The crew vacuumed the LM to remove broken glass from the damaged range/range rate meter. LM checkout was completed as planned.

Based on the first midcourse correction burn test data, it was decided to perform all service propulsion system maneuvers except lunar orbit insertion and transearth injection using bank B only. The insertion and injection maneuvers would be dual bank burns with modified procedures to permit automatic start and shutdown on bank B. The second midcourse correction, using this propulsion system, was made at 073:31:14.81 for 0.91 seconds and changed the velocity by 5.4 ft/sec.

The scientific instrument bay door was jettisoned at 074:06:47.1. The lunar module pilot photographed the jettisoned door and visually observed it slowly tumbling through space away from the CSM and eventually into heliocentric orbit.

At 078:31:46.70, at an altitude of 86.7 n mi above the Moon, the service propulsion engine was fired for 398.36 seconds, inserting the spacecraft into a lunar orbit of 170.1 by 57.7 n mi. The translunar coast had lasted 75 hours 42 minutes 21.37 seconds. During the burn, bank A was shut down 32 seconds before planned cutoff to obtain performance data on bank B for future single bank burns.

Lunar Orbit/Lunar Surface Phase

At 082:39:49:09, a 24.53-second service propulsion system maneuver was performed to establish the descent orbit of 58.5 by 9.6 n mi in preparation for undocking of the LM. A 30.40-second orbit trim maneuver was performed at 095:56:44.70 and adjusted the orbit to 60.3 by 8.8 n mi.

During the 12th lunar revolution on the far side of the Moon at about 100:14, the CSM/LM undocking and separation maneuver was initiated; however, undocking did not occur. The crew and ground control decided that the probe instrumentation LM/CSM umbilical was either loose or disconnected. The command module pilot went into the

tunnel to inspect the connection and found the umbilical plug to be loose. After reconnecting the plug and adjusting

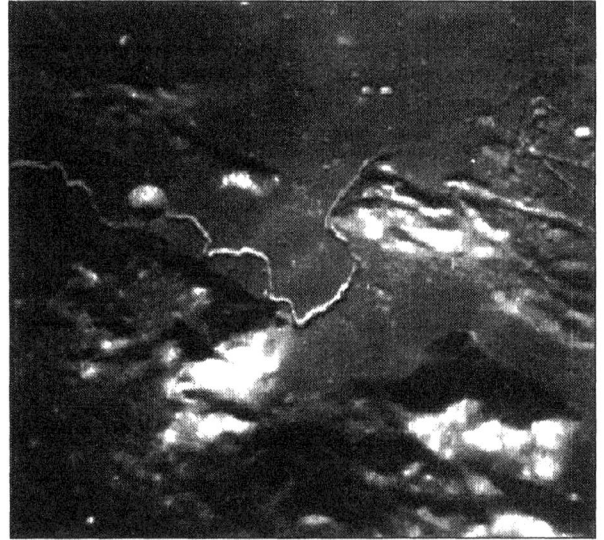

Wide angle view of the target landing site—the Apennine Mountains, adjacent to Hadley Rille (jagged line) (NASA AS15-94-12811).

the spacecraft attitude, undocking and separation were achieved approximately 25 minutes late at 100:39:16.2 at an altitude of 7.4 n mi. A 3.67-second maneuver at 101:38:58.98 circularized the CSM orbit to 65.2 by 54.8 n mi in preparation for the acquisition of scientific data.

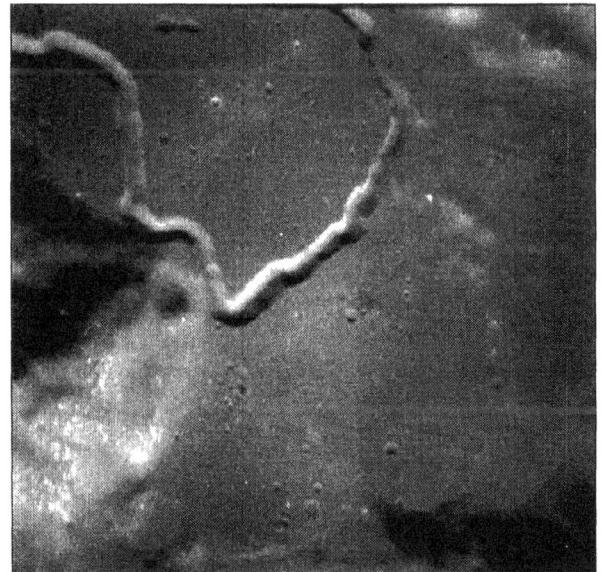

View of planned landing site taken with 250 mm lens (NASA AS15-96-13010).

The powered descent engine firing began at 104:30:09.4 at an altitude of 5.8 n mi and ended 739.2 seconds later, just 0.7 seconds before landing at 22:16:29 GMT (06:16:29 p.m. EDT) on 30 July at 104:42:29.3. The spacecraft landed in the Montes Apenninus (Apennine Mountains), adjacent to Hadley Rille at latitude 26.13222° north and longitude 3.63386° east, and about 1,800 feet from the planned landing point. Approximately 103 seconds of engine firing time remained at landing.

At 106:42:49, two hours after landing, the cabin was depressurized and the commander opened the LM top hatch to photograph and describe the area surrounding the landing site. During this "stand-up EVA" (SEVA), which lasted 33 minutes 7 seconds, he took a series of panoramic photos of the area immediately surrounding the LM landing site.

The first lunar surface extravehicular activity was initiated at 119:39:17 when the cabin of the LM was depressurized.

View of Hadley Delta as seen from the top hatch of the LM during the SEVA period following the LM landing (NASA AS15-87-11748).

On the way down the ladder, the commander deployed the modularized equipment stowage assembly (MESA). The television in the MESA was activated and the pictures of the commander's remaining descent to the lunar surface were excellent. The lunar module pilot then exited to the surface. While the commander removed the television camera from the MESA and deployed it on a tripod, the lunar module pilot collected the contingency sample.

Scott maneuvers LRV at the start of EVA-1 (NASA AS15-85-11471).

At 120:18:31, the crew offloaded the LRV and deployed it 13 minutes later. They unstowed the third Apollo lunar surface experiments package (ALSEP) and other equipment, and configured the LRV for lunar surface operations.

Some problems were experienced in deploying and checking out the LRV but these problems were worked out. During checkout of the LRV, it was found that the front steering mechanism was inoperative. Additionally, there were no readouts on the LRV battery #2 ampere/volt meter. After minor troubleshooting, a decision was made to perform the first extravehicular activity (EVA-1) without the LRV front wheel steering activated.

Irwin works at LRV during EVA-1 (NASA AS15-86-11602).

At 121:44:55, the crew drove the LRV to Elbow Crater, collected and documented samples and gave an enthusiastic and informative commentary on lunar features.

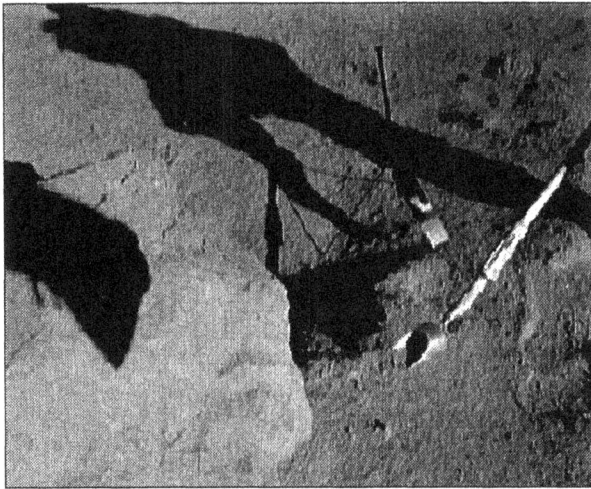

The LMP (large shadow on right) is about to sample fillet on east side of Station 2 boulder during EVA-1. Shadow at left is sample bag held by CDR (NASA (AS15-86-11548).

The mission control center provided television control during various stops. After obtaining additional samples and photographs near St. George Crater, the crew returned to the LM using the LRV navigation system.

Underside of overturned boulder during EVA-1 (NASA AS15-86-11563).

The crew then proceeded to the selected Apollo lunar surface experiments package deployment site, 360 feet west-northwest of the LM. There, the experiments were deployed essentially as planned, except that the second heatflow experiment probe was not emplaced because drilling was more difficult than expected and the hole was not completed.

Irwin at LRV at end of EVA-1 (NASA AS15-86-11603).

The crew entered the LM and the cabin was repressurized at 126:11:59. The first extravehicular activity lasted 6 hours 32 minutes 42 seconds, about 27 minutes less than planned because of higher than anticipated oxygen usage by the commander. The distance traveled in the lunar rover vehicle was 33,800 feet (10.3 km), vehicle drive time was 1 hour 2 minutes, parked time was 1 hour 14 minutes, and an estimated 31.9 pounds (14.5 kg) of samples were collected.

Between the first and second extravehicular periods, the crew spent 16 hours in the LM. The second period began at 142:14:48 when the cabin was depressurized.

After the crew left the LM for the second EVA, they checked out the LRV and prepared it for the second traverse. During the checkout, they recycled the circuit breakers on the vehicle and the front steering became completely operational.

The crew started their traverse at 143:10:43, heading south to the Apennine front, just east of the first traverse. Stops were made at Spur Crater and other points along the base of the front, as well as at Dune Crater on the return trip. Television transmission was very good.

Checkout of LRV prior to EVA-2. Note surface map hanging from steering bar (NASA AS15-82-11200).

The return route closely followed the outbound route. Documented samples, a core sample, and a comprehensive sample were collected, and photographs were taken.

Scott reaches for drill during EVA-2. Solar Wind Spectrometer is in foreground (NASA AS15-87-11847).

During this period, the lunar module pilot performed soil mechanics tasks. The commander also tried to drill for a deep-core sample but terminated the effort because of time constraints.

Irwin uses scoop to make a trench in the lunar soil (NASA AS15-92-12424).

After reaching the LM at 148:32:17, the crew returned to the experiments package site where the commander completed drilling the second hole for the heat flow experiment, emplaced the probe, and collected a core tube sample. The drill core stems were left at the ALSEP site for retrieval during EVA-3.

Rock-strewn "relatively fresh" crater. Apennine Front and Hadley Delta are in background (NASA AS15-82-11082).

Scott on slope of Hadley Delta during EVA-2 (NASA AS15-85-11514).

Scott and Irwin gather lunar samples during EVA-2 (TV still image (NASA S71-41426).

View over station 6a into the Swann Hills (AS15-90-12188).

LM photographed against rolling lunar hills during EVA-2 (NASA AS15-82-11057).

The crew then returned to the LM and deployed the United States flag. The sample container and film were stowed in the LM.

The crew entered the LM and the cabin was repressurized at 149:27:02. The second extravehicular activity period lasted 7 hours 12 minutes 14 seconds. The distance traveled in the lunar rover vehicle was 41,000 feet (12.5 km), vehicle drive time was 1 hour 23 minutes, the vehicle was parked for 2 hour 34 minutes, and an estimated 76.9 pounds (34.9 kg) of samples were collected.

The crew spent almost 14 hours in the LM before the cabin was depressurized for the third extravehicular period at 163:18:14. The third extravehicular activity began 1 hour 45 minutes later than planned due to cumulative changes in the surface activities timeline. Because of this delay and later delays at the ALSEP site, the planned trip to the North Complex was deleted.

Irwin salutes U.S. flag during EVA-3. LM is in background; LRV to the right (NASA AS15-88-11866).

The first stop was the ALSEP site at 164:09:00 to retrieve drill core stem samples left during EVA-2. Two core sections were disengaged and placed in the LRV. The drill and the remaining four sections could not be separated and were left for later retrieval.

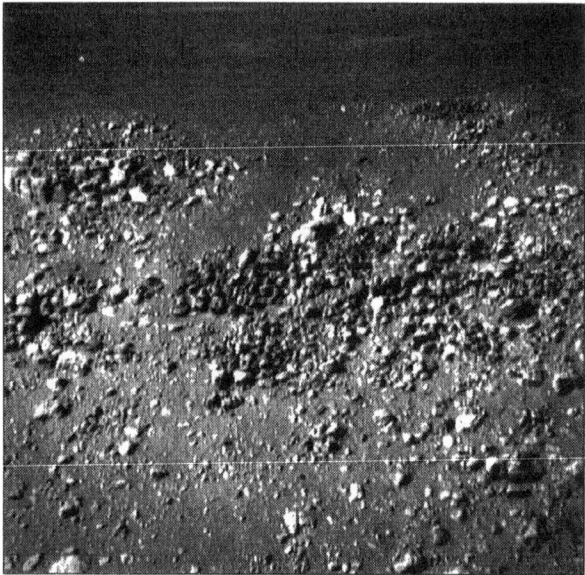

Group of boulders on the west wall of Hadley Rille seen during EVA-3 (NASA AS15-89-12074).

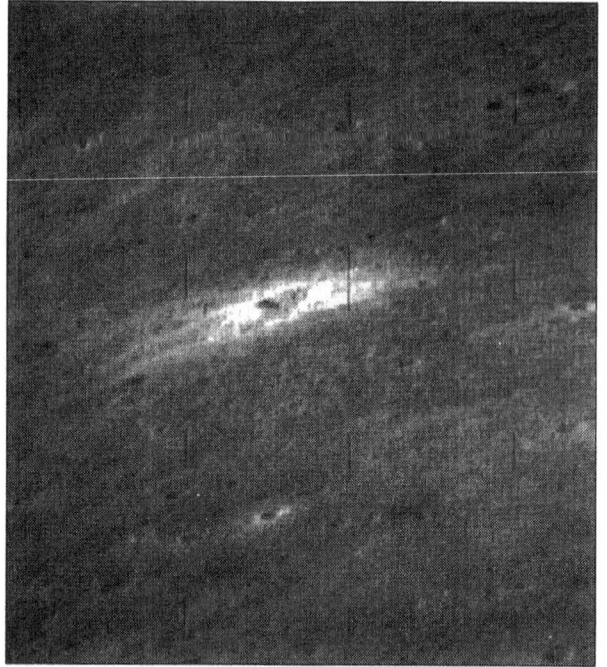

Interesting feature encountered during EVA-3—a white ejecta crater on the east rim of St. George Crater (NASA AS15-89-12164)

The third geologic traverse took a westerly direction and included stops at Scarp Crater, Rim Crater, and "The Terrace" near Rim Crater. Extensive samples and a double-core-tube sample were obtained.

Scott has put his tongs atop station 6a boulder. Notice the LRV right front wheel is off ground (NASA AS15-86-11658).

Blocky-rimmed crater on the left flank of Swann Mountain rises into the background, seen during EVA-3 (NASA AS15-89-12177).

To prove that items of different mass fall at the same speed in zero gravity, Scott drops feather and hammer—and it works—as seen in this TV still (NASA S71-43788).

Photographs were taken of the west wall of Hadley Rille, where exposed layering was observed. The return trip was east toward the LM with a stop at the ALSEP site at 166:43:40 to retrieve the remaining sections of the deep-core sample. One more section was separated, and the remaining three sections were returned in one piece. During sample collecting, the commander tripped over a rock and fell, but experienced no difficulty in getting up.

To honor fallen astronauts and cosmonauts, a plaque and human image were left on the lunar surface (NASA AS15-88-11894).

After returning to the LM, the LRV was unloaded and parked at 167:35:24 for ground-controlled television coverage of the LM ascent. The commander selected a site slightly closer to the LM than planned in order to take advantage of more elevated terrain for better television coverage of the ascent.

Final parking site for the first lunar rover vehicle which will televise the liftoff of the Apollo 15 LM ascent stage (NASA AS15-88-11901).

The third extravehicular period lasted 4 hours 49 minutes 50 seconds. The distance traveled in the lunar rover vehicle was 16,700 feet (5.1 km), vehicle drive time was 35 minutes, the vehicle was parked for 1 hour 22 minutes, and an estimated 60.2 pounds (27.3 kg) of samples were collected. The crew reentered the LM and the cabin was repressurized at 168:08:04, thus ending the Apollo program's fourth piloted exploration of the Moon.

For the mission, the total time spent outside the LM was 18 hours 34 minutes 46 seconds, the total distance traveled in the lunar rover vehicle was 91,500 feet (27.9 km), vehicle drive time was 3 hours 0 minutes, the vehicle was parked during extravehicular activities for 5 hours 10 minutes, and the collected samples totaled 170.44 pounds (77.31 kg; official total in kilograms as determined by the Lunar Receiving Laboratory in Houston). The farthest point traveled from the LM was 16,470 feet.

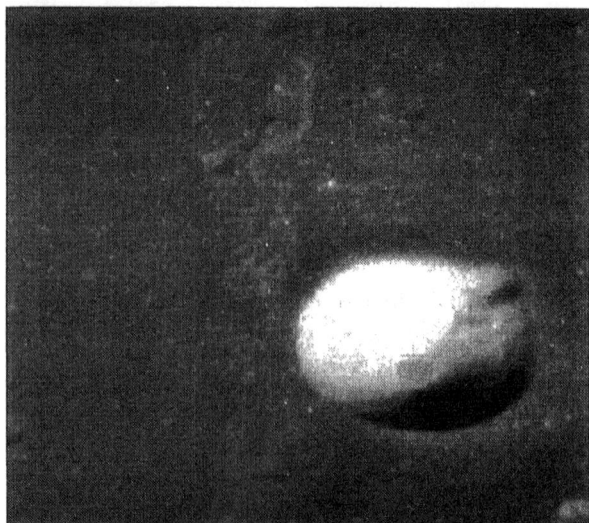

Crater La Hire A, a classic bowl-shaped crater with a ridge to the south (NASA AS15-81-11039).

While the LM was on the surface, the command module pilot completed 34 lunar orbits, conducting scientific instrument module experiments and operating cameras to obtain data concerning the lunar surface and the lunar environment.

Some scientific tasks accomplished during this time were photographing the sunlit lunar surface, gathering data needed for mapping the bulk chemical composition of the lunar surface and for determining the geometry of the Moon along the ground track, visually surveying regions of the Moon to assist in identifying processes that formed geologic features, obtaining lunar atmospheric data, and surveying gamma ray and x-ray sources.

High-resolution photographs were obtained with the panoramic and mapping cameras during the missions. An 18.31-second CSM plane change maneuver had been conducted at 165:11:32.74 and resulted in an orbit of 64.5 by 53.6 n mi.

Craters Aristarchus and Herodotus as seen from the CM (NASA AS15-88-11980).

Oblique view of previous photo—the lunar nearside near northeast ridge of Ocean of Storms (NASA AS15-88-12002).

Ignition of the ascent stage engine for lunar liftoff occurred at 17:11:23 GMT (01:11:23 p.m. EDT) on 2 August 1971 at 171:37:23.2. The LM had been on the lunar surface for 66 hours 54 minutes 53.9 seconds.

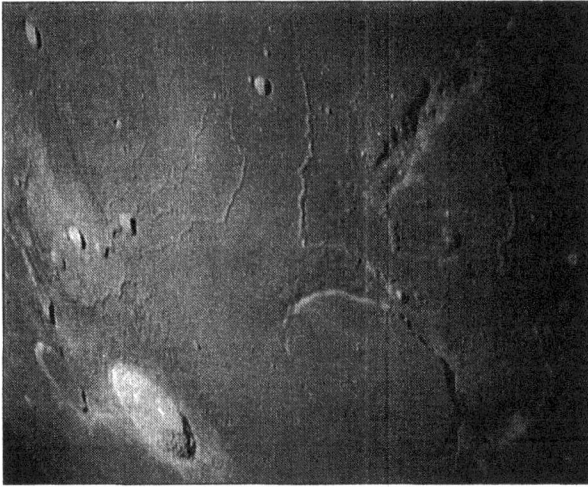

Crater Prinz (left) and Cobra's Head features of Shroter's Valley (NASA AS15-93-12602).

Liftoff of the LM ascent stage as seen from the TV camera mounted on the LRV (NASA S71-41512).

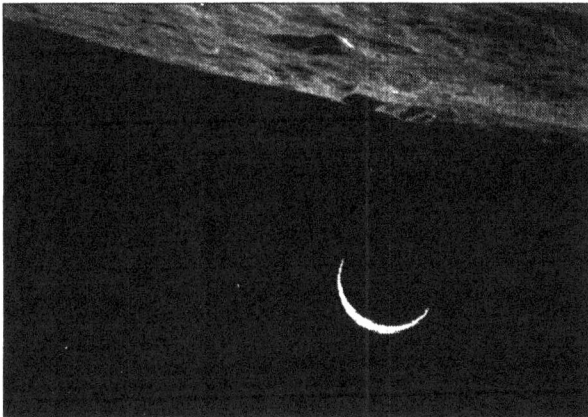

Interesting view of crescent Earthrise as seen from the CM during revolution 70 (NASA AS15-97-13267).

The 431.0-second firing achieved the initial lunar orbit of 42.5 by 9.0 n mi. Several rendezvous sequence maneuvers were required before docking could occur approximately two hours later.

A 2.6-second terminal phase initiate maneuver at 172:29:40.0 adjusted the ascent stage orbit to 64.4 by 38.7 n mi. The ascent stage and the CSM docked at 173:36:25.5 at an altitude of 57.0 n mi. The two craft had been undocked for 72 hours 57 minutes 9.3 seconds.

After transfer of the crew and samples to the CSM, the ascent stage was jettisoned at 179:30.01.4, and the CSM was prepared for transearth injection. Jettison had been delayed one revolution because of difficulty verifying the spacecraft tunnel sealing and astronaut pressure suit integrity.

View of the CM and the Scientific Instrument Module (SIM) bay following rendezvous with the LM ascent stage (NASA AS15-88-11972).

At 181:04:19.8 and 61.5 n mi altitude, the ascent stage was maneuvered to impact the lunar surface by firing the engine to depletion, which occurred 83.0 seconds after ignition. Impact occurred at latitude 26° 21' north and longitude 0° 15' east 03:03:37 GMT on 3 August (11:03:37 p.m. EDT on 2 August) at 181:29:35.8. The impact point was 12.7 n mi (23.5 km) from the planned point and 50 n mi (93 km) west of the Apollo 15 landing site. The impact was recorded by the Apollo 12, 14, and 15 seismic stations.

In preparation for the launch of a subsatellite into lunar orbit, a 3.42-second orbit-shaping maneuver at 221:20:48.02 altered the CSM orbit to 76.0 by 54.3 n mi. The subsatellite was then spring-ejected from the scientific instrument module bay at 222:39:29.1 during the 74th revolution into an orbit of 76.3 by 55.1 n mi at an inclination of -28.7°. The subsatellite was instrumented to measure plasma and energetic-particle fluxes, vector magnetic fields, and subsatellite velocity from which lunar gravitational anomalies could be determined. All systems operated as expected.

Following a 140.90-second maneuver at 67.6 n mi altitude at 223:48:45.84, transearth injection was achieved at 223:51:06.74 at a velocity of 8,272.4 ft/sec after 74 lunar orbits lasting 145:12:41.68.

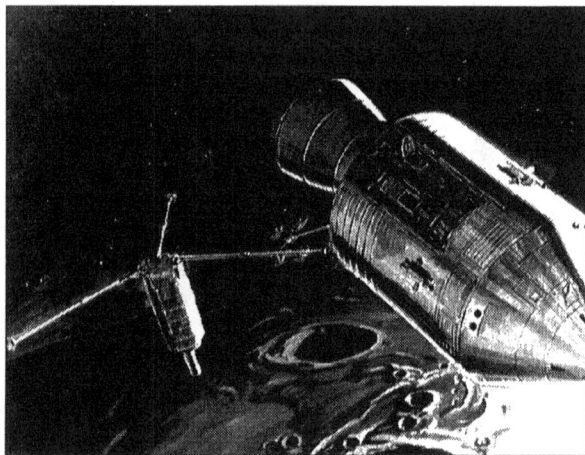

Artist's concept of deployment of the subsatellite deployed into lunar orbit from the SIM bay (NASA S71-39481).

Transearth Phase

At 241:57:12, the command module pilot began a transearth coast extravehicular activity. Television coverage was provided for the 39-minute 7-second extravehicular period during which Worden retrieved panoramic and mapping camera film cassettes from the scientific instrument module bay.

Three excursions were made to the bay. The film cassettes were retrieved during trips one and two. The third trip was used to observe and report the general condition of the instruments, in particular the mapping camera.

The command module pilot reported no evidence of the cause for the mapping camera extend/retract mechanism failure in the extended position and no observable reason for the pan camera velocity/altitude sensor failure. He also

reported that the mass spectrometer boom was not fully retracted. The EVA was completed at 242:36:19. This brought the total extravehicular activity for the mission to 19 hours 46 minutes 59 seconds.

Seismometer readings are studied in Mission Control (NASA S71-41422).

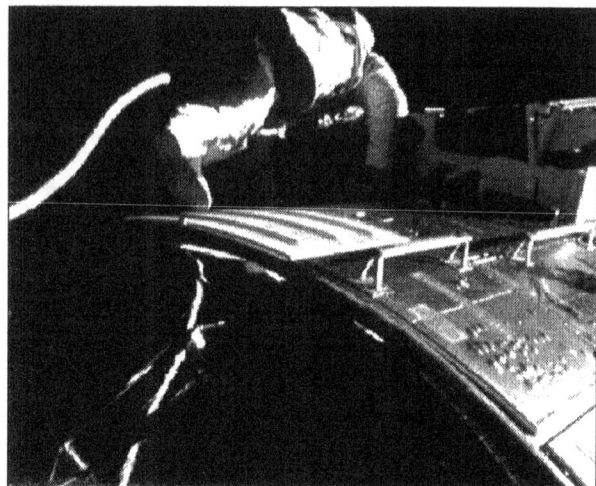

CMP Al Worden retrieves film cassettes from the service module during a transearth EVA (NASA S71-43202).

A 22.30-second midcourse correction of 5.6 ft/sec was performed 291:56:49.91 to put the CSM on a proper track for Earth entry.

Recovery

The service module was jettisoned at 294:43:55.2, and CM entry followed a normal profile. The command module reentered the Earth's atmosphere (400,000 feet altitude) at 294:58:54.7 at a velocity of 36,096 ft/sec, following a transearth coast of 71 hours 7 minutes 48.0 seconds.

The parachute system, with two main parachutes properly inflated and one collapsed, effected splashdown of the CM in the Pacific Ocean at 20:45:53 GMT (04:45:53 p.m. EDT) on 7 August. Mission duration was 295:11:53.0. The impact point was about 1.0 n mi from the target point and 5 n mi from the recovery ship U.S.S. *Okinawa*.

The collapsed parachute contributed to the fastest entry time in the Apollo program, just 778.3 seconds from entry to splashdown. The splashdown site was estimated to be latitude 26.13° north and longitude 158.13° west.

After splashdown, the CM assumed an apex-up flotation attitude. The crew was retrieved by helicopter and was aboard the recovery 39 minutes after splashdown. The CM was recovered 55 minutes later. The estimated CM weight at splashdown was 11,731 pounds, and the estimated distance traveled for the mission was 1,107,945 n mi.

Although one of the three parachutes collapsed prior to CM splashdown, the crew was not harmed (NASA S71-41999).

Apollo 15 crew members are welcomed by family members upon arrival at Ellington AFB (NASA S71-43428).

Conclusions

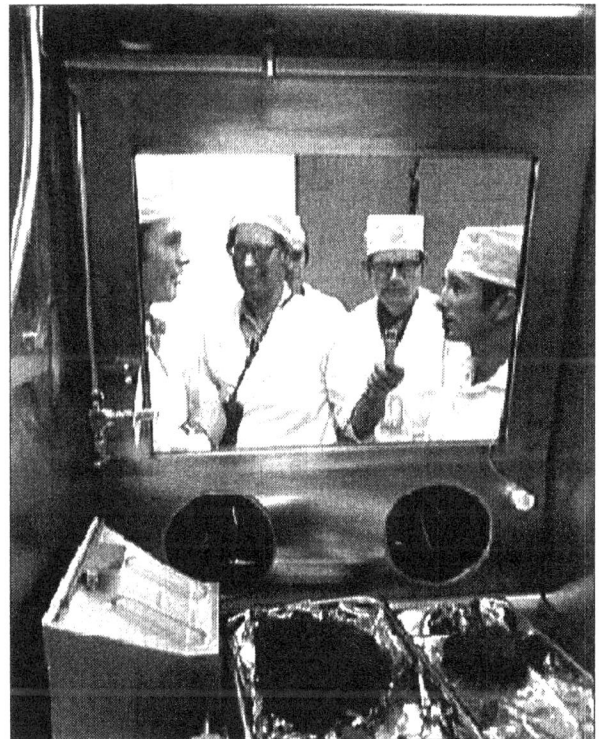

Scott (left) and Irwin join geologists in looking at Apollo 15 rock samples (NASA S71-43203).

The mission accomplished all primary objectives and provided scientists with a large amount of new information concerning the Moon and its characteristics.

Rock sample No. 15415 in the Lunar Receiving Laboratory, Houston (NASA S71-42951).

The Apollo 15 mission was the fourth lunar landing and resulted in the collection of a wealth of scientific information. The Apollo system, in addition to providing a means of transportation, excelled as an operational scientific facility. The following conclusions were made from an analysis of post-mission data:

1. The Apollo 15 mission demonstrated that, with the addition of consumables and the installation of scientific instruments, the CSM is an effective means of gathering scientific data. Real-time data allowed participation by scientists with the crew in planning and making decisions to maximize scientific results.

2. The mission demonstrated that the modified launch vehicle, spacecraft, and life support system configurations could successfully transport larger payloads and safely extend the time spent on the Moon.

3. The modified pressure garment and portable life support systems provided better mobility and extended the lunar surface extravehicular time.

4. The ground-controlled mobile television camera allowed greater real-time participation by Earth-bound scientists and operational personnel during lunar surface extravehicular activity.

5. The practicality of the lunar rover vehicle was demonstrated by greatly increasing load-carrying capability and range of exploration of the lunar surface.

6. The lunar communications relay unit provided the capability for continuous communications en route to and at the extended ranges made possible by the lunar rover vehicle.

7. Landing site visibility was improved by the use of a steeper landing trajectory.

8. Apollo 15 demonstrated that the crew could operate to a greater degree as scientific observers and investigators and rely more on the ground support team for systems monitoring.

9. The value of human space flight was further demonstrated by the unique human capability to observe and think creatively, as shown in the supplementation and redirection of many tasks by the crew to enhance scientific data return.

10. The mission confirmed that, in order to maximize mission success, crews should train with actual flight equipment or equipment with equal reliability.

In an attempt to grow "germ-free" plants, lettuce, tomato, and citrus plants are grown in lunar soil returned by Apollo 15 (NASA S71-51318).

Apollo 15 Objectives

Spacecraft Primary Objectives

1. To perform selenological inspection, survey, and sampling of materials and surface features in a preselected area of the Hadley-Apennine region. *Achieved.*

2. To emplace and activate surface experiments. *Achieved.*

3. To evaluate the capability of the Apollo equipment to provide extended lunar surface stay time, increased extravehicular operations, and surface mobility. *Achieved.*

4. To conduct inflight experiments and photographic tasks from lunar orbit. *Achieved.*

Detailed Objectives

1. Lunar rover vehicle evaluation. *Achieved.*

2. Extravehicular communications with the lunar communications relay unit and ground controlled television assembly. *Achieved.*

3. Extravehicular mobility unit assessment on lunar surface. *Achieved.*

4. Lunar module landing effects evaluation. *Achieved.*

5. Service module orbital photographic tasks. *Achieved.*

6. Command module photographic tasks. *Achieved.*

7. Scientific instrument module thermal data. *Achieved.*

8. Scientific instrument module inspection during extravehicular activity. *Achieved.*

9. Scientific instrument module door jettison evaluation. *Achieved.*

10. Lunar module descent engine performance. *Achieved.*

11. Visual observations from lunar orbit. *Achieved.*

12. Visual light flash phenomenon. *Achieved.*

Experiments

1. Contingency sample collection. *Achieved.*

2. ALSEP V: Apollo Lunar Scientific Experiment Package.

   a. Passive seismic. *Achieved.*

   b. Lunar surface magnetometer. *Achieved.*

   c. Solar wind spectrometer. *Achieved.*

   d. Suprathermal ion detector. *Achieved.*

   e. Heat flow. *Achieved.*

   f. Cold cathode ion gauge. *Achieved.*

3. Lunar geology investigation. *Achieved.*

4. Laser ranging retroreflector. *Achieved.*

5. Solar wind composition. *Achieved.*

6. Gamma ray spectrometer. *Achieved.*

7. X-ray fluorescence. *Achieved.*

8. Alpha particle spectrometer. *Achieved.*

9. S-band transponder (command and service module and lunar module). *Achieved.*

10. Mass spectrometer. *Achieved.*

11. Downlink bistatic radar observations of the Moon. *Achieved.*

12. Apollo window meteoroid. *Achieved.*

13. Ultraviolet photography of the Earth and Moon. *Achieved.*

14. Gegenschein from lunar orbit. *Not achieved. The fourteen 35-mm photographs scheduled for this experiment were not obtained due to an error in the spacecraft photographic attitudes.*

15. Soil mechanics. *Achieved.*

16. Bone mineral measurement. *Achieved.*

17. Lunar dust detector. *Achieved.*

Subsatellite Experiments

1. S-164: S-band transponder. *Achieved.*

2. S-173: Particle shadows/boundary layer. *Achieved.*

3. S-174: Magnetometer. *Achieved.*

Operational Tests

1. For Manned Spacecraft Center.

   a. Lunar gravity measurement using the lunar module primary guidance system. *Achieved.*

   b. Lunar module voice and data relay test. *Achieved.*

2. For Department of Defense/Kennedy Space Center.

   a. Chapel Bell (classified Department of Defense test). *Results classified.*

   b. Radar skin tracking. *Results classified.*

   c. Ionospheric disturbance from missiles. *Results classified.*

d. Acoustic measurement of missile exhaust noise. *Results classified.*

e. Army acoustic test. *Results classified.*

f. Long-focal-length optical system. *Results classified.*

g. Sonic boom measurement. *Results classified.*

Launch Vehicle Objectives

1. To launch on a flight azimuth between 80° and 100° and insert the S-IVB/instrument unit/spacecraft into the planned circular Earth parking orbit. *Achieved.*

2. To restart the S-IVB during either the second or third revolution and inject the S-IVB/instrument unit/spacecraft into the planned translunar trajectory. *Achieved.*

3. To provide the required attitude control for the S-IVB/instrument unit/spacecraft during transposition, docking, and ejection. *Achieved.*

4. To perform an evasive maneuver after ejection of the command and service module/lunar module from the S-IVB/instrument unit. *Achieved.*

5. To attempt to impact the S-IVB/instrument unit on the lunar surface within 350 kilometers (189 nautical miles) of latitude 3.65° south, longitude 7.58° west. *Achieved.*

6. To determine actual impact point within 5.0 kilometers (2.7 nautical miles) and time of impact within one second. *Achieved.*

7. To vent and dump the remaining gases and propellants to safe the S-IVB/instrument unit. *Achieved.*

Apollo 15 Spacecraft History

EVENT	DATE
Individual and combined CM and SM systems test completed at factory.	05 Nov 1969
Saturn S-II stage #10 delivered to KSC.	18 May 1970
Saturn S-IVB stage #510 delivered to KSC.	13 Jun 1970
Saturn V instrument unit #510 delivered to KSC.	26 Jun 1970
Saturn S-IC stage #10 delivered to KSC.	06 Jul 1970
Saturn S-IC stage #10 erected on MLP #3.	08 Jul 1970
Spacecraft/LM adapter #19 delivered to KSC.	08 Jul 1970
Saturn S-II stage #10 erected.	15 Sep 1970
Saturn S-IVB stage #510 erected.	16 Sep 1970
Saturn V instrument unit #510 erected.	17 Sep 1970
LM #10 final engineering evaluation acceptance test at factory.	21 Sep 1970
LM #10 integrated test at factory.	21 Sep 1970
LM ascent stage #10 ready to ship from factory to KSC.	04 Nov 1970
LM ascent stage #10 delivered to KSC.	06 Nov 1970
LM descent stage #10 ready to ship from factory to KSC.	16 Nov 1970
Launch vehicle electrical systems test completed.	17 Nov 1970
Integrated CM and SM systems test completed at factory.	24 Nov 1970
CM #112 and SM #112 ready to ship from factory to KSC.	11 Jan 1971
CM #112 and SM #112 delivered to KSC.	14 Jan 1971
CM #112 and SM #112 mated.	18 Jan 1971
LM ascent stage #10 and descent stage #10 mated.	09 Feb 1971
LM #10 combined systems test completed.	12 Feb 1971
CSM #112 combined systems test completed.	08 Mar 1971
LRV #1 delivered to KSC.	15 Mar 1971
LM #10 altitude tests completed.	06 Apr 1971
CSM #112 altitude tests completed.	09 Apr 1971
Launch vehicle propellant dispersion/malfunction overall test completed.	15 Apr 1971
Launch vehicle service arm overall test completed.	27 Apr 1971
LRV #1 installed.	28 Apr 1971
CSM #112 moved to VAB.	08 May 1971
Spacecraft erected.	08 May 1971
Space vehicle and MLP #3 transferred to launch complex 39A.	11 May 1971
LM #10 combined systems test completed.	17 May 1971
CSM #112 integrated systems test completed.	18 May 1971
CSM #112 electrically mated to launch vehicle.	07 Jun 1971
Space vehicle overall test #1 (plugs in) completed.	09 Jun 1971
LM #8 flight readiness test completed.	10 Jun 1971
Space vehicle flight readiness test completed.	22 Jun 1971
Saturn S-IC stage #10 RP-1 fuel loading completed.	06 Jul 1971
Space vehicle countdown demonstration test (wet) completed.	13 Jul 1971
Space vehicle countdown demonstration test (dry) completed.	14 Jul 1971

Apollo 15 Ascent Phase

Event	GET (hhh:mm:ss)	Altitude (n mi)	Range (n mi)	Earth Fixed Velocity (ft/sec)	Space Fixed Velocity (ft/sec)	Event Duration (deg E)	Geocentric Latitude (deg N)	Longitude (deg E)	Space Fixed Flight Path Angle (deg)	Space Fixed Heading Angle (E of N)
Liftoff	000:00:00.58	0.060	0.000	1.5	1,340.7		28.4470	-80.6041	0.07	90.00
Mach 1 achieved	000:01:05.0	4.224	1.004	1,052.0	2,028.1		28.4497	-80.5854	27.86	87.36
Maximum dynamic pressure	000:01:22.0	7.401	2.970	1,661.1	2,681.3		28.4555	-80.5847	29.80	85.77
S-IC center engine cutoff[2]	000:02:15.96	25.271	25.987	5,518.4	6,708.5	142.5	28.5203	-80.1190	24.217	82.494
S-IC outboard engine cutoff	000:02:39.56	36.947	48.610	7,811.3	9,043.3	166.1	28.5824	-79.6961	21.266	82.129
S-IC/S-II separation[2]	000:02:41.2	37.830	596.012	7,827.6	9,062.2		28.5876	-79.6605	21.021	82.144
S-II outboard engine cutoff	000:09:09.06	95.184	874.532	21,588.4	22,949.6	386.06	29.6810	-63.9910	0.059	89.863
S-II/S-IVB separation[2]	000:09:10.1	95.187	878.126	21,601.2	22,962.5		29.6811	-63.9221	0.047	89.900
S-IVB 1st burn cutoff	000:11:34.67	93.215	1,406.808	24,236.4	25,596.7	141.47	29.2688	-53.8183	0.013	95.149
Earth orbit insertion	000:11:44.67	93.215	1,445.652	24,242.4	25,602.6		29.2052	-53.0807	0.015	95.531

Apollo 15 Earth Orbit Phase

Event	GET (hhh:mm:ss)	Space Fixed Velocity (ft/sec)	Event Duration (sec)	Velocity Change (ft/sec)	Apogee (n mi)	Perigee (n mi)	Period (mins)	Inclination (deg)
Earth orbit insertion	000:11:44.67	25,602.6			91.5	89.6	87.84	29.679
S-IVB 2nd burn ignition	002:50:02.90	25,597.1						
S-IVB 2nd burn cutoff	002:55:53.61	35,603.0	350.71	10,414.7				

Apollo 15 Translunar Phase

Event	GET (hhh:mm:ss)	Altitude (n mi)	Space Fixed Velocity (ft/sec)	Event Duration (sec)	Velocity Change (ft/sec)	Space Fixed Flight Path Angle (deg)	Space Fixed Heading Angle (E of N)
Translunar injection	002:56:03.61	173.679	35,579.1			7.430	73.173
CSM separated from S-IVB	003:22:27.2	4,028.139	24,586.6			46.015	112.493
CSM docked with LM/S-IVB	003:33:49.5	5,985.4	21,811.0			51.66	115.86
CSM/LM ejected from S-IVB	004:18:01.2	12,826.9	16,402.2			61.45	119.20
Midcourse correction ignition	028:40:22.00	114,783.2	4,849.8			77.22	116.83
Midcourse correction cutoff	028:40:22.80	114,784.0	4,845.6	0.80	5.3	77.18	116.76
Midcourse correction ignition	073:31:14.81	12,618.4	3,963.1			-81.08	-139.68
Midcourse correction cutoff	073:31:15.72	12,617.7	3,966.8	0.91	5.4	-81.10	-140.00

[2] Only the commanded time is available for this event.

Apollo 15 Lunar Orbit Phase

Event	GET (hhh:mm:ss)	Altitude (n mi)	Space Fixed Velocity (ft/sec)	Event Duration (sec)	Velocity Change (ft/sec)	Apogee (n mi)	Perigee (n mi)
Lunar orbit insertion ignition	078:31:46.70	86.7	8,188.6				
Lunar orbit insertion cutoff	078:38:25.06	74.1	5,407.5	398.36	3,000.1	170.1	57.7
Descent orbit insertion ignition	082:39:49.09	55.3	5,491.7				
Descent orbit insertion cutoff	082:40:13.62	54.9	5,285	24.53	213.9	58.5	9.6
Descent orbit trim ignition	095:56:44.70	56.4	5,276.9				
Descent orbit trim cutoff	095:57:15.10	50.1	5,314.8	30.40	3.2	60.3	8.8
LM undocking and separation	100:39:16.2	7.4	5,553.6				
CSM orbit circularization ignition	101:38:58.98	57.1	5,276.5				
CSM orbit circularization cutoff	101:39:02.65	55.8	5,352.3	3.67	68.3	65.2	54.8
LM powered descent initiation	104:30:09.4	5.8	5,560.2				
LM powered descent cutoff	104:42:28.6			739.2	6813		
CSM plane change ignition	165:11:32.74	61.8	5,318.1				
CSM plane change cutoff	165:11:51.05	62	5,318.8	18.31	330.6	64.5	53.6
LM lunar liftoff ignition	171:37:23.2	54.8	5,357.1				
LM ascent orbit cutoff	171:44:34.2			431.0	6,059	42.5	9.0
LM terminal phase initiation ignition	172:29:40.0	34.2	5,368.8				
LM terminal phase initiation cutoff	172:29:42.6			2.6	72.7	64.4	38.7
CSM/LM docked	173:36:25.5	57	5,345.8				
LM ascent stage jettisoned	179:30:01.4	57.5	5,342.1				
CSM separation from LM	179:50				2		
LM ascent stage deorbit ignition	181:04:19.8	61.5	5,318.9				
LM ascent stage deorbit cutoff	181:05:42.8	61.8	5,196.0	83.0	200.3		
CSM orbit shaping maneuver ignition	221:20:48.02	53.6	5,362.9				
CSM orbit shaping maneuver cutoff	221:20:51.44	53.7	5,379.2	3.42	66.4	76.0	54.3
Subsatellite deployed	222:39:29.1	62.6	5,331.6			76.3	55.1

Apollo 15 Transearth Phase

Event	GET (hhh:mm:ss)	Altitude (n mi)	Space Fixed Velocity (ft/sec)	Event Duration (sec)	Velocity Change (ft/sec)	Space Fixed Flight Path Angle (deg)	Space Fixed Heading Angle (E of N)
Transearth injection ignition	223:48:45.84	67.6	5,305.9			0.52	-128.90
Transearth injection cutoff	223:51:06.74	71.8	8,272.4	140.90	3,046.8	4.43	-129.08
Midcourse correction ignition	291:56:49.91	25,190.3	11,994.6			-68.47	103.11
Midcourse correction cutoff	291:57:12.21	25,149.3	12,002.4	22.30	5.6	-68.49	103.09
CM/SM separation	294:43:55.2	1,951.8	29,001.7			-36.44	56.65

Apollo 15 Timeline

Event	GET (hhh:mm:ss)	GMT Time	GMT Date
Terminal countdown started.	-028:00:00	23:00:00	24 Jul 1971
Scheduled 9-hour 34-minute hold at T-9 hours.	-009:00:00	18:00:00	25 Jul 1971
Countdown resumed at T-9 hours.	-009:00:00	03:34:00	26 Jul 1971
Scheduled 1-hour hold at T-3 hours 30 minutes.	-003:30:00	09:04:00	26 Jul 1971
Countdown resumed at T-3 hours 30 minutes.	-003:30:00	10:04:00	26 Jul 1971
Guidance reference release.	-000:00:16.939	13:33:43	26 Jul 1971
S-IC engine start command.	-000:00:08.9	13:33:51	26 Jul 1971
S-IC engine ignition (#5).	-000:00:06.5	13:33:53	26 Jul 1971
All S-IC engines thrust OK.	-000:00:01.4	13:33:58	26 Jul 1971
Range zero.	000:00:00.00	13:34:00	26 Jul 1971
All holddown arms released (1st motion) (1.08 g).	000:00:00.3	13:34:00	26 Jul 1971
Liftoff (umbilical disconnected).	000:00:00.58	13:34:00	26 Jul 1971
Tower clearance yaw maneuver started.	000:00:01.68	13:34:01	26 Jul 1971
Yaw maneuver ended.	000:00:09.66	13:34:09	26 Jul 1971
Pitch and roll maneuver started.	000:00:12.21	13:34:12	26 Jul 1971
Roll maneuver ended.	000:00:23.02	13:34:23	26 Jul 1971
Mach 1 achieved.	000:01:05.0	13:35:05	26 Jul 1971
Maximum bending moment (80,000,000 lbf-in).	000:01:20.1	13:35:20	26 Jul 1971
Maximum dynamic pressure (768.58 lb/ft^2).	000:01:22.0	13:35:22	26 Jul 1971
S-IC center engine cutoff command.	000:02:15.96	13:36:16	26 Jul 1971
Pitch maneuver ended.	000:02:36.94	13:36:36	26 Jul 1971
S-IC outboard engine cutoff. Maximum total inertial acceleration (3.97 g).	000:02:39.56	13:36:39	26 Jul 1971
S-IC maximum Earth-fixed velocity.	000:02:40.00	13:36:40	26 Jul 1971
S-IC/S-II separation command.	000:02:41.2	13:36:41	26 Jul 1971
S-II engine start command.	000:02:41.9	13:36:41	26 Jul 1971
S-II ignition.	000:02:43.0	13:36:43	26 Jul 1971
S-II aft interstage jettisoned.	000:03:11.2	13:37:11	26 Jul 1971
Launch escape tower jettisoned.	000:03:15.9	13:37:15	26 Jul 1971
Iterative guidance mode initiated.	000:03:22.62	13:37:22	26 Jul 1971
S-IC apex.	000:04:37.562	13:38:37	26 Jul 1971
S-II maximum total inertial acceleration (1.79 g). S-II center engine cutoff.	000:07:39.56	13:41:39	26 Jul 1971
S-II outboard engine cutoff.	000:09:09.06	13:43:09	26 Jul 1971
S-II maximum Earth-fixed velocity.	000:09:10.00	13:43:10	26 Jul 1971
S-II/S-IVB separation command.	000:09:10.1	13:43:10	26 Jul 1971
S-IVB 1st burn start command.	000:09:10.20	13:43:10	26 Jul 1971
S-IVB 1st burn ignition.	000:09:13.20	13:43:13	26 Jul 1971
S-II apex.	000:09:13.225	13:43:13	26 Jul 1971
S-IC impact (theoretical).	000:09:20.839	13:43:20	26 Jul 1971
S-IVB ullage case jettisoned.	000:09:21.8	13:43:21	26 Jul 1971
S-IVB 1st burn maximum total inertial acceleration (0.65 g). S-IVB 1st burn cutoff.	000:11:34.67	13:45:34	26 Jul 1971
Earth orbit insertion. S-IVB 1st burn maximum Earth-fixed velocity.	000:11:44.67	13:45:44	26 Jul 1971
Orbital navigation started.	000:11:56.3	13:45:56	26 Jul 1971
Maneuver to local horizontal attitude started.	000:13:15.7	13:47:15	26 Jul 1971
S-II impact (theoretical).	000:19:43.912	13:53:43	26 Jul 1971
S-IVB 2nd burn restart preparation.	002:40:24.80	16:14:24	26 Jul 1971
S-IVB 2nd burn restart command.	002:49:54.90	16:23:54	26 Jul 1971
S-IVB 2nd burn ignition.	002:50:02.90	16:24:02	26 Jul 1971
S-IVB 2nd burn cutoff and maximum total inertial acceleration (1.40 g).	002:55:53.61	16:29:53	26 Jul 1971
S-IVB 2nd burn maximum Earth-fixed velocity	002:55:54.00	16:29:54	26 Jul 1971
Translunar injection.	002:56:03.61	16:30:03	26 Jul 1971
Orbital navigation started.	002:58:26.0	16:32:26	26 Jul 1971

Apollo 15 Timeline

Event	GET (hhh:mm:ss)	GMT Time	GMT Date
Maneuver to local horizontal attitude started.	002:58:26.2	16:32:26	26 Jul 1971
Maneuver to transposition and docking attitude started.	003:10:54.6	16:44:54	26 Jul 1971
CSM separated from S-IVB.	003:22:27.2	16:56:27	26 Jul 1971
TV transmission started.	003:25	16:34	26 Jul 1971
CSM docked with LM/S-IVB.	003:33:49.5	17:07:49	26 Jul 1971
TV transmission ended.	003:50	16:34	26 Jul 1971
CSM/LM ejected from S-IVB.	004:18:01.2	17:52:01	26 Jul 1971
S-IVB APS evasive maneuver ignition.	004:40:01.8	18:14:01	26 Jul 1971
S-IVB APS evasive maneuver cutoff.	004:41:22.0	18:15:22	26 Jul 1971
Maneuver to S-IVB LOX dump attitude initiated.	004:49:41.8	18:23:41	26 Jul 1971
S-IVB lunar impact maneuver—CVS venting closed.	004:56:40.6	18:30:40	26 Jul 1971
S-IVB lunar impact maneuver—LOX dump. Start of unplanned velocity increment due to J-2 engine control helium dump.	005:01:20.6	18:35:20	26 Jul 1971
S-IVB lunar impact maneuver—CVS vent opened.	005:01:40.6	18:35:40	26 Jul 1971
S-IVB lunar impact maneuver—LOX dump ended.	005:02:08.7	18:36:08	26 Jul 1971
S-IVB lunar impact maneuver—J-2 engine control helium dump ended.	005:18:51	18:52:51	26 Jul 1971
Maneuver to attitude required for final S-IVB APS burn initiated.	005:27:13.5	19:01:13	26 Jul 1971
S-IVB lunar impact maneuver—1st APS ignition.	005:46:00.7	19:20:00	26 Jul 1971
S-IVB lunar impact maneuver—Start of 1st unplanned velocity increment due to instrument unit thermal control system water valve operations and APS attitude engine reactions.	006:18:00	19:52:00	26 Jul 1971
S-IVB lunar impact maneuver—End of 1st velocity increment due to IU/TCS and APS effects.	006:23:00	19:57:00	26 Jul 1971
S-IVB lunar impact maneuver—Start of 2nd velocity increment due to IU/TCS and APS effects.	006:58:00	20:32:00	26 Jul 1971
S-IVB lunar impact maneuver—End of 2nd velocity increment due to IU/TCS and APS effects.	007:03:00	20:37:00	26 Jul 1971
S-IVB lunar impact maneuver—Start of 3rd velocity increment due to IU/TCS and APS effects.	007:38:00	21:12:00	26 Jul 1971
S-IVB lunar impact maneuver—End of 3rd velocity increment due to IU/TCS and APS effects.	007:43:00	21:17:00	26 Jul 1971
S-IVB lunar impact maneuver—Start of 4th velocity increment due to IU/TCS and APS effects.	008:18:00	21:52:00	26 Jul 1971
S-IVB lunar impact maneuver—End of 4th velocity increment due to IU/TCS and APS effects.	008:23:00	21:57:00	26 Jul 1971
S-IVB lunar impact maneuver—Start of 5th velocity increment due to IU/TCS and APS effects.	008:53:00	22:27:00	26 Jul 1971
S-IVB lunar impact maneuver—End of 5th velocity increment due to IU/TCS and APS effects.	008:58:00	22:32:00	26 Jul 1971
S-IVB lunar impact maneuver—Start of 6th velocity increment due to IU/TCS and APS effects.	009:28:00	23:02:00	26 Jul 1971
S-IVB lunar impact maneuver—End of 6th velocity increment due to IU/TCS and APS effects.	009:33:00	23:07:00	26 Jul 1971
S-IVB lunar impact maneuver—2nd APS ignition.	010:00:01	23:34:01	26 Jul 1971
S-IVB lunar impact maneuver—2nd APS cutoff.	010:01:12	23:35:12	26 Jul 1971
S-IVB 0.3° per second solar heating avoidance roll command.	010:19:22	23:53:22	26 Jul 1971
Midcourse correction ignition.	028:40:22.00	18:14:22	27 Jul 1971
Midcourse correction cutoff.	028:40:22.80	18:14:22	27 Jul 1971
Sextant photography test started.	032:00	21:34	27 Jul 1971
Sextant photography test ended.	032:50	22:24	27 Jul 1971
Preparations for LM ingress.	033:25	22:59	27 Jul 1971
CDR and LMP entered LM for checkout.	033:56	23:30	27 Jul 1971
TV transmission of CM and LM interiors started.	034:55	00:29	28 Jul 1971
TV transmission of CM and LM interiors ended.	035:46	01:20	28 Jul 1971
CDR and LMP entered CM.	036:55	02:29	28 Jul 1971
Visual light flash phenomenon observations started.	051:37	17:11	28 Jul 1971
Visual light flash phenomenon observations ended.	052:33	18:07	28 Jul 1971
LM ingress and housekeeping.	056:26	22:00	28 Jul 1971
CDR and LMP entered LM for checkout.	057:00	22:34	28 Jul 1971
CDR and LMP entered CM.	058:00	23:34	28 Jul 1971
Equigravisphere.	063:55:20	05:29:20	29 Jul 1971
Midcourse correction ignition.	073:31:14.81	15:05:14	29 Jul 1971
Midcourse correction cutoff.	073:31:15.72	15:05:15	29 Jul 1971

Apollo 15 Timeline

Event	GET (hhh:mm:ss)	GMT Time	GMT Date
Scientific instrument module door jettisoned.	074:06:47.1	15:40:47	29 Jul 1971
Lunar orbit insertion ignition (SPS).	078:31:46.70	20:05:46	29 Jul 1971
Lunar orbit insertion cutoff.	078:38:25.06	20:12:25	29 Jul 1971
S-IVB impact on lunar surface.	079:24:41.55	20:58:41	29 Jul 1971
Orbital science photography started.	080:35	22:09	29 Jul 1971
Orbital science photography ended.	080:50	22:24	29 Jul 1971
Terminator photography.	082:00	23:34	29 Jul 1971
Descent orbit insertion ignition (SPS).	082:39:49.09	00:13:49	30 Jul 1971
Descent orbit insertion cutoff.	082:40:13.62	00:14:13	30 Jul 1971
CSM landmark tracking.	083:45	01:19	30 Jul 1971
Terminator photography.	084:35	02:09	30 Jul 1971
TV transmission of landing site started.	095:00	12:34	30 Jul 1971
TV transmission of landing site ended.	095:10	12:44	30 Jul 1971
Descent orbit trim ignition (RCS).	095:56:44.70	13:30:44	30 Jul 1971
Descent orbit trim cutoff.	095:57:15.10	13:31:15	30 Jul 1971
CDR and LMP entered LM for activation, checkout, and platform alignment.	098:00	15:34	30 Jul 1971
CM/LM undocking failure due to loose CM/LM umbilical.	100:14	17:48	30 Jul 1971
LM undocking and separation.	100:39:16.2	18:13:16	30 Jul 1971
CSM orbit circularization ignition (SPS).	101:38:58.98	19:12:59	30 Jul 1971
CSM orbit circularization cutoff.	101:39:02.65	19:13:02	30 Jul 1971
CSM lunar surface landmark tracking.	102:35	20:09	30 Jul 1971
LM landing radar on.	104:25:13.0	21:59:13	30 Jul 1971
LM powered descent engine ignition.	104:30:09.4	22:04:09	30 Jul 1971
LM throttle to full-throttle position.	104:30:35.9	22:04:35	30 Jul 1971
LM manual target (landing site) update.	104:31:44.2	22:05:44	30 Jul 1971
LM pitchover started.	104:33:10.4	22:07:10	30 Jul 1971
LM landing radar range data good.	104:33:26.2	22:07:26	30 Jul 1971
LM landing radar altitude data good.	104:33:38.2	22:07:38	30 Jul 1971
LM landing radar updates enabled.	104:33:50.2	22:07:50	30 Jul 1971
LM throttle down.	104:37:31.1	22:11:31	30 Jul 1971
LM approach phase program selected.	104:39:32.2	22:13:32	30 Jul 1971
LM landing radar antenna to position 2.	104:39:39.0	22:13:39	30 Jul 1971
LM 1st landing point redesignation.	104:39:40.0	22:13:40	30 Jul 1971
LM landing radar switched to low scale.	104:40:13.0	22:14:13	30 Jul 1971
LM attitude hold mode selected.	104:41:08.7	22:15:08	30 Jul 1971
LM landing phase program selected.	104:41:10.2	22:15:10	30 Jul 1971
LM powered descent engine cutoff.	104:42:28.6	22:16:28	30 Jul 1971
LM lunar landing (right side & forward footpad contact).	104:42:29.3	22:16:29	30 Jul 1971
LM final settling.	104:42:31.1	22:16:31	30 Jul 1971
CSM orbital science photography.	106:00	23:34	30 Jul 1971
Stand-up EVA started (Scott).	106:42:49	00:16:49	31 Jul 1971
Stand-up EVA ended.	107:15:56	00:49:56	31 Jul 1971
CSM orbital science photography.	108:00	01:34	31 Jul 1971
CSM orbital science photography.	108:40	02:14	31 Jul 1971
TV transmission started for 1st EVA. CSM bistatic radar test.	110:00	03:34	31 Jul 1971
1st EVA started (LM cabin depressurized).	119:39:17	13:13:17	31 Jul 1971
TV deployed.	119:54:54	13:28:54	31 Jul 1971
Contingency sample collected.	120:00:05	13:34:05	31 Jul 1971
Lunar rover vehicle (LRV) offloaded.	120:18:31	13:52:31	31 Jul 1971

Apollo 15 Timeline

Event	GET (hhh:mm:ss)	GMT Time	GMT Date
LRV deployed.	120:31:33	14:05:33	31 Jul 1971
LRV configured for traverse.	121:24:03	14:58:03	31 Jul 1971
Departed for station 1.	121:44:55	15:18:55	31 Jul 1971
Arrived at station 1. Performed radial sampling, gathered documented samples, and performed panoramic photography.	122:10:46	15:44:46	31 Jul 1971
Departed for station 2.	122:22:36	15:56:36	31 Jul 1971
Arrived at station 2. Gathered samples, obtained a double core tube sample and performed stereopanoramic and 500 mm photography.	122:34:44	16:08:44	31 Jul 1971
CSM deep space measurements.	122:40	16:14	31 Jul 1971
CSM sunrise solar corona photography.	123:05	16:39	31 Jul 1971
Departed for LM.	123:26:02	17:02	31 Jul 1971
Arrived at LM. Offloaded and deployed Apollo lunar surface experiment package (ALSEP), laser ranging retroreflector, and solar wind composition experiment.	123:59:39	17:33:39	31 Jul 1971
CSM sunset solar corona photography.	124:30	18:04	31 Jul 1971
CSM lunar libration photography.	125:00	18:34	31 Jul 1971
1st ALSEP data received on Earth.	125:18:00	18:52	31 Jul 1971
TV transmission ended for 1st EVA.	125:55	19:29	31 Jul 1971
Cold cathode gauge experiment turned on. CSM orbital science photography.	126:00	19:34	31 Jul 1971
1st EVA ended (cabin repressurized).	126:11:59	19:45:59	31 Jul 1971
Heat flow experiment turned on.	126:13	19:47	31 Jul 1971
CSM bistatic radar test.	131:40	01:14	01 Aug 1971
CSM orbital science photography.	142:00	11:34	01 Aug 1971
2nd EVA started (cabin depressurized).	142:14:48	11:48:48	01 Aug 1971
Equipment prepared for LRV traverse.	142:25:04	11:59:04	01 Aug 1971
TV transmission started for 2nd EVA.	142:35	12:09	01 Aug 1971
Departed for station 6.	143:10:43	12:44:43	01 Aug 1971
Arrived at station 6. Gathered samples, obtained a single core tube sample, obtained a special environmental sample from trench, and performed panoramic and 500 mm photography tasks.	143:53:46	13:27:46	01 Aug 1971
CSM Earthshine photography.	144:10:32	13:44:32	01 Aug 1971
Departed for station 6a.	144:58:49	14:32:49	01 Aug 1971
Arrived at station 6a. Gathered samples and performed panoramic photography tasks.	145:01:11	14:35:11	01 Aug 1971
Departed for station 7.	145:22:40	14:56:40	01 Aug 1971
Arrived at station 7. Gathered selected samples, a comprehensive soil sample, and performed panoramic photography.	145:26:25	15:00:25	01 Aug 1971
Departed for station 4.	146:16:09	15:50:09	01 Aug 1971
Arrived at station 4. Gathered samples and performed panoramic photography.	146:28:59	16:02:59	01 Aug 1971
CSM deep space measurements.	146:30	16:04	01 Aug 1971
Departed for LM.	146:45:44	16:19:44	01 Aug 1971
Arrived at LM. Offloaded samples and configured LRV for trip to station 8 (ALSEP site).	147:08:09	16:42:09	01 Aug 1971
Departed for station 8.	147:19:33	16:53:33	01 Aug 1971
CSM orbital science photography.	147:20	16:54	01 Aug 1971
Arrived at station 8. Gathered comprehensive geologic sample, gathered special environmental sample from trench, drilled second heat flow hole and emplaced probe, drilled deep core sample hole, and performed penetrometer experiments.	147:21:15	16:55:15	01 Aug 1971
Departed for LM.	148:31:08	18:05:08	01 Aug 1971
Arrived at LM. Deployed United States flag and started EVA closeout.	148:32:17	18:06:17	01 Aug 1971
CSM zodiacal light photography.	148:40	18:14	01 Aug 1971
CSM orbital science photography.	149:10	18:44	01 Aug 1971
TV transmission ended for 2nd EVA.	149:20	18:54	01 Aug 1971
2nd EVA ended (cabin repressurized).	149:27:02	19:01:02	01 Aug 1971

Apollo 15 Timeline

Event	GET (hhh:mm:ss)	GMT Time	GMT Date
3rd EVA started (LM cabin depressurized).	163:18:14	08:52:14	02 Aug 1971
TV transmission started for 3rd EVA.	163:45	09:19	02 Aug 1971
Departed for ALSEP site.	164:04:13	09:38:13	02 Aug 1971
Arrived at ALSEP site. Recovered deep core sample and photographed LRV operation.	164:09:00	09:43:00	02 Aug 1971
Departed for station 9.	164:48:05	10:22:05	02 Aug 1971
Arrived at station 9. Collected samples and performed panoramic photography tasks.	165:01:22	10:35:22	02 Aug 1971
CSM plane change ignition (SPS).	165:11:32.74	10:45:32	02 Aug 1971
CSM plane change cutoff.	165:11:51.05	10:45:51	02 Aug 1971
Departed for station 9a.	165:16:50	10:50:50	02 Aug 1971
Arrived at station 9a. Gathered extensive samples, obtained a double core tube and performed photographic tasks including 500 mm and stereoscopic panoramic photography.	165:19:26	10:53:26	02 Aug 1971
Departed for station 10.	166:14:25	11:48:25	02 Aug 1971
Arrived at station 10. Gathered samples and performed 500 mm and panoramic photography tasks.	166:16:45	11:50:45	02 Aug 1971
Departed for ALSEP site.	166:28:49	12:02:49	02 Aug 1971
Arrived at ALSEP site. Recovered drilled core sample and performed photographic tasks.	166:43:40	12:17:40	02 Aug 1971
Arrived at LM. EVA closeout procedures started.	166:45:45	12:19:45	02 Aug 1971
Solar wind composition experiment retrieved.	167:10	12:44	02 Aug 1971
Departed for final positioning of LRV to obtain television coverage of LM ascent.	167:32:18	13:06:18	02 Aug 1971
LRV positioned.	167:35:24	13:09:24	02 Aug 1971
3rd EVA ended (LM cabin repressurized).	168:08:04	13:42:04	02 Aug 1971
TV transmission ended for 3rd EVA.	168:20	13:54	02 Aug 1971
CSM Gegenschein photography.	168:30	14:04	02 Aug 1971
LM equipment jettisoned.	169:00	14:34	02 Aug 1971
CSM tracking of LM landing site.	169:30	15:04	02 Aug 1971
Surface television transmission started for lunar liftoff.	171:30	17:04	02 Aug 1971
LM lunar liftoff ignition (LM APS).	171:37:23.2	17:11:23	02 Aug 1971
Surface television transmission ended.	171:40	17:14	02 Aug 1971
Lunar ascent orbit cutoff.	171:44:34.2	17:18:34	02 Aug 1971
Terminal phase initiation ignition.	172:29:40.0	18:03:40	02 Aug 1971
Terminal phase initiation cutoff.	172:29:42.6	18:03:42	02 Aug 1971
TV transmission started.	173:05	18:39	02 Aug 1971
TV transmission ended.	173:10	18:44	02 Aug 1971
Terminal phase finalize.	173:11:07	18:45:07	02 Aug 1971
TV transmission started.	173:35	19:09	02 Aug 1971
CSM/LM docked.	173:36:25.5	19:10:25	02 Aug 1971
TV transmission ended. CDR and LMP prepared to transfer to CSM.	173:40	19:14	02 Aug 1971
Samples and equipment transferred to CSM.	175:00	20:34	02 Aug 1971
CDR and LMP entered CSM and hatch closed.	176:40	22:14	02 Aug 1971
LM ascent stage jettisoned.	179:30:01.4	01:04:01	03 Aug 1971
CSM separation maneuver from LM.	179:50	01:24	03 Aug 1971
LM ascent stage deorbit ignition.	181:04:19.8	02:38:19	03 Aug 1971
LM ascent stage fuel depletion.	181:05:42.8	02:39:42	03 Aug 1971
LM ascent stage impact on lunar surface.	181:29:35.8	03:03:35	03 Aug 1971
Deep space measurements and Gegenschein photography.	195:45	17:19	03 Aug 1971
Ultraviolet photography of lunar maria.	196:35	18:09	03 Aug 1971
Visual light flash phenomenon observations started.	197:00	18:34	03 Aug 1971
Visual observations from lunar orbit.	197:20	18:54	03 Aug 1971
Visual light flash phenomenon observations ended.	198:00	19:34	03 Aug 1971
Orbital science photography.	198:35	20:09	03 Aug 1971
Visual observations from lunar orbit.	199:00	20:34	03 Aug 1971

Apollo 15 Timeline

Event	GET (hhh:mm:ss)	GMT Time	GMT Date
CSM lunar terminator photography.	199:30	21:04	03 Aug 1971
CSM lunar terminator photography.	200:30	22:04	03 Aug 1971
Orbital science photography.	200:50	22:24	03 Aug 1971
Ultraviolet photography of lunar surface.	201:00	22:34	03 Aug 1971
CSM lunar terminator photography.	201:40	23:14	03 Aug 1971
CSM boom photography.	202:20	23:54	03 Aug 1971
CSM lunar terminator photography.	214:05	11:39	04 Aug 1971
Orbital science photography.	214:35	12:09	04 Aug 1971
Deep space measurements.	215:40	13:14	04 Aug 1971
Sunrise solar corona photography.	216:00	13:34	04 Aug 1971
Orbital science photography.	217:00	14:34	04 Aug 1971
CSM lunar terminator photography.	217:20	14:54	04 Aug 1971
CSM lunar terminator photography.	219:20	16:54	04 Aug 1971
Orbit shaping maneuver ignition.	221:20:48.02	18:54:48	04 Aug 1971
Orbit shaping maneuver cutoff.	221:20:51.44	18:54:51	04 Aug 1971
Subsatellite launched.	222:39:29.1	20:13:29	04 Aug 1971
Transearth injection ignition (SPS).	223:48:45.84	21:22:45	04 Aug 1971
Transearth injection cutoff.	223:51:06.74	21:25:06	04 Aug 1971
Moon and star field photography.	224:20	21:54	04 Aug 1971
Corona window calibration.	239:05	12:39	05 Aug 1971
Transearth EVA started (Worden).	241:57:12	15:31:12	05 Aug 1971
Transearth EVA—TV transmission started.	242:00	15:34	05 Aug 1971
Transearth EVA—TV and data acquisition cameras installed and adjusted.	242:02	15:36	05 Aug 1971
Transearth EVA—Camera cassette retrieved.	242:22	15:56	05 Aug 1971
Transearth EVA—TV transmission ended.	242:28	16:02	05 Aug 1971
Transearth EVA—Ingress and hatch closed.	242:33	16:07	05 Aug 1971
Transearth EVA ended.	242:36:19	16:10:19	05 Aug 1971
Visual light flash phenomenon observations started.	264:35	14:09	06 Aug 1971
Visual light flash phenomenon observations ended.	265:35	15:09	06 Aug 1971
Lunar eclipse photography.	269:00	18:34	06 Aug 1971
Sextant photography.	270:00	19:34	06 Aug 1971
Lunar eclipse photography.	271:00	20:34	06 Aug 1971
Contamination photography.	271:50	21:24	06 Aug 1971
Mass spectrometer boom retraction test.	272:45	22:19	06 Aug 1971
Midcourse correction ignition.	291:56:49.91	17:30:49	07 Aug 1971
Midcourse correction cutoff.	291:57:12.21	17:31:12	07 Aug 1971
CM/SM separation.	294:43:55.2	20:17:55	07 Aug 1971
Entry.	294:58:54.7	20:32:54	07 Aug 1971
Communication blackout started.	295:59:13	21:33:13	07 Aug 1971
Communication blackout ended.	295:02:31	20:36:31	07 Aug 1971
Radar contact with CM by established recovery ship.	295:03	20:37	07 Aug 1971
S-band contact with CM established by recovery aircraft.	295:04	20:38	07 Aug 1971
Forward heat shield jettisoned.	295:06:45	20:40:45	07 Aug 1971
Drogue parachute deployed.	295:06:46	20:40:46	07 Aug 1971
VHF recovery beacon contact established with CM by recovery ship and recovery aircraft.	295:07	20:41	07 Aug 1971
Visual sighting of CM established by support helicopters.			
Main parachute deployed.	295:07:34	20:41:34	07 Aug 1971
Splashdown (went to apex-up).	295:11:53.0	20:45:53	07 Aug 1971
Crew aboard recovery helicopter.			
Crew aboard recovery ship.	295:51	21:25	07 Aug 1971
CM aboard recovery ship.	296:46	22:20	07 Aug 1971

APOLLO 16

The Tenth Mission:
The Fifth Lunar Landing

Apollo 16 Summary

(16 April–27 April 1972)

Apollo 16 crew (l. to. r): Ken Mattingly, John Young, Charlie Duke (NASA S72-16660).

Background

Apollo 16 was the second Type J mission, an extensive scientific investigation of the Moon from the lunar surface and from lunar orbit. The vehicles and payload were similar to those of Apollo 15.

The primary objectives were:

- to perform selenological inspection, survey, and sampling of materials and surface features in a preselected area of the Descartes region;

- to emplace and activate surface experiments; and

- to conduct inflight experiments and photographic tasks.

The crew members were Captain John Watts Young (USN), commander; Lt. Commander Thomas Kenneth "Ken" Mattingly, II (USN), command module pilot; and Lt. Colonel Charles Moss Duke, Jr. (USAF), lunar module pilot.

Selected as an astronaut in 1962, Young was making his fourth spaceflight, only the second astronaut to achieve that distinction. He had been pilot of Gemini 3 command pilot of Gemini 10, and command module pilot of Apollo

10, the first test of the LM in lunar orbit and the dress rehearsal for the first piloted landing on the Moon. Born 24 September 1930 in San Francisco, California, Young was 41 years old at the time of the Apollo 16 mission. He received a B.S. in aeronautical engineering from the Georgia Institute of Technology in 1952. His backup for the mission was Fred Wallace Haise, Jr.

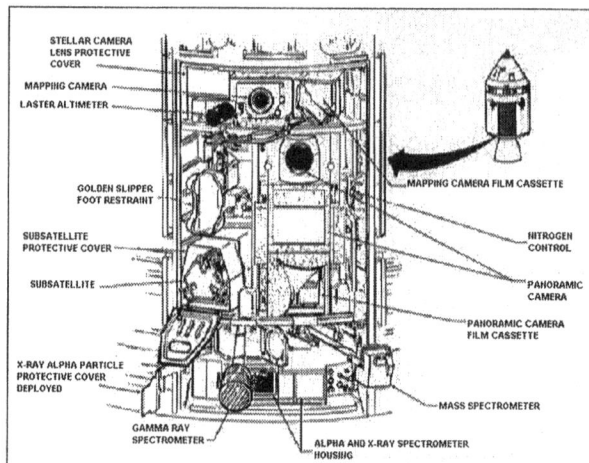

Line drawing of the scientific instrument bay in the Apollo 16 service module (NASA S72-16852).

Mattingly, who had been removed from the command pilot's position one day before the Apollo 13 mission because of his susceptibility to German measles, was making his first spaceflight. Born 17 March 1936 in Chicago, Illinois, Mattingly was 36 years old at the time of the Apollo 16 mission. He received a B.S. in aeronautical engineering from Auburn University in 1958, and was selected as an astronaut in 1966. His backup was Lt. Colonel Stuart Allen Roosa (USAF).

Duke was making his first spaceflight. Born 3 October 1935 in Charlotte, North Carolina, Duke was 36 years old at the time of the Apollo 16 mission. Duke received a B.S. in Naval sciences from the U.S. Naval Academy in 1957 and an M.S. in aeronautics and astronautics from the Massachusetts Institute of Technology in 1964. He was selected as an astronaut in 1966 and his backup was Captain Edgar Dean Mitchell (USN).

The capsule communicators (CAPCOMs) for the mission were Major Donald Herod Peterson (USAF), Major Charles Gordon Fullerton (USAF), Colonel James Benson Irwin (USAF), Haise, Roosa, Mitchell, Major Henry Warren Hartsfield, Jr. (USAF), Anthony Wayne "Tony" England, Ph.D., and Lt. Colonel Robert Franklyn Overmyer (USMC). The support crew consisted of Peterson, England, Hartsfield, and Phillip Kenyon

Chapman. The flight directors were M.P. "Pete" Frank and Philip C. Shaffer (first shift), Eugene F. Kranz and Donald R. Puddy (second shift), and Gerald D. Griffin, Neil B. Hutchinson, and Charles R. Lewis (third shift).

The Apollo 16 launch vehicle was a Saturn V, designated SA-511. The mission also carried the designation Eastern Test Range #1601. The CSM was designated CSM-113, and had the call-sign "Casper." The lunar module was designated LM-11, and had the call-sign "Orion."

Launch Preparations

The terminal countdown was picked up at T-28 hours at 03:54:00 GMT on 14 April. Scheduled holds were initiated at T-9 hours for 9 hours and at T-3 hours 30 minutes for one hour.

At launch time, the Cape Kennedy launch area was experiencing fair weather resulting from a ridge of high pressure extending westward, from the Atlantic Ocean through central Florida. Cumulus clouds covered 20 percent of the sky (base 3,000 feet), the temperature was 88.2° F, the relative humidity was 44 percent, and the barometric pressure was 14.769 lb/in^2. The winds, as measured by the anemometer on the light pole 60.0 feet above ground at the launch site measured 12.2 knots at 269° from true north. The winds, as measured at 530 feet above the launch site, measured 9.9 knots at 256° from true north.

Ascent Phase

Apollo 16 launched from Kennedy Space Center Launch Complex 39, Pad A, at a Range Zero time of 17:54:00 GMT (12:54:00 p.m. EST) on 16 April 1972. The planned launch window extended to 21:43:00 GMT to take advantage of a sun elevation angle on the lunar surface of 11.9°.

Between 000:00:12.7 and 000:00:31.8, the vehicle rolled from a launch pad azimuth of 90° to a flight azimuth of 72.034°. The S-IC engine shut down at 000:02:41.78, followed S-IC/S-II separation, and S-II engine ignition. The S-II engine shut down at 000:09:19.54 followed by separation from the S-IVB, which ignited at 000:09:23.60. The first S-IVB engine cutoff occurred at 000:11:46.21 with deviations from the planned trajectory of only +0.6 ft/sec in velocity; altitude was exactly as planned.

The maximum wind conditions encountered during ascent were 50.7 knots at 257° from true north at 38,880 feet, and a maximum wind shear of 0.0095 sec^{-1} at 44,780 feet.

Parking orbit conditions at insertion, 000:11:56.21 (S-IVB cutoff plus 10 seconds to account for engine tailoff and other transient effects), showed an apogee and perigee of 90.7 by 90.0 n mi, an inclination of 32.542°, a period of 87.84 minutes, and a velocity of 25,605.0 ft/sec. The apogee and perigee were based upon a spherical Earth with a radius of 3,443.934 n mi.

Apollo 16 lifts off from Kennedy Space Center Pad 39A (NASA KSC-72PC-176).

The international designation for the CSM upon achieving orbit was 1972-031A and the S-IVB was designated 1972-031B. After undocking at the Moon, the LM ascent stage would be designated 1972-031C, the descent stage 1972-031E, and the particles and fields subsatellite 1972-031D.

Translunar Phase

After inflight systems checks, the 341.92-second translunar injection maneuver (second S-IVB firing) was performed at 002:33:36.50. The S-IVB engine shut down at 002:39:18.42 and translunar injection occurred ten seconds later, at a velocity of 35,589.9 ft/sec after 1.5 Earth orbits lasting 2 hours 37 minutes 32.21 seconds.

At 003:04:59.0, the CSM was separated from the S-IVB stage, transposed, and docked at 003:21:53.4. The docked spacecraft were ejected from the S-IVB at 003:59:15.1, and an 80.2-second separation maneuver was performed at 004:18:08.3. Color television was transmitted for 18 minutes during the transposition and docking.

At 005:40:07.2, a 54.2-second propulsive force from the S-IVB auxiliary propulsion system targeted the S-IVB for impact on the Moon near the Apollo 12 landing site. As on previous missions, S-IVB impact was desired to produce seismic vibrations that could be used to study the nature of the lunar interior structure. Although launch vehicle systems malfunctions precluded a planned trajectory refinement, the impact point was within the desired area. Loss of S-IVB telemetry prevented establishment of the precise time of impact, making the interpretation of seismic data uncertain. However, it is estimated that the S-IVB impacted the lunar surface at 075:08:04.0. The estimated impact point was latitude 2.1° north and longitude 22.1° west, 173 n mi from the target point, 86 n mi from the Apollo 12 seismometer, 121 n mi from the Apollo 14 seismometer, and 585 n mi from the Apollo 15 seismometer. At impact, the S-IVB weighed 30,805 pounds and was traveling 8,711 ft/sec.

View of North America following translunar injection (NASA AS16-118-18885).

During the CSM/LM docking, light colored particles were noticed coming from the LM area. The particles were unexplained. At 007:18, the crew reported a stream of particles emitting from the LM in the vicinity of aluminum close-out panel 51, which covers the Mylar insulation over reaction control system A. This panel was located below the docking target on the +Z face of the LM ascent stage.

To determine systems status, the crew entered the LM at 008:17 and powered up. All systems were normal and the LM was powered down at 008:52. The CM television was turned on at 008:45 to give the mission control center a view of the particle emission. In order to point the high gain antenna, panel 51 was rotated out of sunlight and a marked decrease was then noted in the quantity of particles. On the television picture, the source of the particles appeared to be a growth of grass-like particles at the base of the panel. The television was turned off at 009:06. Results of the investigation determined that the particles were shredded thermal paint, and that the degraded thermal protection due to the paint shredding would have no effect on subsequent LM operations.

The 45-minute inflight electrophoresis demonstration commenced on schedule at 020:05 and was successful. Ultraviolet photography of the Earth from 58,000 and 117,000 n mi was accomplished as planned.

The only required midcourse correction was made at 030:39:00.66. It lasted 2.01 seconds and was required to ensure proper lunar orbit insertion.

At 038:18:56, the command module computer received an indication that an inertial measurement unit gimbal lock had occurred. The computer correctly downmoded the IMU to "coarse align" mode and set the appropriate alarms. Due to the large number of LM panel particles floating near the spacecraft and blocking the command module pilot's vision of the stars, realignment of the platform was accomplished using the Sun and Moon. It was suspected that the gimbal lock indication was an electrical transient caused by actuation of the thrust vector control enable relay when exiting the IMU alignment program. An erasable software program was uplinked to the crew and entered in the computer. The program would cause the computer to ignore gimbal lock indication during critical periods.

The visual light flash phenomena experiment started at 049:10. Numerous flashes were reported by the crew prior to terminating the experiment at 050:16. The crew also reported the flashes left no after-glow, were instantaneous, and were white.

The second LM housekeeping commenced about 053:30 and was completed at 055:11. All LM system checks were normal. The scientific instrumentation module door was jettisoned at 069:59:01.

At an altitude of 92.9 n mi above the Moon, the service propulsion engine was fired for 374.90 seconds to insert the spacecraft into a lunar orbit of 170.3 by 58.1 n mi. The translunar coast had lasted 71 hours 55 minutes 14.35 seconds.

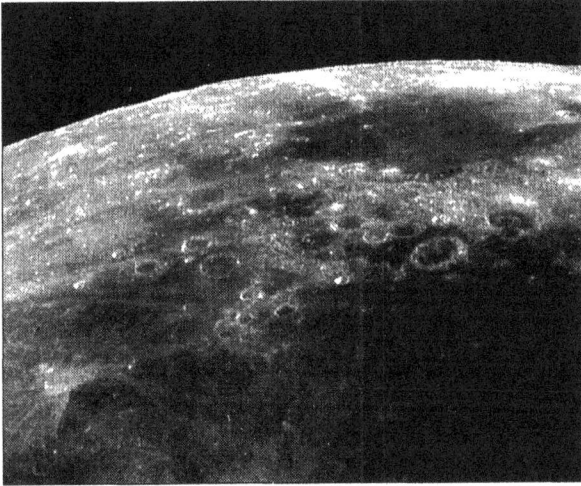

Interesting crater patterns and lunar horizon as seen from the CM (NASA AS16-121-19449).

Lunar Orbit/Lunar Surface Phase

At 078:33:45.04, a 24.35-second service propulsion system maneuver was performed to reach the descent orbit of 58.5 by 10.9 n mi for undocking of the LM.

LM activation started at 093:34, about 11 minutes early. The LM was powered up and all systems were nominal.

After undocking from the CSM, the LM gets a visual inspection from the command module pilot (NASA AS16-118-18894).

Lunar module undocking and separation were performed at 096:13:31, during the 12th revolution. At 103:21:43.08, the service propulsion system was fired for 4.66 seconds to place the CSM in a near-circular lunar orbit of 68.0 by 53.1 n mi in preparation for the acquisition of scientific data.

The CSM as seen from the LM during the twelfth revolution of the Moon (NASA AS16-113-18282).

The CSM was scheduled to perform an orbit circularization maneuver on the 13th lunar revolution at 097:41:44. However, oscillations were detected in a secondary system that controlled the direction of thrust of the service propulsion system engine.

Earthrise over the lunar horizon (NASA AS15-120-19187).

While flight controllers evaluated the problem, the CSM maneuvered into a stationkeeping situation with the LM and prepared either to redock or continue the mission. After 5 hours 45 minutes, tests and analyses showed that the system was still usable and safe; therefore, the vehicles were separated again and the mission continued on a revised timeline. A separation maneuver was performed at 102:30:00, and the 4.66-second CSM circularization maneuver was performed successfully with the primary system at 103:21:43.08.

NASA officials discuss whether to land on the Moon following failure of the circularization maneuver by the CSM (NASA S72-37009).

The 734.0-second powered descent engine firing began at 104:17:25 at an altitude of 10.9 n mi. Landing occurred at 02:23:35 GMT on 21 April (09:23:35 p.m. EST on 20 April) at 104:29:35.

The spacecraft landed in the Plain of Descartes at latitude 8.97301° south and longitude 15.50019° east, 886 feet northwest of the planned landing site. Approximately 102 seconds of engine firing time remained at landing.

LM (black spot) and exhaust plume (white streaking)

Image from Apollo 16 pan camera frame 4623 shows area around LM landing site. Palmetto Crater is at the top. The detail image below clearly shows the LM as a black spot, with white streaking from the descent engine plume. The crater in the center is Spook and to the upper left is Flag Crater.

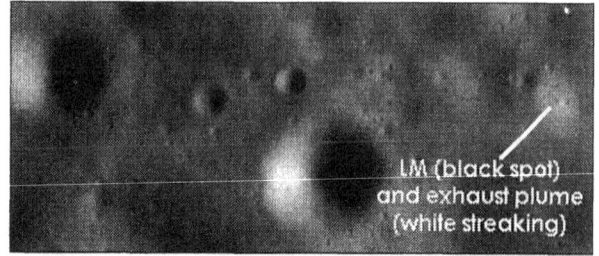

Because the LM had remained in lunar orbit six hours longer than planned, the LM was powered down to conserve electrical power and the first extravehicular activity was delayed in order to provide the crew with a well-deserved sleep period.

The LM cabin was depressurized at 118:53:38 for the first extravehicular activity period. Television coverage of surface activity was delayed until the LRV systems were activated because the LM steerable antenna, used for initial lunar surface television transmission, remained locked in one axis and could not be used.

The lunar surface experiments package was successfully deployed, but the commander accidentally tripped over the electronics cable, breaking it, and rendering the heat flow experiment inoperative.

The ALSEP deployment site as seen during EVA-1 (NASA AS16-113-18347).

This photo, taken during EVA-1, shows the undulating terrain of the landing site, with the LM in the center in the distance. In the foreground is the PSE, with the mortar pack at the left (NASA AS15-113-18359).

After completing their activities at the experiments site, the crew drove the lunar roving vehicle (LRV-2) west to Flag Crater where they made visual observations, photographed items of interest, and collected lunar samples.

Duke also collected lunar samples at Plum Crater during EVA-1 (NASA AS16-114-18423).

The inbound traverse route was just slightly south of the outbound route, and the next stop was Spook Crater. The crew then returned by way of the experiment station to the LM, at which time they deployed the solar wind composition experiment.

At the rim of Plum Crater, Young gathers rock samples (NASA AS16-109-17804)

At the end of a trail of lunar bootprints, Young works at the LRV (NASA AS16-109-17813).

In one of the most famous photographs from the Apollo program, John Young salutes the U.S. flag while "hanging in the air," thanks to the Moon's gravity which is one-sixth that of Earth. The photo was taken during EVA-1 at the Descartes landing site. The LM and lunar rover can be seen to the left (NASA AS16-113-18340).

Several LRV problems occurred during EVA-1. While ascending ridges and traversing very rocky terrain, there was no response at the rear wheels when full throttle was applied. The vehicle continued to move, but the front wheels were digging into the surface.

Following problems with the rover, Young takes it for a "grand prix" test ride during EVA-1 (NASA S72-36970).

Deployed during EVA-1, the ultraviolet camera can be seen in the shadow of the LM. Duke is in the shadows, with the rover and U.S. flag in the background in full sunlight (NASA AS16-114-18439).

The crew entered the LM and the cabin was repressurized at 126:04:40.

The first extravehicular activity lasted 7 hours 11 minutes 2 seconds. The distance traveled in the lunar rover vehicle was 13,800 feet (4.2 km), vehicle drive time was 43 minutes, the vehicle was parked for 3 hours 39 minutes, and an estimated 65.9 pounds (29.9 kg) of samples were collected.

After 16 hours, 30 minutes in the LM, the crew depressurized the cabin at 142:39:35 to begin the second extravehicular period.

After preparing the LRV, the crew headed south-southeast to a mare sampling area near the Cinco Craters on the north slope of Stone Mountain. They then drove in a northwesterly direction, making stops near Stubby and Wreck Craters. The last leg of the traverse was north to the experiments station and the LM.

Later, at station 8, a rear-drive troubleshooting procedure was implemented. During this procedure, a mismatch of power mode switching was identified as the cause of the problem. After a change in switch configuration, the LRV was working properly. An hour and a half later, at stations 9 and 10, the LRV range, bearing, and distance were reported to be inoperative. However, navigation heading was working. When the crew reset the power switches, the navigation system began operating nominally.

Duke begins a photographic pan of the landing site at the start of EVA-2 (NASA AS16-107-17436).

Young aligns the high-gain antenna on the LRV during a stop at station 8 (NASA AS16-108-17670).

View of cosmic ray experiment deployed on the landing gear of the LM (NASA AS16-107-17442).

Young breaks off a piece of rock and takes a soil sample at station 8 (NASA AS16-108-17701).

After the crew arrived at station 10 (LM and ALSEP area), the surface activity was extended about 20 minutes because the crew's consumables usage was lower than predicted. The lunar module pilot then examined the damaged heat flow experiment. Visual inspection revealed that the cable separated at the connector. Results of troubleshooting a model of the experiment at mission control indicated a fix could be accomplished. However it was not attempted because the time required could affect the third EVA.

The period ended with ingress and repressurization of the LM cabin at 150:02:44. During ingress, a two-inch portion of the commander's antenna was broken off, which produced a 15 to 18 db drop in signal strength. Since the commander's backpack radio relayed the lunar module pilot's information to the LM and the lunar communications relay unit for transmission to ground stations, a decision was made later to have the commander use the lunar module pilot's oxygen purge system, which supported the antenna.

Close-up of the RCA television camera affixed to the LRV. The LM is up slope (NASA AS16-115-18549).

The second extravehicular activity lasted 7 hours 23 minutes 9 seconds. The distance traveled in the lunar rover vehicle was 37,100 feet (11.3 km), vehicle drive time was 1 hour 31 minutes, park time was 3 hours 56 minutes, and an estimated 63.9 pounds (29.0 kg) of samples were collected.

The third extravehicular period began 30 minutes early when the cabin was depressurized at 165:31:28, but four stations were deleted because of time limitations.

Full view of LM taken by LMP during EVA-3 (NASA AS16-116-18579).

The crew first drove to the rim of North Ray Crater where photographs were taken and samples gathered, some from House Rock, the largest single rock seen during the extravehicular activities. The extra 30 minutes were used at North Ray Crater.

Young uses the lunar rake during EVA-2 (NASA AS15-110-18020).

Double core tube sample at ALSEP (NASA AS16-115-18557).

They then drove southeast to the second sampling area, Shadow Rock. On completing activities there, the crew drove the vehicle back to the LM, retracing the outbound route.

The third extravehicular activity lasted 5 hours 40 minutes 3 seconds. The distance traveled in the lunar rover vehicle was 37,400 feet (11.4 km), vehicle drive time was 1 hour 12 minutes, the vehicle was parked for 2 hours 26 minutes, and an estimated 78.0 pounds (35.4 kg) of samples were collected.

Block discovered during EVA-2. Note impact impression in soil (NASA AS16-107-17573).

Duke in small boulder field at station 4 sample site (NASA AS16-107-17446).

The crew reentered the LM and the cabin was repressurized at 171:11:31, thus ending the Apollo program's fifth human exploration of the Moon.

For the mission, the total time spent outside the LM was 20 hours 14 minutes 14 seconds, the total distance traveled in the lunar rover vehicle was 88,300 feet (26.9 km), vehi-cle drive time was 3 hours 26 minutes, the vehicle was parked during extravehicular activities for 10 hours 1 minute, and the collected samples totaled 211.00 pounds (95.71 kg) (official total in kilograms as determined by the Lunar Receiving Laboratory in Houston). The farthest point traveled from the LM was 15,092 feet.

Duke examines the surface of House Rock at North Ray Crater (NASA AS16-116-18649).

Duke follows his examination of House Rock by taking soil samples at its base (NASA AS16-116-18653).

While the crew was on the surface, the command module pilot had obtained photographs, measured physical properties of the Moon, and made visual observations. The command module pilot also had made comprehensive deep space measurements, providing scientific data that could be used to validate findings from the Apollo 15 mission. A 7.14-second CSM plane change maneuver was made at 169:05:52.14 and adjusted the orbit to 64.6 by 55.0 n mi.

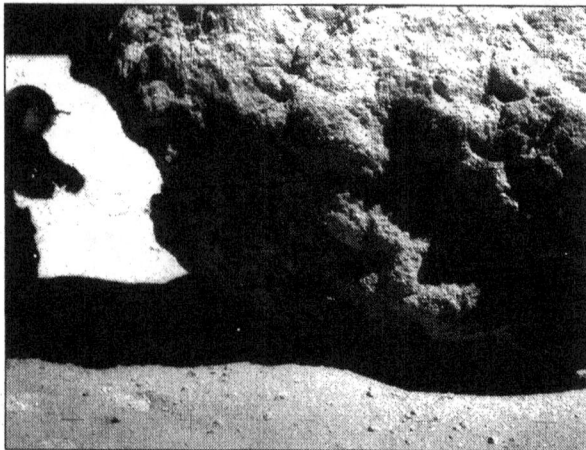

Young examines permanently shadowed area under Shadow Rock, a large boulder at station 13 during EVA-3 (NASA AS16-106-17413).

Close-up of debris-filled small crater at station 11 (NASA AS16-116-18599).

Ignition of the ascent stage engine for lunar liftoff occurred at 01:25:47 GMT on 24 April (at 08:25:47 p.m. EST on 23 April) at 175:31:47.9 and was televised. It had been on the lunar surface for 71 hours 2 minutes 13 seconds.

Photo of station 10 prime rake site before John Young started raking (NASA AS16-117-18826).

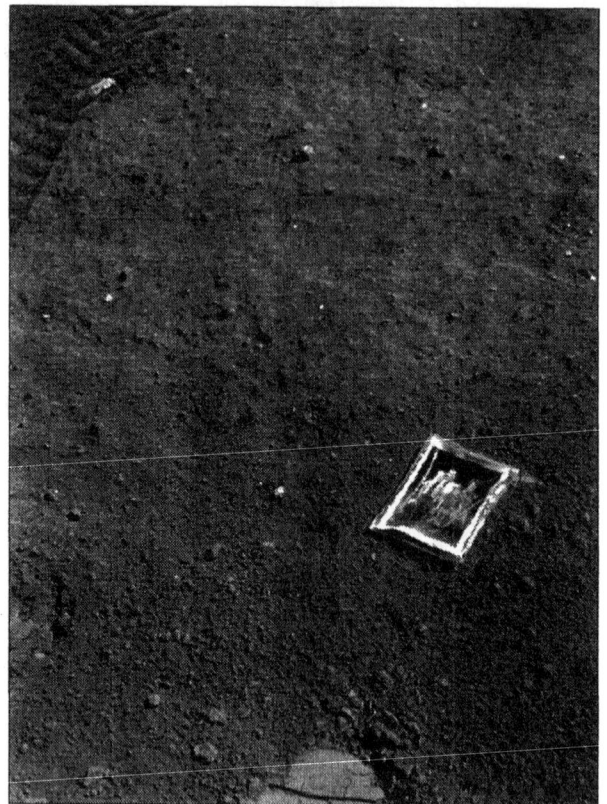

Before entering the LM for the return trip, Duke left a photo of his family on the lunar surface (NASA AS16-117-18841).

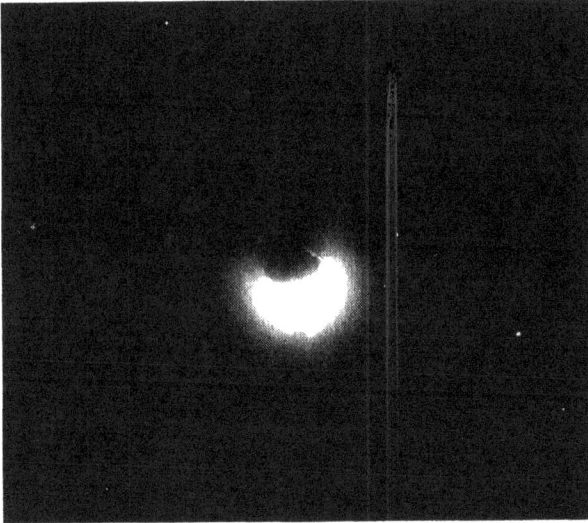

Geocorona, a halo of low density hydrogen around the Earth, photographed with the ultraviolet camera (NASA AS15-123-19650).

Still from television transmission of the LM ascent stage liftoff (NASA S72-35614).

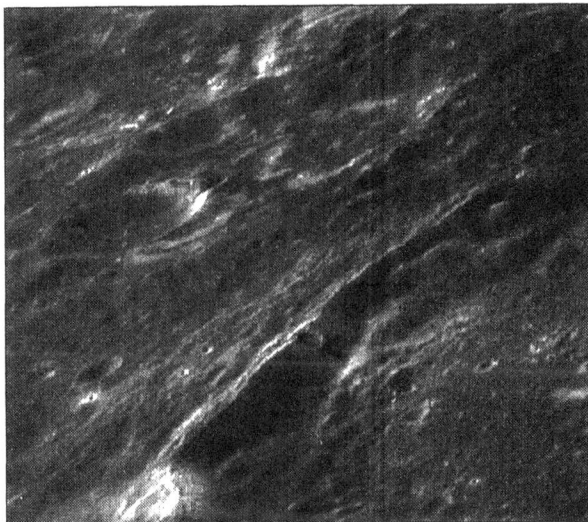

Rim of Guyot crater on the lunar farside (AS16-121-19407).

LM ascent stage, seen against the Sea of Fertility, approaches the CSM following a successful lunar surface expedition (AS16-122-19533).

The 427.7-second firing of the ascent engine placed the vehicle into a 40.2 by 7.9 n mi orbit. Several rendezvous sequence maneuvers were required before docking could occur two hours later. First, a vernier adjustment was made at 175:42:18 at an altitude of 11.2 n mi. Then the terminal phase was initiated with a 2.5 second maneuver at 176:26:05. This maneuver brought the ascent stage to an orbit of 64.2 by 40.1 n mi. Following a nominal rendezvous sequence, the ascent stage docked with the CM at 177:41:18 at an altitude of 65.6 n mi, after being undocked for 81 hours 27 minutes 47 seconds.

After the crew transferred the samples, film, and equipment to the CSM, the ascent stage was jettisoned at 195:00:12 at an altitude of 59.2 n mi. After jettison, the LM lost stability and began tumbling at a rate of about 3° per second. This may have been due to a guidance circuit breaker inadvertently being left open. A maneuver was made at 195:03:13 to separate the CSM from the ascent stage. No deorbit burn maneuver was possible, and the ascent stage remained in lunar orbit for approximately one year.[1]

The mass spectrometer deployment boom stalled during a retract cycle and was, therefore, jettisoned at 195:23:12.

Before the CSM was maneuvered from lunar orbit, a particles and fields subsatellite similar to that launched from Apollo 15 was deployed at 196:02:09 during the 62nd revolution into an orbit of 66 by 52 n mi at an inclination of -110°. The subsatellite was planned to be released during the 73rd revolution into an orbit of 170 by 58 n mi. The subsatellite was instrumented to measure plasma and energetic-particle fluxes, vector magnetic fields, and subsatellite velocity from which lunar gravitational anomalies could be determined. However, as a result of the engine gimbal anomaly earlier in the mission, a planned CSM orbit-shaping maneuver had not been performed before ejection of the subsatellite. As a result, the subsatellite was placed in an orbit with a much shorter lifetime than planned.

It was not possible to activate the subsatellite for about 20 hours after launch because of communications interference resulting from the failure of the ascent stage to deorbit, but this did not interfere with the subsatellite systems. Loss of all tracking and telemetry data occurred at 20:31 GMT on 29 May 1972. Reacquisition of the signal was expected at 22:00 GMT on that day, but was not achieved, and it is believed that the subsatellite struck the far side of the lunar surface during the 425th revolution at longitude 110° east. The lower-than-desired orbit contributed to the short orbital life because the lunar mass concentrations on the front and far sides of the Moon were located relatively near the subsatellite ground track.

The second plane-change maneuver and some orbital science photography were deleted so that transearth injection could be performed 24 hours earlier than originally planned. This decision was made due to the engine problem experienced during the lunar orbit circularization maneuver.

Following a 162.29-second maneuver at 200:21:33.07 at 52.2 n mi, transearth injection was achieved at 200:24:15.36 after 64 lunar orbits lasting 125:49:32.59, at velocity of 8,663.0 ft/sec.

Transearth Phase

Between 202:57 and 203:12, good quality television pictures were transmitted from inside the CM. From 203:29 to 204:12, pictures were broadcast from the LRV camera on the lunar surface.

The first of two midcourse corrections, a 22.6-second 3.4-ft/sec maneuver, was made at 214:35:02.8 to achieve the desired entry interface conditions with Earth.

CMP Mattingly (right) during transearth EVA. The LMP is at left. (NASA S72-37001).

At 218:39:46, the command module pilot began a transearth coast EVA. Television coverage was provided for the 1 hour 23 minute 42 second period, during which Mattingly retrieved film cassettes from the scientific instrument module cameras, visually inspected the equipment, and exposed an experiment for ten minutes to provide data on microbial response to the space environment. This brought the total extravehicular activity for the mission to 22 hours 17 minutes 36 seconds.

A scheduled television press conference started at 243:35 and lasted for 18 minutes. During the conference, the crew gave a brief description of the farside of the Moon. An item of particular interest was the crew's description of Guyot Crater, which appeared to be full of material. The material seemed to have overflowed and spilled down the side of the crater. The crew compared their observations with similar geological formations in Hawaii.

Additional activities during transearth coast included photography for a Skylab program study of the behavior and effects of particles emanating from the spacecraft, and the second light-flash observation session. The second midcourse correction, a 6.4-second maneuver of 1.4 ft/sec, was made at 262:37:20.7.

Recovery

The service module was jettisoned at 265:22:33, and the CM entry followed a normal profile. The command module reentered Earth's atmosphere (400,000 feet altitude) at 265:37:31 at a velocity of 36,090 ft/sec, following a transearth coast of 65 hours 13 minutes 16 seconds.

---

[1] Later analysis indicated that the ascent stage struck the lunar surface before Apollo 17 commenced, but no data were available for substantiation.

While on the drogue parachutes, the CM was viewed on television, and continuous coverage was provided through crew recovery.

The parachute system effected splashdown of the CM in the Pacific Ocean at 19:45:05 GMT (02:45:05 p.m. EST) on 27 April. Mission duration was 265:51:05. The impact point was about 0.3 n mi from the target point and 2.7 n mi from the recovery ship U.S.S. *Ticonderoga*. The splashdown site was estimated to be latitude 0.70° south and longitude 156.22° west. After splashdown, the CM assumed an apex-down flotation attitude, but was successfully returned to the normal flotation position in 4 minutes 30 seconds by the inflatable bag uprighting system.

Apollo 16 CM about to splash down into the central Pacific Ocean (NASA S72-36291).

The crew was retrieved by helicopter and was aboard the recovery ship 37 minutes after splashdown. The CM was recovered 62 minutes later. The estimated CM weight at splashdown was 11,995 pounds, and the estimated distance traveled for the mission was 1,208,746 n mi.

The crew remained aboard the *Ticonderoga* until 17:30 GMT on 29 April, when they were flown to Hickam Air Force Base, Hawaii, where they arrived at 19:21 GMT. They departed by C-141 aircraft for Ellington Air Force Base, Houston, at 20:07 GMT and arrived at 03:40 GMT on 30 April.

The CM arrived in Hawaii at 03:30 GMT on 30 April. At 18:00 GMT on 1 May, it departed for North Island Naval

Air Station, San Diego, for deactivation, where it arrived at 00:00 GMT on 6 May.

Welcome ceremonies aboard the recovery ship U.S.S. *Ticonderoga* (NASA S72-36262).

On May 7, while propellants were being removed from the CM, a tank cart exploded because of overpressurization. Forty-six persons suspected of inhaling toxic fumes were hospitalized, but examination revealed no symptoms of inhalation. The CM was not damaged. An investigation board reported that the ratio of neutralizer to oxidizer being detanked had been too low because of the extra oxidizer retained in the CM tanks as a result of the Apollo 15 parachute anomaly. Changes were made in ground support equipment and detanking procedures to prevent future overpressurization. Deactivation was completed at 00:00 GMT on 11 May. The CM left North Island at 03:00 GMT on 12 May, and was transferred to the North American Rockwell Space Division facility at Downey, California, for postflight analysis. It arrived at 10:30 GMT on 12 May.

Conclusions

The overall performance of the Apollo 16 mission was excellent, with all of the primary mission objectives and most of the detailed objectives being met, although the mission was terminated one day earlier than planned. Experiment data were gathered during lunar orbit, from the lunar surface, and during both the translunar and transearth coast phases for all detailed objectives and experiments except subsatellite tracking for autonomous navigation and the heat flow experiment. Especially significant scientific findings included the first photography obtained of the geocorona in the hydrogen (Lyman alpha) wavelength from outside Earth's atmosphere, and the discovery of two new auroral belts around Earth. The following conclusions were made from an analysis of post-mission data:

1. Lunar dust and soil continued to cause problems with some equipment, although procedural measures were taken and equipment changes and additions were made to control the condition.

2. Loss of the heat flow experiment emphasized that all hardware should be designed for loads accidentally induced by crew movements because of vision and mobility constraints while wearing the pressurized suits.

3. The capability of the S-band omni-directional antenna system to support the overall lunar module mission operations was demonstrated after the failure experienced with the S-band steerable antenna.

4. The performance of the Apollo 16 particles and fields subsatellite showed that the lunar gravitational model was not sufficiently accurate for the orbital conditions that existed to accurately predict the time of impact.

5. The absence of cardiac arrhythmias on this mission was, in part, attributed to a better physiological balance of electrolytes and body fluids resulting from an augmented dietary intake of potassium and a better rest-work cycle that effectively improved the crew's sleep.

6. The ability of the crew and the capability of the spacecraft to land safely in the rough terrain of a lunar highlands region without having high resolution photography prior to the mission was demonstrated. Further, the capability of the lunar roving vehicle to operate under these conditions and on slopes up to 20° was demonstrated.

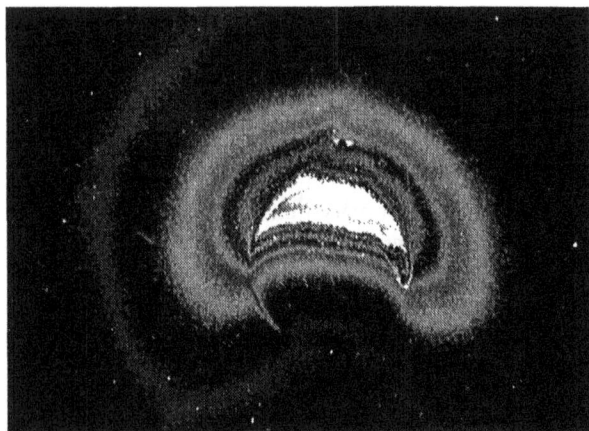

Earth as viewed with far ultraviolet camera (NASA S72-40818).

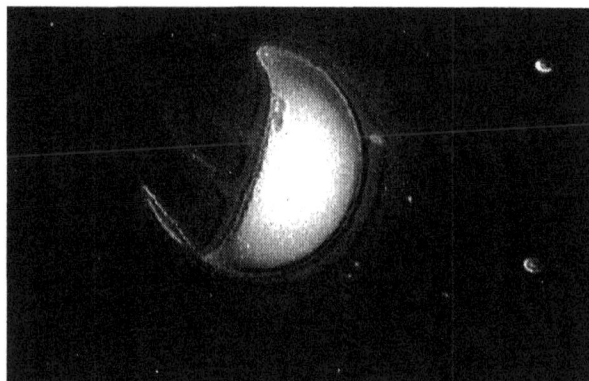

10-minute far ultraviolet exposure of Earth (NASA S72-40821).

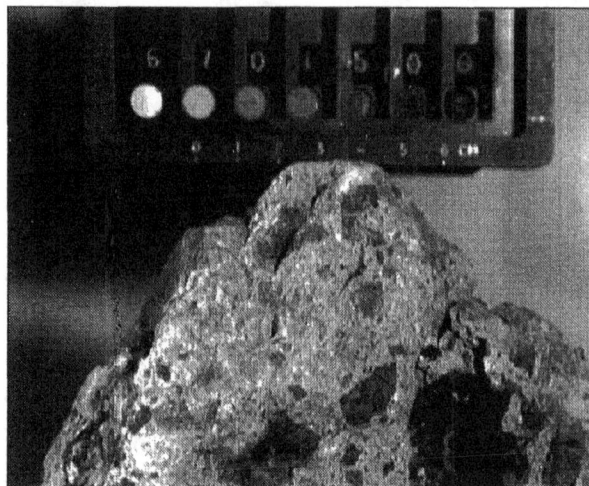

Lunar sample 67015. The white matrix consists of finely fractured and crushed mineral and rock detritus, predominantly feldspar (NASA S72-37216).

Apollo 16 Objectives

Spacecraft Primary Objectives

1. To perform selenological inspection, survey, and sampling of materials and surface features in a preselected area of the Descartes region. *Achieved.*

2. To emplace and activate surface experiments. *Achieved.*

3. To conduct inflight experiments and photographic tasks. *Achieved.*

Detailed Objectives

1. Service module orbital photographic tasks. *Achieved.*

2. Visual light flash phenomenon. *Achieved.*

3. Command module photographic tasks. *Partially achieved. Timeline changes caused data loss.*

4. Visual observations from lunar orbit. *Achieved.*

5. Skylab contamination study. *Partially achieved. Timeline changes caused data loss.*

6. Improved gas/water separator. *Not achieved. Separator failed before it could be evaluated.*

7. Body fluid balance analysis. *Achieved.*

8. Subsatellite tracking for autonomous navigation. *Not achieved. Timeline changes caused data loss.*

9. Improved fecal collection bag. *Achieved.*

10. Skylab food package. *Achieved.*

11. Lunar rover vehicle evaluation. *Achieved.*

Crew Participation Experiments

1. Passive seismic. *Partially achieved. No lunar module ascent stage impact.*

2. Active seismic. *Partially achieved. The fourth mortar was not fired.*

3. Lunar surface magnetometer. *Achieved.*

4. Heat flow. *Not achieved. Electronics package cable broken.*

5. Lunar geology investigation. *Achieved.*

6. Solar wind composition. *Achieved.*

7. Cosmic ray detector (sheets). *Partially achieved. Partial deployment of panel #4.*

8. Gamma ray spectrometer. *Achieved.*

9. X-ray fluorescence. *Achieved.*

10. Alpha particle spectrometer. *Achieved.*

11. S-band transponder (command and service module/lunar module). *Achieved.*

12. Mass spectrometer. *Achieved.*

13. Downlink bistatic radar observations of the Moon. *Achieved.*

14. Ultraviolet photography of Earth and Moon. *Partially achieved. Timeline changes caused data loss.*

15. Gegenschein from lunar orbit. *Achieved.*

16. Soil mechanics. *Partially achieved. No trench was dug due to time constraints.*

17. Far ultraviolet camera/spectroscope. *Achieved.*

18. Portable magnetometer. *Achieved.*

19. Microbial response in space environment. *Achieved.*

Passive Experiments

1. Bone mineral measurement. *Achieved.*

2. Biostack. *Achieved.*

3. Apollo window meteoroid. *Achieved.*

Operational Test

Lunar module voice and data relay. *Achieved.*

Inflight Demonstration

Fluid electrophoresis in space. *Achieved.*

Subsatellite Experiments

1. S-164: S-band transponder. *Achieved.*

2. S-173: Particle shadows/boundary layer. *Achieved.*

3. S-174: Magnetometer. *Achieved.*

Operational Tests for Manned Spacecraft Center and U.S. Department of Defense

1. Chapel Bell (classified, Department of Defense test). *Results classified.*

2. Radar skin tracking. *Results classified.*

3. Ionospheric disturbance from missiles. *Results classified.*

4. Acoustic measurement of missile exhaust noise. *Results classified.*

5. Army acoustic test. *Results classified.*

6. Long-focal-length optical system. *Results classified.*

7. Sonic boom measurement. *Results classified.*

Launch Vehicle Objectives

1. To launch on a flight azimuth between 72° and 100° and insert the S-IVB/instrument unit/spacecraft into the planned circular Earth parking orbit. *Achieved.*

2. To restart the S-IVB during either the second or third revolution and inject the S-IVB/instrument unit/spacecraft into the planned translunar trajectory. *Achieved.*

3. To provide the required attitude control for the S-IVB/instrument unit/spacecraft during transposition, docking, and ejection. *Achieved.*

4. To perform an evasive maneuver after ejection of the command and service module/lunar module from the S-IVB/instrument unit. *Achieved.*

5. To target the S-IVB/instrument stages for impact on the lunar surface at latitude 2.3° south and longitude 31.7° west. *Achieved.*

6. To determine actual impact point within 5.0 kilometers (2.7 n mi) and time of impact within one second. *Not achieved. The desired accuracy was not achieved.*

7. To vent and dump the remaining gases and propellants to safe the S-IVB/instrument unit. *Achieved.*

Apollo 16 Spacecraft History

EVENT	DATE
Saturn S-IVB stage #511 delivered to KSC.	01 Jul 1970
Spacecraft/LM adapter #20 delivered to KSC.	17 Aug 1970
Saturn V instrument unit #511 delivered to KSC.	29 Sep 1970
Saturn S-II stage #11 delivered to KSC.	30 Sep 1970
Individual and combined CM and SM systems test completed at factory.	03 Dec 1970
LM #11 final engineering evaluation acceptance test at factory.	24 Feb 1971
LM #11 integrated test at factory.	24 Feb 1971
Integrated CM and SM systems test completed at factory.	17 Mar 1971
LM descent stage #11 ready to ship from factory to KSC.	01 May 1971
LM descent stage #11 delivered to KSC.	05 May 1971
LM ascent stage #11 ready to ship from factory to KSC.	07 May 1971
LM ascent stage #11 delivered to KSC.	14 May 1971
CM #113 and SM #113 ready to ship from factory to KSC.	26 Jul 1971
CM #113 and SM #113 delivered to KSC.	29 Jul 1971
CM #113 and SM #113 mated.	02 Aug 1971
LRV #2 delivered to KSC.	01 Sep 1971
CSM #113 combined systems test completed.	13 Sep 1971
Saturn S-IC stage #11 delivered to KSC.	17 Sep 1971
Saturn S-IC stage #11 erected on MLP #3.	21 Sep 1971
Saturn V instrument unit #511 delivered to KSC.	29 Sep 1971
Saturn S-II stage #11 erected.	01 Oct 1971
Saturn S-IVB stage #511 erected.	05 Oct 1971
Saturn V instrument unit #511 erected.	06 Oct 1971
Launch vehicle electrical systems test completed.	15 Oct 1971
LM #11 altitude tests completed.	19 Oct 1971
CSM #113 altitude tests completed.	21 Oct 1971
Launch vehicle propellant dispersion/malfunction overall test completed.	08 Nov 1971
LRV #2 installed.	16 Nov 1971
Launch vehicle service arm overall test completed.	18 Nov 1971
CSM #113 moved to VAB.	07 Dec 1971
Spacecraft erected.	08 Dec 1971
Space vehicle and MLP #3 transferred to launch complex 39A.	13 Dec 1971
CSM #113 integrated systems test completed.	03 Jan 1972
LM #11 combined systems test completed.	04 Jan 1972
Space vehicle and MLP #3 returned to VAB.	27 Jan 1972
Space vehicle and MLP #3 returned to launch complex 39A.	09 Feb 1972
CSM #113 integrated systems test repeated.	14 Feb 1972
CSM #113 electrically mated to launch vehicle.	21 Feb 1972
Space vehicle overall test #1 (plugs in) completed.	23 Feb 1972
LM #9 flight readiness test completed.	24 Feb 1972
Space vehicle flight readiness test completed.	02 Mar 1972
Saturn S-IC stage #11 RP-1 fuel loading completed.	20 Mar 1972
Space vehicle countdown demonstration test (wet) completed.	30 Mar 1972
Space vehicle countdown demonstration test (dry) completed.	31 Mar 1972

Apollo 16 Ascent Phase

Event	GET (hhh:mm:ss)	Altitude (n mi)	Range (n mi)	Earth Fixed Velocity (ft/sec)	Space Fixed Velocity (ft/sec)	Event Duration (sec)	Geocentric Latitude (deg N)	Longitude (deg E)	Space Fixed Flight Path Angle (deg)	Space Fixed Heading Angle (E of N)
Liftoff	000:00:00.59	0.060	0.000	0.0	1,340.7		28.4470	-80.6041	0.05	90.00
Mach 1 achieved	000:01:07.5	4.282	1.358	1,076.4	2,075.5		28.4539	-80.5797	26.79	84.51
Maximum dynamic pressure	000:01:26.0	7.755	3.800	1,759.6	2,785.9		28.4670	-80.5359	29.12	81.64
S-IC center engine cutoff[2]	000:02:17.85	24.548	26.821	5,488.2	6,658.8	144.55	28.5847	-80.1207	23.105	76.125
S-IC outboard engine cutoff	000:02:41.78	35.698	49.927	7,753.0	8,961.7	168.5	28.7009	-79.7028	19.914	75.328
S-IC/S-II separation[2]	000:02:43.5	36.560	51.929	7,767.8	8,979.2		28.7109	-79.6666	19.643	75.339
S-II center engine cutoff	000:07:41.77	92.441	592.660	17,039.0	18,357.7	296.57	30.9376	-69.6064	0.116	79.535
S-II outboard engine cutoff	000:09:19.54	93.445	894.079	21,539.3	22,858.7	394.34	31.7737	-63.8100	0.367	82.585
S-II/S-IVB separation[2]	000:09:20.5	93.468	897.389	21,550.4	22,869.8		31.7812	-63.7457	0.358	82.622
S-IVB 1st burn cutoff	000:11:46.21	93.374	1,430.142	24,280.1	25,600.0	142.61	32.5109	-53.2983	0.001	88.496
Earth orbit insertion	000:11:56.21	93.377	1,469.052	24,286.1	25,605.1		32.5262	-52.5300	0.001	88.932

Apollo 16 Earth Orbit Phase

Event	GET (hhh:mm:ss)	Space Fixed Velocity (ft/sec)	Event Duration (sec)	Velocity Change (ft/sec)	Geocentric Latitude (deg N)	Longitude (deg E)	Apogee (n mi)	Perigee (n mi)	Period (mins)	Inclination (deg)
Earth orbit insertion	000:11:56.21	25,605.1			32.5262	-52.5300	91.3	90.0	87.85	32.542
S-IVB 2nd burn ignition	002:33:36.50	25,598.1			-24.5488	137.4789				
S-IVB 2nd burn cutoff	002:39:18.42	35,590.2	341.92	10,389.6	-12.3781	161.7104				32.511

Apollo 16 Translunar Phase

Event	GET (hhh:mm:ss)	Altitude (n mi)	Space Fixed Velocity (ft/sec)	Event Duration (sec)	Velocity Change (ft/sec)	Space Fixed Flight Path Angle (deg)	Space Fixed Heading Angle (E of N)
Translunar injection	002:39:28.42	171.243	35,566.1			7.461	59.524
CSM separated from S-IVB	003:04:59.0	3,870.361	24,824.8			45.397	69.807
CSM/LM ejected from S-IVB	003:59:15.1	12,492.7	16,533.5			61.07	88.39
Midcourse correction ignition	030:39:00.66	119,343.8	4,514.8			76.86	111.56
Midcourse correction cutoff	030:39:02.67	119,345.3	4,508.1	2.01	12.5	76.72	111.50

[2] Data for this event reflects postflight trajectory reconstruction for 36 seconds Ground Elapsed Time.

Apollo 16 Lunar Orbit Phase

Event	GET (hhh:mm:ss)	Altitude (n mi)	Space Fixed Velocity (ft/sec)	Event Duration (sec)	Velocity Change (ft/sec)	Geodetic Latitude (deg N)	Longitude (deg E)	Apogee (n mi)	Perigee (mins)
Lunar orbit insertion ignition	074:28:27.87	93.9	8,105.4			8.15	-166.63		
Lunar orbit insertion cutoff	074:34:42.77	75.3	5,399.2	374.90	2,802	7.12	169.32	170.3	58.1
Descent orbit insertion ignition	078:33:45.04	58.5	5,486.3			8.58	136.02		
Descent orbit insertion cutoff	078:34:09.39	58.4	5,281.9	24.35	209.5	8.58	-137.27	58.5	10.9
LM undocking and separation	096:13:31	33.8	5,417.2			2.37	121.92		
CSM orbit circularization ignition	103:21:43.08	59.2	5,277.8			9.22	-151.98		
CSM orbit circularization cutoff	103:21:47.74	59.1	5,348.7	4.66	81.6	9.23	-151.95	68.0	53.1
LM powered descent initiation	104:17:25	10.944	5,548.8			-8.67	32.73		
LM powered descent cutoff	104:29:39			734	6,703				
CSM plane change ignition	169:05:52.14	58.6	5,349.8			5.60	108.83		
CSM plane change cutoff	169:05:59.28	58.6	5,349.9	7.14	124	5.57	108.50	64.6	55.0
LM lunar liftoff ignition	175:31:47.9								
LM lunar ascent orbit cutoff	175:38:55.7	9.9	5,523.3	427.8	6,054.2	-9.77	5.43	40.2	7.9
LM vernier adjustment	175:42:18	11.2	5,515.2			-10.67	-5.83		
LM terminal phase initiation ignition	176:26:05	40.2	5,351.6			6.88	-147.37		
LM terminal phase initiation cutoff	176:26:07.5			2.5	78.0				
LM terminal phase finalize	177:08:42							64.2	40.1
CSM/LM docked	177:41:18	65.6	5,313.7			-10.53	-55.65		
LM ascent stage jettisoned	195:00:12	59.2	5,347.9						
CSM separation maneuver	195:03:13				2.0	-0.02	-115.98		
Subsatellite launched	196:02:09	58.4	5,349.4			1.13	70.47	66	52

Apollo 16 Transearth Phase

Event	GET (hhh:mm:ss)	Altitude (n mi)	Space Fixed Velocity (ft/sec)	Event Duration (sec)	Velocity Change (ft/sec)	Space Fixed Flight Path Angle (deg)	Space Fixed Heading Angle (E of N)
Transearth injection ignition	200:21:33.07	52.2	5,383.6			0.15	-85.80
Transearth injection cutoff	200:24:15.36	59.7	8,663.0	162.29	3,370.9	5.12	-82.37
Midcourse correction ignition	214:35:02.8	183,668.0	3,806.8			-75.08	165.08
Midcourse correction cutoff	214:35:25.4	183,664.8	3,807.9	22.6	3.4	-80.35	164.99
Midcourse correction ignition	262:37:20.7	25,312.9	12,256.5			-69.02	157.11
Midcourse correction cutoff	262:37:27.1	25,305.2	12,258.3	6.4	1.4	-69.02	157.10

Apollo 16 Timeline

Event	GET (hhh:mm:ss)	GMT Time	GMT Date
Terminal countdown started.	-028:00:00	03:54:00	15 Apr 1972
Scheduled 9-hour hold at T-9 hours.	-009:00:00	22:54:00	15 Apr 1972
Countdown resumed at T-9 hours.	-009:00:00	07:54:00	16 Apr 1972
Scheduled 1-hour hold at T-3 hours 30 minutes.	-003:30:00	13:24:00	16 Apr 1972
Countdown resumed at T-3 hours 30 minutes.	-003:30:00	14:24:00	16 Apr 1972
Guidance reference release.	-000:00:16.963	17:53:43	16 Apr 1972
S-IC engine start command.	-000:00:08.9	17:53:51	16 Apr 1972
S-IC engine ignition (#5).	-000:00:06.7	17:53:53	16 Apr 1972
All S-IC engines thrust OK.	-000:00:01.9	17:53:58	16 Apr 1972
Range zero.	000:00:00.00	17:54:00	16 Apr 1972
All holddown arms released (1st motion) (1.08 g).	000:00:00.3	17:54:00	16 Apr 1972
Liftoff (umbilical disconnected).	000:00:00.59	17:54:00	16 Apr 1972
Tower clearance yaw maneuver started.	000:00:01.7	17:54:01	16 Apr 1972
Yaw maneuver ended.	000:00:10.9	17:54:10	16 Apr 1972
Pitch and roll maneuver started.	000:00:12.7	17:54:12	16 Apr 1972
Roll maneuver ended.	000:00:31.8	17:54:31	16 Apr 1972
Mach 1 achieved.	000:01:07.5	17:55:07	16 Apr 1972
Maximum dynamic pressure (724.72 lb/ft²).	000:01:26.0	17:55:26	16 Apr 1972
Maximum bending moment (71,000,000 lbf-in).	000:01:26.5	17:55:26	16 Apr 1972
S-IC center engine cutoff command.	000:02:17.85	17:56:17	16 Apr 1972
Pitch maneuver ended.	000:02:38.9	17:56:38	16 Apr 1972
S-IC outboard engine cutoff. Maximum total inertial acceleration (3.82 g).	000:02:41.78	17:56:41	16 Apr 1972
S-IC maximum Earth-fixed velocity.	000:02:42.5	17:56:42	16 Apr 1972
S-IC/S-II separation command.	000:02:43.5	17:56:43	16 Apr 1972
S-II engine start command.	000:02:44.2	17:56:44	16 Apr 1972
S-II ignition.	000:02:45.2	17:56:45	16 Apr 1972
S-II aft interstage jettisoned.	000:03:13.5	17:57:13	16 Apr 1972
Launch escape tower jettisoned.	000:03:19.8	17:57:19	16 Apr 1972
Iterative guidance mode initiated.	000:03:24.5	17:57:24	16 Apr 1972
S-IC apex.	000:04:30.973	17:58:31	16 Apr 1972
S-II center engine cutoff. S-II maximum total inertial acceleration (1.74 g).	000:07:41.77	18:01:41	16 Apr 1972
S-IC impact (theoretical).	000:09:07.136	18:03:07	16 Apr 1972
S-II outboard engine cutoff.	000:09:19.54	18:03:19	16 Apr 1972
S-II maximum Earth-fixed velocity.	000:09:20.0	18:03:20	16 Apr 1972
S-II/S-IVB separation command.	000:09:20.5	18:03:20	16 Apr 1972
S-IVB 1st burn start command.	000:09:20.60	18:03:20	16 Apr 1972
S-IVB 1st burn ignition.	000:09:23.60	18:03:23	16 Apr 1972
S-IVB ullage case jettisoned.	000:09:32.3	18:03:32	16 Apr 1972
S-II apex.	000:09:44.122	18:03:44	16 Apr 1972
S-IVB 1st burn cutoff and maximum total inertial acceleration (0.67 g).	000:11:46.21	18:05:46	16 Apr 1972
Earth orbit insertion. S-IVB 1st burn maximum Earth-fixed velocity.	000:11:56.21	18:05:56	16 Apr 1972
Maneuver to local horizontal attitude started.	000:12:07.8	18:06:07	16 Apr 1972
Orbital navigation started.	000:13:26.1	18:07:26	16 Apr 1972
S-II impact (theoretical).	000:20:02.390	18:14:02	16 Apr 1972
S-IVB 2nd burn restart preparation.	002:23:58.60	20:17:58	16 Apr 1972
S-IVB 2nd burn restart command.	002:33:28.50	20:27:28	16 Apr 1972
S-IVB 2nd burn ignition.	002:33:36.50	20:27:36	16 Apr 1972
S-IVB 2nd burn cutoff and maximum total inertial acceleration (1.42 g).	002:39:18.42	20:33:18	16 Apr 1972
S-IVB safing procedures started.	002:39:19.1	20:33:19	16 Apr 1972
S-IVB 2nd burn maximum Earth-fixed velocity.	002:39:20.0	20:33:20	16 Apr 1972
Translunar injection.	002:39:28.42	20:33:28	16 Apr 1972

Apollo 16 Timeline

Event	GET (hhh:mm:ss)	GMT Time	GMT Date
Maneuver to local horizontal attitude and orbital navigation started.	002:41:50.3	20:35:50	16 Apr 1972
Maneuver to transposition and docking attitude started.	002:54:19.3	20:48:19	16 Apr 1972
CSM separated from S-IVB.	003:04:59.0	20:58:59	16 Apr 1972
TV transmission started.	003:10	21:04	16 Apr 1972
CSM docked with LM/S-IVB.	003:21:53.4	21:15:53	16 Apr 1972
TV transmission ended.	003:28	21:22	16 Apr 1972
CSM/LM ejected from S-IVB.	003:59:15.1	21:53:15	16 Apr 1972
TV transmission started.	004:10	22:04	16 Apr 1972
S-IVB yaw maneuver to attain attitude for evasive maneuver.	004:10:01	22:04:01	16 Apr 1972
S-IVB APS evasive maneuver ignition.	004:18:08.3	22:12:08	16 Apr 1972
S-IVB APS evasive maneuver cutoff.	004:19:28.5	22:13:28	16 Apr 1972
TV transmission ended.	004:20	22:14	16 Apr 1972
Maneuver to S-IVB LOX dump attitude initiated.	004:27:48.4	22:21:48	16 Apr 1972
Alternate (second) maneuver to LOX dump attitude.	004:31:09	22:25:09	16 Apr 1972
S-IVB lunar impact maneuver—CVS vent opened.	004:34:47.1	22:28:47	16 Apr 1972
S-IVB lunar impact maneuver—LOX dump started.	004:39:27.1	22:33:27	16 Apr 1972
S-IVB lunar impact maneuver—CVS vent closed.	004:39:47.1	22:33:47	16 Apr 1972
S-IVB lunar impact maneuver—LOX dump ended.	004:40:15.1	22:34:15	16 Apr 1972
Maneuver to attitude required for final S-IVB APS burn initiated.	005:30:37.2	23:24:37	16 Apr 1972
S-IVB lunar impact maneuver—APS ignition.	005:40:07.2	23:34:07	16 Apr 1972
S-IVB lunar impact maneuver—APS cutoff.	005:41:01.4	23:35:01	16 Apr 1972
S-IVB lunar impact maneuver—3-axis tumble command initiated.	005:55:06.2	23:49:06	16 Apr 1972
Command to inhibit instrument unit flight control computer to leave S-IVB in 3-axis tumble mode.	005:55:37	23:49:37	16 Apr 1972
Crew reported stream of particles coming from LM.	007:18	01:12	17 Apr 1972
Unscheduled crew transfer to LM for system checks.	008:17	02:11	17 Apr 1972
TV transmission to give Mission Control a view of the particle emissions started.	008:45	02:39	17 Apr 1972
LM powered down.	008:52	02:46	17 Apr 1972
TV transmission ended.	009:06	03:00	17 Apr 1972
Electrophoresis demonstration started.	025:05	18:59	17 Apr 1972
Electrophoresis demonstration ended.	025:50	19:44	17 Apr 1972
Loss of S-IVB tracking data precluded exact determination of impact time and location within mission objectives.	027:09:59	21:03:59	17 Apr 1972
Midcourse correction ignition (SPS).	030:39:00.66	00:33	18 Apr 1972
Midcourse correction cutoff.	030:39:02.67	00:33:02	18 Apr 1972
LM pressurized.	032:30	02:24	18 Apr 1972
CDR and LMP entered LM for housekeeping and communication checkout.	033:00	02:54	18 Apr 1972
CDR and LMP entered CM.	035:00	04:54	18 Apr 1972
False gimbal lock indication.	038:18:56	08:12:56	18 Apr 1972
Visual light flash phenomenon observations started.	049:10	19:04	18 Apr 1972
Visual light flash phenomenon observations ended.	050:16	20:10	18 Apr 1972
CDR and LMP entered LM for housekeeping.	053:30	23:24	18 Apr 1972
CDR and LMP entered CM.	055:11	01:05	19 Apr 1972
Skylab food test.	056:30	02:24	19 Apr 1972
Equigravisphere.	059:19:45	05:13:45	19 Apr 1972
Scientific instrument module door jettisoned.	069:59:01	15:53:01	19 Apr 1972
Lunar orbit insertion ignition (SPS).	074:28:27.87	20:22:27	19 Apr 1972
Lunar orbit insertion cutoff.	074:34:42.77	20:28:42	19 Apr 1972
S-IVB impact on lunar surface.	075:08:04.0	21:02:04	19 Apr 1972
Descent orbit insertion ignition (SPS).	078:33:45.04	00:27:45	20 Apr 1972
Descent orbit insertion cutoff.	078:34:09.39	00:28:09	20 Apr 1972

Apollo 16 Timeline

Event	GET (hhh:mm:ss)	GMT Time	GMT Date
CSM landmark tracking.	079:30	01:24	20 Apr 1972
Solar monitor door/tie-down release.	080:10	02:04	20 Apr 1972
CDR and LMP entered LM.	092:50	14:44	20 Apr 1972
LM activation and system checks.	093:34	15:28	20 Apr 1972
Terminator photography.	094:40	16:34	20 Apr 1972
LM undocking and separation.	096:13:31	18:07:31	20 Apr 1972
CSM landmark tracking.	096:40	18:34	20 Apr 1972
CSM checkout indicated no rate feedback and SPS engine gimbal position indicator showed yaw oscillations. Planned circularization maneuver at 097:41:44 not performed.	097:40	19:34	20 Apr 1972
Rendezvous (CSM active).	100:00	21:54	20 Apr 1972
LM separation from CSM.	102:30:00	00:24	21 Apr 1972
CSM and LM platforms realigned.	102:40	00:34	21 Apr 1972
CSM orbit circularization ignition (SPS).	103:21:43.08	01:15:43	21 Apr 1972
CSM orbit circularization cutoff.	103:21:47.74	01:15:47	21 Apr 1972
LM powered descent engine ignition (LM DPS).	104:17:25	02:11:25	21 Apr 1972
LM throttle to full-throttle position.	104:17:53	02:11:53	21 Apr 1972
LM manual target (landing site) update.	104:19:16	02:13:16	21 Apr 1972
CSM landmark tracking.	104:20	02:14:20	21 Apr 1972
LM landing radar velocity data good.	104:20:38	02:14:38	21 Apr 1972
LM landing radar range data good.	104:21:24	02:15:24	21 Apr 1972
LM landing radar updates enabled.	104:21:54	02:15:54	21 Apr 1972
LM landing point redesignation phase entered.	104:24:14	02:18:14	21 Apr 1972
LM throttle down.	104:24:54	02:18:54	21 Apr 1972
LM landing radar antenna to position 2.	104:26:50	02:20:50	21 Apr 1972
LM approach phase program selected and pitchover.	104:26:52	02:20:52	21 Apr 1972
LM 1st landing point redesignation.	104:27:20	02:21:20	21 Apr 1972
LM landing radar switched to low scale.	104:27:32	02:21:32	21 Apr 1972
LM attitude hold mode selected.	104:28:37	02:22:37	21 Apr 1972
LM landing phase program selected.	104:28:42	02:22:42	21 Apr 1972
LM lunar landing.	104:29:35	02:23:35	21 Apr 1972
LM powered descent engine cutoff.	104:29:39	02:23:39	21 Apr 1972
Mission clock updated (000:11:48.00 added).	118:06:31	16:00:31	21 Apr 1972
CSM terminator photography.	118:20	16:14	21 Apr 1972
1st EVA started (LM cabin depressurized).	118:53:38	16:47:38	21 Apr 1972
Lunar roving vehicle (LRV) offloaded.	119:25:29	17:19:29	21 Apr 1972
LRV deployed.	119:32:44	17:26:44	21 Apr 1972
Far ultraviolet camera/spectroscope deployed.	119:54:01	17:48:01	21 Apr 1972
TV transmission started for 1st EVA.	120:05:40	17:59:40	21 Apr 1972
United States flag deployed.	120:15	18:09	21 Apr 1972
Apollo lunar surface experiments package (ALSEP) offloaded.	120:21:35	18:15:35	21 Apr 1972
CSM terminator photography.	120:30	18:24	21 Apr 1972
CSM Gum nebula photography.	121:20	19:14	21 Apr 1972
ALSEP deployed, deep core sample gathered, and LRV configured for traverse.	122:55:23	20:49:23	21 Apr 1972
Departed for station 1.	122:58:02	20:52:02	21 Apr 1972
CSM zodiacal photography.	123:00	20:54	21 Apr 1972
CDR reported bright flash on lunar surface.	123:09:40	21:03:40	21 Apr 1972
Arrived at station 1. Performed radial sampling, gathered rake and documented samples, and performed panoramic and stereographic photography.	123:23:54	21:17:54	21 Apr 1972
Departed for station 2.	124:14:32	22:08:32	21 Apr 1972
Arrived at station 2. Performed a lunar portable magnetometer measurement, gathered samples and performed panoramic and 500 mm photography.	124:21:10	22:15:10	21 Apr 1972

Apollo 16 Timeline

Event	GET (hhh:mm:ss)	GMT Time	GMT Date
Departed for ALSEP site (station 3/10).	124:48:07	22:42:07	21 Apr 1972
Arrived at station 3/10. Performed "grand prix" with LRV, retrieved core sample, armed the active seismic experiment mortar package, and departed for LM.	124:54:14	22:48:14	21 Apr 1972
Arrived at LM. Deployed solar wind composition experiment, gathered samples, performed photography, and started EVA closeout.	125:05:09	22:59:09	21 Apr 1972
Solar wind composition experiment deployed.	125:07:00	23:01	21 Apr 1972
TV transmission ended for 1st EVA.	125:35	23:29	21 Apr 1972
1st EVA ended (LM cabin repressurized).	126:04:40	23:58:40	21 Apr 1972
CSM ultraviolet photography.	126:20	00:14	22 Apr 1972
CSM Gegenschein calibration.	127:00	00:54	22 Apr 1972
CSM orbital science visual observations.	128:00	01:54	22 Apr 1972
LM crew debriefing.	128:20	02:14	22 Apr 1972
CSM terminator photography.	128:30	02:24	22 Apr 1972
CSM orbital science visual observations.	129:25	03:19	22 Apr 1972
CSM orbital science photography.	130:00	03:54	22 Apr 1972
CSM terminator photography.	131:20	05:14	22 Apr 1972
2nd EVA started (LM cabin depressurized).	142:39:35	16:33:35	22 Apr 1972
LRV prepared for traverse.	142:49:29	16:43:29	22 Apr 1972
CSM Gegenschein photography.	142:30	16:24	22 Apr 1972
TV transmission started for 2nd EVA.	142:55	16:49	22 Apr 1972
CSM Gegenschein photography.	142:30	16:24	22 Apr 1972
Departed for station 4.	143:31:40	17:25:40	22 Apr 1972
Arrived at station 4. Performed penetrometer measurements, gathered samples, obtained a double core tube sample, gathered a soil trench sample, and performed 500 mm and panoramic photography.	144:07:26	18:01:26	22 Apr 1972
CSM deep space measurement.	144:45	18:39	22 Apr 1972
Departed for station 5.	145:05:16	18:59:16	22 Apr 1972
Arrived at station 5. Gathered samples, performed lunar portable magnetometer measurement, and performed panoramic photography.	145:10:05	19:04:05	22 Apr 1972
CSM orbital science photography.	145:35	19:29	22 Apr 1972
Departed for station 6.	145:58:40	19:52:40	22 Apr 1972
CSM orbital science visual observations.	146:05	19:59	22 Apr 1972
Arrived at station 6. Gathered samples and performed panoramic photography.	146:06:37	20:00:37	22 Apr 1972
Departed for station 8 (station 7 deleted).	146:29:18	20:23:18	22 Apr 1972
Arrived at station 8. Gathered samples, obtained a double core tube sample, and performed panoramic photography.	146:40:19	20:34:19	22 Apr 1972
CSM terminator photography.	147:15	21:09	22 Apr 1972
Departed for station 9.	147:48:15	21:42:15	22 Apr 1972
Arrived at station 9. Gathered samples, obtained single core tube sample, and performed panoramic photography.	147:53:12	21:47:12	22 Apr 1972
Departed for station 10.	148:29:45	22:23:45	22 Apr 1972
Arrived at station 10. Gathered samples, performed penetrometer measurements, obtained a double core tube sample, and performed panoramic photography.	148:54:16	22:48:16	22 Apr 1972
CSM solar corona photography.	149:05	22:59	22 Apr 1972
Departed for LM.	149:21:17	23:15:17	22 Apr 1972
Arrived at LM and started EVA activity closeout.	149:23:24	23:17:24	22 Apr 1972
TV transmission ended for 2nd EVA.	149:40	23:34	22 Apr 1972
2nd EVA ended (LM cabin repressurized).	150:02:44	23:56:44	22 Apr 1972
CSM photography of mass spectrometer boom.	153:05	02:59	23 Apr 1972
CSM orbital science visual observations.	153:40	03:34	23 Apr 1972
CSM terminator photography.	154:20	04:14	23 Apr 1972

Event	GET (hhh:mm:ss)	GMT Time	GMT Date
CSM orbital science photography.	155:05	04:59	23 Apr 1972
CSM bistatic radar test started.	155:20	05:14	23 Apr 1972
CSM bistatic radar test ended.	156:00	05:54	23 Apr 1972
CSM mass spectrometer retraction test started.	165:30	15:24	23 Apr 1972
3rd EVA started (LM cabin depressurized).	165:31:28	15:25:28	23 Apr 1972
LRV prepared for traverse.	165:43:29	15:37:29	23 Apr 1972
TV transmission started for 3rd EVA.	165:40	15:34	23 Apr 1972
CSM mass spectrometer retraction test ended.	166:00	15:54	23 Apr 1972
Departed for station 11.	166:09:13	16:03:13	23 Apr 1972
Arrived at station 11. Gathered samples, performed 500 mm and panoramic photography.	166:44:50	16:38:50	23 Apr 1972
CSM solar camera photography.	166:50	16:44	23 Apr 1972
CSM orbital science visual observations.	167:50	17:44	23 Apr 1972
Departed for station 13.	168:09:46	18:03:46	23 Apr 1972
Arrived at station 13. Gathered samples, performed lunar portable magnetometer measurement and performed panoramic photography.	168:17:39	18:11:39	23 Apr 1972
Departed for station 10 prime.	168:46:33	18:40:33	23 Apr 1972
CSM plane change ignition (SPS).	169:05:52.14	18:59:52	23 Apr 1972
CSM plane change cutoff.	169:05:59.28	18:59:59	23 Apr 1972
Arrived at station 10 prime. Gathered samples, obtained a double core tube sample, and performed 500 mm and panoramic photography.	169:15:38	19:09:38	23 Apr 1972
LRV driven to LM. Samples gathered. EVA closeout started.	169:51:48	19:45:48	23 Apr 1972
Solar wind composition experiment retrieved.	170:12:00	20:06	23 Apr 1972
Departed for LRV final parking area.	170:23:06	20:17:06	23 Apr 1972
Arrived at final parking area. Performed two lunar portable magnetometer measurements, gathered samples and continued EVA closeout.	170:27:09	20:21:09	23 Apr 1972
Film from far ultraviolet camera/spectroscope retrieved.	171:01:42	20:55:42	23 Apr 1972
CSM Gegenschein photography.	171:00	20:54	23 Apr 1972
TV transmission ended for 3rd EVA. CSM deep space measurement.	171:10	21:04	23 Apr 1972
3rd EVA ended (LM cabin repressurized).	171:11:31	21:05:31	23 Apr 1972
LM equipment jettisoned.	172:15	22:09	23 Apr 1972
TV transmission started.	175:15	01:09	24 Apr 1972
LM lunar liftoff ignition (LM APS).	175:31:47.9	01:25:47	24 Apr 1972
Lunar ascent orbit cutoff.	175:38:55.7	01:32:55	24 Apr 1972
TV transmission ended.	175:40	01:34	24 Apr 1972
Vernier adjustment.	175:42:18	01:36:18	24 Apr 1972
TV transmission started.	176:18	02:12	24 Apr 1972
TV transmission ended.	176:25	02:19	24 Apr 1972
Terminal phase initiation ignition (LM APS).	176:26:05	02:20:05	24 Apr 1972
Terminal phase initiation cutoff.	176:26:07.5	02:20:07	24 Apr 1972
LM 1st midcourse correction.	176:35	02:29	24 Apr 1972
LM 2nd midcourse correction.	176:50	02:44	24 Apr 1972
Terminal phase finalize.	177:08:42	03:02:42	24 Apr 1972
CSM/LM docked.	177:41:18	03:35:18	24 Apr 1972
Transfer and stowing of equipment and samples started.	178:15	04:09	24 Apr 1972
Mass spectrometer deployed.	178:40	04:34	24 Apr 1972
Transfer and stowing of equipment and samples ended.	180:00	05:54	24 Apr 1972
Transfer of items to LM ascent stage started.	192:00	17:54	24 Apr 1972
Transfer of items to LM ascent stage ended. LM ascent stage activated.	192:30	18:24	24 Apr 1972
Maneuver to LM jettison attitude.	192:55	18:49	24 Apr 1972
Hatch closeout.	194:30	20:24	24 Apr 1972
LM prepared for jettison.	194:35	20:29	24 Apr 1972

Apollo 16 Timeline

Event	GET (hhh:mm:ss)	GMT Time	GMT Date
LM ascent stage jettisoned.	195:00:12	20:54:12	24 Apr 1972
LM tumbling started.			
LM tumbling ended.			
CSM separation maneuver.	195:03:13	20:57:13	24 Apr 1972
Mass spectrometer boom jettisoned.	195:23:12	21:17:12	24 Apr 1972
Subsatellite launched.	196:02:09	21:56:09	24 Apr 1972
Sunrise solar corona photography.	196:40	22:34	24 Apr 1972
Transearth injection ignition (SPS).	200:21:33.07	02:15:33	25 Apr 1972
Transearth injection cutoff.	200:24:15.36	02:18:15	25 Apr 1972
X-Ray spectrometer—Scorpius X-1 observation started.	201:31	03:25	25 Apr 1972
X-Ray spectrometer—Scorpius X-1 observation ended.	202:11	04:05	24 Apr 1972
Mission clock updated (024:34:12 added).	202:18:12	04:12:12	25 Apr 1972
TV transmission from CM started.	202:57	04:51	25 Apr 1972
TV transmission from CM ended.	203:12	05:06	25 Apr 1972
TV transmission from lunar surface (LRV camera) started.	203:29	05:23	25 Apr 1972
TV transmission from lunar surface ended.	204:12	06:06	25 Apr 1972
Midcourse correction ignition.	214:35:02.8	16:29:02	25 Apr 1972
Midcourse correction cutoff.	214:35:25.4	16:29:25	25 Apr 1972
Transearth EVA started (Mattingly).	218:39:46	20:33:46	25 Apr 1972
TV transmission started for transearth EVA.	218:40	20:34	25 Apr 1972
Installation of television camera and data acquisition cameras started.	218:50	20:44	25 Apr 1972
Camera cassette retrieval and scientific instrument module inspection.	219:10	21:04	25 Apr 1972
Microbial response in space environment experiment.	219:30	21:24	25 Apr 1972
TV transmission ended for transearth EVA.	219:49	21:43	25 Apr 1972
Ingress and hatch closing started.	219:50	21:44	25 Apr 1972
Transearth EVA ended.	220:03:28	21:57:28	25 Apr 1972
X-Ray spectrometer—Cygnus X-1 observation started.	221:01	22:55	25 Apr 1972
X-Ray spectrometer—Cygnus X-1 observation ended.	224:01	01:55	26 Apr 1972
X-Ray spectrometer—Scorpius X-1 observation started.	224:21	02:15	26 Apr 1972
Contamination control.	226:10	04:04	26 Apr 1972
X-Ray spectrometer—Scorpius X-1 observation ended.	226:51	04:45	26 Apr 1972
Apollo 15 subsatellite reactivated.	226:50	04:44	26 Apr 1972
Visual light flash phenomenon observations started.	238:00	15:54	26 Apr 1972
Visual light flash phenomenon observations ended.	239:00	16:54	26 Apr 1972
X-Ray spectrometer—Scorpius X-1 observation started.	242:21	20:15	26 Apr 1972
Televised press conference started.	243:35	21:29	26 Apr 1972
Televised press conference ended.	243:53	21:47	26 Apr 1972
Jet firing test.	245:00	22:54	26 Apr 1972
X-Ray spectrometer—Scorpius X-1 observation ended.	245:51	23:45	26 Apr 1972
Skylab contamination photography started.	245:30	23:24	26 Apr 1972
Skylab contamination photography ended.	247:00	00:54	27 Apr 1972
X-ray spectrometer—Cygnus X-1 observation started.	248:51	02:45	27 Apr 1972
X-ray spectrometer—Cygnus X-1 observation ended.	251:51	05:45	27 Apr 1972
Midcourse correction ignition.	262:37:20.7	16:31:20	27 Apr 1972
Midcourse correction cutoff.	262:37:27.1	16:31:27	27 Apr 1972
Earth ultraviolet photography.	263:00	16:54	27 Apr 1972
CM/SM separation.	265:22:23	19:16:23	27 Apr 1972
Entry.	265:37:31	19:31:31	27 Apr 1972
Communication blackout started.	265:37:47	19:31:47	27 Apr 1972
Radar contact with CM by recovery ship.	265:40	19:34	27 Apr 1972
Communication blackout ended.	265:41:01	19:35:01	27 Apr 1972

Apollo 16 Timeline

Event	GET (hhh:mm:ss)	GMT Time	GMT Date
Forward heat shield jettisoned.	265:45:25	19:39:25	27 Apr 1972
Drogue parachute deployed	265:45:26	19:39:26	27 Apr 1972
Visual contact with CM established by recovery forces.	265:45	19:39	27 Apr 1972
Main parachute deployed.	265:46:16	19:40:16	27 Apr 1972
VHF recovery beacon contact with CM established by recovery ship.	265:46	19:40	27 Apr 1972
Voice contact with CM established by recovery ship.	265:47	19:41	27 Apr 1972
Splashdown (went to apex-down).	265:51:05	19:45:05	27 Apr 1972
CM returned to apex-up position.	265:55:30	19:49:30	27 Apr 1972
Swimmers deployed to CM.	265:56	19:50	27 Apr 1972
flotation collar inflated.	266:06	20:00	27 Apr 1972
Hatch opened for crew egress.	266:10	20:04	27 Apr 1972
Crew aboard recovery helicopter.	266:22	20:16	27 Apr 1972
Crew aboard recovery ship.	266:28	20:22	27 Apr 1972
CM aboard recovery ship.	267:30	21:24	27 Apr 1972
1st sample flight departed recovery ship.	305:51	11:45	29 Apr 1972
1st sample flight arrived in Hawaii.	308:20	14:14	29 Apr 1972
1st sample flight departed Hawaii.	309:09	15:03	29 Apr 1972
Flight crew departed recovery ship.	311:36	17:30	29 Apr 1972
Flight crew arrived in Hawaii.	313:27	19:21	29 Apr 1972
Flight crew departed Hawaii.	314:13	20:07	29 Apr 1972
1st sample flight arrived in Houston.	316:38	22:32	29 Apr 1972
CM arrived in Hawaii.	321:36	03:30	30 Apr 1972
Flight crew arrived in Houston.	321:46	03:40	30 Apr 1972
CM departed Hawaii.	360:06	18:00	01 May 1972
CM arrived at North Island, San Diego.	462:06	00:00	06 May 1972
Explosive failure of ground support equipment decontamination unit tank during deactivation of nitrogen tetroxide portion of CM RCS.			07 May 1972
CM deactivated.	606:06	00:00	11 May 1972
CM departed San Diego.	609:06	03:00	12 May 1972
CM arrived at contractor's facility in Downey, CA.	616:36	10:30	12 May 1972
Final telemetry from subsatellite (just before impact on lunar surface).	1034:37	20:31	29 May 1972

APOLLO 17

The Eleventh Mission:
The Sixth Lunar Landing

Apollo 17 Summary

(7 December–19 December 1972)

Apollo 17 crew (l. to r.): Jack Schmitt, Gene Cernan (seated), Ron Evans (NASA S72-50438).

Background

Apollo 17 was the third Type J mission, an extensive scientific investigation of the Moon on the lunar surface and from lunar orbit. Although the spacecraft and launch vehicle were similar to those for Apollo 15 and 16, some experiments were unique to this mission. It was also the final piloted lunar landing mission of the Apollo program.

The primary objectives were:

• to perform selenological inspection, survey, and sampling of materials and surface features in a preselected area of the Taurus-Littrow region;

• to emplace and activate surface experiments; and

• to conduct inflight experiments and photographic tasks.

The targeted landing site was the Taurus-Littrow region, selected because of the certainty of acquiring highlands material, the potential for superior orbital coverage, and for better use of the LRV.

The crew members were Captain Eugene Andrew "Gene" Cernan, (USN), commander; Commander Ronald Ellwin Evans (USN), command module pilot; and Harrison Hagan "Jack" Schmitt, Ph.D., lunar module pilot.

[1] Evans died of a heart attack on 7 April 1990 in Scottsdale, Arizona.

Selected as an astronaut in 1963, Cernan was making his third spaceflight. He had been pilot of Gemini 9-A and lunar module pilot of Apollo 10, the first test of the LM in lunar orbit and the dress rehearsal for the first piloted landing on the Moon. Born 14 March 1934 in Chicago, Illinois, Cernan was 38 years old at the time of the Apollo 17 mission. He received a B.S. in electrical engineering from Purdue University in 1956 and an M.S. in aeronautical engineering from the U.S. Naval Postgraduate School in 1963. His backup for the mission was Captain John Watts Young (USN).

Evans and Schmitt were making their first spaceflights. Born 10 November 1933 in St. Francis, Kansas, Evans was 39 years old at the time of the mission. He received a B.S. in electrical engineering from the University of Kansas in 1956 and a M.S. in aeronautical engineering from the U.S. Naval Postgraduate School in 1964, and he was selected as an astronaut in 1966.[1] His backup was Lt. Colonel Stuart Allen Roosa (USAF).

A geologist, Schmitt was the first true scientist to explore the Moon. Born 3 July 1935 in Santa Rita, New Mexico, he was 37 years old at the time of the Apollo 17 mission. Schmitt received a B.S. in science from the California Institute of Technology in 1957 and a Ph.D. in geology from Harvard University in 1964. He was selected as an astronaut in 1965. His backup was Colonel Charles Moss Duke, Jr. (USAF).

The capsule communicators (CAPCOMs) for the mission were Major Charles Gordon Fullerton (USAF), Lt. Colonel Robert Franklyn Overmyer (USMC), Robert Alan Ridley Parker, Ph. D., Joseph Percival Allen IV, Ph. D., Captain Alan Bartlett Shepard, Jr. (USN), Commander Thomas Kenneth "Ken" Mattingly, II (USN), Duke, Roosa, and Young. The support crew were Overmyer, Parker, and Fullerton. The flight directors were Gerald D. Griffin (first shift), Eugene F. Kranz and Neil B. Hutchinson (second shift), and M.P. "Pete" Frank and Charles R. Lewis (third shift).

The Apollo 17 launch vehicle was a Saturn V, designated SA-512. The mission also carried the designation Eastern Test Range #1701. The CSM was designated CSM-114, and had the call-sign "America." The lunar module was designated LM-12, and had the call-sign "Challenger."

Launch Preparations

The terminal countdown was picked up at T-28 hours on at 12:53:00 GMT on 5 December 1972. Scheduled holds were initiated at T-9 hours for nine hours and at T-3 hours 30 minutes for one hour.

The launch countdown proceeded smoothly until 2 minutes 47 seconds before the scheduled launch, when the Terminal Countdown Sequencer failed to issue the S-IVB LOX tank pressurization command. As a result, an automatic hold command was issued at T-30 seconds which lasted 1 hour 5 minutes 11 seconds. The countdown was recycled to T-22 minutes, but was held again at T-8 minutes to resolve the sequencer corrective action. This hold lasted 1 hour 13 minutes 19 seconds The countdown was then picked up at T-8 minutes and proceeded smoothly to launch. The delays totaled 2 hours 40 minutes.

During the evening launch of Apollo 17, the Cape Kennedy area was experiencing mild temperatures with gentle surface winds. These conditions resulted from a warm moist air mass covering most of Florida. This warm air was separated from an extremely cold air mass over the rest of the south by a cold front oriented northeast-southwest and passing through the Florida panhandle. Surface winds in the Cape Kennedy area were light and northwesterly. The maximum wind belt was located north of Florida, giving less intense wind flow aloft over the Cape Kennedy area. At launch time, stratocumulus clouds covered 20 percent of the sky (base 2,600 feet) and cirrus clouds covered 50 percent (base 26,000 feet); the temperature was 70.0° F; the relative humidity was 93 percent; and the barometric pressure was 14.795 lb/in^2. The winds, as measured by the anemometer on the light pole 60.0 feet above ground at the launch site measured 8.0 knots at 5° from true north. The winds, as measured at 530 feet above the launch site, measured 10.5 knots at 335° from true north.

Ascent Phase

Apollo 17 was launched from Kennedy Space Center Launch Complex 39, Pad A, at a Range Zero time of 05:33:00 GMT (12:33:00 a.m. EST) on 7 December 1972. The planned launch window was 02:53:00 GMT to 06:31:00 GMT on 7 December to take advantage of a sun elevation angle on the lunar surface of 13.3°.

Between 000:00:12.9 and 000:00:14.3, the vehicle rolled from a launch pad azimuth of 90° to a flight azimuth of 91.504°. The S-IC engine shut down at 000:02:41.20, followed by S-IC/S-II separation, and S-II engine ignition. The S-II engine shut down at 000:09:19.66 followed by separation from the S-IVB, which ignited at 000:09:23.80. The first S-IVB engine cutoff occurred at 000:11:42.65, with

deviations from the planned trajectory of only +1.0 ft/sec in velocity and only -0.1 n mi in altitude.

The maximum wind conditions encountered during ascent were 87.6 knots at 311° from true north at 38,945 feet, and a maximum wind shear of 0.0177 sec^{-1} at 26,164 feet.

Apollo 17 lifts off from Kennedy Space Center Pad 39A (NASA S72-55482).

Parking orbit conditions at insertion, 000:11:52.65 (S-IVB cutoff plus 10 seconds to account for engine tailoff and other transient effects), showed an apogee and perigee of 90.3 by 90.0 n mi, an inclination of 28.526°, a period of 87.83 minutes, and a velocity of 25,604.0 ft/sec. The apogee and perigee were based upon a spherical Earth with a radius of 3,443.934 n mi.

The international designation for the CSM upon achieving orbit was 1972-096A and the S-IVB was designated 1972-096B. After undocking at the Moon, the LM ascent stage would be designated 1972-096C and the descent stage 1972-096D.

Translunar Phase

After inflight systems checks, the 351.04-second translunar injection maneuver (second S-IVB firing) was performed at 003:12:36.60. The S-IVB engine shut down at 003:18:27.64 and translunar injection occurred ten seconds later at a velocity of 35,579.4 ft/sec after two Earth orbits lasting 3 hours 6 minutes 44.99 seconds.

View of Earth during translunar flight. This photo is unique because it was the only Apollo lunar mission from which the crew could see the Earth's South Pole (NASA AS17-148-22726).

At 003:42:27.6, the CSM was separated from the S-IVB stage, transposed, and docked at 003:57:10.7. During docking, there were indications of a ring latch malfunction. The LM was pressurized, the hatch removed, and troubleshooting revealed that the handles for latches 7, 9, and 10 were not locked. All were manually set and the docked spacecraft were ejected from the S-IVB at 004:45:02.3. A 79.9-second separation maneuver was performed at 005:03:01.1.

The S-IVB tanks were vented at 006:09:59.8, and the auxiliary propulsion system was fired for 98.2 seconds to target the S-IVB for a lunar impact. A second, 102.4-second maneuver was performed at 011:14:59.8.

The S-IVB impacted the lunar surface at 086:59:40.99. The impact point was latitude 4.33° south and longitude 12.37° west, 84 n mi from the target point, 182 n mi from the Apollo 12 seismometer, 84 n mi from the Apollo 14 seismometer, 559 n mi from the Apollo 15 seismometer, and 460 n mi from the Apollo 16 seismometer. The impact was recorded by all four instruments. At impact, the S-IVB weighed 30,712 pounds and was traveling 8,346 ft/sec.

The 2-hour 40-minute launch delay caused ground controllers to modify Apollo 17's trajectory so that it would arrive at the Moon at the originally scheduled time. They shortened the translunar coast time by having the crew make a 1.73-second 10.5 ft/sec midcourse correction at 035:29:59.91.

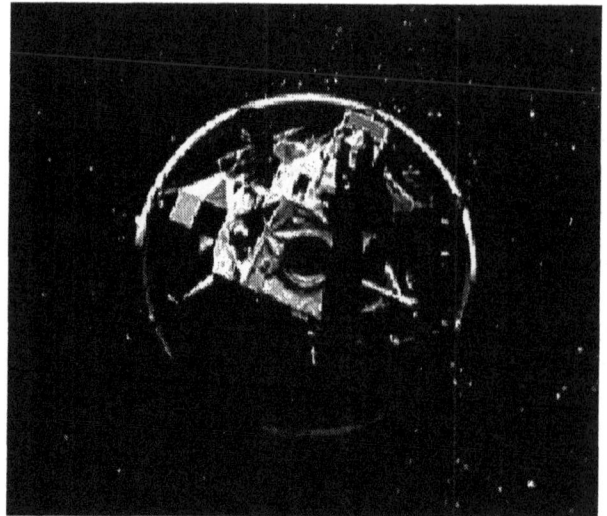

View of LM inside S-IVB stage following separation from the CSM (NASA AS17-148-22688).

The commander and lunar module pilot transferred to the LM at 040:10. At ingress, it was discovered that #4 docking latch was not properly latched. The command module pilot moved the latch handle between 30° and 45°, disengaging the hook from the docking ring. After discussion with ground control, it was decided to curtail further action on the latch until the second LM activation. The remainder of the LM housekeeping was nominal and the LM was closed out at 042:11.

The heat flow and convection demonstrations were conducted as planned. The first demonstration began at 042:55 and was performed with the spacecraft in attitude hold while the second run was accomplished with the spacecraft in the passive thermal control mode. The demonstrations produced satisfactory results, and were concluded at 046:00.

The second LM housekeeping session commenced at 059:59 and was completed at 062:16. All LM systems checks were nominal. During the LM housekeeping period, the command module pilot performed troubleshooting on the docking latch #4 problem experienced during the first session. Following instructions from the ground controllers, he stroked the latch handle and succeeded in cocking the latch. The latch was left in the cocked position for the CSM/LM rendezvous.

At 068:19, a one-hour visual light flash phenomenon observation was conducted by the crew. They reported seeing light flashes ranging from bright to dull.

The scientific instrument module bay door was jettisoned at 081:32:40.

At 086:14:22.60, at an altitude of 76.8 n mi above the Moon, the service propulsion engine was fired for 393.16 seconds to insert the spacecraft into a lunar orbit of 170.0 by 52.6 n mi. The translunar coast had lasted 83 hours 2 minutes 18.11 seconds.

Lunar Orbit/Lunar Surface Phase

At 090:31:37.43, a 22.27-second service propulsion system maneuver was performed and lowered the spacecraft to the descent orbit of 59.0 by 14.5 n mi in preparation for undocking of the LM.

The CSM/LM combination was retained in this orbit 17 hours before the spacecraft were undocked and separated by a 3.4-second maneuver at 107:47:56 at an altitude of 47.2 n mi, while in an orbit of 61.5 by 11.5 n mi. After undocking, a 3.80-second maneuver at 109:17:28.92 circularized the CSM orbit to 70.0 by 54.0 n mi.

The second LM descent orbit insertion maneuver, performed for 21.5 seconds at 109:22:42, lowered the orbit to 59.6 by 6.2 n mi. The 725-second powered descent maneuver was initiated from this orbit at 110:09:53 at an altitude of 8.7 n mi.

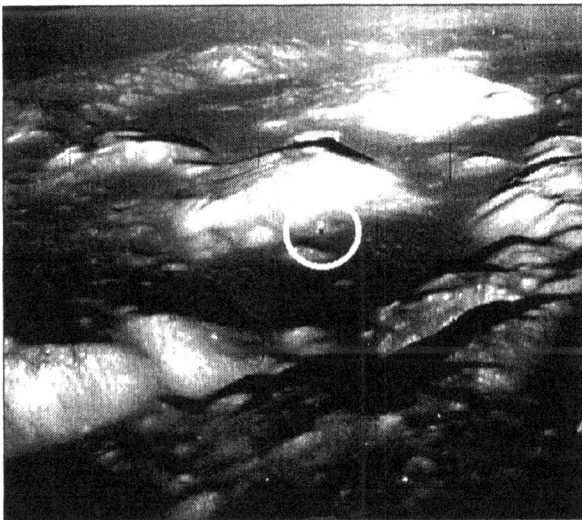

CSM (inside circle) barely seen against the Taurus Littrow landing site (NASA AS17-147-22465).

Landing occurred at 19:54:57 GMT (02:54:57 p.m. EST) on 11 December at 110:21:58. The spacecraft landed in the Taurus-Littrow region at latitude 20.19080° north and longitude 30.77168° east, within 656 feet of the planned landing point. Approximately 117 seconds of engine firing time remained at landing.

The first extravehicular activity began at 114:21:49 with the depressurization of the LM cabin. After exiting to the surface, the crew offloaded the lunar roving vehicle (LRV-3) at 114:51:10.

Cernan checks out LRV during EVA-1 and prior to loading it with equipment (NASA AS17-147-22526).

After deploying the LRV, and prior to traversing to the ALSEP site, the commander inadvertently knocked the right rear fender extension off the LRV. The extension was subsequently secured to the fender with tape. Later during EVA-1, the extension came off and showered the crew and the LRV with a great deal of lunar dust.

Following an LRV test drive the crew gathered samples and performed panoramic photography.

The crew deployed the U.S. flag at 115:40:58 and offloaded the ALSEP package at 115:58:30. Following several traverse gravimeter readings, the ALSEP was deployed 607 feet (185 m) west-northwest of the LM.

At the ALSEP site, at 118:35:27, Cernan drilled two holes for heat flow experiment probes and a deep core hole.

Cernan drives the LRV by the LM during EVA-1 (NASA AS17-147-22527).

Cernan salutes U.S. flag during EVA-1 (NASA AS17-134-20380).

Schmitt takes his turn posing with the flag during EVA-1. Note the Earth at the top of the figure (NASA AS17-134-20384).

Schmitt collects lunar rake samples during EVA-1 (NASA AS17-134-20425).

Panorama of Schmitt, SEP transmitter, LRV, LM, Geophone Rock, and ALSEP during EVA-1 (NASA AS17-134-20435).

At 119:56:47, the crew departed for the surface electrical properties experiment, with a stop to deploy a seismic profiling explosive charge.

The crew entered the LM and the cabin was repressurized at 121:33:42. The first EVA lasted 7 hours 11 minutes 53 seconds. The distance traveled in the lunar rover vehicle was 10,800 feet (3.3 km), vehicle drive time was 33 minutes, and an estimated 31.5 pounds (14.3 kg) of samples were collected.

The second extravehicular activity began 80 minutes late, with cabin depressurization at 137:55:06.

Prior to starting the EVA traverse, ground controllers sent instructions for improvising a replacement for the lost fender extension. A rig of four maps, taped together and held in position by two clamps from portable utility lights, made an excellent substitute for the extension.

Flight team discusses repairs to the damaged LRV fender that occurred during EVA-1 (NASA S72-55170).

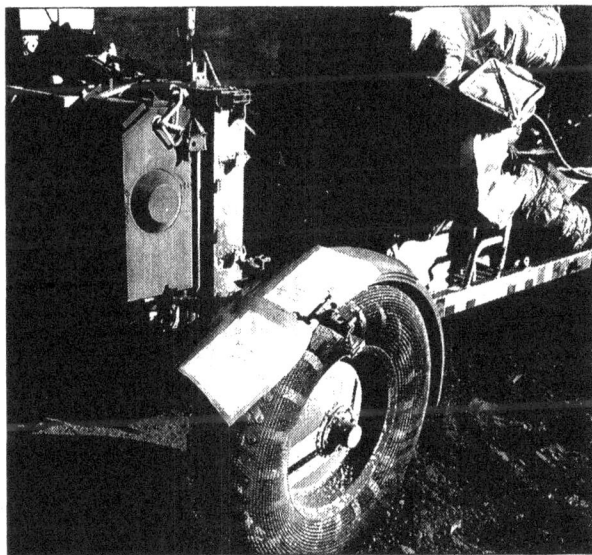

This figure from EVA-2 shows the makeshift repair to the LRV. (NASA AS17-137-20979).

The crew loaded the LRV and departed for the surface electrical properties experiment site at 138:44:02. During the traverse, the extravehicular plan was modified to allow more time at points of geological interest.

Schmitt uses a "lunar scoop" to retrieve soil samples at station 5 during EVA-2 (NASA AS17-145-22157).

The crew deployed three explosive packages in support of the lunar seismic profiling experiment, made seven traverse gravimeter measurements, gathered numerous samples, and completed their 500 mm and panoramic photographic tasks.

Schmitt enjoys the Moon's one-sixth Earth gravity as he searches for rock samples during EVA-2 (NASA AS17-145-22165).

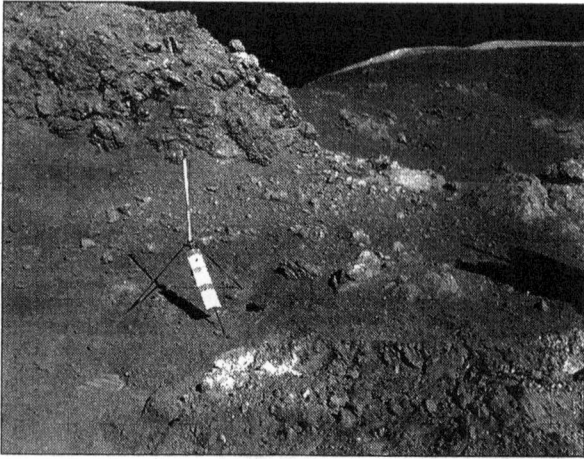

View of the orange soil found at station 4 at the rim of Shorty Crater during EVA-2 (NASA AS17-137-20990).

An orange-colored material, believed to be of volcanic origin, was found at station 4 (Shorty Crater).

The crew entered the LM and the cabin was repressurized at 145:32:02. The second extravehicular activity lasted 7 hours 36 minutes 56 seconds. The distance traveled in the lunar rover vehicle was 66,600 feet (20.3 km), vehicle drive time was 2 hours 25 minutes, and an estimated 75.2 pounds (34.1 kg) of samples were collected.

After a 15 hour 30-minute period in the LM, the cabin was depressurized at 160:52:48 for the third EVA, about 50 minutes later than planned.

A view of "Tracy's Rock," named for Cernan's daughter, and Henry Crater at station 6. The LRV is parked at an outcrop of rocks and near the shadow of the large boulder (NASA AS17-140-21493).

Schmitt, with gnomon in hand, stands to the left of "Tracy's Rock," a large split boulder. (NASA AS17-140-21496).

Photo taken at station 8 of a small boulder before it was rolled over so soil samples could be taken (NASA AS17-146-22365).

Specific sampling objectives were accomplished and nine traverse gravimeter measurements were made, as well as additional 500 mm and panoramic photography.

The surface electrical properties experiment was terminated because the receiver temperature was increasing to a level that could have affected the data tape. Consequently, the tape recorder was removed on the way back to the LM.

The cosmic ray experiment and the lunar neutron probe experiment were retrieved at 161:20:17, and several seismic profiling charges were deployed.

Schmitt holds a scoop over the small boulder seen in the previous image, after it was rolled over (NASA AS17-146-22371).

Many 500 mm panoramic images were taken during EVA-3. This one, at Station 6, shows the LM, in the center, surrounded by the rolling hills of the landing site (NASA AS17-139-21203).

The third extravehicular activity lasted 7 hours 15 minutes 8 seconds. The distance traveled in the lunar rover vehicle was 39,700 feet (12.1 km), vehicle drive time was 1 hour 31 minutes, and an estimated 136.7 pounds (62.0 kg) of samples were collected.

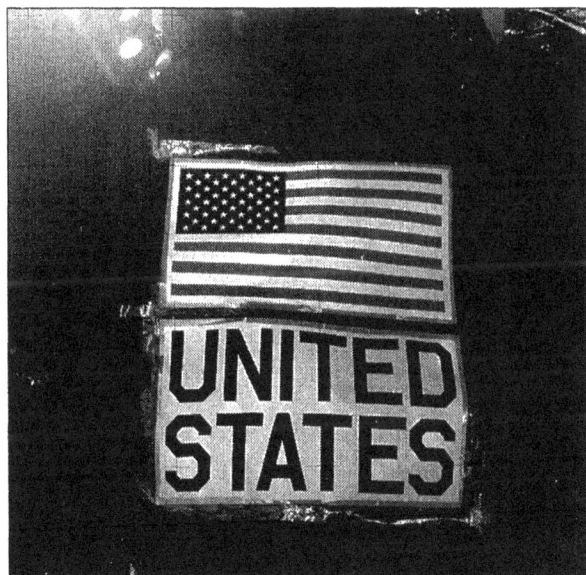

Interesting image of the hole left by a core tube sample taken during EVA-3. The regular shape was characteristic of the soil at Taurus-Littrow, which proved to be very stable when penetrated by the core tubes (NASA AS17-146-22295).

Close-up view of the U.S. flag and "United States" banner displayed on the outside of the LM (NASA AS17-134-20469).

The crew entered the LM, and, following equipment jettison, the cabin was repressurized at 168:07:56, thus ending the Apollo program's sixth and final human exploration of the Moon.

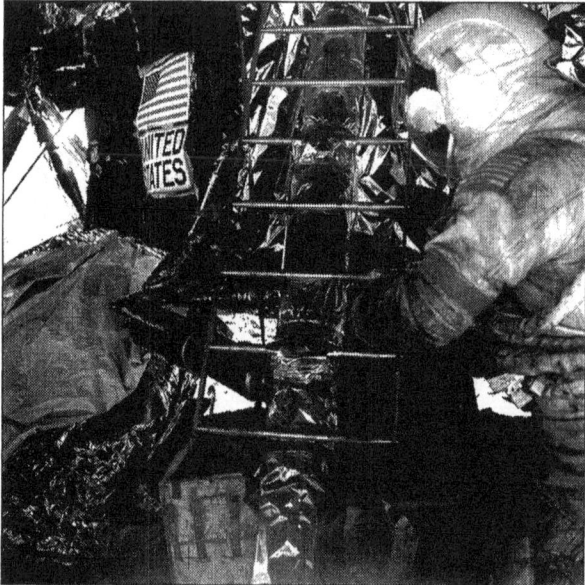

Cernan prepares to mount the LM ladder. Commemorative plaque can barely be seen above the third rung from the bottom (NASA AS17-134-20482).

The only geologist to visit the lunar surface, Schmitt smiles inside the LM following the final EVA of the Apollo program (NASA 134-20530).

For the mission, the total time spent outside the LM was 22 hours 3 minutes 57 seconds, the total distance traveled in the lunar rover vehicle was 117,000 feet (35.7 km), vehicle drive time was 4 hours 29 minutes, and the collected samples totaled 243.65 pounds (110.52 kg, official total in kilograms as determined by the Lunar Receiving Laboratory in Houston). The farthest point traveled from the LM was 24,180 feet. Good quality television transmissions were received during all three EVA's.

View of craters Eratosthenese and Copernicus from CM (NASA AS17-145-22285).

Numerous science activities were conducted in lunar orbit while the surface was being explored. In addition to the panoramic camera, the mapping camera, and the laser altimeter (which were used on previous missions), three new experiments were included in the service module.

An ultraviolet spectrometer measured lunar atmospheric density and composition, an infrared radiometer mapped the thermal characteristics of the Moon, and a lunar sounder acquired data on the subsurface structure.

The CSM orbit did not decay as predicted while the LM was on the Moon. Consequently, a 37.50-second orbital trim maneuver was performed at 178:54:05.45 to lower the orbit to 67.3 by 62.5 n mi. In addition, a planned 20.05-second plane change maneuver was made at 179:53:53.83 in preparation for rendezvous and resulted in an orbit of 62.8 by 62.5 n mi.

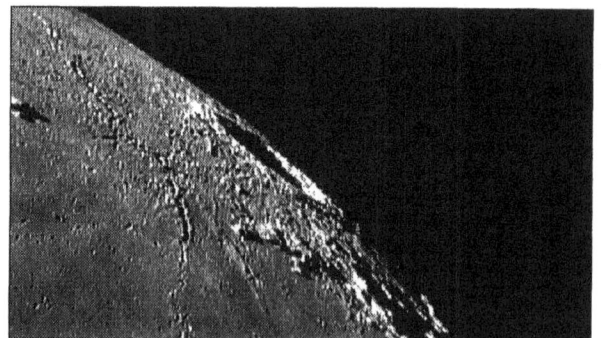

Interesting oblique view of crater Copernicus as seen from lunar orbit (NASA AS17-145-22287).

Ignition of the ascent stage engine for lunar liftoff occurred at 05:54:37 GMT (22:54:37 p.m. EST) on 14 December at 185:21:37. The LM had been on the lunar surface for 74 hours 59 minutes 40 seconds.

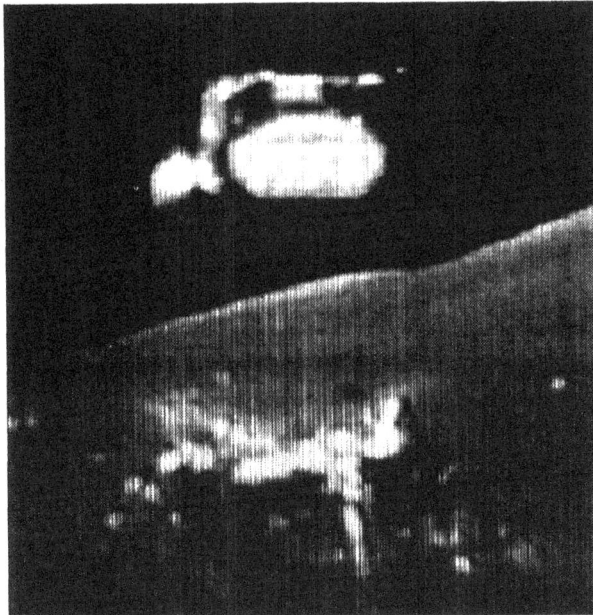

The LM ascent stage lifts off from the lunar surface as seen by the television camera mounted on the LRV (NASA S72-55421).

A clear view of the Scientific Instrument Module bay as seen from the approaching LM. CMP Evans would later do a spacewalk to retrieve film and a camera from the bay (NASA AS-145-22257).

The LM ascent stage approaches the CM for docking (NASA AS17-149-22857).

The nose of the CM, with docking probe, as seen just before linkup (NASA AS17-145-22273).

The 441-second maneuver was made to achieve the initial lunar orbit of 48.5 by 9.1 n mi. Several rendezvous sequence maneuvers were required before docking could occur two hours later. A 10-second vernier adjustment maneuver at 185:32:12 adjusted the orbit to 48.5 by 9.4 n mi. Finally, the 3.2-second terminal phase initiation at 186:15:58 brought the ascent stage to an orbit of 64.7 by 48.5 n mi.

The ascent stage and the CSM docked at 187:37:15 at an altitude of 60.6 n mi. The two spacecraft had been undocked for 79 hours 49 minutes 19 seconds.

After transfer of the crew and samples to the CSM, the ascent stage was jettisoned at 191:18:31, and the CSM was prepared for transearth injection. The ascent stage was then maneuvered by remote control to strike the lunar surface.

A 12-second maneuver was made at 191:23:31 to separate the CSM from the ascent stage, and resulted in an orbit of 63.9 by 61.2 n mi. A 116-second deorbit firing at 60.5 n mi altitude depleted the ascent stage propellants by 193:00:10. Impact occurred at latitude 19° 57' 58" north and longitude 30° 29' 23" east at 193:17:21. The impact point was 0.94 n mi (1.75 km) from the planned point and 5.35 n mi (9.9 km) southwest of the Apollo 17 landing site. The impact was recorded by the Apollo 12, 14, 15, and 16 seismic stations.

Explosive packages placed by the crew on the lunar surface were detonated at 210:15:35 and 212:45:01. Both events were picked up by the lunar seismic profiling geophones, and the resulting flash and dust from the second explosion were seen on television.

The television assembly and lunar communications relay unit failed to operate when attempts were made to command the camera on at 218:20, 235:04, and 235:13. It was later determined that the relay unit experienced an over-temperature failure.

Following a 143.69-second maneuver at 234:02:09.18 at an altitude of 62.1 n mi, transearth injection was achieved at 234:04:32.87, at a velocity of 8,374.3 ft/sec, after 75 lunar orbits lasting 147 hours 43 minutes 37.11 seconds. The crew had spent an additional day in lunar orbit performing scientific experiments.

Transearth Phase

Two more explosive packages were detonated (235:09:52 and 238:12:50), and the geophones received strong signals.

At 254:54:40, the command module pilot began a 1-hour 5-minute 44-second transearth coast extravehicular activity, televised to Earth, during which he retrieved the lunar sounder film, panoramic camera, and mapping camera cassettes in three trips to the scientific instrument module bay. This brought the total extravehicular activity for the mission to 23 hours 9 minutes 41 seconds.

Three final explosive packages were detonated at 257:43:56, 259:12:02, and 262:34:29, and were detected by the lunar surface geophones.

During the remainder of transearth flight, the crew performed another light-flash experiment, and operated the infrared radiometer and ultraviolet spectrometer. One mid-course correction was required, a 9-second 2.1-ft/sec maneuver at 298:38:01.

Evans performs a transearth EVA to retrieve items from the SIM bay (NASA AS17-152-23374).

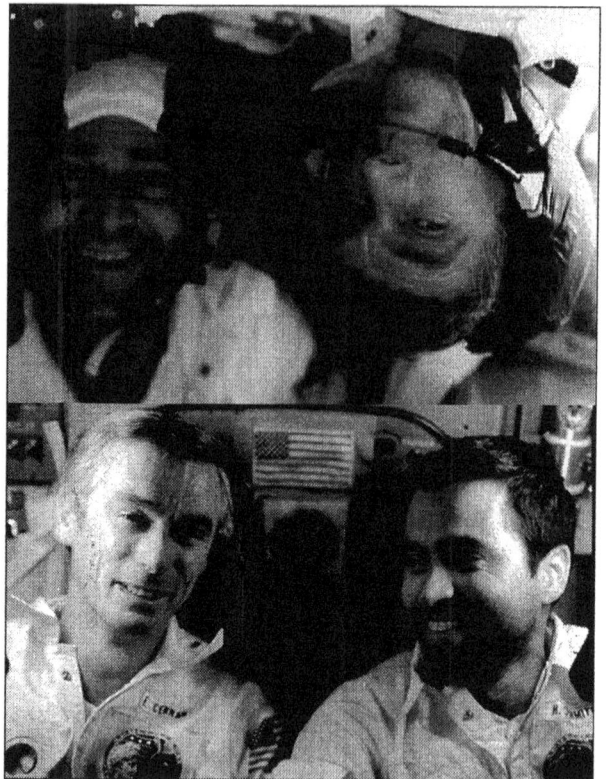

Cernan and Evans, and Cernan and Schmitt, enjoy the final trip home from the Moon (NASA AS17-162-24053 and 24149).

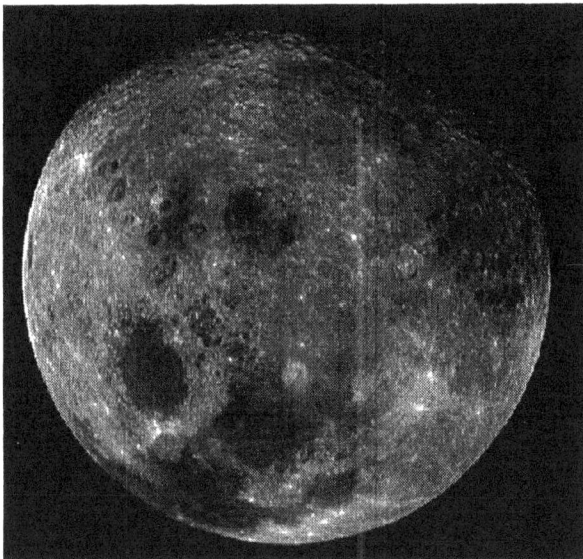

One last look at the Moon during transearth coast (NASA AS17-152-23312).

Recovery

The service module was jettisoned at 301:23:49, and the CM entry followed a normal profile. The command module reentered Earth's atmosphere (400,000 feet altitude) at 301:38:38 at a velocity of 36.090.3 ft/sec, following a transearth coast of 67 hours 34 minutes 05 seconds.

The parachute system effected splashdown of the CM in the Pacific Ocean at 19:24:59 GMT (02:24:59 p.m. EST) on 19 December. Mission duration was 301:51:59. The impact point was about 1.0 n mi from the target point and 3.5 n mi from the recovery ship U.S.S. *Ticonderoga*.

The splashdown site was estimated to be latitude 17.88° south and longitude 166.11° west. After splashdown, the CM assumed an apex-up flotation attitude. The crew was retrieved by helicopter and was aboard the recovery ship 52 minutes after splashdown. The CM was recovered 71 minutes later. The estimated CM weight at splashdown was 12,120 pounds, and the estimated distance traveled for the mission was 1,291,299 n mi.

The crew departed the *Ticonderoga* at 00:38 GMT on 21 December and arrived in Houston at 15:50 GMT. The CM was sent for deactivation to North Island Naval Air Station, San Diego, where it arrived at 19:30 GMT on 27 December. Deactivation was completed at 22:00 GMT on 30 December. The CM left North Island at 19:00 GMT on 2 January, and was delivered to the North American Rockwell Space Division facility in Downey, California, for postflight analysis. It arrived at 22:00 GMT.

The final Apollo mission nears splashdown (NASA S72-55834).

The Apollo 17 crew arrives aboard the recovery ship U.S.S. *Ticonderoga* after retrieval by helicopter (NASA S72-55937).

On the first anniversary of their mission, Cernan and Schmitt (c. and r., respectively) present a U.S. flag that went to the Moon with them to flight controllers in Houston. Chief of the flight Control Division, Gene Kranz looks on (NASA S73-38346).

Conclusions

All facets of the Apollo 17 mission were conducted with skill, precision, and relative ease because of experienced personnel and excellent performance of equipment. The following conclusions were made from an analysis of post-mission data:

1. The Apollo 17 mission was the most productive and trouble-free piloted mission, and represented the culmination of continual advancements in hardware, procedures, training, planning, operations, and scientific experiments.

2. The Apollo 17 mission demonstrated the practicality of training scientists to become qualified astronauts while retaining their expertise and scientific knowledge.

3. Stars and the horizon were not visible during night launches, therefore out-of-the-window alignment techniques could not be used for attitude reference.

4. The dynamic environment within the cabin during the early phases of the launch made system troubleshooting or corrective actions by the crew impractical. Therefore, either the ground control or automation should be relied upon for system troubleshooting and, in some cases, corrective actions.

5. As a result of problems on this and other missions, further research was needed to increase the dependability of mechanisms used to extend and retract equipment repeatedly in the space environment.

Lunar sample 72255 (NASA S72-16007).

Apollo 17 Objectives

Spacecraft Primary Objectives

1. To perform selenological inspection, survey, and sampling of materials and surface features in a preselected area of the Taurus-Littrow region. *Achieved.*

2. To emplace and activate surface experiments. *Achieved.*

3. To conduct inflight experiments and photographic tasks. *Achieved.*

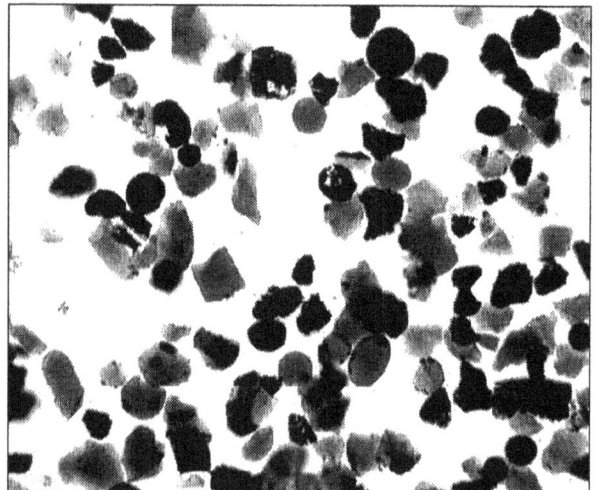

Micrograph of orange soil particles discovered on the lunar surface (NASA S73-15171).

View of lunar rock sample 76055 (NASA S72-15713).

Detailed Objectives

1. To obtain (service module) lunar surface photographs and altitude data from lunar orbit. *Achieved.*

2. To obtain data on the visual light flash phenomenon. *Achieved.*

3. To obtain (command module) photographs of lunar surface features of scientific interest and photographs of low brightness astronomical and terrestrial sources. *Achieved.*

4. To record visual observations (from lunar orbit) of particular lunar surface features and processes. *Achieved.*

5. To obtain data on Apollo spacecraft-induced contamination (Skylab contamination study). *Achieved.*

6. To obtain data on whole body metabolic gains or losses, together with associated endocrinological controls (food compatibility assessment). *Achieved.*

7. To obtain data on the use of the protective pressure garment. *Achieved.*

Experiments

1. ALSEP V: Apollo Lunar Surface Experiments Package.

   a. S-037: Heat flow experiment. *Achieved.*

   b. S-202: Lunar ejecta and meteorites experiment. *Partially achieved. Operation was restricted during lunar day due to overheating.*

   c. S-203: Lunar seismic profiling experiment. *Achieved.*

   d. S-205: Lunar atmospheric composition experiment. *Achieved.*

   e. S-207: Lunar surface gravimeter experiment. *Partially achieved. Data obtained in the seismic and free oscillation channels only.*

2. Collect and document samples, and study lunar surface geology. *Achieved.*

3. Cosmic ray detector (sheets) experiment. *Achieved.*

4. S-band transponder experiment (command and service module/lunar module). *Achieved.*

5. Far ultraviolet spectrometer experiment. *Achieved.*

6. Infrared scanning radiometer experiment. *Achieved.*

7. Traverse gravimeter experiment. *Achieved.*

8. Surface electrical properties experiment. *Achieved.*

9. Lunar sounder experiment. *Achieved.*

10. Lunar neutron probe experiment. *Achieved.*

Inflight Demonstration

Heat flow and convection. *Achieved.*

Passive Objectives

1. Long-term lunar surface exposure. *Achieved.*

2. S-160: Gamma ray spectrometer. *Achieved.*

3. S-176: Apollo window meteoroid. *Achieved.*

4. S-200: Soil mechanics. *Achieved.*

5. M-211: Biostack IIA. *Achieved.*

6. M-212: Biocore. *Achieved.*

Operational Tests for Manned Spacecraft Center/Department of Defense

1. Chapel Bell (classified Department of Defense test). *Results classified.*

2. Radar skin tracking. *Results classified.*

3. Ionospheric disturbance from missiles. *Results classified.*

4. Acoustic measurement of missile exhaust noise. *Results classified.*

5. Army acoustic test. *Results classified.*

6. Long-focal-length optical system. *Results classified.*

7. Sonic boom measurement. *Results classified.*

8. Skylab Medical Mobile Laboratory. *Results classified.*

Launch Vehicle Objectives

1. To launch on a flight azimuth between 72° and 100° and insert the S-IVB/instrument unit/spacecraft into the planned circular Earth parking orbit. *Achieved.*

2. To restart the S-IVB during either the first or second opportunity over the Atlantic and inject the S-IVB/instrument unit/spacecraft into the planned translunar trajectory. *Achieved.*

3. To provide the required attitude control for the S-IVB/instrument unit/spacecraft during transposition, docking, and ejection. *Achieved.*

4. To perform an evasive maneuver after ejection of the command and service module/lunar module from the S-IVB/instrument unit. *Achieved.*

5. To attempt to impact the S-IVB/instrument unit on the lunar surface within 350 kilometers (189 nautical miles) of latitude 7° south, longitude 8° west. *Achieved.*

6. To determine actual impact point within 5.0 kilometers (2.7 nautical miles) and time of impact within one second. *Achieved.*

7. To vent and dump the remaining gases and propellants to safe the S-IVB/instrument unit. *Achieved.*

Apollo 17 Spacecraft History

EVENT	DATE
Saturn S-II stage #12 delivered to KSC.	27 Oct 1970
Saturn S-IVB stage #512 delivered to KSC.	21 Dec 1970
Individual and combined CM and SM systems test completed at factory.	08 May 1971
LM #12 final engineering evaluation acceptance test at factory.	23 May 1971
LM #12 integrated test at factory.	23 May 1971
LM ascent stage #12 ready to ship from factory to KSC.	14 Jun 1971
LM descent stage #12 ready to ship from factory to KSC.	14 Jun 1971
LM ascent stage #12 delivered to KSC.	16 Jun 1971
LM descent stage #12 delivered to KSC.	17 Jun 1971
Integrated CM and SM systems test completed at factory.	02 Aug 1971
CM #114 and SM #114 ready to ship from factory to KSC.	17 Mar 1972
CM #114 and SM #114 delivered to KSC.	24 Mar 1972
Spacecraft/LM adapter #21 delivered to KSC.	24 Mar 1972
CM #114 and SM #114 mated.	28 Mar 1972
CSM #114 combined systems test completed.	09 May 1972
Saturn S-IC stage #12 delivered to KSC.	11 May 1972
Saturn S-IC stage #12 erected on MLP #3.	15 May 1972
LM ascent stage #12 and descent stage #12 mated.	18 May 1972
Saturn S-II stage #12 erected.	19 May 1972
LRV #3 delivered to KSC.	02 Jun 1972
LM #12 combined systems test completed.	07 Jun 1972
Saturn S-IVB instrument unit #512 delivered to KSC.	07 Jun 1972
CSM #114 altitude tests completed.	19 Jun 1972
Saturn S-IVB instrument unit #512 erected.	20 Jun 1972
Saturn S-IVB stage #512 erected.	23 Jun 1972
Launch vehicle electrical systems test completed.	12 Jul 1972
LM #12 altitude tests completed.	25 Jul 1972
Launch vehicle propellant dispersion/malfunction overall test completed.	01 Aug 1972
Launch vehicle service arm overall test completed.	11 Aug 1972
LRV #3 installed.	13 Aug 1972
CSM #114 moved to VAB.	22 Aug 1972
Spacecraft erected.	23 Aug 1972
Spacecraft moved to VAB.	24 Aug 1972
Space vehicle and MLP #3 transferred to launch complex 39A.	28 Aug 1972
LM #12 combined systems test completed.	06 Sep 1972
CSM #114 integrated systems test completed.	11 Sep 1972
LM #10 flight readiness test completed.	04 Oct 1972
CSM #114 electrically mated to launch vehicle.	11 Oct 1972
Space vehicle overall test #1 (plugs in) completed.	12 Oct 1972
Space vehicle overall test completed.	17 Oct 1972
Space vehicle flight readiness test completed.	20 Oct 1972
Saturn S-IC stage #12 RP-1 loading completed.	10 Nov 1972
Space vehicle countdown demonstration test (wet) completed.	20 Nov 1972
Space vehicle countdown demonstration test (dry) completed.	21 Nov 1972

Apollo 17 Ascent Phase

Event	GET (hhh:mm:ss)	Altitude (n mi)	Range (n mi)	Earth Fixed Velocity (ft/sec)	Space Fixed Velocity (ft/sec)	Event Duration (sec)	Geocentric Latitude (deg N)	Longitude (deg E)	Space Fixed Flight Path Angle (deg)	Space Fixed Heading Angle (E of N)
Liftoff	000:00:00.63	0.060	0.000	1.1	1,340.6		28.4470	-80.6041	0.05	90.00
Mach 1 achieved	000:01:07.5	4.315	1.265	1,076.7	2,085.8		28.4465	-80.5082	26.91	90.29
Maximum dynamic pressure	000:01:22.5	6.992	3.071	1,611.1	2,650.5		28.4457	-80.5460	28.89	91.04
S-IC center engine cutoff[2]	000:02:19.30	25.388	27.795	5,646.8	6,862.7	146.2	28.4329	-80.0781	23.199	91.355
S-IC outboard engine cutoff	000:02:41.20	35.900	49.145	7,757.4	9,012.1	168.1	28.4211	-79.6741	20.4285	91.718
S-IC/S-II separation[2]	000:02:42.9	36.776	51.112	7,778.4	9,036.1		28.4200	-79.6369	20.151	91.741
S-II center engine cutoff	000:07:41.21	93.420	591.254	17,064.6	18,439.6	296.61	27.5754	-69.4919	-0.058	97.647
S-II outboard engine cutoff	000:09:19.66	93.182	895.010	21,559.1	22,933.5	395.06	26.7251	-63.8908	0.254	100.395
S-II/S-IVB separation[2]	000:09:20.6	93.195	898.234	21,567.7	22,942.1		26.7147	-63.8314	0.244	100.424
S-IVB 1st burn cutoff	000:11:42.65	92.082	1,417.476	24,225.0	25,598.0	138.85	24.7139	-54.4952	0.00118	104.718
Earth orbit insertion	000:11:52.65	92.057	1,456.314	24,230.9	25,603.9		24.5384	-53.8107	0.0003	105.021

Apollo 17 Earth Orbit Phase

Event	GET (hhh:mm:ss)	Altitude (n mi)	Space Fixed Velocity (ft/sec)	Event Duration (sec)	Velocity Change (ft/sec)	Apogee (n mi)	Perigee (n mi)	Period (mins)	Inclination (deg)
Earth orbit insertion	000:11:52.65	92.057	25,603.9			90.3	90.0	87.83	28.526
S-IVB 2nd burn ignition	003:12:36.60	96.417	22,589.4						
S-IVB 2nd burn cutoff	003:18:27.64	162.127	35,579.5	351.04	10,376				28.466

Apollo 17 Translunar Phase

Event	GET (hhh:mm:ss)	Altitude (n mi)	Space Fixed Velocity (ft/sec)	Event Duration (sec)	Velocity Change (ft/sec)	Space Fixed Flight Path Angle (deg)	Space Fixed Heading Angle (E of N)
Translunar injection	003:18:37.64	169.401	35,555.3			7.379	118.110
CSM separated from S-IVB	003:42:27.6	3,566.842	25,344.9			44.177	102.769
CSM/LM ejected from S-IVB	004:45:02.3	13,393.6	16,012.8			61.80	83.485
Midcourse correction ignition	035:29:59.91	128,217.7	4,058.1			76.40	66.71
Midcourse correction cutoff	035:30:01.64	128,246.9	4,066.8	1.7	10.5	76.48	66.84

[2] Data for this event reflects postflight trajectory reconstruction for 36 seconds Ground Elapsed Time.

Apollo 17 Lunar Orbit Phase

Event	GET (hhh:mm:ss)	Altitude (n mi)	Space Fixed Velocity (ft/sec)	Event Duration (sec)	Velocity Change (ft/sec)	Apogee (n mi)	Perigee (n mi)
Lunar orbit insertion ignition	086:14:22.60	76.8	8,110.2				
Lunar orbit insertion cutoff	086:20:55.76	51.2	5,512.1	393.16	2,988	170	52.6
1st descent orbit insertion ignition	090:31:37.43	51.1	5,512.7				
1st descent orbit insertion cutoff	090:31:59.70	50.9	5,322.1	22.27	197	59	14.5
CSM/LM separation initiated	107:47:56	47.2	5,342.8				
CSM/LM separation cutoff	107:47:59.4			3.4	1	61.5	11.5
CSM orbit circularization ignition	109:17:28.92	58.6	5,279.9				
CSM orbit circularization cutoff	109:17:32.72	58.8	5,349.9	3.80	70.5	70	54
LM 2nd descent orbit insertion ignition	109:22:42	59.6	5,274.5				
LM 2nd descent orbit insertion cutoff	109:23:03.5	59.6	5,267.0	21.5	7.5	59.6	6.2
LM powered descent initiation	110:09:53	8.7	5,550.3				
LM powered descent cutoff	110:21:58			721	6,698		
CSM orbital trim ignition	178:54:05.45	64.9	5,315.1				
CSM orbital trim cutoff	178:54:42.95			37.50	9.2	67.3	62.5
CSM plane change cutoff	179:54:13.88	60.5	5,341.1	20.05	366	62.8	62.5
LM lunar liftoff ignition	185:21:37						
LM ascent orbit cutoff	185:28:58	8	5,542.3	441	6,075.7	48.5	9.1
LM vernier adjustment initiated	185:32:12	9.4	5,534.7				
LM vernier adjustment cutoff	185:32:22			10	10.0	48.5	9.4
LM terminal phase initiation ignition	186:15:58	44.6	5,333.3				
LM terminal phase initiation cutoff	186:16:01.2			3.2	53.8	64.7	48.5
CSM/LM docked	187:37:15	60.6	5,341.7				
LM ascent stage jettisoned	191:18:31	60.6	5,343.4				
CSM separation cutoff	191:23:43			12	2.0	63.9	61.2
LM ascent stage deorbit ignition	192:58:14	60.5	5,343.7				
LM ascent stage deorbit cutoff	193:00:10	58.9	5,130.1	116	286.0		

Apollo 17 Transearth Phase

Event	GET (hhh:mm:ss)	Altitude (n mi)	Space Fixed Velocity (ft/sec)	Event Duration (sec)	Velocity Change (ft/sec)	Space Fixed Flight Path Angle (deg)	Space Fixed Heading Angle (E of N)
Transearth injection ignition	234:02:09.18	62.1	5,337.1			-0.18	257.32
Transearth injection cutoff	234:04:32.87	63.1	8,374.3	143.69	3,046.3	2.46	259.47
Midcourse correction ignition	298:38:01	25,016.3	12,021.1			-68.43	34.63
Midcourse correction cutoff	298:38:10	24,999.7	12,025.8	9	2.1	-68.42	34.63

Apollo 17 Timeline

Event	GET (hhh:mm:ss)	GMT Time	GMT Date
Terminal countdown started.	-028:00:00	12:53:00	05 Dec 1972
Scheduled 9-hour hold at T-9 hours.	-009:00:00	07:53:00	06 Dec 1972
Countdown resumed at T-9 hours.	-009:00:00	16:53:00	06 Dec 1972
Scheduled 1-hour hold at T-3 hours 30 minutes.	-003:30:00	22:23:00	06 Dec 1972
Countdown resumed at T-3 hours 30 minutes.	-003:30:00	23:23:00	06 Dec 1972
Terminal Countdown Sequencer (TCS) failed to issue the S-IVB LOX pressurization command.	-000:02:47	02:50:13	07 Dec 1972
Unscheduled but automatic 1-hour 5-minute 11-second hold at T-30 seconds due to TCS failure.	-000:00:30	02:52:30	07 Dec 1972
Countdown recycled to T-22 minutes.	-000:22:00	03:57:41	07 Dec 1972
Unscheduled 1-hour 13-minute 19-second hold at T-8 minutes to resolve TCS corrective action.	-000:08:00	04:11:41	07 Dec 1972
Countdown resumed at T-8 minutes.	-000:08:00	05:25:00	07 Dec 1972
Guidance reference release.	-000:00:16.960	05:32:43	07 Dec 1972
S-IC engine start command.	-000:00:08.9	05:32:51	07 Dec 1972
S-IC engine ignition (#5).	-000:00:06.9	05:32:53	07 Dec 1972
All S-IC engines thrust OK.	-000:00:01.6	05:32:58	07 Dec 1972
Range zero.	000:00:00.00	05:33:00	07 Dec 1972
All holddown arms released (1st motion) (1.08 g).	000:00:00.24	05:33:00	07 Dec 1972
Liftoff (umbilical disconnected).	000:00:00.63	05:33:00	07 Dec 1972
Tower clearance yaw maneuver started.	000:00:01.7	05:33:01	07 Dec 1972
Yaw maneuver ended.	000:00:09.7	05:33:09	07 Dec 1972
Pitch and roll maneuver started.	000:00:12.9	05:33:12	07 Dec 1972
Roll maneuver ended.	000:00:14.3	05:33:14	07 Dec 1972
Mach 1 achieved.	000:01:07.5	05:34:07	07 Dec 1972
Maximum bending moment (96,000,000 lbf-in).	000:01:19	05:34:19	07 Dec 1972
Maximum dynamic pressure (701.75 lb/ft²).	000:01:22.5	05:34:22	07 Dec 1972
S-IC center engine cutoff command.	000:02:19.30	05:35:19	07 Dec 1972
Pitch maneuver ended.	000:02:40.1	05:35:40	07 Dec 1972
S-IC outboard engine cutoff. Maximum total inertial acceleration (3.87 g).	000:02:41.20	05:35:41	07 Dec 1972
S-IC maximum Earth-fixed velocity.	000:02:42.0	05:35:42	07 Dec 1972
S-IC/S-II separation command.	000:02:42.9	05:35:42	07 Dec 1972
S-II engine start command.	000:02:43.6	05:35:43	07 Dec 1972
S-II ignition.	000:02:44.6	05:35:44	07 Dec 1972
S-II aft interstage jettisoned.	000:03:12.9	05:36:12	07 Dec 1972
Launch escape tower jettisoned (planned time, actual time not recorded).	000:03:19	05:36:19	07 Dec 1972
Iterative guidance mode initiated.	000:03:24.1	05:36:24	07 Dec 1972
S-IC apex.	000:04:33.689	05:37:33	07 Dec 1972
S-II center engine cutoff. Maximum total inertial acceleration (1.74 g).	000:07:41.21	05:40:41	07 Dec 1972
S-IC impact (theoretical).	000:09:11.708	05:42:11	07 Dec 1972
S-II outboard engine cutoff.	000:09:19.66	05:42:19	07 Dec 1972
S-II/S-IVB separation command. S-II maximum Earth-fixed velocity.	000:09:20.6	05:42:20	07 Dec 1972
S-IVB 1st burn start command.	000:09:20.70	05:42:20	07 Dec 1972
S-IVB 1st burn ignition.	000:09:23.80	05:42:23	07 Dec 1972
S-IVB ullage case jettisoned.	000:09:32.4	05:42:32	07 Dec 1972
S-II apex.	000:09:34.527	05:42:34	07 Dec 1972
S-IVB 1st burn cutoff and maximum total inertial acceleration (0.67 g).	000:11:42.65	05:44:42	07 Dec 1972
Earth orbit insertion.	000:11:52.65	05:44:52	07 Dec 1972
S-IVB 1st burn maximum Earth-fixed velocity.	000:11:52.7	05:44:52	07 Dec 1972
Maneuver to local horizontal attitude started.	000:12:04.4	05:45:04	07 Dec 1972
S-II impact (theoretical).	000:19:56.947	05:52:56	07 Dec 1972
S-IVB 2nd burn restart preparation.	003:02:58.60	08:35:58	07 Dec 1972
S-IVB 2nd burn restart command.	003:12:28.60	08:45:28	07 Dec 1972
S-IVB 2nd burn ignition.	003:12:36.60	08:45:36	07 Dec 1972

Apollo 17 Timeline

Event	GET (hhh:mm:ss)	GMT Time	GMT Date
S-IVB 2nd burn cutoff and maximum total inertial acceleration (1.41 g).	003:18:27.64	08:51:27	07 Dec 1972
S-IVB 2nd burn maximum Earth-fixed velocity.	003:18:28.5	08:51:28	07 Dec 1972
Translunar injection.	003:18:37.64	08:51:37	07 Dec 1972
S-IVB safing procedure—CVS opened.	003:18:28.3	08:51:28	07 Dec 1972
Maneuver to local horizontal attitude and orbital navigation started.	003:20:59.6	08:53:59	07 Dec 1972
Maneuver to transposition and docking attitude started.	003:33:28.9	09:06:28	07 Dec 1972
CSM separated from S-IVB.	003:42:27.6	09:15:27	07 Dec 1972
TV transmission started.	003:50	09:23	07 Dec 1972
CSM docked with LM/S-IVB.	003:57:10.7	09:30:10	07 Dec 1972
TV transmission ended.	004:10	09:43	07 Dec 1972
CSM/LM ejected from S-IVB.	004:45:02.3	10:18:02	07 Dec 1972
S-IVB APS evasive maneuver ignition.	005:03:01.1	10:36:01	07 Dec 1972
S-IVB APS evasive maneuver cutoff (estimated).	005:04:21.0	10:37:21	07 Dec 1972
S-IVB lunar impact maneuver—CVS opened.	005:19:39.8	10:52:39	07 Dec 1972
S-IVB lunar impact maneuver—LOX dump started.	005:24:20.2	10:57:20	07 Dec 1972
S-IVB lunar impact maneuver—CVS closed.	005:24:40.0	10:57:40	07 Dec 1972
S-IVB lunar impact maneuver—LOX dump ended.	005:25:07.9	10:58:07	07 Dec 1972
Maneuver to attitude for 1st S-IVB APS lunar impact burn.	006:02:15	11:35:15	07 Dec 1972
S-IVB lunar impact maneuver—1st APS ignition command.	006:09:59.8	11:42:59	07 Dec 1972
S-IVB lunar impact maneuver—1st APS cutoff command.	006:11:38.0	11:44:38	07 Dec 1972
Maneuver to S-IVB solar heating attitude.	006:17:44	11:50:44	07 Dec 1972
Maneuver to attitude for 2nd S-IVB APS lunar impact burn.	011:02:40	16:35:40	07 Dec 1972
S-IVB lunar impact maneuver—2nd APS ignition command.	011:14:59.8	16:47:59	07 Dec 1972
S-IVB lunar impact maneuver—2nd APS cutoff command.	011:16:42.0	16:49:42	07 Dec 1972
S-IVB 3-axis tumble mode initiated.	011:31:42	17:04:42	07 Dec 1972
S-IVB passive thermal control maneuver.	011:31:50	17:04:50	07 Dec 1972
Command to inhibit instrument unit flight control computer to leave the S-IVB in 3-axis tumble mode.	011:32:12.5	17:05:12	07 Dec 1972
Midcourse correction ignition (SPS).	035:29:59.91	17:02:59	08 Dec 1972
Midcourse correction cutoff.	035:30:01.64	17:03:01	08 Dec 1972
Maneuver to LM checkout attitude.	039:05	20:38	08 Dec 1972
Preparations for intravehicular transfer.	039:20	20:53	08 Dec 1972
LM pressurization started.	039:30	21:03	08 Dec 1972
CDR and LMP entered LM for housekeeping and communications check.	040:10	21:43	08 Dec 1972
LM closeout.	042:11	23:44	08 Dec 1972
Heat flow and convection demonstration started.	042:55	00:28	09 Dec 1972
Heat flow and convection demonstration ended.	043:45	01:18	09 Dec 1972
Heat flow and convection demonstration started.	045:20	02:53	09 Dec 1972
Heat flow and convection demonstration ended.	046:00	03:33	09 Dec 1972
LM pressurization started.	059:30	17:03	09 Dec 1972
CDR and LMP entered LM for telemetry checkout.	059:59	17:32	09 Dec 1972
CDR and LMP entered CM.	060:35	18:08	09 Dec 1972
Mission clock updated (002:40:00 added).	065:00	22:33	09 Dec 1972
Apollo light flash phenomenon experiment started.	065:39	23:12	09 Dec 1972
Apollo light flash phenomenon experiment ended.	066:39	00:12	10 Dec 1972
Equigravisphere.	070:37:45	04:10:45	10 Dec 1972
Scientific instrument module door jettisoned.	081:32:40	15:05:40	10 Dec 1972
Inflight science phase of mission initiated with turn-on of Far Ultraviolet Spectrometer.	083:26	16:59	10 Dec 1972
Ultraviolet photography of dark Moon.	084:50	18:23	10 Dec 1972
Lunar orbit insertion ignition (SPS).	086:14:22.60	19:47:22	10 Dec 1972
Lunar orbit insertion cutoff.	086:20:55.76	19:53:55	10 Dec 1972

Apollo 17 Timeline

Event	GET (hhh:mm:ss)	GMT Time	GMT Date
S-IVB impact on lunar surface.	086:59:40.99	20:32:40	10 Dec 1972
Terminator photography.	087:05	20:38	10 Dec 1972
Orbital science visual observations.	087:15	20:48	10 Dec 1972
Orbital science photography.	088:00	21:33	10 Dec 1972
1st descent orbit insertion ignition (SPS).	090:31:37.43	00:04:37	11 Dec 1972
1st descent orbit insertion cutoff.	090:31:59.70	00:04:59	11 Dec 1972
Landmark observations.	090:50	00:23	11 Dec 1972
CDR and LMP entered LM.	105:02	14:35	11 Dec 1972
CSM/LM separation maneuver initiated (RCS).	107:47:56	17:20:56	11 Dec 1972
CSM/LM separation maneuver cutoff.	107:47:59.4	17:20:59	11 Dec 1972
CSM orbit circularization ignition (SPS).	109:17:28.92	18:50:28	11 Dec 1972
CSM orbit circularization cutoff.	109:17:32.72	18:50:32	11 Dec 1972
2nd descent orbit insertion ignition (LM RCS).	109:22:42	18:55:42	11 Dec 1972
2nd descent orbit insertion cutoff.	109:23:03.5	18:56:03	11 Dec 1972
CSM landmark tracking started.	109:40	19:13	11 Dec 1972
LM powered descent engine ignition (DPS).	110:09:53	19:42:53	11 Dec 1972
LM throttle to full-throttle position.	110:10:21	19:43:21	11 Dec 1972
LM manual target (landing site) update.	110:11:25	19:44:25	11 Dec 1972
LM landing radar velocity data good.	110:13:28	19:46:28	11 Dec 1972
LM landing radar range data good.	110:14:06	19:47:06	11 Dec 1972
LM landing radar updates enabled.	110:14:32	19:47:32	11 Dec 1972
LM throttle down.	110:17:19	19:50:19	11 Dec 1972
LM approach phase program selected.	110:19:15	19:52:15	11 Dec 1972
LM landing radar antenna to position 2.	110:19:16	19:52:16	11 Dec 1972
LM 1st landing point redesignation.	110:19:26	19:52:26	11 Dec 1972
LM landing radar switched to low scale.	110:19:54	19:52:54	11 Dec 1972
LM landing phase program selected.	110:20:51	19:53:51	11 Dec 1972
LM lunar landing and powered descent engine cutoff.	110:21:58	19:54:58	11 Dec 1972
CSM landmark tracking ended.	111:20	20:53	11 Dec 1972
1st EVA started (LM cabin depressurized).	114:21:49	23:54:49	11 Dec 1972
CSM orbital science visual observations.	114:45	00:18	12 Dec 1972
Lunar roving vehicle (LRV) offloaded.	114:51:10	00:24:10	12 Dec 1972
LRV deployed, test drive performed and documented with photography, gathered samples and performed 500 mm and panoramic photography.	115:13:50	00:46:50	12 Dec 1972
CSM orbital science photography.	115:15	00:48	12 Dec 1972
United States flag deployed and documented with photographs and stereo photography.	115:40:58	01:13:58	12 Dec 1972
Traverse gravimeter experiment reading obtained.	115:50:51	01:23:51	12 Dec 1972
Cosmic ray experiment deployed.	115:54:40	01:27:40	12 Dec 1972
Apollo lunar surface experiment (ALSEP) package offloaded.	115:58:30	01:31:30	12 Dec 1972
Traverse gravimeter experiment reading obtained.	116:06:01	01:39:01	12 Dec 1972
Traverse gravimeter experiment reading obtained.	116:11:54	01:44:54	12 Dec 1972
Traverse gravimeter experiment reading obtained.	116:46:17	02:19:17	12 Dec 1972
CSM orbital science photography.	117:10	02:43	12 Dec 1972
1st ALSEP data received on Earth.	117:21:00	02:54	12 Dec 1972
Heat flow experiment turned on.	117:29	03:02	12 Dec 1972
ALSEP deployment completed and documented with photographs and panoramic photography.	118:07:43	03:40:43	12 Dec 1972
CSM terminator photography.	118:10	03:43	12 Dec 1972
Lunar seismic profiling experiment (S-203) turned on. CSM Earthshine photography started.	118:25	03:58	12 Dec 1972
Deep core sample obtained and lunar neutron probe experiment deployed.	118:35:27	04:08:27	12 Dec 1972
Traverse gravimeter experiment reading obtained.	118:43:08	04:16:08	12 Dec 1972
CSM Earthshine photography ended.	118:50	04:23	12 Dec 1972

Apollo 17 Timeline

Event	GET (hhh:mm:ss)	GMT Time	GMT Date
Departed for station 1.	119:11:02	04:44:02	12 Dec 1972
Arrived at station 1 and deployed seismic profiling experiment explosive charge 6, obtained traverse gravimeter experiment reading and documented rake samples and performed panoramic photography.	119:24:02	04:57:02	12 Dec 1972
Lunar surface gravimeter experiment (S-207) activated.	119:50	05:23	12 Dec 1972
Departed for surface electrical properties experiment site with a stop to deploy seismic profiling experiment explosive charge 7, and performed panoramic photography.	119:56:47	05:29:47	12 Dec 1972
Arrived at surface electrical properties experiment site. Deployed antennas and the transmitter, gathered samples and performed documentary and panoramic photograph traverse gravimeter experiment reading obtained.	120:11:02	05:44:02	12 Dec 1972
Departed for LM.	120:33:39	06:06:39	12 Dec 1972
Arrived at LM and started EVA activity closeout.	120:36:15	06:09:15	12 Dec 1972
Traverse gravimeter experiment reading obtained.	121:16:37	06:49:37	12 Dec 1972
Traverse gravimeter experiment reading obtained.	121:21:11	06:54:11	12 Dec 1972
1st EVA ended (LM cabin repressurized).	121:33:42	07:06:42	12 Dec 1972
CSM zodiacal light photography.	130:35	16:08	12 Dec 1972
CSM orbital science photography.	134:00	19:33	12 Dec 1972
CSM solar corona photography.	134:50	20:23	12 Dec 1972
CSM orbital science visual observations.	137:00	22:33	12 Dec 1972
2nd EVA started (LM cabin depressurized).	137:55:06	23:28:06	12 Dec 1972
Traverse gravimeter experiment reading obtained.	138:04:08	23:37:08	12 Dec 1972
TV transmission started for 2nd EVA.	138:05	23:38	12 Dec 1972
LRV loaded for traverse and a traverse gravimeter experiment reading obtained.	138:39:00	00:12	13 Dec 1972
Departed for surface electrical properties experiment site.	138:44:02	00:17:02	13 Dec 1972
Arrived at surface electrical properties experiment site. Activated experiment, gathered samples, and performed panoramic photography.	138:47:05	00:20:05	13 Dec 1972
Departed for station 2 with four short stops—one to deploy seismic profiling experiment explosive charge 4, and three to gather en route samples.	138:51:43	00:24:43	13 Dec 1972
CSM orbital science photography and visual observations.	139:45	01:18	13 Dec 1972
Arrived at station 2. Traverse gravimeter experiment reading obtained, gathered samples including a rake sample, and performed documentary and panoramic photography.	140:01:30	01:34:30	13 Dec 1972
Departed for station 3 with one stop to obtain a traverse gravimeter experiment reading, gather samples and perform panoramic and 500 mm photography.	141:07:25	02:40:25	13 Dec 1972
Arrived at station 3. Traverse gravimeter experiment reading obtained, gathered samples including a double core-tube sample and a rake sample, and performed panoramic and 500 mm photography.	141:48:38	03:21:38	13 Dec 1972
CSM terminator photography.	142:05	03:38	13 Dec 1972
Departed for station 4 with two short stops to gather en route samples.	142:25:56	03:58:56	13 Dec 1972
Arrived at station 4. Traverse gravimeter experiment reading obtained, gathered samples including a trench sample and a double core-tube sample, and performed documentary and panoramic photography.	142:42:57	04:15:57	13 Dec 1972
Departed for station 5 with one stop to deploy seismic profiling experiment explosive charge, gather samples, and perform panoramic photography.	143:19:03	04:52:03	13 Dec 1972
Arrived at station 5. Traverse gravimeter experiment reading obtained, gathered samples, and performed documentary and panoramic photography.	143:45:15	05:18:15	13 Dec 1972
Departed for the LM with a short stop to deploy seismic profiling experiment explosive charge 8 documented with photographs, and a stop at the ALSEP site to allow the LMP to relevel the lunar surface gravimeter experiment.	144:15:58	05:48:58	13 Dec 1972
Arrived at the LM and started EVA closeout.	144:32:24	06:05:24	13 Dec 1972
TV transmission ended for 2nd EVA.	144:55	06:28	13 Dec 1972
Traverse gravimeter experiment reading obtained.	145:19:24	06:52:24	13 Dec 1972

Apollo 17 Timeline

Event	GET (hhh:mm:ss)	GMT Time	GMT Date
2nd EVA ended (LM cabin repressurized).	145:32:02	07:05:02	13 Dec 1972
CSM orbital science photography.	154:40	16:13	13 Dec 1972
CSM terminator photography.	156:50	18:23	13 Dec 1972
3rd EVA started (LM cabin depressurized).	160:52:48	22:25:48	13 Dec 1972
Zodiacal light photography.	160:55	22:28	13 Dec 1972
Traverse gravimeter experiment reading obtained.	161:02:40	22:35:40	13 Dec 1972
TV transmission started for 3rd EVA.	161:15	22:48	13 Dec 1972
LRV loaded for traverse, and panoramic and 500 mm photography performed.	161:16:15	22:49:15	13 Dec 1972
Traverse gravimeter experiment reading obtained.	161:19:45	22:52:45	13 Dec 1972
Cosmic ray experiment retrieved.	161:20:17	22:53:17	13 Dec 1972
Departed for surface electrical properties experiment site.	161:36:31	23:09:31	13 Dec 1972
Arrived at surface electrical properties experiment site. Activated the experiment, gathered samples, and performed documentary photography.	161:39:07	23:12:07	13 Dec 1972
Departed for station 6 with two short stops to gather en route samples.	161:42:36	23:15:36	13 Dec 1972
CSM orbital science photography.	161:50	23:23	13 Dec 1972
Arrived at station 6. Traverse gravimeter experiment reading obtained, gathered samples including a single core-tube sample, a rake sample, and performed documentary, panoramic, and 500 mm photography.	162:11:24	23:44:24	13 Dec 1972
Departed for station 7.	163:22:10	00:55:10	14 Dec 1972
Arrived at station 7. Gathered samples and performed documentary and panoramic photography.	163:29:05	01:02:05	14 Dec 1972
CSM orbital science visual observations.	163:30	01:03	14 Dec 1972
Departed for station 8 with one short stop to gather en route samples.	163:51:09	01:24:09	14 Dec 1972
Arrived at station 8. Two traverse gravimeter experiment readings obtained, gathered samples including rake and trench samples, and performed documentary and panoramic photography.	164:07:40	01:40:40	14 Dec 1972
Departed for station 9.	164:55:33	02:28:33	14 Dec 1972
Arrived at station 9. Seismic profiling experiment explosive charge 5 deployed, two traverse gravimeter readings obtained, gathered samples including a trench sample and a double core-tube sample, and performed documentary, panoramic and 500 mm photography. Removed data storage electronics assembly from surface electrical properties receiver.	165:13:10	02:46:10	14 Dec 1972
Departed for the LM with two short stops - one to gather en route samples and the other to deploy seismic profiling experiment explosive charge 2 and perform documentary and panoramic photography.	166:09:25	03:42:25	14 Dec 1972
Arrived at LM and started EVA closeout.	166:37:51	04:10:51	14 Dec 1972
Traverse gravimeter experiment reading obtained.	166:55:09	04:28:09	14 Dec 1972
final traverse gravimeter experiment reading obtained.	167:11:11	04:44:11	14 Dec 1972
ALSEP photography completed.	167:33:58	05:06:58	14 Dec 1972
Lunar neutron probe experiment retrieved..	167:36:43	05:09:43	14 Dec 1972
LRV positioned to monitor LM ascent.	167:39:57	05:12:57	14 Dec 1972
Seismic profiling experiment explosive charge 3 deployed.	167:44:41	05:17:41	14 Dec 1972
TV transmission ended for 3rd EVA. Equipment jettisoned.	167:45	05:18	14 Dec 1972
3rd EVA ended (LM cabin repressurized).	168:07:56	05:40:56	14 Dec 1972
Orbital trim maneuver ignition (RCS).	178:54:05.45	16:27:05	14 Dec 1972
Orbital trim maneuver cutoff.	178:54:42.95	16:27:42	14 Dec 1972
CSM plane change ignition (RCS).	179:53:53.83	17:26:53	14 Dec 1972
CSM plane change cutoff.	179:54:13.88	17:27:13	14 Dec 1972
1st equipment jettison from LM.	180:15	17:48	14 Dec 1972
CSM zodiacal light photography.	182:20	19:53	14 Dec 1972

Apollo 17 Timeline

Event	GET (hhh:mm:ss)	GMT Time	GMT Date
CSM landmark tracking.	182:40	20:13	14 Dec 1972
2nd equipment jettison from LM.	183:24	20:57	14 Dec 1972
CSM landmark tracking.	183:00	20:33	14 Dec 1972
TV transmission for lunar liftoff started.			
LM lunar liftoff ignition (LM APS).	185:21:37	22:54:37	14 Dec 1972
TV transmission for lunar liftoff ended.			
LM ascent orbit cutoff.	185:28:58	23:01:58	14 Dec 1972
Vernier adjustment maneuver initiated (LM RCS).	185:32:12	23:05:12	14 Dec 1972
Vernier adjustment maneuver cutoff.	185:32:22	23:05:22	14 Dec 1972
Terminal phase initiation ignition (LM APS).	186:15:58	23:48:58	14 Dec 1972
Terminal phase initiation cutoff.	186:16:01.2	23:49:01	14 Dec 1972
LM midcourse corrections.	186:30	00:03	14 Dec 1972
CSM/LM docked.	187:37:15	01:10:15	15 Dec 1972
Transfer, stowing of equipment and samples started.	188:00	01:33	15 Dec 1972
Transfer, stowing of equipment and samples ended.	190:05	03:38	15 Dec 1972
CDR and LMP entered CM.	190:10	03:43	15 Dec 1972
LM closeout.	190:30	04:03	15 Dec 1972
LM ascent stage jettisoned.	191:18:31	04:51:31	15 Dec 1972
Separation maneuver initiated.	191:23:31	04:56:31	15 Dec 1972
Separation maneuver cutoff.	191:23:43	04:56:43	15 Dec 1972
LM ascent stage deorbit ignition.	192:58:14	06:31:14	15 Dec 1972
LM ascent stage deorbit cutoff.	193:00:10	06:33:10	15 Dec 1972
LM ascent stage impact on lunar surface.	193:17:21	06:50:21	15 Dec 1972
Terminator photography.	206:20	19:53	15 Dec 1972
Orbital science visual observations.	206:40	20:13	15 Dec 1972
Orbital science visual observations.	207:10	20:43	15 Dec 1972
Explosive package detonated on lunar surface.	210:15:35	23:58:35	15 Dec 1972
Explosive package detonated on lunar surface.	212:45:01	02:28:01	16 Dec 1972
Orbital science photography.	213:10	02:43	16 Dec 1972
Terminator photography.	215:20	04:53	16 Dec 1972
Terminator photography.	231:20	20:53	16 Dec 1972
Transearth injection ignition (SPS).	234:02:09.18	23:35:09	16 Dec 1972
Transearth injection cutoff.	234:04:32.87	23:37:32	16 Dec 1972
TV transmission started.	234:10	23:43	16 Dec 1972
TV transmission ended.	234:35	00:08	17 Dec 1972
Ultraviolet spectrometer of Lyman Alpha region started.	235:00	00:33	17 Dec 1972
Explosive package detonated on lunar surface.	235:09:52	00:42:52	17 Dec 1972
Ultraviolet spectrometer of Lyman Alpha region ended.	236:00	01:33	17 Dec 1972
Ultraviolet spectrometer of Earth.	236:05	01:38	17 Dec 1972
Ultraviolet spectrometer of Moon.	237:15	02:48	17 Dec 1972
Ultraviolet spectrometer of Moon off.	238:00	03:33	17 Dec 1972
Explosive package detonated on lunar surface.	238:12:50	03:45	17 Dec 1972
Ultraviolet spectrometer on for passive thermal control galactic scan.	239:40	05:13	17 Dec 1972
TV transmission started for transearth EVA.			
CM cabin depressurization and hatch opening for transearth EVA.			
Transearth EVA started (Evans).	254:54:40	20:27:40	17 Dec 1972
Installation of television camera and data acquisition camera started.	255:00	20:33	17 Dec 1972
Panoramic film cassette retrieved.	255:23	20:56	17 Dec 1972
Mapping camera film cassette retrieved.	255:36	21:09	17 Dec 1972
CM hatch closed.	255:40	21:13	17 Dec 1972
TV transmission ended for transearth EVA.	255:42	21:15	17 Dec 1972

Apollo 17 Timeline

Event	GET (hhh:mm:ss)	GMT Time	GMT Date
Transearth EVA ended.	256:00:24	21:33:24	17 Dec 1972
Ultraviolet coma cluster observation.	257:00	22:33	17 Dec 1972
Explosive package detonated on lunar surface.	257:43:56	23:16:56	17 Dec 1972
Explosive package detonated on lunar surface.	259:12:02	01:45:02	18 Dec 1972
Ultraviolet Alpha ERI measurements.	260:30	02:03	18 Dec 1972
Ultraviolet passive thermal control measurements for Alpha ERI and Alpha GRU.	261:20	02:53	18 Dec 1972
Ultraviolet passive thermal control for galactic scan.	262:30	04:03	18 Dec 1972
Explosive package detonated on lunar surface.	262:34:29	04:07:29	18 Dec 1972
Ultraviolet dark north observation.	274:30	16:03	18 Dec 1972
Apollo light flash observation and investigation.	277:10	18:43	18 Dec 1972
Ultraviolet spectrometer of Virgo cluster.	279:10	20:43	18 Dec 1972
Ultraviolet spectrometer viewing dark south.	280:50	22:23	18 Dec 1972
TV transmission for inflight press conference started.	281:20	22:53	18 Dec 1972
TV transmission ended.	281:47	23:20	18 Dec 1972
Ultraviolet spectrometer of Spica.	283:45	01:18	19 Dec 1972
Ultraviolet passive thermal control for galactic scan.	285:30	03:03	19 Dec 1972
Midcourse correction ignition (RCS).	298:38:01	16:11:01	19 Dec 1972
Midcourse correction cutoff.	298:38:10	16:11:10	19 Dec 1972
Inflight science phase of mission ended with turn-off of Far Ultraviolet Spectrometer.	299:20	16:53	19 Dec 1972
CM/SM separation.	301:23:49	18:56:49	19 Dec 1972
Entry.	301:38:38	19:11:38	19 Dec 1972
Communication blackout started.	301:38:55	19:11:55	19 Dec 1972
Radar contact with CM established by recovery ship.	301:41	19:14	19 Dec 1972
Communication blackout ended.	301:42:15	19:15:15	19 Dec 1972
Forward heat shield jettisoned.	301:46:20	19:19:20	19 Dec 1972
Drogue parachute deployed.	301:46:22	19:19:22	19 Dec 1972
Visual contact with CM established by recovery ship and photo helicopter.	301:47	19:20	19 Dec 1972
Main parachute deployed.	301:47:13	19:20:13	19 Dec 1972
VHF recovery beacon contact with CM established by recovery ship.	301:48	19:21	19 Dec 1972
Voice contact with CM established by recovery ship.	301:49	19:22	19 Dec 1972
Splashdown (went to apex-up).	301:51:59	19:24:59	19 Dec 1972
Swimmers deployed to CM.	302:02	19:35	19 Dec 1972
Flotation collar inflated.	302:08	19:41	19 Dec 1972
Hatch opened for crew egress.	302:21	19:54	19 Dec 1972
Crew aboard recovery helicopter.	302:33	20:06	19 Dec 1972
Crew aboard recovery ship.	302:44	20:17	19 Dec 1972
CM aboard recovery ship.	303:55	21:28	19 Dec 1972
1st sample flight departed recovery ship.	323:52	17:25	20 Dec 1972
1st sample flight arrived in Hawaii.	330:27	00:00	21 Dec 1972
Flight crew departed recovery ship.	331:05	00:38	21 Dec 1972
1st sample flight departed Hawaii.	333:37	03:10	21 Dec 1972
1st sample flight arrived in Houston.	340:43	10:16	21 Dec 1972
Flight crew arrived in Houston.	346:17	15:50	21 Dec 1972
CM arrived at North Island Naval Air Station, San Diego.	493:57	19:30	27 Dec 1972
CM deactivated.	568:27	22:00	30 Dec 1972
CM departed San Diego.	637:27	19:00	02 Jan 1973
CM arrived at contractor's facility in Downey, CA.	640:27	22:00	02 Jan 1973

APOLLO BY THE NUMBERS

NASA

APOLLO

Statistical Tables

General Background[1]

	Apollo 7	Apollo 8	Apollo 9	Apollo 10	Apollo 11	Apollo 12	Apollo 13	Apollo 14	Apollo 15	Apollo 16	Apollo 17
Mission Information											
Mission Type	C	C prime	D	F	G	H-1	H-2	H-3	J-1	J-2	J-3
Purpose	CSM manned flight demonstration.	CSM manned flight demonstration.	Lunar module manned flight demonstration.	Lunar module manned flight demonstration.	Manned lunar landing demonstration.	Precision manned lunar landing demonstration and systematic lunar exploration.	Precision manned lunar landing demonstration and systematic lunar exploration.	Precision manned lunar landing demonstration and systematic lunar exploration.	Extensive scientific investigation of moon on lunar surface and from lunar orbit.	Extensive scientific investigation of moon on lunar surface and from lunar orbit.	Extensive scientific investigation of moon on lunar surface and from lunar orbit.
Trajectory Type	Earth Orbital	Lunar Orbital	Earth Orbital	Lunar Orbital	Lunar Landing	Lunar Landing	Lunar Landing	Lunar Landing	Lunar Landing	Lunar Landing	Lunar Landing
Payload Description	Block II CSM, adapter, and LES.	Block II CSM, lunar module, adapter, and LES.	Block II CSM, lunar module, adapter, and LES.	Block II CSM, lunar module, adapter, and LES.	Block II CSM, lunar module, adapter, and LES.	Block II CSM, lunar module, adapter, and LES.	Block II CSM, lunar module, adapter, and LES.	Block II CSM, lunar module, adapter, and LES.	Block II CSM, lunar module, adapter, and LES.	Block II CSM, lunar module, adapter, and LES.	Block II CSM, lunar module, adapter, and LES.
Launch Information											
Launch Site	Cape Kennedy	KSC	KSC	KSC	KSC	KSC	KSC	KSC	KSC	KSC	KSC
Launch Complex	Complex 34	Complex 39A	Complex 39A	Complex 39B	Complex 39A	Complex 39A	Complex 39A	Complex 39A	Complex 39A	Complex 39A	Complex 39A
Geodetic Latitude (deg N)	28.521963	28.608422	28.608422	28.627306	28.608422	28.608422	28.608422	28.608422	28.608422	28.608422	28.608422
Geocentric Latitude (deg N)	28.3608	28.4470	28.4470	28.4658	28.4470	28.4470	28.4470	28.4470	28.4470	28.4470	28.4470
Longitude (deg E)	-80.561141	-80.604133	-80.604133	-80.620869	-80.604133	-80.604133	-80.604133	-80.604133	-80.604133	-80.604133	-80.604133
Range Zero[2]											
KSC Date	11 Oct 1968	21 Dec 1968	03 Mar 1969	18 May 1969	16 Jul 1969	14 Nov 1969	11 Apr 1970	31 Jan 1971	26 Jul 1971	16 Apr 1972	07 Dec 1972
KSC Time	11:02:45 a.m.	07:51:00 a.m.	11:00:00 a.m.	12:49:00 p.m.	09:32:00 a.m.	11:22:00 a.m.	02:13:00 p.m.	04:03:02 p.m.	09:34:00 a.m.	12:54:00 p.m.	12:33:00 a.m.
KSC Time Zone	EDT	EST	EST	EDT	EDT	EST	EST	EST	EDT	EST	EST
GMT Date	11 Oct 1968	21 Dec 1968	03 Mar 1969	18 May 1969	16 Jul 1969	14 Nov 1969	11 Apr 1970	31 Jan 1971	26 Jul 1971	16 Apr 1972	07 Dec 1972
GMT Time	15:02:45	12:51:00	16:00:00	16:49:00	13:32:00	16:22:00	19:13:00	21:03:02	13:34:00	17:54:00	05:33:00
Actual GMT Liftoff Time	15:02:45.36	12:51:00.67	16:00:00.67	16:49:00.58	13:32:00.63	16:22:00.68	19:13:00.61	21:03:02.57	13:34:00.58	17:54:00.59	05:33:00.63
Selected Durations											
Ascent to Orbit (sec)	626.76	694.98	674.66	713.76	709.33	703.91	759.83	710.56	704.67	716.21	712.65
Earth Orbit	259:42:59	002:44:30.53	240:32:55.5	002:27:26.82	002:38:23.70	002:41:30.03	002:28:07.32	002:22:42.68	002:44:18.94	002:27:32.21	003:06:44.99
Revolutions	163.0	1.5	151.0	1.5	1.5	1.5	1.5	1.5	1.5	1.5	2.0
Translunar Coast	—	066:16:21.8	—	073:22:29.5	073:05:34.87	080:38:01.67	—	079:28:18.30	075:42:21.45	071:55:14.35	083:02:18.12
Time on Lunar surface	—	—	—	—	021:36:21	031:31:12	—	033:30:31	066:54:54	071:02:13	074:59:39
Lunar Orbit	—	020:10:13.0	—	061:43:23.6	059:30:25.79	088:58:11.52	—	066:35:39.99	145:12:41.68	125:49:32.59	147:43:37.11
Revolutions	—	10	—	31	30	45	—	34	74	64	75
CSM/LM Undocked	—	—	006:22:50	008:10:05	027:51:00.0	037:42:17.9	—	039:45:08.9	072:57:09.3	081:27:47	079:49:19
Transearth Coast	—	057:23:32.5	—	054:09:40.8	059:36:52.0	071:52:51.96	—	067:09:13.8	071:07:48	065:13:16	067:34:05
CM Earth Entry (sec from 400,000 ft to Splashdown)	937.0	869.2	1,004	869	929	846	835	853	778	814	801
Mission Duration	260:09:03	147:00:42.0	241:00:54	192:03:23	195:18:35	244:36:25	142:54:41	216:01:58.1	295:11:53.0	265:51:05	301:51:59

1 Compiled from mission reports, launch vehicle reports, and other sources

2 Range Zero was the integral second before liftoff.

Crew Information—Earth Orbit and Lunar Orbit Missions[3]

	Apollo 7	Apollo 8	Apollo 9	Apollo 10
Commander	Walter Marty Schirra, Jr.	Frank Frederick Borman, II	James Alton McDivitt	Thomas Patten Stafford
Date of Birth	12 Mar 1923	14 Mar 1928	10 Jun 1929	17 Sep 1930
Place of Birth	Hackensack, NJ	Gary, IN	Chicago, IL	Weatherford, OK
Age On Launch Date	45	40	39	38
Status	Captain	Colonel	Colonel	Colonel
	USN	USAF	USAF	USAF
Year Selected Astronaut	1959	1962	1962	1962
Prior Space Flights	MA-8, GT-6A	GT-7	GT-4	GT-6A, GT-9A
Backup	Thomas Patten Stafford	Neil Alden Armstrong	Charles Conrad, Jr.	Leroy Gordon Cooper, Jr.
Status	Colonel	Civilian	Commander	Colonel
	USAF	NASA	USN	USAF
Command Module Pilot	Donn Fulton Eisele	James Arthur Lovell, Jr.	David Randolph Scott	John Watts Young
Date of Birth	23 Jun 1930	25 Mar 1928	06 Jun 1932	24 Sep 1930
Place of Birth	Columbus, OH	Cleveland, OH	San Antonio, TX	San Francisco, CA
Date of Death	01-Dec-87	—	—	—
Place of Death	Tokyo, Japan	—	—	—
Age On Launch Date	38	40	36	38
Status	Major	Captain	Colonel	Commander
	USAF	USN	USAF	USN
Year Selected Astronaut	1963	1962	1963	1962
Prior Space Flights	None	GT-7, GT-12	GT-8	GT-3, GT-10
Backup	John Watts Young	Edwin Eugene Aldrin, Jr.	Richard Francis Gordon, Jr.	Donn Fulton Eisele
Status	Commander	Colonel	Commander	Lt. Colonel
	USN	USAF	USN	USAF
Lunar Module Pilot	Ronnie Walter Cunningham	William Alison Anders	Russell Louis Schweickart	Eugene Andrew Cernan
Date of Birth	16 Mar 1932	17 Oct 1933	25 Oct 1935	14 Mar 1934
Place of Birth	Creston, IA	Hong Kong	Neptune, NJ	Chicago, IL
Age On Launch Date	36	35	33	35
Status	Civilian	Major	Civilian	Commander
		USAF		USN
Year Selected Astronaut	1963	1963	1963	1963
Prior Space Flights	None	None	None	GT-9A
Backup	Eugene Andrew Cernan	Fred Wallace Haise, Jr.	Alan LaVern Bean	Edgar Dean Mitchell
Status	Commander	Civilian	Commander	Commander
	USN	NASA	USN	USN

[3] Compiled from press kits and mission reports, and *Who's Who in Space* (Cassutt).

Crew Information—Lunar Landing Missions[4]

	Apollo 11	Apollo 12	Apollo 13	Apollo 14	Apollo 15	Apollo 16	Apollo 17
Commander	Neil Alden Armstrong	Charles Conrad, Jr.	James Arthur Lovell, Jr.	Alan Bartlett Shepard, Jr.	David Randolph Scott	John Watts Young	Eugene Andrew Cernan
Date of Birth	05 Aug 1930	02 Jun 1930	25 Mar 1928	18 Nov 1923	06 Jun 1932	24 Sep 1930	14 Mar 1934
Place of Birth	Wapakoneta, OH	Philadelphia, PA	Cleveland, OH	East Derry, NH	San Antonio, TX	San Francisco, CA	Scottsdale, AZ
Date of Death	—	—	—	21-07-98	—	—	—
Place of Death	—	—	—	Monterey, CA	—	—	—
Age On Launch Date	38	39	42	47	39	41	38
Status	Civilian	Commander, USN	Captain, USN	Captain, USN	Colonel, USAF	Captain, USN	Captain, USN
Year Selected Astronaut	1962	1962	1962	1959	1963	1962	1963
Prior Space Flights	GT-8	GT-5, GT-11	GT-7, GT-12, Apollo 8	MR-3	GT-8, Apollo 9	GT-3, GT-10, Apollo 10	GT-9A, Apollo 10
Backup	James Arthur Lovell, Jr.	David Randolph Scott	John Watts Young	Eugene Andrew Cernan	Richard Francis Gordon, Jr.	Fred Wallace Haise, Jr.	John Watts Young
Status	Captain, USN	Colonel, USAF	Commander, USN	Captain, USN	Captain, USN	Civilian, NASA	Captain, USN
Command Module Pilot	Michael Collins	Richard Francis Gordon, Jr.	John Leonard Swigert, Jr.	Stuart Allen Roosa	Alfred Merrill Worden	Thomas Kenneth Mattingly, II	Ronald Ellwin Evans
Date of Birth	31 Oct 1930	05 Oct 1929	30 Aug 1931	16 Aug 1933	07 Feb 1932	17 Mar 1936	10 Nov 1933
Place of Birth	Rome, Italy	Seattle, WA	Denver, CO	Durango, CO	Jackson, MI	Chicago, IL	St Francis, KS
Date of Death	—	—	27 Dec 82	12 Dec 94	—	—	07 Apr 1990
Place of Death	—	—	Washington, DC	Washington, DC	—	—	Scottsdale, AZ
Age On Launch Date	38	40	38	37	39	36	39
Status	Lt. Colonel, USAF	Commander, USN	Civilian	Major, USAF	Major, USAF	Lt. Commander, USN	Commander, USN
Year Selected Astronaut	1963	1963	1966	1966	1966	1966	1966
Prior Space Flights	GT-10	GT-11	None	None	None	None	None
Backup	William Alison Anders	Alfred Merrill Worden	Thomas Kenneth Mattingly, II	Ronald Ellwin Evans	Vance DeVoe Brand	Stuart Allen Roosa	Stuart Allen Roosa
Status	Lt. Colonel, USAF	Major, USAF	Lt. Commander, USN	Commander, USN	Civilian, NASA	Lt. Colonel, USAF	Lt. Colonel, USAF
Lunar Module Pilot	Edwin Eugene Aldrin, Jr.	Alan LaVern Bean	Fred Wallace Haise, Jr.	Edgar Dean Mitchell	James Benson Irwin	Charles Moss Duke, Jr.	Harrison Hagan Schmitt
Date of Birth	20 Jan 1930	15 Mar 1932	14 Nov 1933	17 Sep 1930	17 Mar 1930	03 Oct 1935	03 Jul 1935
Place of Birth	Montclair, NJ	Wheeler, TX	Biloxi, MS	Hereford, TX	Pittsburgh, PA	Charlotte, NC	Santa Rita, NM
Date of Death	—	—	—	—	08 Aug 91	—	—
Place of Death	—	—	—	—	Glenwood Springs, CO	—	—
Age On Launch Date	39	37	36	40	41	36	37
Status	Colonel, Sc. D., USAF	Commander, USN	Civilian	Commander, Sc. D., USN	Lt. Colonel, USAF	Lt. Colonel, USAF	Civilian, Ph. D.
Year Selected Astronaut	1963	1963	1966	1966	1966	1966	1965
Prior Space Flights	GT-12	None	None	None	None	None	None
Backup	Fred Wallace Haise, Jr.	James Benson Irwin	Charles Moss Duke, Jr.	Joe Henry Engle	Harrison Hagan Schmitt	Edgar Dean Mitchell	Charles Moss Duke, Jr.
Status	Civilian, NASA	Lt. Colonel, USAF	Major, USAF	Lt. Colonel, USAF	Civilian, NASA	Captain, USN	Colonel, USAF

4 Compiled from press kits and mission reports, and "Who's Who in Space" (Cassutt).

Apportionment of Training According to Mission Type[5]

Training Category	Missions Before 1st Lunar Landing (Apollo 7–10)		Early Lunar Landing Missions (Apollo 11–14)		Final Lunar Landing Missions (Apollo 15–17)	
	Hours	% of Total	Hours	% of Total	Hours	% of Total
Simulators	11,511	36	15,029	56	11,413	45
Special Purpose	4,023	13	5,379	220	9,246	36
Procedures	7,924	25	2,084	8	1,265	5
Briefings	5,894	18	3,070	11	2,142	9
Spacecraft Tests	2,576	8	1,260	5	1,255	5
Total	31,928	100	26,822	100	25,321	100

Apollo Training Exercises[6]

Exercise	Apollo 7	Apollo 8	Apollo 9	Apollo 10	Apollo 11	Apollo 12	Apollo 13	Apollo 14	Apollo 15	Apollo 16	Apollo 17
Lunar Surface Activity Simulations (Sessions)											
Surface Operations			—	—	20	31	42	43	91	47	341
Operations Before/After EVA			—	—	10	4	11	18	20	20	93
Total Per Mission			—	—	30	35	53	61	111	67	434
Geology Field Trips[7]			—	—	1	4	7	7	12	18	13
Integrated Crew/Ground Mission Simulations (Days)											
Command Module Simulator	18	14	—	—	6 (1)	10	13	12 (3)	13 (6)	16 (5)	13 (2)
Lunar Module Simulator	0	0	—	—	4	3	5	5 (2)	5	7 (1)	6
Command Module and Lunar Module Simulators	0	0	—	—	7	12	9	12 (1)	7	10	9
Total Per Mission	18	14	—	—	17 (1)	25	27	29 (6)	25 (6)	33 (6)	28 (2)

5 *Apollo Program Summary Report* (JSC-09423), pps. 6-20 to 6-23. Includes participation of Mission Control Center personnel. Numbers in parentheses indicate simulations accomplished by follow-on or support crew members.

6 Ibid.

7 Each field trip lasted from one to seven days.

Capsule Communicators (CAPCOMS)[8]

Apollo 7

Col. Thomas Patten Stafford, USAF
Lt. Cdr. Ronald Ellwin Evans, USN
Maj. William Reid Pogue, USAF
John Leonard Swigert, Jr.
Cdr. John Watts Young, USN
Cdr. Eugene Andrew Cernan, USN

Apollo 8

Lt. Col. Michael Collins, USAF
Lt. Cdr. Thomas Kenneth Mattingly, II, USN
Maj. Gerald Paul Carr, USMC
Neil Alden Armstrong
Col. Edwin Eugene Aldrin, USAF/Sc.D.
Vance DeVoe Brand
Fred Wallace Haise, Jr.

Apollo 9

Maj. Stuart Allen Roosa, USAF
Lt. Cdr. Ronald Ellwin Evans, USN
Maj. Alfred Merrill Worden, USAF
Cdr. Charles Conrad, Jr., USN
Cdr. Richard Francis Gordon, Jr., USN
Cdr. Alan LaVern Bean, USN

Apollo 10

Maj. Charles Moss Duke, Jr., USAF
Maj. Joe Henry Engle, USAF
Maj. Jack Robert Lousma, USMC
Lt. Cdr. Bruce McCandless, II, USN

Apollo 11

Maj. Charles Moss Duke, Jr., USAF
Lt. Cdr. Ronald Ellwin Evans, USN
Lt. Cdr. Bruce McCandless, II, USN
Capt. James Arthur Lovell, Jr., USN
Lt. Col. William Alison Anders, USAF
Lt. Cdr. Thomas Kenneth Mattingly, II, USN
Fred Wallace Haise, Jr.
Don Leslie Lind, Ph.D.
Owen Kay Garriott, Jr., Ph.D.
Harrison Hagan Schmitt, Ph.D.

Apollo 12

Lt. Col. Gerald Paul Carr, USMC
Edward George Gibson, Ph.D.
Cdr. Paul Joseph Weitz, USN
Don Leslie Lind, Ph.D.
Col. David Randolph Scott, USAF
Maj. Alfred Merrill Worden, USAF
Lt. Col. James Benson Irwin, USAF

Civilian Backup CAPCOMS
Dickie K. Warren
James O. Rippey
James L. Lewis
Michael R. Wash

Apollo 13

Cdr. Joseph Peter Kerwin, USN/MD/MC
Vance DeVoe Brand
Maj. Jack Robert Lousma, USMC
Cdr. John Watts Young, USN
Lt. Cdr. Thomas Kenneth Mattingly, II, USN

Apollo 14

Maj. Charles Gordon Fullerton, USAF
Lt. Cdr. Bruce McCandless, II, USN
Fred Wallace Haise, Jr.
Lt. Cdr. Ronald Ellwin Evans, USN

Apollo 15

Joseph Percival Allen, IV, Ph.D.
Maj. Charles Gordon Fullerton, USAF
Karl Gordon Henize, Ph.D.
Cdr. Edgar Dean Mitchell, USN/Sc. D.
Robert Alan Ridley Parker, Ph.D.
Harrison Hagan Schmitt, Ph.D.
Capt. Alan Bartlett Shepard, Jr., USN
Capt. Richard Francis Gordon, Jr., USN
Vance DeVoe Brand

Apollo 16

Maj. Donald Herod Peterson, USAF
Maj. Charles Gordon Fullerton, USAF
Col. James Benson Irwin, USAF
Fred Wallace Haise, Jr.
Lt. Col. Stuart Allen Roosa, USAF
Cdr. Edgar Dean Mitchell, USN
Maj. Henry Warren Hartsfield, Jr., USAF
Anthony Wayne England, Ph.D.
Lt. Col. Robert Franklyn Overmyer, USMC

Apollo 17

Maj. Charles Gordon Fullerton, USAF
Lt. Col. Robert Franklyn Overmyer
Robert Alan Ridley Parker, Ph.D.
Joseph Percival Allen, IV, Ph.D.
Capt. Alan Bartlett Shepard, Jr., USN
Cdr. Thomas Kenneth Mattingly, II, USN
Col. Charles Moss Duke, Jr., USAF
Lt. Col. Stuart Allen Roosa, USAF
Capt. John Watts Young, USN

8 Derived from various documents and memoranda in Rice University archives. Military ranks for astronauts who are not also backups are implied from available information and B. Hello (Rockwell) memo, 10 December 1969.

Support Crews[9]

Apollo 7

Lt. Cdr. Ronald Ellwin Evans, USN
Maj. William Reid Pogue, USAF
John Leonard Swigert, Jr.

Apollo 8

Vance DeVoe Brand
Lt. Cdr. Thomas Kenneth Mattingly, II, USN
Maj. Gerald Paul Carr, USMC

Apollo 9

Maj. Jack Robert Lousma, USMC
Lt. Cdr. Edgar Dean Mitchell, USN/Sc.D.
Maj. Alfred Merrill Worden, USAF

Apollo 10

Maj. Charles Moss Duke, Jr., USAF
Maj. Joe Henry Engle, USAF
Lt. Col. James Benson Irwin, USAF

Apollo 11

Lt. Cdr. Thomas Kenneth Mattingly, II, USN
Lt. Cdr. Ronald Ellwin Evans, USN
Maj. William Reid Pogue, USAF
John Leonard Swigert, Jr.

Apollo 12

Maj. Gerald Paul Carr, USMC
Cdr. Paul Joseph Weitz, USN
Edward George Gibson, Ph.D.

Apollo 13

Maj. Jack Robert Lousma, USMC
Vance DeVoe Brand
Maj. William Reid Pogue, USAF

Apollo 14

Lt. Cdr. Bruce McCandless, II, USN
Lt. Col. William Reid Pogue, USAF
Maj. Charles Gordon Fullerton, USAF
Phillip Kenyon Chapman, Sc.D.

Apollo 15

Karl Gordon Henize, Ph.D.
Joseph Percival Allen, IV, Ph.D.
Robert Alan Ridley Parker, Ph.D.

Apollo 16

Maj. Donald Herod Peterson, USAF
Anthony Wayne England, Ph.D.
Maj. Henry Warren Hartsfield, Jr., USAF
Philip Kenyon Chapman, Sc.D.

Apollo 17

Lt. Col. Robert Franklyn Overmyer, USMC
Robert Alan Ridley Parker, Ph.D.
Maj. Charles Gordon Fullerton, USAF

[9] Compiled from various documents and memoranda in the Rice University archives. For Apollo 7, Bill Pogue replaced Maj. Edward Galen Givens, Jr., USAF, who was killed in an automobile accident in Pearland, TX on 6 June 1967. Military ranks are implied from available information and B. Hello (Rockwell) memo, 10 December 1969.

Flight Directors[10]

Apollo 7

	Director
Shift #1	Glynn S. Lunney
Shift #2	Eugene F. Kranz
Shift #3	Gerald D. Griffin

Apollo 8

	Director
Shift #1	Clifford E. Charlesworth
Shift #2	Glynn S. Lunney
Shift #3	Milton L. Windler

Apollo 9

	Director
Shift #1	Eugene F. Kranz
Shift #2	Gerald D. Griffin
Shift #3	M.P. "Pete" Frank III

Apollo 10

	Director
Shift #1	Glynn S. Lunney
Shift #2	Gerald D. Griffin
Shift #3	Milton L. Windler
	M.P. "Pete" Frank III

Apollo 11

	Director
Shift #1	Clifford E. Charlesworth
	Gerald D. Griffin
Shift #2	Eugene F. Kranz
Shift #3	Glynn S. Lunney

Apollo 12

	Director
Shift #1	Gerald D. Griffin
Shift #2	M.P. "Pete" Frank III
Shift #3	Clifford E. Charlesworth
Shift #4	Milton L. Windler

Apollo 13

	Director
Shift #1	Milton L. Windler
Shift #2	Gerald D. Griffin
Shift #3	Eugene F. Kranz
Shift #4	Glynn S. Lunney

Apollo 14

	Director
Shift #1	M.P. "Pete" Frank III
Shift #2	Glynn S. Lunney
Shift #3	Milton L. Windler
Shift #4	Gerald D. Griffin
	Glynn S. Lunney

Apollo 15

	Director
Shift #1	Gerald D. Griffin
Shift #2	Milton L. Windler
Shift #3	Glynn S. Lunney
	Eugene F. Kranz

Apollo 16

	Director
Shift #1	M.P. "Pete" Frank III
	Philip C. Shaffer
Shift #2	Eugene F. Kranz
	Donald R. Puddy
Shift #3	Gerald D. Griffin
	Neil B. Hutchinson
	Charles R. Lewis

Apollo 17

	Director
Shift #1	Gerald D. Griffin
	Eugene F. Kranz
Shift #2	Neil B. Hutchinson
	M.P. "Pete" Frank III
Shift #3	Charles R. Lewis

---

10 Compiled from various documents and memoranda in the Rice University archives.

Apollo Space Vehicle Configuration

S-IB (Apollo 7)

- Reached 1.640 million pounds of thrust at liftoff
- Accelerated total space vehicle to ~7,620 fps (inertial/space-fixed)
- Reached ~33 nautical miles in ~ 2.5 minutes

S-IC

- Reached 7.650 million pounds of thrust at liftoff
- Accelerated total space vehicle to ~7,880 fps (inertial/space-fixed)
- Reached ~58 nautical miles in ~ 2.5 minutes

S-II interstage

- Interfaced first and second stages
- Housed second stage engines
- Provided ullage for S-II engine start

S-II

- Accelerated vehicle from ~7,880 fps to ~ 22,850 fps in ~370 seconds.
- Achieved altitude of ~101 nautical miles
- Housed S-II retro-rocket mounting

S-IVB Interstage

- Provided structural transition from diameter of S-II to S~IVB
- Housed S-IVB engine
- Had attitude control about 3 axed and +X ullage with APS, up to 505 seconds of burn time

S-IVB

- Increased inertial/space-fixed velocity from 7,620 fps to 25,553 fps in 470 sec to accomplish orbit (Apollo 7)
- Increased inertial/space-fixed velocity from 22,850 fps to 25,568 fps in 154 sec to accomplish orbit (all other flights)
- Accelerated space vehicle to ~35,500 fps for TLI (all except Apollo 7)

Instrument Unit

- Provided launch vehicle guidance; navigation; control signals; telemetry; command communications; tracking; EDS rates and display activation timing and stage functional sequencing

Spacecraft/Lunar Module Adapter

- Housed and supported the LM, aerodynamically enclosed, supported LM
- Provided the structural electrical interface between spacecraft and launch vehicle
- Provided diameter transition from S-IVB to CSM
- Allowed LM extraction

Lunar Module Descent Stage

- Provided velocity change for lunar deorbit and lunar landing (throttleable)
- Protected ascent stage from landing damage
- Provided ascent stage/descent stage staging
- Provided LM ascent stage launch pad
- Stowed lunar scientific equipment

Lunar Module Ascent Stage

- Provided mission life support for two crew members
- Contained secondary command control and communications
- Computed and performed lunar landing abort, launch, rendezvous and docking with CSM
- Facilitated CM, LM ingress/egress inter- and extra-vehicular activities
- Maneuvered about and along three axes in the near-lunar environment

Service Module

- Provided velocity change for course correction, lunar orbit insertion, transearth injection and CSM aborts
- Provided attitude control and translation
- Supplemented environmental, electrical power and reaction control requirements of CM

Command Module

- Provided mission life support for three crew members
- Provided inertial/space-fixed navigation
- Provided command control and communication center
- Provided attitude control about three axes
- Acted as a limited lifting body
- Provided CM-LM ingress/egress for inter- and extra-vehicular activity

Launch Escape System

- Transported CM away from space vehicle (and mainland) during launch abort
- Oriented CM attitude for launch abort descent
- Jettisoned safely as required
- Sensed flight dynamics
- Provided CM thermal protection

Designations[11]

	Apollo 7	Apollo 8	Apollo 9	Apollo 10	Apollo 11	Apollo 12	Apollo 13	Apollo 14	Apollo 15	Apollo 16	Apollo 17
Call-Signs											
Command Module	Apollo 7	Apollo 8	Gumdrop	Charlie Brown	Columbia	Yankee Clipper	Odyssey	Kitty Hawk	Endeavour	Casper	America
Lunar Module	—	—	Spider	Snoopy	Eagle	Intrepid	Aquarius	Antares	Falcon	Orion	Challenger
NASA/Contractor Designations											
Space Vehicle	AS-205	AS-503	AS-504	AS-505	AS-506	AS-507	AS-508	AS-509	AS-510	AS-511	AS-512
Launch Vehicle	SA-205	SA-503	SA-504	SA-505	SA-506	SA-507	SA-508	SA-509	SA-510	SA-511	SA-512
Launch Vehicle Type	Saturn 1B	Saturn V	Saturn V	Saturn V	Saturn V	Saturn V	Saturn V	Saturn V	Saturn V	Saturn V	Saturn V
Launch Vehicle 1st Stage	S-1B-5	S-IC-3	S-IC-4	S-IC-5	S-IC-6	S-IC-7	S-IC-8	S-IC-9	S-IC-10	S-IC-11	S-IC-12
Launch Vehicle 2nd Stage	S-IVB-205	S-II-3	S-II-4	S-II-5	S-II-6	S-II-7	S-II-8	S-II-9	S-II-10	S-II-11	S-II-12
Launch Vehicle 3rd Stage	—	S-IVB-503	S-IVB-504	S-IVB-505	S-IVB-506	S-IVB-507	S-IVB-508	S-IVB-509	S-IVB-510	S-IVB-511	S-IVB-512
Instrument Unit	S-IU-205	S-IU-503	S-IU-504	S-IU-505	S-IU-506	S-IU-507	S-IU-508	S-IU-509	S-IU-510	S-IU-511	S-IU-512
Spacecraft/LM Adapter	SLA-5	SLA-11A	SLA-12A	SLA-13A	SLA-14	SLA-15	SLA-16	SLA-17	SLA-19	SLA-20	SLA-21
Command Module	CM-101	CM-103	CM-104	CM-106	CM-107	CM-108	CM-109	CM-110	CM-112	CM-113	CM-114
Service Module	SM-101	SM-103	SM-104	SM-106	SM-107	SM-108	SM-109	SM-110	SM-112	SM-113	SM-114
Lunar Module	—	Lunar Module Test Article (LTA-B)	LM-3	LM-4	LM-5	LM-6	LM-7	LM-8	LM-10	LM-11	LM-12
Lunar Roving Vehicle	—	—	—	—	—	—	—	—	LRV-1	LRV-2	LRV-3
VAB High Bay	—	1	3	2	1	3	1	3	3	3	3
Firing Room	—	1	2	3	1	2	1	2	1	1	1
Mobile Launcher Platform	—	MLP-1	MLP-2	MLP-3	MLP-1	MLP-2	MLP-3	MLP-2	MLP-3	MLP-3	MLP-3
Computer Programs	[Not found]	Colossus	Colossus, Sundance	Colossus 2, Luminary 1	Colossus 2A, Luminary 1A	Colossus 2C, Luminary 1B	Colossus 2D, Luminary 1C	Colossus 2E, Luminary 1D	Colossus 3, Luminary 1E	Colossus 3, Luminary 1F	Colossus 3, Luminary 1G
Eastern Test Range Number	66	170	9025	920	5307	2793	3381	7194	7744	1601	1701
International Designations											
CSM	1968-089A	1968-118A	1969-018A	1969-043A	1969-059A	1969-099A	1970-029A	1971-008A	1971-063A	1972-031A	1972-096A
S-IVB Stage	1968-089B	1968-118B	1969-018B	1969-043B	1969-059B	1969-099B	1970-029B	1971-008B	1971-063B	1972-031B	1972-096B
LM Ascent Stage [12]	—	—	1969-018C	1969-043D	1969-059C	1969-099C	1970-029C	1971-008C	1971-063C	1972-031C	1972-096C
LM Descent Stage	—	—	1969-018D	1969-043C	1969-059D	1969-099D	1970-029C	1971-008D	1971-063E	1972-031E	1972-096D
Lunar Subsatellite	—	—	—	—	—	—	—	—	1971-063D	1972-031D	—
NORAD Designations											
CSM	03486	03626	03769	03941	04039	04225	04371	04900	05351	06000	06300
S-IVB Stage	03487	03627	03770	03943	04040	04226	04372	04904	05352	06001	06301
LM Ascent Stage	—	—	03771	03949	04041	04246	—	04905	05366	06005	06307
LM Descent Stage	—	—	03780	03948	—	—	—	—	05377	06009	—
Lunar Subsatellite	—	—	—	—	—	—	—	—	—	—	—

11 Compiled from RAE Table of Earth Satellites 1957-1986; press kits; mission implementation plans; Saturn V flight evaluation reports; Apollo Program Summary Report; Stages to Saturn: A Technological History of the Apollo/Saturn Launch Vehicles; and other sources.

12 Ascent and descent stages for Apollo 13 remained as one piece until Earth entry.

Launch Vehicle/Spacecraft Key Facts[13]

	Apollo 7	Apollo 8	Apollo 9	Apollo 10	Apollo 11	Apollo 12	Apollo 13	Apollo 14	Apollo 15	Apollo 16	Apollo 17
First Stage (S-IB)											
Contractor	Chrysler	—	—	—	—	—	—	—	—	—	—
Diameter, base, ft	21.500	—	—	—	—	—	—	—	—	—	—
Diameter, top, ft	21.667	—	—	—	—	—	—	—	—	—	—
Height, ft	80.200	—	—	—	—	—	—	—	—	—	—
Engines, type/number	H-1/8	—	—	—	—	—	—	—	—	—	—
Fuel	RP-1	—	—	—	—	—	—	—	—	—	—
Oxidizer	LO_2	—	—	—	—	—	—	—	—	—	—
Rated thrust each engine, lbf	200,000	—	—	—	—	—	—	—	—	—	—
Rated thrust total, lbf	1,600,000	—	—	—	—	—	—	—	—	—	—
Thrust at 35 to 38 sec, lbf	1,744,400	—	—	—	—	—	—	—	—	—	—
First Stage (S-IC)											
Contractor	—	Boeing	Boeing	Boeing	Boeing	Boeing	Boeing	Boeing	Boeing	Boeing	Boeing
Diameter, base, ft	—	33.000	33.000	33.000	33.000	33.000	33.000	33.000	33.000	33.000	33.000
Diameter, top, ft	—	33.000	33.000	33.000	33.000	33.000	33.000	33.000	33.000	33.000	33.000
Height, ft	—	138.030	138.030	138.030	138.030	138.030	138.030	138.030	138.030	138.030	138.030
Engines, type/number	—	F-1/5	F-1/5	F-1/5	F-1/5	F-1/5	F-1/5	F-1/5	F-1/5	F-1/5	F-1/5
Fuel	—	RP-1	RP-1	RP-1	RP-1	RP-1	RP-1	RP-1	RP-1	RP-1	RP-1
Oxidizer	—	LO_2	LO_2	LO_2	LO_2	LO_2	LO_2	LO_2	LO_2	LO_2	LO_2
Rated thrust each engine, lbf	—	1,500,000	1,522,000	1,522,000	1,522,000	1,522,000	1,522,000	1,522,000	1,522,000	1,522,000	1,522,000
Rated thrust total, lbf	—	7,500,000	7,610,000	7,610,000	7,610,000	7,610,000	7,610,000	7,610,000	7,610,000	7,610,000	7,610,000
Thrust at 35 to 38 sec, lbf	—	7,560,000	7,576,000	7,536,000	7,552,000	7,594,000	7,560,000	7,504,000	7,558,000	7,620,000	7,599,000
Second Stage (S-II)											
Contractor	—	North American Rockwell	North American Rockwell	North American Rockwell	North American Rockwell	North American Rockwell	North American Rockwell	North American Rockwell	North American Rockwell	North American Rockwell	North American Rockwell
Diameter, ft	—	33.000	33.000	33.000	33.000	33.000	33.000	33.000	33.000	33.000	33.000
Height, ft	—	81.500	81.500	81.500	81.500	81.500	81.500	81.500	81.500	81.500	81.500
Engines, type/number	—	J-2/5	J-2/5	J-2/5	J-2/5	J-2/5	J-2/5	J-2/5	J-2/5	J-2/5	J-2/5
Fuel	—	LH_2	LH_2	LH_2	LH_2	LH_2	LH_2	LH_2	LH_2	LH_2	LH_2
Oxidizer	—	LO_2	LO_2	LO_2	LO_2	LO_2	LO_2	LO_2	LO_2	LO_2	LO_2
Rated thrust each engine, lbf	—	225,000	230,000	230,000	230,000	230,000	230,000	230,000	230,000	230,000	230,000
Rated thrust total, lbf	—	1,125,000	1,150,000	1,150,000	1,150,000	1,150,000	1,150,000	1,150,000	1,150,000	1,150,000	1,150,000
Thrust, ESC+61 sec, lbf	—	1,143,578	1,155,611	1,159,477	1,155,859	1,161,534	1,160,767	1,164,464	1,169,662	1,163,534	1,156,694
Thrust, OECO, lbf	—	865,302	730,000	642,068	625,751	611,266	635,725	580,478	548,783	787,380	787,009

13 Compiled from Saturn launch vehicle flight evaluation reports. Thrust for S-IC stage is at sea level and for the S-II and S-IVB stages is at altitude. Thrust listed at "35 to 38 sec," "Engine Start Command (ESC) + 61 seconds", and at Outboard Engine Cutoff (OECO) is actual thrust as flown.

Launch Vehicle/Spacecraft Key Facts

	Apollo 7	Apollo 8	Apollo 9	Apollo 10	Apollo 11	Apollo 12	Apollo 13	Apollo 14	Apollo 15	Apollo 16	Apollo 17
Third Stage (S-IVB)											
Contractor	McDonnell Douglas	McDonnell Douglas	McDonnell Douglas	McDonnell Douglas	McDonnell Douglas	McDonnell Douglas	McDonnell Douglas	McDonnell Douglas	McDonnell Douglas	McDonnell Douglas	McDonnell Douglas
Diameter, ft (base)	33.000	33.000	33.000	33.000	33.000	33.000	33.000	33.000	33.000	33.000	33.000
Diameter, ft (top)	21.667	21.667	21.667	21.667	21.667	21.667	21.667	21.667	21.667	21.667	21.667
Height, ft	58.400	58.630	58.630	58.630	58.630	58.630	58.630	58.630	58.630	58.630	58.630
Engines, type/number	J-2/1	J-2/1	J-2/1	J-2/1	J-2/1	J-2/1	J-2/1	J-2/1	J-2/1	J-2/1	J-2/1
Fuel	—	LH_2	LH_2	LH_2	LH_2	LH_2	LH_2	LH_2	LH_2	LH_2	LH_2
Oxidizer	—	LO_2	LO_2	LO_2	LO_2	LO_2	LO_2	LO_2	LO_2	LO_2	LO_2
Rated thrust total, lbf	200,000	230,000	230,000	230,000	230,000	230,000	230,000	230,000	230,000	230,000	230,000
Thrust, lbf - 1st burn	207,802	202,678	232,366	204,965	202,603	206,956	199,577	201,572	202,965	206,439	205,797
Thrust, lbf - 2nd burn	—	201,777	203,568	204,712	201,061	207,688	198,536	201,738	203,111	206,807	205,608
Thrust, lbf - 3rd burn	—	—	199,516	—	—	—	—	—	—	—	—
Instrument Unit (IU)											
Contractor	IBM	IBM	IBM	IBM	IBM	IBM	IBM	IBM	IBM	IBM	IBM
Diameter, ft	21.667	21.667	21.667	21.667	21.667	21.667	21.667	21.667	21.667	21.667	21.667
Height, ft	3.000	3.000	3.000	3.000	3.000	3.000	3.000	3.000	3.000	3.000	3.000
Service Module (SM)											
Contractor	North American Rockwell	North American Rockwell	North American Rockwell	North American Rockwell	North American Rockwell	North American Rockwell	North American Rockwell	North American Rockwell	North American Rockwell	North American Rockwell	North American Rockwell
Diameter, ft	12.833	12.833	12.833	12.833	12.833	12.833	12.833	12.833	12.833	12.833	12.833
Height (with engine bell), ft	24.583	24.583	24.583	24.583	24.583	24.583	24.583	24.583	24.583	24.583	24.583
Height (engine bell), ft	9.750	9.750	9.750	9.750	9.750	9.750	9.750	9.750	9.750	9.750	9.750
Fairing, ft	24.583	24.583	24.583	24.583	24.583	24.583	24.583	24.583	24.583	24.583	24.583
Main structure, ft	1.917	1.917	1.917	1.917	1.917	1.917	1.917	1.917	1.917	1.917	1.917
SPS nozzle structure	12.917	12.917	12.917	12.917	12.917	12.917	12.917	12.917	12.917	12.917	12.917
Weight, lb	19,730	51,258	36,159	51,371	51,243	51,105	51,105	51,744	54,063	54,044	54,044
Weight, dry, lb									13,470	13,450	13,450
Propellant, lb									40,593	40,594	40,594
Rated Thrust, SPS engine, lbf	20,500	20,500	20,500	20,500	20,500	20,500	20,500	20,500	20,500	20,500	20,500
Spacecraft/LM Adapter											
Contractor	Grumman	Grumman	Grumman	Grumman	Grumman	Grumman	Grumman	Grumman	Grumman	Grumman	Grumman
Minimum diameter, ft	12.833	12.833	12.833	12.833	12.833	12.833	12.833	12.833	12.833	12.833	12.833
Maximum diameter, ft	21.667	21.667	21.667	21.667	21.667	21.667	21.667	21.667	21.667	21.667	21.667
Height, ft	28.000	27.999	27.999	27.999	27.999	27.999	27.999	27.999	27.999	27.999	27.999
Upper jettisonable panels, ft	21.129	21.208	21.208	21.208	21.208	21.208	21.208	21.208	21.208	21.208	21.208
Lower fixed panels, ft	6.871	6.791	6.791	6.791	6.791	6.791	6.791	6.791	6.791	6.791	6.791
Lunar Module (LM)											
Contractor	Grumman	Grumman	Grumman	Grumman	Grumman	Grumman	Grumman	Grumman	Grumman	Grumman	Grumman
Overall											
Width, ft	—	—	31.000	31.000	31.000	31.000	31.000	31.000	31.000	31.000	31.000
Height, ft	—	—	22.917	22.917	22.917	22.917	22.917	22.917	22.917	22.917	22.917
Footpad diameter, ft	—	—	3.083	3.083	3.083	3.083	3.083	3.083	3.083	3.083	3.083
Sensing probe length, ft	—	—	5.667	5.667	5.667	5.667	5.667	5.667	5.667	5.667	5.667
Weight (lb)	—	(LTA) 19,900	32,034	30,735	33,278	33,562	33,493	33,685	36,238	36,237	36,262

Launch Vehicle/Spacecraft Key Facts

	Apollo 7	Apollo 8	Apollo 9	Apollo 10	Apollo 11	Apollo 12	Apollo 13	Apollo 14	Apollo 15	Apollo 16	Apollo 17
LM Descent Stage											
Diameter, ft	—	—	14.083	14.083	14.083	14.083	14.083	14.083	14.083	14.083	14.083
Height, ft	—	—	10.583	10.583	10.583	10.583	10.583	10.583	10.583	10.583	10.583
Weight, dry, lb[14]	—	—	4,265	4,703	4,483	4,875	4,650	4,716	6,179	6,083	6,155
Maximum rated thrust, lb	—	—	9,870	9,870	9,870	9,870	9,870	9,870	9,870	9,870	9,870
LM Ascent Stage											
Diameter, ft	—	—	14.083	14.083	14.083	14.083	14.083	14.083	14.083	14.083	14.083
Height, ft	—	—	12.333	12.333	12.333	12.333	12.333	12.333	12.333	12.333	12.333
Cabin volume, cu ft	—	—	235	235	235	235	235	235	235	235	235
Habitable volume, cu ft	—	—	160	160	160	160	160	160	160	160	160
Crew compartment height, ft	—	—	7.667	7.667	7.667	7.667	7.667	7.667	7.833	7.833	7.833
Crew compartment depth, ft	—	—	3.500	3.500	3.500	3.500	3.500	3.500	3.500	3.500	3.500
Weight, dry, lb	—	—	5,071	4,781	4,804	4,760	4,668	4,691	4,690	4,704	4,729
Maximum rated thrust, lb	—	—	2,524	1,650	3,218	3,224	N/A	3,218.2	3,225.6	3,224.7	3,234.8
Lunar Roving Vehicle (LRV)											
Contractor	—	—	—	—	—	—	—	—	Boeing	Boeing	Boeing
Length, ft	—	—	—	—	—	—	—	—	10.167	10.167	10.167
Width, ft	—	—	—	—	—	—	—	—	6.000	6.000	6.000
Wheel base, ft	—	—	—	—	—	—	—	—	7.500	7.500	7.500
Weight, lb	—	—	—	—	—	—	—	—	462	462	462
Payload capacity, lb	—	—	—	—	—	—	—	—	1,080	1,080	1,080
Command Module (CM)											
Contractor	North American Rockwell	North American Rockwell	North American Rockwell	North American Rockwell	North American Rockwell	North American Rockwell	North American Rockwell	North American Rockwell	North American Rockwell	North American Rockwell	North American Rockwell
Diameter, ft	12.833	12.833	12.833	12.833	12.833	12.833	12.833	12.833	12.833	12.833	12.833
Height, ft	11.417	11.417	11.417	11.417	11.417	11.417	11.417	11.417	11.417	11.417	11.417
Docking probe cone, ft	2.583	2.583	2.583	2.583	2.583	2.583	2.583	2.583	2.583	2.583	2.583
Main structure, ft	6.750	6.750	6.750	6.750	6.750	6.750	6.750	6.750	6.750	6.750	6.750
Aft/heat shield, ft	2.083	2.083	2.083	2.083	2.083	2.083	2.083	2.083	2.083	2.083	2.083
Weight, lb	12,659	12,392	12,405	12,277	12,250	12,365	12,365	12,831	12,831	12,874	12,874
Habitable volume, cu ft	210	210	210	210	210	210	210	210	210	210	210
Launch Escape System (LES)											
Contractor	North American Rockwell	North American Rockwell	North American Rockwell	North American Rockwell	North American Rockwell	North American Rockwell	North American Rockwell	North American Rockwell	North American Rockwell	North American Rockwell	North American Rockwell
Diameter, ft	4.000	4.000	4.000	4.000	4.000	4.000	4.000	4.000	4.000	4.000	4.000
Height, ft	33.460	33.460	33.460	33.460	33.460	33.460	33.460	33.460	33.460	33.460	33.460
Rocket motors (1 each)											
Thrust, LES, lb	155,000	147,000	147,000	147,000	147,000	147,000	147,000	147,000	147,000	147,000	147,000
Thrust, pitch control motor, lb	3,000	2,400	2,400	2,400	2,400	2,400	2,400	2,400	2,400	2,400	2,400
Thrust tower jettison motor, lb	33,000	31,500	31,500	31,500	31,500	31,500	31,500	31,500	31,500	31,500	31,500
Total Vehicle											
Height (ft)	223.488	363.013	363.013	363.013	363.013	363.013	363.013	363.013	363.013	363.013	363.013

14 LM ascent and descent stages, LRV and CM dry weights are as published in mission press kits. All other weights are actual "as flown."

Launch Windows[15]

	Apollo 7	Apollo 8	Apollo 9	Apollo 10	Apollo 11	Apollo 12	Apollo 13	Apollo 14	Apollo 15	Apollo 16	Apollo 17
Launch Window Opening											
KSC Date	11 Oct 1968	21 Dec 1968	03 Mar 1969	18 May 1969	16 Jul 1969	14 Nov 1969	11 Apr 1970	31 Jan 1971	26 Jul 1971	16 Apr 1972	06 Dec 1972
KSC Time	11:00:00 a.m.	07:50:22 a.m.	11:00:00 a.m.	12:49:00 p.m.	09:32:00 a.m.	11:22:00 a.m.	02:13:00 p.m.	03:23:00 p.m.	09:34:00 a.m.	12:54:00 p.m.	09:53:00 p.m.
Time Zone	EST	EST	EST	EDT	EDT	EST	EST	EST	EDT	EST	EST
GMT Date	11 Oct 1968	21 Dec 1968	03 Mar 1969	18 May 1969	16 Jul 1969	14 Nov 1969	11 Apr 1970	31 Jan 1971	26 Jul 1971	16 Apr 1972	07-Dec-72
GMT Time	16:00:00	12:50:22	16:00:00	16:49:00	13:32:00	16:22:00	19:13:00	20:23:00	13:34:00	17:54:00	02:53:00
Launch Window Closing											
KSC Date	11 Oct 1968	21 Dec 1968	03 Mar 1969	18 May 1969	16 Jul 1969	14 Nov 1969	11 Apr 1970	31 Jan 1971	26 Jul 1971	16 Apr 1972	07 Dec 1972
KSC Time	03:00:00 p.m.	12:31:40 p.m.	02:15:00 p.m.	05:09:00 p.m.	01:54:00 p.m.	02:28:00 p.m.	05:36:00 p.m.	07:12:00 p.m.	12:11:00 p.m.	04:43:00 p.m.	01:31:00 a.m.
Time Zone	EST	EST	EST	EDT	EDT	EST	EST	EST	EDT	EST	EST
GMT Date	11 Oct 1968	21 Dec 1968	03 Mar 1969	18 May 1969	16 Jul 1969	14 Nov 1969	11 Apr 1970	01 Feb 1971	26 Jul 1971	16 Apr 1972	07 Dec 1972
GMT Time	20:00:00	17:31:40	19:15:00	21:09:00	17:54:00	19:28:00	22:36:00	00:12:00	16:11:00	21:43:00	06:31:00
Window Duration											
H:MM:SS	4:00:00	4:41:18	3:15:00	4:20:00	4:22:00	3:06:00	3:23:00	3:49:00	3:37:00	3:49:00	3:38:00
Minutes	240	281	195	260	262	186	203	229	217	229	218
Targeted Lunar Sun											
Elevation Angle (deg)	—	6.74	—	11.0	10.8	5.1	10.0	10.3	12.0	11.9	13.3

15 Compiled from press kits, mission implementation plans, and mission reports.

Launch Weather[16]

	Apollo 7	Apollo 8	Apollo 9	Apollo 10	Apollo 11	Apollo 12	Apollo 13	Apollo 14	Apollo 15	Apollo 16	Apollo 17
Surface Observations											
Pressure (lb/in.)	14.765	14.804	14.642	14.779	14.798	14.621	14.676	14.652	14.788	14.769	14.795
Temperature (°F)	82.9	59.0	67.3	80.1	84.9	68.0	75.9	71.1	85.6	88.2	70.0
Relative Humidity	65%	88%	61%	75%	73%	92%	57%	86%	68%	44%	93%
Dew Point (°F)	70	56	53	72	75	65	60	67	74	62.6	68.0
Visibility (s mi)	11.5	9.9	9.9	11.2	9.9	3.7	9.9	9.9	9.9	9.9	6.8
Surface Wind Conditions											
1st Level Wind Site (ft)	64.0	60.0	60.0	60.0	60.0	60.0	60.0	60.0	60.0	60.0	60.0
1st Level Wind Speed (ft/sec)	33.5	18.7	22.6	32.2	10.8	22.3	20.7	16.4	16.7	20.7	13.5
1st Level Wind Direction (deg)	090	348	160	142	175	280	105	255	156	269	005
2nd Level Wind Site (ft.)	N/R	N/R	N/R	N/R	N/R	N/R	N/R	530.0	530.0	530.0	530.0
2nd Level Wind Speed (ft/sec)	N/R17	N/R	N/R	N/R	N/R	N/R	N/R	27.9	17.7	16.7	17.7
2nd Level Wind Direction (deg)	N/R	N/R	N/R	N/R	N/R	N/R	N/R	275	158	256	335
Cloud Coverage											
1st Level Cover	30%	40%	70%	40%	10%	100%/rain	40%	70%	70%	20%	20%
1st Level Type	Cumulonimbus	Cirrus	Stratocumulus	Cumulus	Cumulus	Stratocumulus	Altocumulus	Cumulus	Cirrus	Cumulus	Stratocumulus
1st Level Altitude (ft)	2,100	N/R	3,500	2,200	2,400	2,100	19,000	4,000	25,000	3,000	26,000
2nd Level Cover	—	—	100%	20%	20%	—	100%	20%	—	—	50%
2nd Level Type	—	—	Altostratus	Altocumulus	Altocumulus	—	Cirrostratus	Altocumulus	—	—	Cirrus
2nd Level Altitude (ft)	—	—	9,000	11,000	15,000	—	26,000	8,000	—	—	26,000
3rd Level Cover	—	—	—	100%	90%	—	—	—	—	—	—
3rd Level Type	—	—	—	Cirrus	Cirrostratus	—	—	—	—	—	—
3rd Level Altitude (ft)	—	—	—	Unknown	Unknown	—	—	—	—	—	—
Maximum Wind Speed/Ascent											
Speed (ft/sec)	136.2	150.9	250.0	154	203	256	246	207	249.3	85.6	252.6
Altitude (ft)	172,000	108,300	38,480	295,276	183,727	180,446	256,562	193,570	182,900	38,880	145,996
Maximum Dynamic Pressure											
Ground Elapsed Time (sec)	75.5	78.9	85.5	82.6	83.0	81.1	81.3	81.0	82.0	86.0	82.5
Max q (lb/ft²)	665.60	776.938	630.73	694.232	735.17	682.95	651.63	655.8	768.58	726.81	701.75
Altitude (ft)	39,903	44,062	45,138	43,366	44,512	42,133	40,876	40,398	44,971	47,122	42,847

16 Compiled from Saturn launch vehicle reports, trajectory reconstruction reports, and *Summary of Atmospheric Data Observations For 155 Flights of MSFC/ABMA Related Aerospace Vehicles.*

17 This measurement not used or not recorded at launch time.

Launch Weather[18]

	Apollo 7	Apollo 8	Apollo 9	Apollo 10	Apollo 11	Apollo 12	Apollo 13	Apollo 14	Apollo 15	Apollo 16	Apollo 17
Maximum Wind Conditions in the High Dynamic Pressure Region											
Altitude (ft)	44,500	49,900	38,480	46,520	37,400	46,670	44,540	43,270	45,110	38,880	39,945
Wind Speed (ft/sec)	51.1	114.1	250.0	139.4	31.6	156.1	182.5	173.2	61.1	85.#	147.9
Wind Direction (deg)	309	284	264	270	297	245	252	255	063	25`	311
Maximum Wind Components											
Pitch Plane - Pitch (ft/sec)	51.8	102.4	244.4	133.9	24.9	154.9	182.4	173.2	-58.4	85._	114.2
Pitch Plane - Altitude (ft)	36,800	49,500	38,390	45,280	36,680	46,670	44,540	43,720	45,030	38,880	39,945
Yaw Plane - Yaw (ft/sec)	51.5	74.1	71.2	61.4	23.3	-64.0	49.2	81.7	24.0	41.#	95.8
Yaw Plane - Altitude (ft)	47,500	51,800	37,500	48,720	39,530	44,780	42,750	33,460	44,040	50,85#	37,237
Maximum Shear Value: (D h=1000 m) (sec^{-1})											
Pitch Plane Shear (sec^{-1})	0.0113	0.0103	0.0248	0.0203	0.0077	0.0183	0.0166	0.0201	0.0110	0.009#	0.0177
Pitch Plane Altitude (ft)	48,100	52,500	49,700	50,200	48,490	46,750	50,610	43,720	36,830	44,78#	26,164
Yaw Plane Shear (sec^{-1})	0.0085	0.0157	0.0254	0.0125	0.0056	0.0178	0.0178	0.0251	0.0071	0.011-	0.0148
Yaw Plane Altitude (ft)	46,500	57,800	48,160	50,950	33,790	47,820	45,850	38,880	47,330	50,85#	34,940
Maximum % Density Deviations											
Negative Deviation From PAFB6319	-0.1	-0.7	-6.1	-1.0	-0.2	-7.6	-2.8	-5.0	None	-0.#	-0.0
Altitude (n mi)	4.32	4.32	7.56	4.32	4.45	8.50	7.69	7.69	None	4.8#	0.00
+ Positive Deviation from IRA63	+1.3	+3.3	None	+3.3	+4.4	+1.2	+0.5	None	+4.2	+4.#	+1.7
Altitude (n mi)	5.80	8.50	None	7.56	7.69	5.67	8.64	None	7.56	8.6-	7.02

18 Compiled from Saturn launch vehicle reports, trajectory reconstruction reports, and *Summary of Atmospheric Data Observations For 155 Flights of MSFC/ABMA Related Aerospace Vehicles.*

19 Patrick Air Force Base Reference Atmosphere, 1963.

Apollo Program Budget Appropriations ($000)[20]

	1960	1961	1962	1963	1964	1965	1966	1967	1968	1969	1970	1971	1972	1973	Program Total
Advanced Technical Development Studies	$100	$1,000	$0	$0	$0	$0	$0	$0	$0	$0	$0	$0	$0	$0	$1,100
Orbital Flight Tests	$0	$0	$63,900	$0	$0	$0	$0	$0	$0	$0	$0	$0	$0	$0	$63,900
Biomedical Flight Tests	$0	$0	$16,550	$0	$0	$0	$0	$0	$0	$0	$0	$0	$0	$0	$16,550
High-Speed Reentry Tests	$0	$0	$27,550	$0	$0	$0	$0	$0	$0	$0	$0	$0	$0	$0	$27,550
Spacecraft Development	$0	$0	$52,000	$0	$0	$0	$0	$0	$0	$0	$0	$0	$0	$0	$52,000
Instrumentation & Scientific Equipment	$0	$0	$0	$11,500	$0	$0	$0	$0	$0	$0	$0	$0	$0	$0	$11,500
Operational Support	$0	$0	$0	$2,500	$0	$0	$0	$0	$0	$0	$0	$0	$0	$0	$2,500
Little Joe II Development	$0	$0	$0	$8,800	$0	$0	$0	$0	$0	$0	$0	$0	$0	$0	$8,800
Supporting Development	$0	$0	$0	$3,000	$0	$0	$0	$0	$0	$0	$0	$0	$0	$0	$3,000
Command and Service Modules	$0	$0	$0	$345,000	$545,874	$577,834	$615,000	$560,400	$455,300	$346,000	$282,821	$0	$0	$0	$3,728,229
Lunar Module	$0	$0	$0	$123,100	$135,000	$242,600	$310,800	$472,500	$399,600	$326,000	$231,433	$0	$0	$0	$2,241,033
Guidance & Navigation	$0	$0	$0	$32,400	$91,499	$81,038	$115,000	$76,654	$113,000	$43,900	$33,866	$0	$0	$0	$587,357
Integration, Reliability, & Checkout	$0	$0	$0	$0	$60,699	$24,763	$34,400	$29,975	$66,600	$65,100	$0	$0	$0	$0	$281,537
Spacecraft Support	$0	$0	$0	$0	$43,503	$83,663	$95,400	$110,771	$60,500	$121,800	$170,764	$0	$0	$0	$686,401
Saturn C-1	$0	$0	$0	$90,864	$0	$0	$0	$0	$0	$0	$0	$0	$0	$0	$90,864
Saturn I	$0	$0	$0	$0	$187,077	$40,265	$800	$0	$0	$0	$0	$0	$0	$0	$228,142
Saturn IB	$0	$0	$0	$0	$146,817	$262,690	$274,185	$236,600	$146,600	$41,347	$0	$0	$0	$0	$1,108,239
Saturn V	$0	$0	$0	$0	$763,382	$964,924	$1,177,320	$1,135,500	$998,900	$534,453	$484,439	$189,059	$142,458	$26,300	$6,416,835
Engine Development	$0	$0	$0	$0	$166,000	$166,300	$134,095	$49,800	$18,700	$0	$0	$0	$0	$0	$534,895
Apollo Mission Support	$0	$0	$0	$0	$133,101	$170,542	$210,385	$243,900	$296,800	$0	$0	$0	$0	$0	$1,054,728
Manned Space Flight Operations	$0	$0	$0	$0	$0	$0	$0	$0	$0	$546,400	$422,728	$314,963	$307,450	$0	$1,591,541
Advanced Development	$0	$0	$0	$0	$0	$0	$0	$0	$0	$0	$0	$11,500	$12,500	$0	$24,000
Flight Modules	$0	$0	$0	$0	$0	$0	$0	$0	$0	$0	$0	$245,542	$55,033	$0	$300,575
Science Payloads	$0	$0	$0	$0	$0	$0	$0	$0	$0	$0	$60,094	$106,194	$52,100	$0	$218,388
Ground Support	$0	$0	$0	$0	$0	$0	$0	$0	$0	$0	$0	$46,411	$31,659	$0	$78,070
Spacecraft	$0	$0	$0	$0	$0	$0	$0	$0	$0	$0	$0	$0	$0	$50,400	$50,400
Apollo Program	**$100**	**$1,000**	**$160,000**	**$617,164**	**$2,272,952**	**$2,614,619**	**$2,967,385**	**$2,916,200**	**$2,556,000**	**$2,025,000**	**$1,686,145**	**$913,669**	**$601,200**	**$76,700**	**$19,408,134**
NASA Total	$523,375	$964,000	$16,717,500	$3,674,115	$3,974,979	$4,270,695	$4,511,644	$4,175,100	$3,970,000	$3,193,559	$3,113,765	$2,555,000	$2,507,700	$2,509,900	$56,661,332
Apollo Share of Total Budget	**>1%**	**>1%**	**1%**	**17%**	**57%**	**61%**	**66%**	**70%**	**64%**	**63%**	**54%**	**36%**	**24%**	**3%**	**34%**

20 *The Apollo Spacecraft: A Chronology*, volumes I through IV.

Call Signs[21]

Mission	Command Module	Lunar Module
Apollo 7	"Apollo 7".	None.
Apollo 8	"Apollo 8."	None.
Apollo 9	"Gumdrop." Derived from the appearance of the spacecraft when transported on Earth. During shipment, it was covered in blue wrappings giving appearance of a wrapped gumdrop.	"Spider," derived from its bug-like configuration.
Apollo 10	"Charlie Brown," from a character in the comic strip Peanuts(c) drawn by Charles L. Schulz. As in the comic, the CM "Charlie Brown" would be the guardian of the LM "Snoopy".	"Snoopy," after the beagle character in the same comic strip. The name referred to the fact that the LM would be "snooping" around the lunar surface in low orbit. Also, at the Manned Spacecraft Center, Snoopy was a symbol of quality performance. Employees who did outstanding work were awarded a silver Snoopy pin.
Apollo 11	"Columbia," after Jules Verne's mythical moonship, "Columbiad," and the close relationship of the word to our Nation's origins.	"Eagle," after the eagle selected for the mission insignia.
Apollo 12	"Yankee Clipper," selected from names submitted by employees of the command module prime contractor.	"Intrepid," selected from names submitted by employees of the lunar module prime contractor.
Apollo 13	"Odyssey," reminiscent of the long voyage of Odysseus of Greek mythology.	"Aquarius," after the Egyptian god Aacquarius, the water carrier. Aquarius brought fertility and therefore life and knowledge to the Nile Valley, as the Apollo 13 crew hoped to bring knowledge from the Moon.
Apollo 14	"Kitty Hawk," the site of the Wright brothers' first flight.	"Antares," for the star on which the LM oriented itself for lunar landing.
Apollo 15	"Endeavour," for the ship which carried Captain James Cook on his 18th-century scientific voyages.	"Falcon," named for the USAF Academy mascot by Apollo 15's all-Air Force crew.
Apollo 16	"Casper," named for a cartoon character, "Casper the Friendly Ghost," because the white Teflon suits worn by the crew looked shapeless on television screens.	"Orion," for a constellation, because the crew would depend on star sightings to navigate in cislunar space.
Apollo 17	"America," as a tribute and a symbol of thanks to the American people who made the Apollo program possible.	"Challenger," indicative of the challenges of the future, beyond the Apollo program.

21 Excerpted and reworked from *Astronaut Mission Patches and Spacecraft Callsigns*, by Dick Lattimer; unpublished draft in JSC History Office; *Space Patches From Mercury to the Space Shuttle*; and various NASA documents.

Mission Insignias[22]

Apollo 7

Symbolizing the Earth-orbital nature of the mission, a command and service module combination circled the globe, trailing an ellipse of orange flame. The background was navy blue, symbolizing the depth of space. In the center was Earth, with the North and South American continents appearing against light blue oceans. The crew's names appeared in an arc at the bottom. A Roman numeral VII appeared in the Pacific region of the globe.

Apollo 8

The shape of the insignia symbolized the Apollo command module. The red figure 8 circled Earth and the Moon, representing not only the number of the mission but the translunar and transearth trajectories.

Apollo 9

Orbiting near the command module, the lunar module symbolized the first flight of the spacecraft that would take humans to the lunar surface on future flights. A Saturn V launch vehicle was depicted at the left. The crew names appeared around the top edge of the insignia, and the mission name, Apollo IX, appeared along the bottom. The 'D' in McDivitt had a red interior, identifying Apollo 9 as the "D" mission in the Apollo series.

Apollo 10

Shaped like a shield, the design of the insignia was based more on mechanics than on mission goals. The large Roman numeral 'X' identified the mission, and was three-dimensional to give the effect of sitting on the Moon. The command module circled the Moon as the LM fired the descent engine for its low pass over the surface. Earth appeared in the background. Although Apollo 10 did not land, the prominence of the 'X' indicated the mission would make a significant contribution to the Apollo program.

Apollo 11

The American eagle, symbolic of the United States, was about to land on the Moon. In its talons, an olive branch indicated that the crew "came in peace for all mankind." Earth, the place from which the crew came and would return safely in order to fulfill President John F. Kennedy's challenge to the nation, rested on a field of black, representing the vast unknown of space.

Apollo 12

An American clipper ship and the blue and gold motif signified that the crew was all-Navy and symbolically related the era of the clipper ship to the era of space flight. As the clipper ship brought foreign shores closer to the United States, and marked the increased utilization of the seas by this Nation, spacecraft have opened the way to the other planets, Apollo 12 marked the increased utilization of space-based on knowledge gained in earlier missions. The portion of the Moon shown represented the Ocean of Storms area in which Apollo would land. The four stars represented the crew and C.C. Williams, who would have been the lunar module pilot had he not died in an aircraft accident.

Apollo 13

Apollo, the Sun god of Greek mythology, was represented as the Sun, with three horses driving his chariot across the surface of the Moon, symbolizing how the Apollo flights have extended the light of knowledge to all humankind. The Latin phrase "Ex Luna, Scientia" means "From the Moon, Knowledge."

Apollo 14

The Apollo 14 insignia featured the astronaut insignia approaching the Moon and leaving a comet trail from the liftoff point on Earth. The mission name and crew name appeared in the border.

Apollo 15

Three stylized birds, or symbols of flight, representing the Apollo 15 crew, were superimposed over an artist's concept of the landing site, next to the Hadley Rille at the foot of the Lunar Apennines. Beneath the symbols, a formation on the lunar surface formed a 'XV' signifying the mission number. Two of the birds flew closer to the surface, representing the two crew members who actually landed.

Apollo 16

Resting on a gray field representing the lunar surface, the American eagle and red, white, and blue striped shield paid tribute to the people of the United States. Crossing the shield while orbiting the Moon was a gold NASA vector. Sixteen stars, representing the mission number, and the crew names, appeared on a blue border, outlined in gold.

Apollo 17

The insignia was dominated by the image of the Greek Sun god Apollo. Suspended in space behind the head of Apollo was an American eagle. The red bars of the eagle's wing represented the bars in the U.S. flag. The three white stars symbolized the crew members. The background was deep blue space. Within it were the Moon, Saturn and a spiral galaxy. The Moon was partially overlaid by the eagle's wing suggesting it is a celestial body humans have visited and conquered. The thrust of the eagle and the gaze of Apollo to the right toward Saturn and the galaxy implied that human goals in space will someday include the planets and perhaps the stars. The colors of the insignia were red, white, and blue, the colors of our flag, with the addition of gold to symbolize the golden age of space flight.

22 Excerpted and reworked from *Astronaut Mission Patches and Spacecraft Callsigns*, by Dick Lattimer, unpublished draft in JSC History Office; *Space Patches From Mercury to the Space Shuttle*; and various NASA documents.

Ground Ignition Weights[23]

Weights In Pounds Mass	Apollo 7	Apollo 8	Apollo 9	Apollo 10	Apollo 11	Apollo 12	Apollo 13	Apollo 14	Apollo 15	Apollo 15	Apollo 17
Ground Ignition Time Relative to Range Zero (sec)	-2.988	-6.585	-6.3	-6.4	-6.4	-6.5	-6.7	-6.5	-6.5	-6?	-6.9
S-IB stage, dry	84,530	—	—	—	—	—	—	—	—	—	—
S-IB stage, fuel	276,900	—	—	—	—	—	—	—	—	—	—
S-IB stage, oxidizer	631,300	—	—	—	—	—	—	—	—	—	—
S-IB stage, other	1,182	—	—	—	—	—	—	—	—	—	—
S-IB stage, total	993,912	—	—	—	—	—	—	—	—	—	—
S-IB/S-IVB interstage, dry	5,543	—	—	—	—	—	—	—	—	—	—
Retromotor Propellant	1,061	—	—	—	—	—	—	—	—	—	—
S-IC stage, dry	—	305,232	294,468	293,974	287,531	287,898	287,899	287,310	286,208	287,85?	287,356
S-IC stage, fuel	—	1,357,634	1,431,678	1,423,254	1,424,889	1,424,287	1,431,384	1,428,561	1,410,798	1,439,89?	1,431,921
S-IC stage, oxidizer	—	3,128,034	3,301,203	3,302,827	3,305,786	3,310,199	3,304,734	3,312,769	3,312,030	3,311,22?	3,314,388
S-IC stage, other	—	6,226	5,508	5,491	5,442	5,442	5,401	5,194	4,283	5,39?	5,395
S-IC stage, total	—	4,797,126	5,032,857	5,025,546	5,023,648	5,027,826	5,029,418	5,033,834	5,013,319	5,044,37?	5,039,060
S-IC/S-II interstage, dry	—	12,436	11,591	11,585	11,477	11,509	11,454	11,400	9,083	10,09?	9,975
S-II stage, dry	—	88,500	84,312	84,273	79,714	80,236	77,947	78,120	78,908	80,36?	80,423
S-II stage, fuel	—	793,795	821,504	823,325	819,050	825,406	836,741	837,484	837,991	846,17?	844,094
S-II stage, oxidizer	—	154,907	158,663	158,541	158,116	157,986	159,931	159,232	158,966	160,55?	160,451
S-II stage, other	—	1,426	1,188	1,250	1,260	1,250	1,114	1,051	1,082	9?	934
S-II stage, total	—	1,038,628	1,065,667	1,067,389	1,058,140	1,064,878	1,075,733	1,075,887	1,076,947	1,088,00?	1,085,902
S-II/S-IVB interstage, dry	—	8,731	7,998	8,045	8,076	8,021	8,081	8,060	8,029	8,05?	8,019
S-IVB stage, dry	21,852	25,926	25,089	25,680	24,852	25,064	25,097	25,030	25,198	25,09?	25,040
S-IVB stage, fuel	39,909	43,395	43,709	43,388	43,608	43,663	43,657	43,546	43,674	43,77?	43,752
S-IVB stage, oxidizer	193,330	192,840	189,686	192,089	192,497	190,587	191,890	190,473	195,788	195,37?	195,636
S-IVB stage, other	1,432	1,626	1,667	1,684	1,656	1,873	1,673	1,687	1,655	1,64?	1,658
S-IVB stage, total	256,523	263,787	260,151	262,841	262,613	261,187	262,317	260,736	266,315	265,84?	266,086
Total Instrument Unit	4,263	4,842	4,281	4,267	4,275	4,277	4,502	4,505	4,487	4,50?	4,470
Spacecraft/Lunar Module Adapter	3,943	3,951	4,012	3,969	3,951	3,960	3,947	3,962	3,964	3,9?	3,961
LM (LTA Apollo 8)	—	19,900	32,034	30,735	33,278	33,562	33,493	33,685	36,238	36,23?	36,262
Command and Service Module	32,495	63,531	59,116	63,560	63,507	63,559	63,795	64,448	66,925	66,9?	66,942
Total Launch Escape System	8,874	8,890	8,869	8,936	8,910	8,963	8,991	9,027	9,108	9,1?	9,104
Total Spacecraft (CSM)	45,312	96,272	104,031	107,200	109,646	110,044	110,226	111,122	116,235	116,31?	116,269
Total Vehicle	1,306,614	6,221,823	6,486,577	6,486,873	6,477,875	6,487,742	6,501,733	6,505,548	6,494,415	6,537,23?	6,529,784

23 Actual weights at S-IC stage ignition, compiled from Saturn launch vehicle flight evaluation reports. Weights to do not add to vehicle totals due to truncated data in reports.

Ascent Data[24]

	Apollo 7	Apollo 8	Apollo 9	Apollo 10	Apollo 11	Apollo 12	Apollo 13	Apollo 14	Apollo 15	Apollo 16	Apollo 17
Pre-Staging											
Pad Azimuth (deg East of North)	100	90.0	90.0	90.0	90.0	90.0	90.0	90.0	90.0	90.0	90.0
Flight Azimuth (deg East of North)	72	72.124	72.0	72.028	72.058	72.029	72.043	75.558	80.088	72.034	91.503
Mach 1 - GET (sec)	62.15	61.48	68.2	66.8	66.3	66.1	68.4	68.0	65.0	67.5	67.5
Mach 1 Altitude (ft)	25,034	24,128	25,781	25,788	25,736	25,610	26,697	26,355	25,663	26,019	26,221
Maximum Bending Moment - GET (sec)	73.1	74.7	79.4	84.6	91.5	77.5	76	76	80.1	86.5	79
Maximum Bending Moment (lbf-in)	7,546,000	60,000,000	86,000,000	88,000,000	33,200,000	37,000,000	69,000,000	116,000,000	80,000,000	71,000,000	96,000,000
Maximum q - GET (sec)	75.5	78.90	85.5	82.6	83.0	81.1	81.3	81.0	82.0	86.0	82.5
Maximum q Altitude (ft)	39,903	44,062	45,138	43,566	44,512	42,133	40,876	40,398	44,971	47,122	42,847
Maximum q (lbf/f²)	665.60	776.938	630.73	694.232	735.17	682.95	651.63	655.8	768.58	726.81	701.75
S-IC Stage Burn (S-IB Apollo 7)											
Duration (sec)	147.31	160.41	169.06	168.03	168.03	168.2	170.3	170.6	166.1	168.5	168.1
Maximum Total Inertial Acceleration - GET (sec)	140.10	153.92	162.84	161.71	161.71	161.82	163.70	164.18	159.56	161.78	161.20
Maximum Total Inertial Acceleration - (ft/sec²)	137.76	127.46	123.75	126.21	126.67	125.79	123.36	122.90	127.85	122.90	124.51
Maximum Total Inertial Acceleration - (g)	4.28	3.96	3.85	3.92	3.94	3.91	3.83	3.82	3.97	3.82	3.87
Maximum Earth-Fixed Velocity - GET (sec)	144.6	154.47	163.45	161.96	162.30	162.18	164.10	164.59	160.00	162.5	162.0
Maximum Earth-Fixed Velocity (ft/sec)	6,490.1	7,727.36	7,837.89	7,835.76	7,882.9	7,852.0	7,820.9	7,774.9	7,387.6	7,779.5	7,790.0
Apex - GET (sec)	259.4	266.54	266.03	266.87	269.1	275.6	271.7	271.2	277.562	270.973	273.689
Apex - Altitude (n mi)	64.4	64.69	59.23	60.61	62.1	66.4	63.1	62.9	68.8	63.1	64.9
Apex - Range (n mi)	132.6	175.70	172.37	172.90	176.8	181.4	176.0	174.5	182.9	174.8	177.2
S-II Stage Burn											
Duration (sec)	—	367.85	371.06	388.59	384.22	389.14	426.64	392.55	386.06	394.34	395.06
Maximum Total Inertial Acceleration - GET (sec)	—	524.14	536.31	460.69	460.70	460.83	537.00	463.17	459.56	461.77	461.21
Maximum Total Inertial Acceleration - (ft/sec²)	—	59.71	64.34	58.46	58.53	58.79	53.31	58.10	57.58	56.00	56.00
Maximum Total Inertial Acceleration - (g)	—	1.86	2.00	1.82	1.82	1.83	1.66	1.81	1.79	1.74	1.74
Maximum Earth-Fixed Velocity GET (sec)	—	524.90	536.45	553.50	549.00	553.20	593.50	560.07	550.00	560.0	560.6
Maximum Earth-Fixed Velocity (ft/sec)	—	21,068.14	21,441.11	21,317.81	21,377.0	21,517.8	21,301.6	21,574.5	21,601.4	21,550.9	21,567.6
Apex - GET (sec)	—	560.34	593.58	597.21	587.0	581.7	632.2	600.2	553.225	584.122	574.527
Apex - Altitude (n mi)	—	104.21	102.50	102.31	101.9	103.2	103.0	102.4	95.2	93.7	93.3
Apex - Range (n mi)	—	934.06	1,026.36	1,035.06	1,005.9	985.3	1,098.8	1,032.2	888.9	978.7	946.2
S-IVB First Burn											
Duration (sec)	469.79	156.69	123.84	146.95	147.13	137.31	152.93	137.16	141.47	142.61	138.85
Maximum Total Inertial Acceleration - GET (sec)	616.9	685.08	664.74	703.84	699.41	693.99	750.00	700.66	694.67	706.21	702.65
Maximum Total Inertial Acceleration (ft/sec²)	82.22	23.10	25.72	22.60	22.08	22.21	21.85	21.62	21.00	21.59	21.46
Maximum Total Inertial Acceleration (g)	2.56	0.72	0.80	0.70	0.69	0.69	0.68	0.67	0.65	0.67	0.67
Maximum Earth-Fixed Velocity - GET (sec)	619.3	685.50	674.66	703.84	709.33	703.91	750.50	710.56	704.67	716.21	712.70
Maximum Earth-Fixed Velocity (ft/sec)	24,208.4	24,244.26	24,246.39	24,240.09	24,243.8	24,242.3	24,243.1	24,221.8	24,242.4	24,286.1	24,231.0
S-IVB Second Burn											
Duration (sec)	—	317.72	62.06	343.06	346.83	341.14	350.85	350.84	350.71	341.92	351.04
Maximum Total Inertial Acceleration - GET[25]	—	002:55:55.61	004:46:57.68	002:39:10.66	002:50:03.11	002:53:04.02	002:41:37.23	002:34:23.34	002:55:53.61	002:39:18.42	003:18:27.64
Maximum Total Inertial Acceleration (ft/sec²)	—	49.77	39.90	47.90	46.65	47.74	46.23	46.55	45.01	45.64	45.44
Maximum Total Inertial Acceleration (g)	—	1.55	1.24	1.49	1.45	1.48	1.44	1.45	1.40	1.42	1.41
Maximum Earth-Fixed Velocity - GET	—	002:55:56.00	004:46:58.20	002:39:11.30	002:50:03.50	002:53:04.32	002:41:37.80	002:34:23.67	002:55:54.00	002:39:20.0	003:18:28.5
Maximum Earth-Fixed Velocity (ft/sec)	—	34,178.74	26,432.58	34,251.67	34,230.3	34,063.0	34,231.0	34,194.9	34,236.9	34,269.0	34,202.4
S-IVB Third Burn											
Duration (sec)	—	—	242.06	—	—	—	—	—	—	—	—
Maximum Total Inertial Acceleration - GET	—	—	006:08:53.00	—	—	—	—	—	—	—	—
Maximum Total Inertial Acceleration (ft/sec²)	—	—	54.40	—	—	—	—	—	—	—	—
Maximum Total Inertial Acceleration (g)	—	—	1.69	—	—	—	—	—	—	—	—
Maximum Earth-Fixed Velocity - GET	—	—	006:11:23.50	—	—	—	—	—	—	—	—
Maximum Earth-Fixed Velocity (ft/sec)	—	—	29,923.49	—	—	—	—	—	—	—	—

24 Compiled from Saturn V launch vehicle flight evaluation reports, Apollo/Saturn V postflight trajectory reports, and mission reports. Segments do not add to totals due to rounding in the Saturn reports.

25 GET is expressed as hours:minutes:seconds (hhh:mm:ss) for the S-IVB second and third burns.

Earth Orbit Data[26]

	Apollo 7	Apollo 8	Apollo 9	Apollo 10	Apollo 11	Apollo 12	Apollo 13	Apollo 14	Apollo 15	Apollo 16	Apollo 17
Earth Orbit Insertion											
Insertion - GET (sec)	626.76	694.98	674.66	713.76	709.33	703.91	759.83	710.56	704.67	716.2	712.65
Altitude (ft)	748,439	627,819	626,777	627,869	626,909	626,360	628,710	626,364	566,387	567,371	559,348
Surface Range (n mi)	1,121.743	1,430.363	1,335.515	1,469.790	1,460.697	1,438.608	1,572.300	1,444.989	1,445.652	1,469.0?2	1,456.314
Earth Fixed Velocity (ft/sec)	24,208.5	24,242.9	24,246.39	24,244.3	24,243.9	24,242.3	24,242.1	24,221.6	24,242.4	24,28?.1	24,230.9
Space-Fixed Velocity (ft/sec)	25,553.2	25,567.06	25,569.78	25,567.88	25,567.9	25,565.9	25,566.1	25,565.8	25,602.6	25,60?.0	25,603.9
Geocentric Latitude (deg N)	31.4091	32.4741	32.4599	32.5303	32.5027	31.5128	32.5249	31.0806	29.2052	32.5?62	24.5384
Geodetic Latitude (deg N)	31.58	32.6487	32.629	32.700	32.672	32.6823	32.6945	31.2460	29.3650	32.6?63	24.6805
Longitude (deg E)	-61.2293	-53.2923	-55.1658	-52.5260	-52.6941	-53.1311	-50.4902	-52.9826	-53.0807	-52.5?00	-53.8107
Space-Fixed Flight Path Angle (deg)	0.005	0.0006	-0.0058	-0.0049	0.012	-0.014	0.005	-0.003	0.015	0.?01	0.003
Space-Fixed Heading Angle (deg E of N)	86.32	88.532	87.412	89.933	88.848	88.580	90.148	91.656	95.531	88?932	105.021
Apogee (n mi)	152.34	99.99	100.74	100.32	100.4	100.1	100.3	100.1	91.5	91.3	90.3
Perigee (n mi)	123.03	99.57	99.68	99.71	98.9	97.8	99.3	98.9	89.6	90.0	90.0
Period (min)	89.55	88.19	88.20	88.20	88.18	88.16	88.19	88.18	87.84	87.85	87.83
Inclination (deg)	31.608	32.509	32.552	32.546	32.521	32.540	32.547	31.120	29.679	?2.542	28.526
Descending Node (deg)		42.415	45.538	123.132	123.088	123.126	123.084	117.455	109.314	1?3.123	86.978
Eccentricity	0.0045	0.00006	0.000149	0.000086	0.00021	0.00032	0.0001	0.0002	0.0003	?.0002	0.0000
Earth Orbit—Revolutions	163.0	1.5	151.0	1.5	1.5	1.5	1.5	1.5	1.5	1.5	2.0
Earth Orbit Duration	259:42:59.24	002:44:30.53	240:32:55.54	002:27:26.82	002:38:23.70	002:41:30.03	002:28:27.32	002:22:42.68	002:44:18.94	002:2?:32.21	003:06:44.99

26 Compiled from Saturn V launch vehicle flight evaluation reports, Apollo/Saturn V postflight trajectory reports and mission reports.

Saturn Stage Earth Impact[27]

	Apollo 7	Apollo 8	Apollo 9	Apollo 10	Apollo 11	Apollo 12	Apollo 13	Apollo 14	Apollo 15	Apollo 16	Apollo 17
S-IB Impact											
GET (sec)	560.2	—	—	—	—	—	—	—	—	—	—
Surface Range (n mi)	265.002	—	—	—	—	—	—	—	—	—	—
Geodetic Latitude (deg N)	29.7605	—	—	—	—	—	—	—	—	—	—
Longitude (deg E)	-75.7183	—	—	—	—	—	—	—	—	—	—
S-IC Impact											
GET (sec)	—	540.410	536.436	539.12	543.7	554.5	546.9	546.2	560.389	547.136	551.708
Surface Range (n mi)	—	353.462	346.635	348.800	357.1	365.200	355.300	351.700	368.800	351.600	356.6
Geodetic Latitude (deg N)	—	30.2040	30.1830	30.188	30.212	30.273	30.177	29.835	29.4200	30.207	28.219
Longitude (deg E)	—	-74.1090	-74.238	-74.207	-74.038	-73.895	-74.0650	-74.0420	-73.6530	-74.147	-73.8780
S-II Impact											
GET (sec)	—	1,145.106	1,205.346	1,217.89	1,213.7	1,221.6	1,258.1	1,246.3	1,143.912	1,202.390	1,146.947
Surface Range (n mi)	—	2,245.913	2,413.198	2,389.290	2,371.8	2,404.4	2,452.600	2,462.100	2,261.3	2,312.000	2292.800
Geodetic Latitude (deg N)	—	31.8338	31.4618	31.522	31.535	31.465	31.320	29.049	26.975	31.726	20.056
Longitude (deg E)	—	-37.2774	-34.0408	-34.512	-34.844	-34.214	-33.2890	-33.567	-37.924	-35.990	-39.6040
S-IVB Earth Impact											
GET	162:27:15	—	—	—	—	—	—	—	—	—	—
SC Date	18 Oct 1968	—	—	—	—	—	—	—	—	—	—
GMT Date	18 Oct 1968	—	—	—	—	—	—	—	—	—	—
KSC Time	05:30 a.m.	—	—	—	—	—	—	—	—	—	—
Time Zone	EDT	—	—	—	—	—	—	—	—	—	—
GMT Time	09:30 GMT	—	—	—	—	—	—	—	—	—	—
Latitude (deg N)	-8.90	—	—	—	—	—	—	—	—	—	—
Longitude (deg E)	081.6	—	—	—	—	—	—	—	—	—	—

[27] Theoretical impacts compiled from Saturn V launch vehicle flight evaluation reports, and Apollo/Saturn V postflight trajectory reports. Impact date is same as launch date except for S-IVB stage, as indicated.

Launch Vehicle Propellant Usage[28]

	Apollo 7				Apollo 8				Apollo 9				Apollo 10			
	Burn Start	Burn End	Change	Burn Rate (lb/sec)	Burn Start	Burn End	Change	Burn Rate (lb/sec)	Burn Start	Burn End	Change	Burn Rate (lb/sec)	Burn Start	Burn End	Change	Burn Rate (lb/sec)
S-IB Burn (sec)	-2.988	144.32	147.31	—	—	—	—	—	—	—	—	—	—	—	—	—
Oxidizer (LOX), lb	631,300	3,231	628,069	4,263.6	—	—	—	—	—	—	—	—	—	—	—	—
Fuel (RP-1), lb	276,900	4,728	272,172	1,847.6	—	—	—	—	—	—	—	—	—	—	—	—
Total, lb	908,200	7,959	900,241	6,111.3	—	—	—	—	—	—	—	—	—	—	—	—
S-IC Burn (sec)	—	—	—	—	-6.585	153.82	160.41	—	-6.3	162.76	169.06	—	-6.4	161.63	168.03	—
Oxidizer (LOX), lb	—	—	—	—	3,128,034	46,065	3,081,969	19,213.7	3,301,203	45,230	3,255,973	19,259.3	3,302,827	40,592	3,262,235	19,414.6
Fuel (RP-1)	—	—	—	—	1,357,634	26,622	1,331,012	8,297.8	1,431,678	42,390	1,389,288	8,217.7	1,423,254	28,537	1,394,717	8,300.4
Total, lb	—	—	—	—	4,485,668	72,687	4,412,981	27,511.5	4,732,881	87,620	4,645,261	27,477.0	4,726,081	69,129	4,656,952	27,715.0
S-II Burn (sec)	—	—	—	—	156.19	524.04	367.85	—	165.16	536.22	371.06	—	164.05	552.64	388.59	—
Oxidizer (LOX), lb	—	—	—	—	793,795	5,169	788,626	2,143.9	821,504	3,230	818,274	2,205.2	823,325	3,536	819,789	2,109.7
Fuel (LH₂), lb	—	—	—	—	154,907	4,514	150,393	408.8	158,663	3,381	155,282	418.5	158,541	4,622	153,919	396.1
Total, lb	—	—	—	—	948,702	9,683	939,019	2,552.7	980,167	6,611	973,556	2,623.7	981,866	8,158	973,708	2,505.7
S-IVB 1st Burn (sec)	146.97	616.76	469.79	—	528.29	684.98	156.69	—	540.82	664.66	123.84	—	556.81	703.76	146.95	—
Oxidizer (LOX), lb	193,330	1,671	191,659	408.0	192,840	132,220	60,620	386.9	189,686	133,421	56,265	454.3	192,089	133,883	58,206	396.1
Fuel (LH₂), lb	39,909	2,502	37,407	79.6	43,395	30,678	12,717	81.2	43,709	32,999	10,710	86.5	43,388	31,564	11,824	80.5
Total, lb	233,239	4,173	229,066	487.6	236,235	162,898	73,337	468.0	233,395	166,420	66,975	540.8	235,477	165,447	70,030	476.6
S-IVB 2nd Burn (sec)	—	—	—	—	10,237.79	10,555.51	317.72	—	17,155.54	17,217.60	62.06	—	9,207.52	9,550.58	343.06	—
Oxidizer (LOX), lb	—	—	—	—	131,975	8,064	123,911	390.0	132,988	109,298	23,690	381.7	133,471	5,274	128,197	373.7
Fuel (LH₂), lb	—	—	—	—	28,358	2,759	25,599	80.6	29,369	24,476	4,893	78.8	29,116	2,177	26,939	78.5
Total, lb	—	—	—	—	160,333	10,823	149,510	470.6	162,357	133,774	28,583	460.6	162,587	7,451	155,136	452.2
S-IVB 3rd Burn (sec)	—	—	—	—	—	—	—	—	22,039.26	22,281.32	242.06	—	—	—	—	—
Oxidizer (LOX), lb	—	—	—	—	—	—	—	—	108,927	34,051	74,876	309.3	—	—	—	—
Fuel (LH₂), lb	—	—	—	—	—	—	—	—	23,520	8,951	14,569	60.2	—	—	—	—
Total, lb	—	—	—	—	—	—	—	—	132,447	43,002	89,445	369.5	—	—	—	—
Oxidizer-Fuel Ratio																
S-IB Stage	2.280	—	2.308	—	—	—	—	—	—	—	—	—	—	—	—	—
S-IC Stage	—	—	—	—	2.304	—	2.316	—	2.306	—	2.344	—	2.321	—	2.339	—
S-II Stage	—	—	—	—	5.124	—	5.244	—	5.178	—	5.270	—	5.193	—	5.326	—
S-IVB Stage 1st burn	4.844	—	5.124	—	4.444	—	4.767	—	4.340	—	5.254	—	4.427	—	4.923	—
S-IVB Stage 2nd burn	—	—	—	—	4.654	—	4.840	—	4.528	—	4.842	—	4.584	—	4.759	—
S-IVB Stage 3rd burn	—	—	—	—	—	—	—	—	4.631	—	5.139	—	—	—	—	—

28 All times are referenced to Range Zero; all other values represent actual usage, in pounds mass. Sources are the Saturn V launch vehicle flight evaluation reports and *Results of the Fifth Saturn IB Vehicle Test Flight (Apollo 7)*.

Launch Vehicle Propellant Usage[29]

	Apollo 11				Apollo 12				Apollo 13				Apollo 14			
	Burn Start	Burn End	Change	Burn Rate (lb/sec)	Burn Start	Burn End	Change	Burn Rate (lb/sec)	Burn Start	Burn End	Change	Burn Rate (lb/sec)	Burn Start	Burn End	Change	Burn Rate (lb/sec)
S-IC Burn (sec)	-6.4	161.63	168.03	—	-6.5	161.74	168.24	—	-6.7	163.60	170.30	—	-6.5	164.10	170.60	—
Oxidizer (LOX), lb	3,305,786	39,772	3,266,014	19,437.1	3,310,199	42,093	3,268,106	19,425.3	3,304,734	38,921	3,265,813	19,176.8	3,312,769	42,570	3,270,199	19,168.8
Fuel (RP-1), lb	1,424,889	30,763	1,394,126	8,296.9	1,424,287	36,309	1,387,978	8,250.0	1,431,384	27,573	1,403,811	8,243.2	1,428,561	32,312	1,396,249	8,184.3
Total, lb	4,730,675	70,535	4,660,140	27,734.0	4,734,486	78,402	4,656,084	27,675.2	4,736,118	66,494	4,669,624	27,420.0	4,741,330	74,882	4,666,448	27,353.2
S-II Burn (sec)	164.00	548.22	384.22	—	163.20	552.34	389.14	—	166.00	592.64	426.64	—	166.50	559.05	392.55	—
Oxidizer (LOX), lb	819,050	3,536	815,514	2,122.5	825,406	3,536	821,870	2,112.0	836,741	3,533	833,208	1,953.0	837,484	2,949	834,535	2,125.9
Fuel (LH₂), lb	158,116	10,818	147,298	383.4	157,986	4,610	153,376	394.1	159,931	4,532	155,399	364.2	159,232	3,232	156,000	397.4
Total, lb	977,166	14,354	962,812	2,505.9	983,392	8,146	975,246	2,506.2	996,672	8,065	988,607	2,317.2	996,716	6,181	990,535	2,523.3
S-IVB 1st Burn (sec)	552.20	699.33	147.13	—	556.60	693.91	137.31	—	596.90	749.83	152.93	—	563.40	700.56	137.16	—
Oxidizer (LOX), lb	192,497	135,144	57,353	389.8	190,587	135,909	54,678	398.2	191,890	132,768	59,122	386.6	190,473	136,815	53,658	391.2
Fuel (LH₂), lb	43,608	31,736	11,872	80.7	43,663	32,346	11,317	82.4	43,657	31,455	12,202	79.8	43,546	32,605	10,941	79.8
Total, lb	236,105	166,880	69,225	470.5	234,250	168,255	65,995	480.6	235,547	164,223	71,324	466.4	234,019	169,420	64,599	471.0
S-IVB 2nd Burn (sec)	9,856.20	10,203.03	346.83	—	10,042.80	10,383.94	341.14	—	9,346.30	9,697.15	350.85	—	8,912.40	9,263.24	350.84	—
Oxidizer (LOX), lb	134,817	5,350	129,467	373.3	135,617	4,659	130,958	383.9	132,525	3,832	128,693	366.8	136,551	5,812	130,739	372.6
Fuel (LH₂), lb	29,324	2,112	27,212	78.5	29,804	2,109	27,695	81.2	29,367	1,963	27,404	78.1	30,428	2,672	27,756	79.1
Total, lb	164,141	7,462	156,679	451.7	165,421	6,768	158,653	465.1	161,892	5,795	156,097	444.9	166,979	8,484	158,495	451.8
Oxidizer-Fuel Ratio																
S-IB Stage	—		—	—	—		—	—	—		—	—	—		—	—
S-IC Stage	2.320		2.343	—	2.324		2.355	—	2.309		2.326	—	2.319		2.342	—
S-II Stage	5.180		5.536	—	5.225		5.359	—	5.232		5.362	—	5.260		5.350	—
S-IVB Stage 1st burn	4.414		4.831	—	4.365		4.831	—	4.395		4.845	—	4.374		4.904	—
S-IVB Stage 2nd burn	4.597		4.758	—	4.550		4.729	—	4.513		4.696	—	4.488		4.710	—
S-IVB Stage 3rd burn	—		—	—	—		—	—	—		—	—	—		—	—

29 All times are referenced to Range Zero; all other values represent actual usage, in pounds mass. Sources are the Saturn V launch vehicle flight evaluation reports.

Launch Vehicle Propellant Usage[30]

	Apollo 15 Burn Start	Apollo 15 Burn End	Apollo 15 Change	Burn Rate (lb/sec)	Apollo 16 Burn Start	Apollo 16 Burn End	Apollo 16 Change	Burn Rate (lb/sec)	Apollo 17 Burn Start	Apollo 17 Burn End	Apollo 17 Change	Burn Rate (lb/sec)	Program Totals Burn Start	Program Totals Burn End	Program Totals Change	Program Totals Burn Rate (lb/sec)
S-IC Burn (sec)	-6.5	159.56	166.06	—	-6.7	161.78	168.48	—	-6.9	161.20	168.10	—	—	—	1677.31	—
Oxidizer (LOX), lb	3,312,030	31,135	3,280,895	19,757.3	3,311,226	34,028	3,277,198	19,451.6	3,314,388	36,479	3,277,909	19,499.8	32,903,196	396,885	32,506,311	19,380.1
Fuel (RP-1), lb	1,410,798	27,142	1,383,656	8,332.3	1,439,894	31,601	1,408,293	8,358.8	1,431,921	26,305	1,405,616	8,361.8	14,204,300	309,554	13,894,746	8,284.0
Total, lb	4,722,828	58,277	4,664,551	28,089.6	4,751,120	65,629	4,685,491	27,810.4	4,746,309	62,784	4,683,525	27,861.5	47,107,496	706,439	46,401,057	27,664.1
S-II Burn (sec)	163.00	549.06	386.06	—	165.20	559.54	394.34	—	164.60	559.66	395.06	—	—	—	3,895.51	—
Oxidizer (LOX), lb	837,991	3,109	834,882	2,162.6	846,157	3,141	843,016	2,137.8	844,094	3,137	840,957	2,128.7	8,285,547	34,876	8,250,671	2,118.0
Fuel (LH$_2$), lb	158,966	4,022	154,944	401.3	160,551	2,884	157,667	399.8	160,451	3,024	157,427	398.5	1,587,344	45,639	1,541,705	395.8
Total, lb	996,957	7,131	989,826	2,563.9	1,006,708	6,025	1,000,683	2,537.6	1,004,545	6,161	998,384	2,527.2	9,872,891	80,515	9,792,376	2,513.8
S-IVB 1st Burn (sec)	553.20	694.67	141.47	—	563.60	706.21	142.61	—	563.80	702.65	138.85	—	—	—	1,424.94	—
Oxidizer (LOX), lb	195,788	140,293	55,495	392.3	195,372	138,937	56,435	395.7	195,636	140,047	55,589	400.4	1,926,858	1,359,437	567,421	398.2
Fuel (LH$_2$), lb	43,674	32,416	11,258	79.6	43,727	32,081	11,646	81.7	43,752	32,685	11,067	79.7	436,119	320,565	115,554	81.1
Total, lb	239,462	172,709	66,753	471.9	239,099	171,018	68,081	477.4	239,388	172,732	66,656	480.1	2,362,977	1,680,002	682,975	479.3
S-IVB 2nd Burn (sec)	10,202.90	10,553.61	350.71	—	9,216.50	9,558.42	341.92	—	11,556.60	11,907.64	351.04	—	—	—	3,156.17	—
Oxidizer (LOX), lb	139,665	4,273	135,392	386.1	138,532	3,869	134,663	393.8	139,879	4,219	135,660	386.5	1,356,020	154,650	1,201,370	380.6
Fuel (LH$_2$), lb	29,799	1,722	28,077	80.1	29,968	2,190	27,778	81.2	30,050	2,212	27,838	79.3	295,583	44,392	251,191	79.6
Total, lb	169,464	5,995	163,469	466.1	168,500	6,059	162,441	475.1	169,929	6,431	163,498	465.8	1,651,603	199,042	1,452,561	460.2
Oxidizer/Fuel Ratio																
S-IB Stage	—				—				—						2.308	
S-IC Stage	2.348				2.300				2.315						2.339	
S-II Stage	5.272				5.270				5.261						5.352	
S-IVB Stage 1st burn	4.483				4.468				4.471						4.910	
S-IVB Stage 2nd burn	4.687				4.623				4.655						4.783	
S-IVB Stage 3rd burn	—				—				—						5.139	

30 All times are referenced to Range Zero; all other values represent actual usage, in pounds mass. Sources are the Saturn V launch vehicle flight evaluation reports.

Translunar Injection[31]

	Apollo 8	Apollo 10	Apollo 11	Apollo 12	Apollo 13	Apollo 14	Apollo 15	Apollo 16	Apollo 17
GET	002:56:05.51	002:39:20.58	002:50:13.03	002:53:13.94	002:41:47.15	002:34:33.24	002:56:03.61	002:39:28.42	003:18:37.64
KSC Date	21 Dec 1968	18 May 1969	16 Jul 1969	14 Nov 1969	11 Apr 1970	31 Jan 1971	26 Jul 1971	16 Apr 1972	07 Dec 1972
GMT Date	21 Dec 1968	18 May 1969	16 Jul 1969	14 Nov 1969	11 Apr 1970	31 Jan 1971	26 Jul 1971	16 Apr 1972	07 Dec 1972
KSC Time	10:47:05 a.m.	03:28:20 p.m.	12:22:13 p.m.	02:15:13 p.m.	04:54:47 p.m.	06:37:35 p.m.	12:30:03 p.m.	03:33:28 p.m.	03:51:37 a.m.
Time Zone	EST	EDT	EDT	EST	EST	EST	EDT	EST	EST
GMT Time	15:47:05	19:28:20	16:22:13	19:15:13	21:54:47	23:37:35	16:30:03	20:33:28	08:51:37
Altitude (ft)	1,137,577	1,093,217	1,097,229	1,209,284	1,108,555	1,090,930	1,055,296	1,040,493	1,029,299
Altitude (n mi)	187.221	179.920	180.581	199.023	182.445	179.544	173.679	171.243	169.401
Earth Fixed Velocity (ft/sec)	34,140.1	34,217.2	34,195.6	34,020.5	34,195.3	34,151.5	34,202.2	34,236.6	34,168.3
Space-Fixed Velocity (ft/sec)	35,505.41	35,562.96	35,545.6	35,389.8	35,538.4	35,511.6	35,579.1	35,566.1	35,555.3
Geocentric Latitude (deg N)	21.3460	-13.5435	9.9204	16.0791	-3.8635	-19.4388	24.8341	-11.9117	4.6824
Geodetic Latitude (deg N)	21.477	-13.627	9.983	16.176	-3.8602	-19.554	24.9700	-11.9881	4.7100
Longitude (deg E)	-143.9242	159.9201	-164.8373	-154.2798	167.2074	141.7312	-142.1295	162.4820	-53.1190
Flight Path Angle (deg)[32]	7.897	7.379	7.367	8.584	7.635	7.480	7.430	7.461	7.379
Heading Angle (deg E of N)	67.494	61.065	60.073	63.902	59.318	65.583	73.173	59.524	118.110
Inclination (deg)	30.636	31.698	31.383	30.555	31.817	30.834	29.696	32.511	28.466
Descending Node (deg)	38.983	123.515	121.847	120.388	122.997	117.394	108.439	122.463	86.042
Eccentricity	0.97553	0.97834	0.97696	0.96966	0.9772	0.9722	0.9760	0.9741	0.97
C3 (ft/sec)	-15,918,930	-14,084,265	-14,979,133	-19,745,586	-14,814,090	-18,096,135	-15,643,934	-16,881,439	-18,152,226

31 Compiled from Saturn V launch vehicle flight evaluation reports and mission reports.

32 Flight path angle and heading angle are 'space-fixed' for these measurements.

S-IVB Solar Trajectory[33]

	Apollo 8	Apollo 9	Apollo 10	Apollo 11	Apollo 12
S-IVB Closest Approach To Moon					
GET	069:58:55.2	—	078:51:03.6	078:42	085:48
KSC Date	24 Dec 1968	—	21 May 1969	19 Jul 1969	18 Nov 1969
GMT Date	24 Dec 1968	—	21 May 1969	19 Jul 1969	18 Nov 1969
KSC Time	05:49:55 a.m.	—	07:40 p.m.	04:14 p.m.	01:10 a.m.
KSC Time Zone	EST	—	EDT	EDT	EST
GMT Time	10:49:55	—	23:40	20:14	06:10
Lunar Radius (n mi)	1,620	—	2,619	2,763	4,020
Altitude Above Lunar Surface (n mi)	681	—	1,680	1,825	3,082
Velocity Increase Due To Lunar Gravity (n mi/sec)	0.79	—	0.459	0.367	0.296
S-IVB Solar Orbit Conditions					
Semi-Major Axis (n mi)	77,130,000	74,848,893	77,740,000	77,260,000	—
Eccentricity	—	0.07256	—	—	—
Aphelion (n mi)	79,770,000	80,280,052	82,160,000	82,000,000	—
Perihelion (n mi)	74,490,000	69,417,732	73,330,000	72,520,000	—
Inclination (deg)	23.47	24,390	23.46	0.3836	—
Period (days)	340.8	325.8	344.88	342	—

33 Compiled from Saturn V launch vehicle flight evaluation reports.

S-IVB Lunar Impact[34]

	Apollo 13	Apollo 14	Apollo 15	Apollo 16	Apollo 17
S-IVB Lunar Impact					
GET	077:56:39.7	082:37:52.17	079:24:41.55	075:08:04.0	086:59:40.99
KSC Date	14 Apr 1970	04 Feb 1971	29 Jul 1971	19 Apr 1972	10 Dec 1972
GMT Date	15 Apr 1970	04 Feb 1971	29 Jul 1971	19 Apr 1972	10 Dec 1972
KSC Time	08:09:39 p.m.	02:40:54 a.m.	04:58:41 p.m.	04:02:04 p.m.	03:42:40 p.m.
Time Zone	EST	EST	EDT	EST	EST
GMT Time	01:09:39	07:40:54	20:58:41	21:02:04	20:32:40
Weight (lbm)	29,599	30,836	30,880	30,805	30,712
Velocity (ft/sec)	8,461	8,343	8,455	8,711	8,346
Energy (ergs)	4.63×10^{17}	4.52×10^{17}	4.61×10^{17}	4.59×10^{17}	4.71×10^{17}
Angle From Horizontal (deg)	76	69	62	~79	55
Heading Angle (deg N to W)	100.6	75.7	83.46	104.7	83
S-IVB Lunar Impact—Tumble Rate (deg/sec)	12	1	1	—	—
Selenographic Latitude (deg N)	-2.75	-8.09	-1.51	1.3	-4.21
Selenographic Longitude (deg E)	-27.86	-26.02	-11.81	-23.8	-12.31
Crater Diameter (calculated) (ft)	134.8	133.9	134.8		
Crater Diameter (measured) (ft)	135.0	129.6	—		
Distance To Target (n mi)	35.4	159	83	173	84
Distance To Seismic Stations (n mi)					
Apollo 12	73	93	192	71	183
Apollo 14	—	—	99	131	85
Apollo 15	—	—	—	593	557
Apollo 16	—	—	—	—	459
Azimuth To Seismic Stations (deg)					
Apollo 12	274	207	083	355	096
Apollo 14	—	—	069	308	096
Apollo 15	—	—	—	231	209
Apollo 16	—	—	—	—	278

[34] Compiled from Saturn V launch vehicle flight evaluation reports, preliminary science reports, and mission reports. Apollo 16 data based on seismic data due to loss of S-IVB tracking prior to impact. Impact times are estimates for when impact occurred on the Moon, not when signal received on Earth, a method used by other sources.

LM Lunar Landing[35]

	Apollo 10[36]	Apollo 11	Apollo 12	Apollo 13[37]	Apollo 14	Apollo 15	Apollo 16	Apollo 17
LM Lunar Landing Conditions								
PDI Burn Duration (sec)	—	756.39	717.0	—	764.61	739.2	734	721
Hover Time Remaining (sec)	—	45	103	—	68	103	102	117
Landing Site	Sea of Tranquility	Sea of Tranquility	Ocean of Storms	Fra Mauro	Fra Mauro	Hadley-Apennine	Plains of Descartes	Taurus-Littrow
Targeted Latitude (deg N)	0.7333°	0.6833°	2.9833°	-3.6167°	-3.6719°	26.0816°	-9.0002°	20.1639°
Targeted Longitude (deg E)	23.6500°	23.7167°	-23.4000°	-17.5500°	17.4627	3.6583°	15.5164°	30.7495°
Actual Landing Latitude (deg N)	—	0.67408°	-3.01239°	—	-3.64530°	26.13222°	-8.97301°	20.19080°
Actual Landing Longitude (deg E)	—	23.47297°	-23.42157°	—	-17.47136°	3.63386°	15.50019°	30.77168°
GET	—	102:45:39.9	110:32:36.2	—	108:15:11.40	104:42:31.1	104:29:35.1	110:21:58
KSC Date	—	20 Jul 1969	19 Nov 1969	—	05 Feb 1971	30 Jul 1971	20 Apr 1972	11 Dec 1972
GMT Date	—	20 Jul 1969	19 Nov 1969	—	05 Feb 1971	30 Jul 1971	21 Apr 1972	11 Dec 1972
KSC Time	—	04:17:39 p.m.	01:54:36 a.m.	—	04:18:13 a.m.	06:16:29 p.m.	09:23:35 p.m.	02:54:58 p.m.
Time Zone	—	EDT	EST	—	EST	EDT	EST	EST
GMT Time	—	20:17:39	06:54:36	—	09:18:13	22:16:29	02:23:35	19:54:58
Sun Angle (deg)	11.0	10.8	5.1	18.5	10.3	12.2	119	13.0
LM Surface Angle (deg)	—	4.5° tilt east; yaw 13° south	3° pitch up, 3.8° roll left	—	1° pitch down; 6.9° roll right; 1.4° yaw left in tilt of 11° from horizontal	6.9° pitch up; 8.6° roll left resulting	0° roll, 2.3° pitch up, slight yaw south	4 to 5° pitch up, 0° roll, near 0° yaw
LM Distance To Target (ft)		22,500 ft W of landing ellipse center	535 ft NW of Surveyor III	—	55 ft N; 165 ft E	1,800 ft NW	668 ft N; 197 ft W	656 ft
Distance To Seismic Stations (n mi)								
Apollo 12		—	—	—	98	641	641	[Not found]
Apollo 14		—	98	—	—	591	544	[Not found]
Apollo 15		—	641	—	591	—	604	[Not found]
Apollo 16		—	641	—	544	604	—	[Not found]
Azimuth To Seismic Stations (deg)								
Apollo 12		—	—	—	96	40	100	[Not found]
Apollo 14		—	276	—	—	33	101	[Not found]
Apollo 15		—	226	—	218	—	160	[Not found]
Apollo 16		—	276	—	277	342	—	[Not found]

35 Compiled from mission reports and summary science reports. Actual landing site coordinates based on International Astronomical Union (IAU) Mean Earth Polar Axis coordinate system as described in the Journal of Geophysical Research, vol. 105, pages 20,227 to 20,280; 2000.

36 Although not planned as a lunar landing mission, Apollo 10 flew over the area to be targeted by the first lunar landing mission.

37 Data is for intended landing site; mission aborted.

LM Descent Stage Propellant Status[38]

Weight (lbm)	Apollo 9	Apollo 10	Apollo 11	Apollo 12	Apollo 13	Apollo 14	Apollo 15	Apollo 16	Apollo 17
Loaded									
Fuel	6,977	7,009.5	6,975	7,079	7,083.6	7,072.8	7,537.6	7,530.4	7,521.7
Oxidizer	11,063	11,209.2	11,209	11,350	11,350.9	11,344.4	12,023.9	12,028.9	12,042.5
Total	18,040	18,218.7	18,184	18,429	18,434.5	18,417.2	19,561.5	19,559.3	19,564.2
Consumed									
Fuel	4,127	295.0	6,724	6,658	3,225.5	6,812.8	7,058.3	7,105.4	7,041.3
Oxidizer	6,524	470.0	10,690	10,596	5,117.4	10,810.4	11,315.0	11,221.9	11,207.6
Total	10,651	765.0	17,414	17,254	8,342.9	17,623.2	18,373.3	18,327.3	18,248.9
Remaining at Cutoff									
Fuel	—	—	251	421	—	260.0	479	425	480.0
Oxidizer	—	—	519	754	—	534.0	709	807	835.0
Total	—	—	770	1,175	—	794.0	1,188	1,232	1,315.0
Usable at Cutoff									
Fuel	—	—	216	386	—	228.0	433	396	455.0
Oxidizer	—	—	458	693	—	400.0	622	732	770.0
Total	—	—	674	1,079	—	628.0	1,055	1,128	1,225.0
Remaining at Cutoff (No Landing)									
Fuel	2,850	6,714.5	—	—	3,858.1	—	—	—	—
Oxidizer	4,539	10,739.2	—	—	6,233.5	—	—	—	—
Total	7,389	17,453.7	—	—	10,091.6	—	—	—	—

[38] Compiled from mission reports.

LM Ascent Stage Propellant Status[39]

Weight (lbm)	Apollo 9	Apollo 10	Apollo 11	Apollo 12	Apollo 14	Apollo 15	Apollo 16	Apollo 17
Loaded								
Fuel	1,626	981	2,020	2,012	2,007.0	2,011.4	2,017.8	2,026.9
Oxidizer	2,524	1,650	3,218	3,224	3,218.2	3,225.6	3,224.7	3,234.8
Total	4,150	2,631	5,238	5,236	5,225.2	5,237.0	5,242.5	5,261.7
Transferred from RCS								
Fuel	—	—	—	—	—	—	16.0	—
Oxidizer	—	—	—	—	—	—	44.0	—
Total	—	—	—	—	—	—	60.0	—
Consumed by RCS								
Fuel	22	13.9	23	31	—	—	—	—
Oxidizer	44	28.0	46	62	—	—	—	—
Total	66	41.9	69	93	—	—	—	—
Consumed by APS Prior to Jettison								
Fuel	31	67	1,833	1,831	—	—	—	—
Oxidizer	59	108	2,934	2,943	—	—	—	—
Total	90	175	4,767	4,774	—	—	—	—
Remaining at Jettison								
Fuel	—	—	164	150	128.0	118.0	164.0	108.9
Oxidizer	—	—	238	219	204.2	173.0	257.7	175.6
Total	—	—	402	369	332.2	291.0	421.7	284.5
Consumed at Fuel Depletion								
Fuel	—	13	—	—	—	—	—	—
Oxidizer	—	106	—	—	—	—	—	—
Total	—	119	—	—	—	—	—	—
Consumed at Oxidizer Depletion								
Fuel	68	—	—	—	—	—	—	—
Oxidizer	0	—	—	—	—	—	—	—
Total	68	—	—	—	—	—	—	—
Total Consumed								
Fuel	1,558	887	1,856	1,862	1,879.0	1,893.4	1,869.8	1,918.0
Oxidizer	2,524	1,408	2,980	3,005	3,014.0	3,052.6	3,011.0	3,059.2
Total	4,082	2,295	4,836	4,867	4,893.0	4,946.0	4,880.8	4,977.2

[39] Compiled from mission reports.

LM Ascent and Ascent Stage Lunar Impact[40]

	Apollo 11	Apollo 12	Apollo 14	Apollo 15	Apollo 16[41]	Apollo 17
LM Ascent						
GET	124:22:00.79	142:03:47.78	141:45:40	171:37:23.2	175:31:47.9	185:21:37
KSC Date	21 Jul 1969	20 Nov 1969	06 Feb 1971	02 Aug 1971	23 Apr 1972	14 Dec 1972
GMT Date	21 Jul 1969	20 Nov 1969	06 Feb 1971	02 Aug 1971	24 Apr 1972	14 Dec 1972
KSC Time	01:54:00 p.m.	09:25:47 a.m.	01:48:42 p.m.	01:11:23 p.m.	08:25:47 p.m.	05:54:37 p.m.
KSC Time Zone	EDT	EST	EST	EDT	EST	EST
GMT Time	17:54:00	14:25:47	18:48:42	17:11:23	01:25:47	22:54:37
LM Ascent Stage Lunar Impact						
GET	—	149:55:16.4	147:42:23.4	181:29:35.8	—	193:17:21
KSC Date	—	20 Nov 1969	06 Feb 1971	02 Aug 1971	—	15 Dec 1972
GMT Date	—	20 Nov 1969	07 Feb 1971	03 Aug 1971	—	15 Dec 1972
KSC Time	—	05:17:16 p.m.	07:45:25 p.m.	11:03:35 p.m.	—	01:50:21 a.m.
Time Zone	—	EST	EST	EDT	—	EST
GMT Time	—	22:17:16	00:45:25	03:03:35	—	06:50:21
Selenocentric Latitude (deg N)	—	-3.94000	-3.42000	26.35583	—	19.96611
Selenocentric Longitude (deg E)	—	-21.20000	-19.67000	0.25000	—	30.48972
Selenocentric Latitude	—	3° 56' 24" S	3° 25' 12" S	26° 21' 21" N	—	19° 57' 58" N
Selenocentric Longitude	—	21° 12' 00" W	19° 40' 01" W	0° 15' 00" E	—	30° 29' 23" E
Velocity (ft/sec)	—	5,512	5,512	5,577	—	5,479
Mass (lbm)	—	5,254	5,077	5,258	—	4,982
LM Ascent Stage Lunar Impact Energy (ergs)	—	3.36×10^{16}	3.25×10^{16}	3.43×10^{16}	—	3.15×10^{16}
Angle From Horizontal (deg)	—	-3.7	-3.6	-3.2	—	[Not found]
Heading Angle (deg)	—	305.85	282	284	—	283
Crater Diameter (calculated) (ft)	—	29.9	29.6	30.2	—	[Not found]
Crater Diameter (measured) (ft)	—	—	—	—	—	[Not found]
Distance To Target (n mi)	—	35	7	12	—	0.7
Distance to LM Descent Stage Landing Site (n mi)	—	41.0	36	50	—	4.7
Distance to Apollo 17 Landing Site (n mi)	—	—	—	—	—	4.7
Distance to Seismic Stations (n mi)						
Apollo 12	—	39	62	610	—	945
Apollo 14	—	—	36	566	—	863
Apollo 15	—	—	—	50	—	416
Apollo 16	—	—	—	—	—	532
Azimuth to Seismic Stations (deg)						
Apollo 12	—	112	096	036	—	064
Apollo 14	—	—	276	029	—	061
Apollo 15	—	—	—	276	—	098
Apollo 16	—	—	—	—	—	027

40 Compiled from Saturn V launch vehicle flight evaluation report and mission report for each flight. Actual landing site coordinates based on International Astronomical Union (IAU) Mean Earth Polar Axis coordinate system as described in the Journal of Geophysical Research, vol. 105, pages 20,227 to 20,280; 2000.

41 Deorbit maneuver was not possible and LM ascent stage remained in lunar orbit for about one year. No impact information is available.

Extravehicular Activity[42]

	Apollo 9	Apollo 11	Apollo 12	Apollo 14	Apollo 15	Apollo 16	Apollo 17
Earth Orbit EVA							
1st EVA Participant	Scott	—	—	—	—	—	—
1st EVA Duration	01:01	—	—	—	—	—	—
2nd EVA Participant	Schweickart	—	—	—	—	—	—
2nd EVA Duration	01:07:00	—	—	—	—	—	—
2nd EVA Duration Outside LM	00:47:01	—	—	—	—	—	—
LM Stand-Up EVA							
Participant	—	—	—	—	Scott	—	—
Duration	—	—	—	—	00:33:07	—	—
First Surface EVA							
Duration	—	02:31:40	03:56:03	04:47:50	06:32:42	07:11:02	07:11:53
Total Distance Traveled (n mi)	—	0.5	0.5	0.5	5.6	2.3	1.8
LRV Ride Time	—	—	—	—	01:02	00:43	00:33
LRV Park Time	—	—	—	—	01:14	03:39	—
Total LRV Time	—	—	—	—	02:16	04:22	—
Samples Collected (lbm)[43]	—	47.51	36.82	45.19	31.97	65.92	31.53
Second Surface EVA							
Duration	—	—	03:49:15	04:34:41	07:12:14	07:23:09	07:36:56
Total Distance Traveled (n mi)	—	—	0.7	1.6	6.7	6.1	11.0
LRV Ride Time	—	—	—	—	01:23	01:31	02:25
LRV Park Time	—	—	—	—	02:34	03:56	—
Total LRV Time	—	—	—	—	03:57	05:27	—
Samples Collected (lbm)	—	—	38.80	49.16	76.94	63.93	75.18
Third Surface EVA							
Duration	—	—	—	—	04:49:50	05:40:03	07:15:08
Total Distance Traveled (n mi)	—	—	—	—	2.8	6.2	6.5
LRV Ride Time,	—	—	—	—	00:35	01:12	01:31
LRV Park Time	—	—	—	—	01:22	02:26	—
Total LRV Time	—	—	—	—	01:57	03:38	—
Samples Collected (lbm)	—	—	—	—	60.19	78.04	136.69
Total Lunar Surface EVA							
Total Duration	—	02:31:40	07:45:18	09:22:31	18:34:46	20:14:14	22:03:57
Total Distance Traveled (n mi)	—	0.5	1.2	2.2	15.1	14.5	19.3
Total Samples Collected (lbm)	—	47.51	75.73	93.21	170.44	211.00	243.65
Total LRV Ride Time	—	—	—	—	3:00	03:26	04:29
Total LRV Park Time	—	—	—	—	05:10	10:01	—
Total LRV Time	—	—	—	—	08:10	13:27	—
Maximum Distance Traveled From LM (ft)	—	200[44]	1,350[45]	4,770[46]	16,470	15,092[47]	25,029
Transearth EVA							
Participant	—	—	—	—	Worden	Mattingly	Evans
Duration	—	—	—	—	00:39:07	01:23:42	01:05:44

42 Compiled from mission reports. Durations represent time from cabin depressurization to cabin pressurization.

43 Returned sample weights provided by Lunar Sample Curator, NASA Johnson Space Center.

44 Apollo 11 Preliminary Science Report (SP-214), p. 44.

45 Apollo 12 Preliminary Science Report, p. 26 (measured from map).

46 Skylab A Chronology (SP-4011), pps. 420-421 for Apollo 14, Apollo 15 and Apollo 17.

47 Measured from map in Apollo 16 Preliminary Science Report (SP-315).

Lunar Surface Experiments Package Arrays and Status[48]

Experiment	Principal Investigator	Apollo 11	Apollo 12	Apollo 14	Apollo 15	Apollo 16	Apollo 17
Array		EASEP	ALSEP A	ALSEP C	ALSEP A-2	ALSEP D	ALSEP E
Design Life (days)		14	365	365	365	365	730
Date Commanded Off			30 Sep 1977	Failed Jan 1976	30 Sep 1977	30 Sep 1977	30 Sep 1977
Passive Seismic Experiment	Gary Latham, University of Texas	X	X	X	X	X	
Laser-Ranging Retroreflector	J. E. Faller, Wesleyan University	100 corner		100 corner	300 corner	X	—
Lunar Surface Magnetometer	Palmer Dyal, Ames Research Center / Charles Sonett, University of Arizona	X	X	X			
Solar Wind Spectrometer (Exposure)	Conway W. Snyder, Jet Propulsion Laboratory	1 hr 17 min[49]	18 hr 42 min	21 hr 0 min	41 hr 8 min	45 hr 5 min	
Suprathermal Ion Detector Experiment	John Freeman, Rice University		X	X	X		
Heat Flow Experiment	Mark Langseth, Lamont-Doherty Geological Observatory, Columbia University				X	X	X
Charged-Particle Lunar Environment Experiment	D. Reasoner, Rice University			X			
Cold-Cathode Gage Experiment	Francis Johnson, University of Texas		X	X	X		
Active Seismic Experiment	Robert Kovach, Stanford University			X		X	
Lunar Seismic Profiling Experiment	Robert Kovach, Stanford University						X
Lunar Surface Gravimeter	Joseph Weber, University of Maryland						X
Lunar Mass Spectrometer	John H. Hoffman, University of Texas						X
Lunar Ejecta Meteoroid Experiment	Otto Berg, Goddard Space Flight Center						X
Dust Detector	James Bates, Manned Spacecraft Center	X	X	X	X		

48 *Apollo Lunar Surface Experiments Package (ALSEP): Five Years of Lunar Science and Still Going Strong*, Bendix Aerospace.
49 JSC-09423, p. 3-54.

Lunar Surface Experiments[50]

Designation	Experiment	Apollo 11	Apollo 12	Apollo 14	Apollo 15	Apollo 16	Apollo 17
M-515	Lunar Dust Detector		X	X	X		
S-031	Passive Seismic Experiment	X	X	X	X		
S-033	Active Seismic Experiment			X			
S-034	Lunar Surface Magnetometer		X		X		
S-035	Solar Wind Spectrometer		X		X		
S-036	Suprathermal Ion Detector		X	X	X		
S-037	Heat Flow Experiment			X	X	X	X
S-038	Charged Particle Lunar Environment			X			
S-058	Cold Cathode Ion Gauge		X	X	X		
S-059	Lunar Field Geology	X	X	X	X	X	X
S-078	Laser Ranging Retroreflector	X		X	X		
S-080	Solar Wind Composition	X	X	X	X	X	
S-151	Cosmic-Ray Detection (helmets)	X					
S-152	Cosmic-Ray Detector (sheets)	X					
S-184	Lunar Surface Close-up (photography)		X			X	
S-198	Portable Magnetometer			X		X	
S-199	Lunar Gravity Traverse			X	X	X	X
S-200	Soil Mechanics			X	X	X	X
S-201	Far-Ultraviolet Camera/Spectroscope					X	
S-202	Lunar Ejecta and Meteorites						X
S-203	Lunar Seismic Profiling						X
S-204	Surface Electrical Properties						X
S-205	Lunar Atmospheric Composition						X
S-207	Lunar Surface Gravimeter						X
S-229	Lunar Neutron Probe						X
—	Lunar sample Analysis	X	X	X	X	X	X
—	Surveyor III Analysis		X				
—	Long-term Lunar Surface Exposure						X

50 *Project Apollo:* NASA Facts.

Lunar Surface Experiments[51]

Central Station

The heart of the experiment package, provided the radio frequency link to Earth for telemetering data, command/control, and power distribution to the experiments.

Early Apollo Scientific Experiment Package (EASEP)

Flown on Apollo 11 only, this experiment package was powered by solar energy and contained an abbreviated set of experiments. It continued to return data for 71 days.

Active Seismic Experiments

Used an astronaut-activated thumper device and mortar firing explosive charges to generate seismic signals. This experiment used geophone seismic listening devices to determine lunar structure to depths of about 1,000 feet.

Heat Flow Experiment

Probes containing temperature sensors were implanted in holes to depths of 8 feet to measure the near-surface temperature gradient and thermal conductivity from which heat flow from the lunar interior could be determined.

Lunar Mass Spectrometer

Used a magnetic deflection mass spectrometer to identify lunar atmospheric components and their relative abundance.

Lunar Seismic Profiling Experiment

Flown on Apollo 17 only, this experiment was an advanced version of the Active Seismic Experiment. It used four geophones to detect seismic signals generated by eight explosive charges weighing from about .10 to 6.5 pounds. The charges were deployed at distances up to 2 n mi from the Lunar Module and were detonated by timers after the Lunar Module departed. Lunar structure to depths of 1.5 n mi was measured. Used in a listening mode, the experiment continued to provide data on Moon/thermal quakes and meteoroid impacts beyond its planned lifetime.

Solar Wind Spectrometer

Measured interaction between the Moon and the solar wind by sensing flow-direction and energies of both electrons and positive ions. Results showed that solar wind plasma measurements on the lunar surface are indistinguishable from simultaneous plasma measurements made by nearby satellites.

Suprathermal Ion Detector

Provided information on the energy and mass spectra of positive ions near the lunar surface. Evidence of prompt ionization and acceleration of gases generated on the Moon was found in the return data.

Charged Particle Lunar Environment

Measured the fluxes of charged particles, both electrons and ions, having energies from 50 to 50,000 electron volts. The instrument measured plasma particles originating in the Sun and low-energy particle flux in the magnetic tail of Earth.

Laser Ranging Retroreflector

The retroreflector bounced laser pulses back to Earth ground stations to provide data for precise measurements of the Earth-Moon distance to determine Earth wobble about its axis, continental drift, lunar librations, etc. Arrays of 100 retroreflecting corners were flown on Apollos 11 and 14, and an array of 300 corners was flown on Apollo 15.

Lunar Surface Magnetometer

Measured the intrinsic remnant lunar magnetic field and the magnetic response of the Moon to large-scale solar and terrestrial magnetic fields. The electrical conductivity of the lunar interior was also determined from measurements of the Moon's response to magnetic field step-transients. Three boom-mounted sensors measured mutually-orthogonal components of the field.

Lunar Ejecta and Meteorites Experiment

Three separate detectors which measured energy, speed, and direction of dust particles. Oriented east, west, and up. The dust particles measured were meteorites, secondary ejecta from meteorites, and, possibly, lunar surface particles levitated and accelerated by lunar surface phenomena.

Cold Cathode Ion Gauge

A separate experiment combined in an integrated package with the Suprathermal Ion Detector. It determined the density of neutral gas particles in the lunar atmosphere.

Passive Seismic Experiment

Detected Moonquakes and meteoroid impacts to enable scientists to determine the Moon's internal composition.

Radioisotope Thermoelectric Generator

Supplied about 70 watts of electrical power for continuous day-night operation.

Lunar Surface Gravimeter

Measured and sensed changes in the vertical component of lunar gravity, using a spring mass suspension. It also provided data on the lunar tides.

[51] *Apollo Lunar Surface Experiments Package (ALSEP): Five Years of Lunar Science and Still Going Strong*, Bendix Aerospace.

Lunar Orbit Experiments[52]

Designation	Experiment	Apollo 8	Apollo 11	Apollo 12	Apollo 14	Apollo 15	Apollo 16	Apollo 17
S-151	Cosmic Ray Detector (Helmets)	X						
S-158	Multispectral Photography			X			X	
S-160	Gamma-Ray Spectrometer					X	X	X
S-161	X-Ray Fluorescence					X	X	
S-162	Alpha-Particle Spectrometer					X	X	
S-164	S-Band Transponder (CSM/LM)				X	X	X	X
S-164	S-Band Transponder (Subsatellite)					X	X	
S-165	Mass Spectrometer					X	X	X
S-169	Far-Ultraviolet Spectrometer						X	
S-170	Bistatic Radar				X	X	X	X
S-171	Infrared Scanning Radiometer							X
S-173[53]	Particle Shadows/Boundary Layer					X	X	
S-174	Magnetometer					X	X	
S-176	Command Module Window Meteoroid				X	X	X	X
S-177	Ultraviolet Photography, Earth and Moon					X	X	
S-178	Gegenschein from Lunar Orbit				X	X		X
S-209	Lunar Sounder							X
—	Candidate Exploration Sites							
—	CM Orbital Science Photography				X		X	X
—	CM Photographic Tasks				X			
—	Dim Light Photography				X			
—	Lunar Mission Photography From CM	X		X	X			
—	Selenodetic Reference Point Update			X	X			
—	SM Orbital Photographic Tasks[54]				X	X	X	X
—	Transearth Lunar Photography						X	X
—	Visual Observations From Lunar Orbit					X	X	X

52 *Project Apollo: NASA Facts.*

53 Experiments S-173 and S-174 were Particles and Fields Subsatellite experiments.

54 Included panoramic camera photography, mapping camera photography, and laser altimetry. Also supported geologic objectives.

Geology and Soil Mechanics Tools and Equipment[55]

Item	Apollo 11	Apollo 12	Apollo 14	Apollo 15	Apollo 16	Apollo 17
Apollo Lunar Surface Hand Tools						
Hammer	1	1	1	1	1	1
Large Scoop	1	1	1	0	0	0
Adjustable Scoop	0	0	0	1	1	1
Extension Handle	1	1	1	1	2	2
Gnomon	1	1	1	1	1	1
Tongs	1	1	1	1	2	2
Adjustable Trenching Tool	0	0	1	0	0	0
Rake	0	0	0	1	1	1
Core Tubes	2	4	6	0	0	0
Core Tube Caps	2	1	0	0	0	0
Drive Tubes (Lower)	0	0	0	5	5	5
Drive Tubes (Upper)	0	0	0	4	4	4
Drive Tube Cape and Bracket Assembly	0	0	0	3	5	5
Drive Tube Tool Assembly	1	0	0	0	1	1
Spring Scale	1	1	0	0	0	0
Sample Scale	0	0	1	1	1	1
Tool Carrier	0	0	0	1	1	0
Sample Return Container	2	2	2	2	2	2
Bags and Special Containers						
Small Sample Bags	5	0	0	0	0	0
Documented Sample Bags (15-Bag Dispenser)	1	3	1	0	0	0
Documented Sample Bags (20-Bag Dispenser)	0	0	0	6	7	8
Documented Sample Bags (35-Bag Dispenser)	0	1	2	0	0	0
Round Documented Sample Bag	0	0	0	0	0	48
Protective Padded Sample Bag	0	0	0	0	2	0
Documented Sample Weigh Bag	2	4	4	0	0	0
Sample Collection Bag	0	0	0	2	2	2
Gas Analysis Sample Container	1	1	0	0	0	0
Core Sample Vacuum Container	0	1	3	3	1	1
Solar Wind Composition Bag	2	1	1	0	0	0
Magnetic Shield Sample Container	0	1	1	0	0	0
Extra Sample Collection Bags	0	0	0	4	6	6
Organic Control Sample	0	1	2	2	2	0
Lunar Surface Sampler (Beta Cloth)	0	0	0	0	1	0
Lunar Surface Sampler (Velvet)	0	0	0	0	1	0
Lunar Roving Vehicle Soil Sampler	0	0	0	0	0	1
Magnetic Sample Assembly	0	0	0	0	1	0
Tether Hook	1	1	1	0	0	0
Lunar Surface Drill	0	0	0	1	1	1
Core Stem With Bit	0	0	0	1	1	1
Core Stems Without Bit	0	0	0	5	5	5
Core Stem Cap and Retainer Assembly	0	0	0	2	2	2
Self-Recording Penetrometer	0	0	0	1	1	0

55 JSC-09423, p. 3-27.

Lunar Subsatellites[56]

	Apollo 15	Apollo 16
Designations		
International	1971-063D	1972-031D
NORAD	05377	06009
Deploy Conditions		
GET	222:39:29.1	196:02:02
KSC Date	04 Aug 1971	24 Apr 1972
GMT Date	04 Aug 1971	24 Apr 1972
KSC Time	04:13:29 p.m.	04:56:09 p.m.
KSC Time Zone	EDT	EST
GMT Time	20:13:29	21:56:09
Weight (lbs)	78.5	90
Apogee (n mi)	76.3	66
Perigee (n mi)	55.1	52
Inclination (deg)	-28.7	-11
Period (min)	120	120
Flight Path Angle (deg)	-0.60	-0.41
Heading Angle (deg)	-41.78	-79.43
Weight (lbm)	79	93
Status	Selenocentric orbit, 1984	Impacted lunar surface
	Data for last telemetry	
GET (hh:mm)	[Unknown]	1,034:37
KSC Date	[Unknown]	29 May 1972
GMT Date	30 Jul 1971	29 May 1972
KSC Time	[Unknown]	03:31 p.m. EDT
GMT Time	[Unknown]	20:31
Revolutions	[Unknown]	425
Lunar Impact Latitude (deg N)	[Unknown]	[Unknown]
Lunar Impact Longitude (deg E)	[Unknown]	110

[56] Compiled from *Apollo 15 Preliminary Science Report* (SP-289) and *Apollo 16 Preliminary Science Report* (SP-315) and mission reports.

Entry, Splashdown, and Recovery[57]

	Apollo 7	Apollo 8	Apollo 9	Apollo 10	Apollo 11	Apollo 12	Apollo 13	Apollo 14	Apollo 15	Apollo 16	Apollo 17[58]
Earth Entry											
Velocity (ft/sec)	25,846.4	36,221.1	25,894	36,314	36,194.4	36,116.618	36,210.6	36,170.2	36,096.4	36,196.1	36,090.3
Maximum Entry Velocity (ft/sec)	25,955	36,303	25,989	36,397	36,277						
Maximum g	3.33	6.84	3.35	6.78	6.56	6.57	5.56	6.76	6.23	7.19	6.49
Range (n mi)	1,594	1,292	1,835	1,295	1,497	1,250	1,250	1,234	1,184	1,190	1,190
Geodetic Latitude (deg N)	-29.92	20.83	33.52	-23.60	-3.19	-13.80	-28.23	-36.36	14.23	-19.87	0.71
Longitude (deg E)	92.62	-179.89	-99.05	174.39	171.96	173.52	173.44	165.80	-175.02	-162.13	-173.34
Flight Path Angle (deg E of N)	-2.0720	-6.50	-1.74	-6.54	-6.48	-6.48	-6.269	-6.370	-6.51	-6.55	-6.49
Heading Angle (deg)	87.47	121.57	99.26	71.89	50.18	98.16	77.21	70.84	52.06	21.08	156.53
Lift To Drag Ratio		0.300		0.305	0.300	0.309	0.291	0.280	0.290	0.286	0.290
Max. Heating Rate (BTU/ft²/sec)		296		296	286	285	271	310	289	346	346
Total Heating Load (BTU/ft²)		26,140		25,728	26,482	26,224	25,710	27,111	25,881	27,939	27,939
Duration (sec)	937.0	869.2	1,003.8	868.5	929.3	845.9	835.3	852.8	778.3	814.0	801.0
Avg. Radiation Skin Dose (Rads)[59]	0.16	0.16	0.20	0.48	0.18	0.58	0.24	1.14	0.30	0.51	0.55
Earth Splashdown											
GET	260:09:03	147:00:42.0	241:00:54	192:03:23	195:18:35	244:36:25	142:54:41	216:01:58.1	295:11:53.0	265:51:05	301:51:59
KSC Date	22 Oct 1968	27 Dec 1968	13 Mar 1969	26 May 1969	24 Jul 1969	24 Nov 1969	17 Apr 1970	09 Feb 1971	07 Aug 1971	27 Apr 1972	19 Dec 1972
GMT Date	22 Oct 1968	27 Dec 1968	13 Mar 1969	26 May 1969	24 Jul 1969	24 Nov 1969	17 Apr 1970	09 Feb 1971	07 Aug 1971	27 Apr 1972	19 Dec 1972
KSC Time	07:11:48 a.m.	10:51:42 a.m.	12:00:54 p.m.	12:52:23 a.m.	12:50:35 p.m.	03:58:25 p.m.	01:07:41 p.m.	04:05:00 p.m.	04:45:53 p.m.	02:45:05 p.m.	02:24:59 p.m.
Time Zone	EDT	EST	EST	EDT	EDT	EST	EST	EST	EDT	EST	EST
GMT Time	11:11:48	15:51:42	17:00:54	16:52:23	16:50:35	20:58:25	18:07:41	21:05:00	20:45:53	19:45:05	19:24:59
Splashdown Site (Ocean)	Atlantic	Pacific	Atlantic	Pacific	Pacific	Pacific	Pacific	Pacific	Pacific	Pacific	Pacific
Latitude (deg N)	27.63	8.10	23.22	-15.07	13.30	-15.78	-21.63	-27.02	26.13	-0.70	-17.88
Longitude (deg E)	-64.15	-165.00	-67.98	-164.65	-169.15	-165.15	-165.37	-172.67	-158.13	-156.22	-166.11
CM Weight (lbm)	11,409	10,977	11,094	10,901	10,873	11,050	11,133	11,481.2	11,731	11,995	12,120
Distance To Target (n mi)	1.9	1.4	2.7	1.3	1.7	2.0	1.0	0.6	1.0	3.0	1.0
Distance To Recovery Ship (n mi)	7	2.6	3	2.9	13	3.91	3.5	3.8	5	2.7	3.5
Distance Traveled (n mi)	3,953,842	504,006	3,664,820	721,250	828,743	828,134	541,103	1,000,279	1,107,945	1,208,746	1,291,299
Maximum Distance Traveled From Earth (n mi)	244.2	203,752.37	275.0	215,548	210,391	(not found)	216,075	(not found)	(not found)	(not found)	(not found)

57 Compiled from mission reports, USN Historical Office data, *Apollo Program Summary Report* (JSC-09423) and other sources.

58 Some Apollo 17 entry phase data are preflight predictions because actual data were not obtained.

59 *Space Physiology & Medicine*, SP-447.

Entry, Splashdown, and Recovery

	Apollo 7	Apollo 8	Apollo 9	Apollo 10	Apollo 11	Apollo 12	Apollo 13	Apollo 14	Apollo 15	Apollo 16	Apollo 17
Splashdown Weather											
1st Level Cloud Type	Light rain showers	Scattered clouds	30%	10%	—	—	Broken	High Scattered	Scattered	Scattered	Scattered
1st Level Cloud Cover (ft)	600 (overcast)	2,000	2,000	2,000	—	—	2,000	2,000	2,000	2,000	3,000
2nd Level Cloud Type	—	Overcast	Broken	20%	—	—	—	—	—	—	—
2nd Level Cloud Cover (ft)	—	9,000	9,000	7,000	—	—	—	—	—	—	—
Visibility (n mi)	2	10	10	10	—	10	10	10	10	10	10
Wind Speed (ft/sec)	27	32	15	8	27	—	—	—	—	—	—
Wind Speed (knots)	16	19	9	5	16	15	10	15	10	10	10
Wind Direction (deg from True N)	260	70	200	100	—	68	—	—	—	110	130
Air Temperature (F)	74	—	79	—	—	—	—	—	—	—	—
Water Temperature (F)	81	82	76	85	—	—	—	—	—	—	—
Wave Height (ft)	3	6	7	3	3	3, with 15 ft swells	4	4	3	4	2 to 3
Wave Direction (deg from True N)	260	110	340	—	—	—	—	—	—	—	—
Spacecraft Recovery											
Flotation Attitude	Inverted	Inverted	Upright	Upright	Inverted	Inverted	Upright	Upright	Upright	Inverted	Upright
Minutes To Upright	12.0	6.0	0.0	0.0	7.6	4.5	0.0	0.0	0.0	4.5	0.0
Minutes To CM Pickup	111	148	132	96	188	108	88	124	94	99	123
Launch Site Pickup Time	09:03 a.m.	01:20 p.m.	02:13 p.m.	02:28 p.m.	03:58 p.m.	05:45 p.m.	02:36 p.m.	06:09 p.m.	06:20 p.m.	04:24 p.m.	04:28 p.m.
Time Zone	EST	EST	EST	EDT	EDT	EST	EST	EST	EDT	EST	EST
GMT Pickup Time	13:03	18:20	19:13	18:28	19:58	22:45	19:36	23:09	22:20	21:28	21:28
Crew Recovery											
Minutes To Crew Pickup	56	88	49	39	63	60	45	48	39	37	52
Launch Site Pickup Time	08:08 a.m.	12:20 p.m.	12:50 p.m.	01:31 p.m.	01:53 p.m.	04:57 p.m.	01:53 p.m.	04:53 p.m.	05:25 p.m.	03:33 p.m.	03:17 p.m.
Time Zone	EST	EST	EST	EDT	EDT	EST	EST	EST	EDT	EST	EST
GMT Pickup Time	12:08	17:20	17:50	17:31	17:53	21:57	18:53	21:53	21:25	20:22	20:17
Recovery Ship	Essex (CVS-9)	Yorktown (CVS-10)	Guadalcanal (LPH-7)	Princeton (LPH-5)	Hornet (CVS-12)	Hornet (CVS-12)	Iwo Jima (LPH-2)	New Orleans (LPH-11)	Okinawa (LPH-3)	Ticonderoga (CVS-14)	Ticonderoga (CVS-14)
Commanding Officer (Captain)	John A. Harkins	John G. Fifield	Roy M. Sudduth	Carl M. Cruise	Carl J. Seiberlich	Carl J. Seiberlich	Leland E. Kirkemo	Robert W. Carius	Andrew F. Huff	Frank T. Hamler	Frank T. Hamler
Recovery Forces[60]											
Navy Ships Deployed	9	12	6	8	5	—	4	5	4	4	3
Atlantic Ocean	4	6	3	4	3	3	2	3	2	1	1
Pacific Ocean	5	6	3	4	2	2	2	2	2	3	2
Aircraft Deployed	31	43	29	30	31	26	22	19	17	17	15
Navy	8	21	7	10	13	9	8	5	5	6	5
Air Force	23	22	22	20	18	17	14	14	12	11	10

60 JSC-09423, p. 7-13

Selected Mission Weights (lbs)[61]

	Apollo 7	Apollo 8	Apollo 9	Apollo 10	Apollo 11	Apollo 12	Apollo 13	Apollo 14	Apollo 15	Apollo 16	Apollo 17
CSM/LM at EOI	36,419	87,382	95,231	98,273	100,756.4	101,126.9	101,261.2	102,083.6	107,142	107,226	107,161
CSM/LM at Separation	—	—	—	94,063	96,566.6	—	—	—	—	—	—
CSM/LM at Transposition & Docking	—	—	91,055	94,243	96,767.5	97,119.8	97,219.4	98,037.2	103,105	103,175	103,167
CSM at Transposition & Docking	—	—	58,925	63,560	63,473.0	63,535.6	63,720.3	64,388.0	66,885	66,923	66,893
LM at Transposition & Docking	—	—	32,130	30,683	33,294.5	33,584.2	33,499.1	33,649.2	36,220	36,252	36,274
CSM/LM at 1st MCC Ignition	—	63,307	—	93,889	96,418.2	96,870.6	97,081.5	97,901.5	—	—	—
CSM/LM at 1st MCC Cutoff	—	—	—	93,413	96,204.2	96,401.2	96,851.1	—	—	—	—
CSM/LM Before Cryogenic Tank Anomaly	—	—	—	—	—	—	96,646.9	—	—	—	—
CSM/LM After Cryogenic Tank Anomaly	—	—	—	—	—	—	96,038.7	—	—	—	—
CSM/LM at 2nd MCC Ignition	—	62,845	—	—	—	—	95,959.9	97,104.1	—	—	—
CSM/LM at 2nd MCC Cutoff	—	—	—	—	—	—	95,647.1	—	—	—	—
CSM at TEI Ignition	—	45,931	—	37,254	36,965.7	34,130.6	95,424.0	34,554.4	35,899	38,697	36,394
CSM at TEI Cutoff	—	32,008	—	26,172	26,792.7	25,724.5	87,456.0	24,631.9	—	—	—
CSM at 3rd MCC Ignition	—	—	—	—	—	—	87,325.3	—	—	—	—
CSM at 3rd MCC Cutoff	—	—	—	—	—	—	87,263.3	—	—	—	—
CSM/LM at LOI Ignition	—	62,827	—	93,319	96,061.1	96,261.1	—	97,033.1	102,589	102,642	102,639
CSM/LM at LOI Cutoff	—	46,743	—	69,429	72,037.6	72,335.6	—	71,823.0	76,329	77,647	76,540
CSM/LM at Circularization Ignition	—	46,716	—	69,385	72,019.9	72,243.7	—	71,768.8	76,278	77,595	76,354
CSM/LM at Circularization Cutoff	—	—	—	68,455	70,905.9	71,028.4	—	—	—	—	—
CSM/LM at Descent Orbit Insertion	—	—	—	68,238	70,760.3	70,897.3	—	70,162.3	74,460	76,590	74,762
CSM/LM at Separation for Lunar Landing	—	—	—	—	—	—	—	—	—	—	—
CSM at Separation for Lunar Landing	—	—	—	37,072	37,076.8	36,911.8	—	36,036.4	37,742	39,847	37,991
LM at Separation for Lunar Landing	—	—	—	31,166	33,683.5	33,985.5	—	34,125.9	36,718	36,743	36,771
LM at Powered Descent Initiation	—	—	—	—	33,669.6	33,971.8	—	34,067.8	36,634	36,617	36,686
LM at Descent Orbit Insertion Ignition	—	—	—	31,137	33,401.6	33,719.3	—	—	—	—	—
LM at Descent Orbit Insertion Cutoff	—	—	—	30,903	—	—	—	—	—	—	—
LM at Lunar Landing	—	—	—	—	16,153.2	16,564.2	—	16,371.7	18,175	18,208	18,305
CSM at Plane Change	—	—	—	—	—	—	—	35,610.4	37,219	38,994	37,464
CSM at Circularization Ignition	—	—	—	—	—	—	—	35,996.3	37,716	39,595	37,960
LM at Phasing Ignition	—	—	—	30,824	—	—	—	—	—	—	—
LM at Phasing Cutoff	—	—	—	30,283	—	—	—	—	—	—	—
LM at Fuel Depletion	—	—	5,616	5,243	—	—	—	—	—	—	—
CSM/LM Ascent Stage at Docking	—	—	36,828	44,930	42,585.4	41,071.8	—	39,906.8	41,754	44,318	41,914
CSM at Docking	—	—	26,895	36,995	36,847.4	35,306.2	—	34,125.5	35,928	38,452	36,036
LM Ascent Stage at Lunar Liftoff	—	—	—	8,077	10,749.6	10,779.6	—	10,779.8	10,915	10,949	10,997
LM Ascent Stage at Orbit Insertion for Docking	—	—	—	8,273	5,928.6	5,965.6	—	5,917.8	5,985	6,001	6,042
LM Ascent Stage at Terminal Phase Initiation	—	—	—	8,052	5,881.5	5,885.9	—	5,880.1	5,965	5,972	5,970
LM Ascent Stage After Staging	—	—	—	—	—	—	—	—	—	—	—
LM Ascent Stage at Coelliptic Sequence Initiation	—	—	—	—	—	—	—	—	—	—	—
LM Ascent Stage at Docking	—	—	9,933	7,935	5,738.0	5,765.6	—	5,781.3	5,826	5,866	5,878
CSM at After Post-Docking Jettison	—	—	27,139	—	37,100.5	35,622.9	—	34,596.3	36,407	38,992	36,619
LM Ascent Stage After Post-Docking Jettison	—	—	—	7,663	5,462.5	5,436.5	—	5,307.6	5,325	5,306	5,277
CSM (CSM/LM) at Subsatellite Jettison	—	—	—	—	—	—	—	—	36,019	38,830	—
CSM (CSM/LM) at 4th MCC Ignition	—	—	—	—	—	—	87,132.1	—	—	—	—
CSM at 4th MCC Cutoff	—	—	—	—	—	—	87,101.5	—	—	—	—
CSM at Pre-Entry Separation	23,435	31,768	24,183	25,095	26,656.5	25,444.2	—	24,375.0	26,323	27,225	26,659
CSM/LM Before CSM/LM Separation	—	—	—	—	—	—	87,057.3	—	—	—	—
CM/LM After CSM/LM Separation	—	—	—	—	—	—	37,109.7	—	—	—	—
SM After Pre-Entry Separation	11,071	19,589	11,924	12,957	14,549.1	13,160.7	—	11,659.9	13,358	14,199	13,507
CM After Pre-Entry Separation	12,364	12,179	12,259	12,138	12,107.4	12,283.5	12,367.6	12,715.1	12,965	13,026	13,152
CM at Entry	12,356	12,171	12,257	12,137	12,095.5	12,275.5	12,361.4	12,703.5	12,953	13,015	13,140
CM at Drogue Deployment	11,936	11,712	11,839	11,639	11,603.7	11,785.7	11,869.4	—	—	—	—
CM at Main Parachute Deployment	11,855	11,631	11,758	11,558	11,318.9	11,496.1	11,579.8	12,130.8	12,381	12,442	12,567
CM at Landing	11,409	10,977	11,094	10,901	10,873.0	11,050.2	11,132.9	11,481.2	11,731	11,995	12,120

61 Compiled from mission reports. Apollo 7 did not have a LM. Apollo 13 includes CSM and LM until separation before Earth entry.

Command Module Cabin Temperature History (°F)[62]

Mission	Apollo 7	Apollo 8	Apollo 9	Apollo 10	Apollo 11	Apollo 12	Apollo 13	Apollo 14	Apollo 15	Apollo 16	Apollo 17
Launch	70	65	65	75	70	70	70	70	70	70	70
Average	70	72	70	73	63	67	64	74	69	70	69
High	79	81	72	80	73	80	71	77	81	80	81
Low	64	61	65	64	55	58	58	60	59	57	61
Reentry	65	61	67	58	55	60	75	59	59	57	62

62 *Biomedical Results of Apollo*, SP-368, p. 133. All temperatures are in Fahrenheit, measured at the inlet to the heat exchanger.

Accumulated Time in Space During Apollo Missions[63]

	Apollo 7	Apollo 8	Apollo 9	Apollo 10	Apollo 11	Apollo 12	Apollo 13	Apollo 14	Apollo 15	Apollo 16	Apollo 17	Flight Time (sec)	Flight Time (hh:mm:ss)
Mission Duration (hh:mm:ss)	260:09:03	147:00:42	241:00:54	192:03:23	195:18:35	244:36:25	142:54:41	216:01:58	295:11:53	265:51:05	301:51:59		
Mission Duration (sec)	936,543	529,242	867,654	691,403	703,115	880,585	514,481	777,718	1,062,713	957,065	1,086,719		
David Randolph Scott			867,654						1,062,713			1,930,367	536:12:47
Eugene Andrew Cernan				691,403							1,086,719	1,778,122	493:55:22
John Watts Young				691,403						957,065		1,648,468	457:54:28
Ronald Ellwin Evans											1,086,719	1,086,719	301:51:59
Harrison Hagan Schmitt											1,086,719	1,086,719	301:51:59
James Benson Irwin									1,062,713			1,062,713	295:11:53
Alfred Merrill Worden									1,062,713			1,062,713	295:11:53
James Arthur Lovell, Jr.		529,242					514,481					1,043,723	289:55:23
Charles Moss Duke, Jr.										957,065		957,065	265:51:05
Thomas Kenneth Mattingly, II										957,065		957,065	265:51:05
Ronnie Walter Cunningham	936,543											936,543	260:09:03
Donn Fulton Eisele	936,543											936,543	260:09:03
Walter Marty Schirra, Jr.	936,543											936,543	260:09:03
Alan LaVern Bean						880,585						880,585	244:36:25
Charles Conrad, Jr.						880,585						880,585	244:36:25
Richard Francis Gordon, Jr.						880,585						880,585	244:36:25
James Alton McDivitt			867,654									867,654	241:00:54
Russell Louis Schweickart			867,654									867,654	241:00:54
Edgar Dean Mitchell								777,718				777,718	216:01:58
Stuart Allen Roosa								777,718				777,718	216:01:58
Alan Bartlett Shepard, Jr.								777,718				777,718	216:01:58
Edwin Eugene Aldrin, Jr.					703,115							703,115	195:18:35
Neil Alden Armstrong					703,115							703,115	195:18:35
Michael Collins					703,115							703,115	195:18:35
Thomas Patten Stafford				691,403								691,403	192:03:23
William Alison Anders		529,242										529,242	147:00:42
Frank Frederick Borman, II		529,242										529,242	147:00:42
Fred Wallace Haise, Jr.							514,481					514,481	142:54:41
John Leonard Swigert, Jr.							514,481					514,481	142:54:41
Total Seconds From Liftoff	2,809,629	1,587,726	2,602,962	2,074,209	2,109,345	2,641,755	1,543,443	2,333,154	3,188,139	2,871,195	3,260,157	27,021,714	
Total Time In Space (hh:mm:ss)	780:27:09	441:02:06	723:02:42	576:10:09	585:55:45	733:49:15	428:44:03	648:05:54	885:35:39	797:33:15	905:35:57	7,506:01:54	7,506:01:54

[63] Calculated.

Apollo Medical Kits[64]

	Apollo 7	Apollo 8	Apollo 9	Apollo 10	Apollo 11	Apollo 12	Apollo 13	Apollo 14	Apollo 15	Apollo 16	Apollo 17
Command Module Medical Kit											
Methylcellulose eye drops (0.25%)	2/1	2/2	2/0	2/0	2/0	2/0	2/0	2/0	1/0	2/0	1/0
Tetrahydrozoline HCl (Visine)	—	—	—	—	—	—	—	—	—	—	1/1
Compress Bandage	2/0	2/0	2/0	2/0	2/0	2/0	2/0	2/0	2/0	2/0	2/0
Band-Aids®	12/2	12/0	12/0	12/0	12/0	12/0	12/0	12/0	12/0	12/0	12/0
Antibiotic ointment	1/1	1/0	1/0	1/0	1/0	2/0	2/0	2/0	2/0	2/1	2/1
Skin cream	1/0	1/1	1/1	1/0	1/0	1/0	1/0	1/0	1/0	1/1	1/0
Demerol injectors (90 mg)	3/0	3/0	3/0	3/0	3/0	3/0	3/0	3/0	3/0	—	—
Marezine injectors	3/0	3/0	3/0	3/0	3/0	3/0	3/0	3/0	3/0	—	—
Marezine tablets (50 mg)	24/3	24/1	24/4	12/0	—	—	—	—	—	—	—
Dexedrine tablets (5 mg)	12/1	12/0	12/0	12/0	12/0	12/0	12/1	12/0	12/0	12/0	12/0
Darvon compound capsules (60 mg)	12/2	18/0	18/0	18/0	18/0	18/0	12/1	18/0	18/0	18/0	18/0
Actifed® tablets (60 mg)	24/24	60/0	60/12	60/2	60/0	60/18	60/0	60/0	60/0	60/0	60/1
Lomotil tablets	24/8	24/3	24/1	24/13	24/2	24/0	24/1	24/0	24/0	24/0	48/5
Nasal emollient	1/0	2/1	2/1	1/0	1/0	1/0	1/0	1/0	1/0	1/0	1/0
Aspirin tablets (5 gr)	72/48	72/8	72/2	72/16	72/Unk	72/6	72/30	72/0	72/0	72/0	72/0
Tetracycline (250 mg)	24/02	24/0	24/0	15/0	—	—	—	—	60/0	60/0	60/0
Ampicillin	—	60/0	60/0	45/0	60/0	60/0	60/0	60/0	60/0	60/0	60/0
Seconal® capsules (100 mg)	—	21/1	21/10	21/0	21/0	21/6	21/0	—	21/0	21/3	21/16
Seconal® capsules (50 mg)	—	12/7	—	—	—	—	—	—	—	—	—
Nose drops (Afrin™)	—	3/0	3/1	3/0	3/0	3/1	3/0	3/1	3/0	3/0	3/3
Benadryl® (50 mg)	—	8/0	—	—	—	—	—	—	—	—	—
Tylenol® (325 mg)	—	14/7	—	—	—	—	—	—	—	—	—
Bacitracin eye ointment	—	—	1/0	—	—	—	—	—	—	—	—
Scopolamine (0.3 mg)—Dexedrine (5 mg capsules)	—	—	—	—	12/6	12/0	12/2	12/0	12/0	12/0	12/1
Mylicon tablets	—	—	—	—	40/0	40/0	40/0	40/0	40/0	40/0	40/0
Opthaine	—	—	—	—	—	—	1/0	1/0	1/0	1/0	1/0
Multi-Vitamins	—	—	—	—	—	—	—	20/0	—	—	—
Auxiliary Medications											
Pronestyl	—	—	—	—	—	—	—	—	—	80/0	80/0
Lidocaine	—	—	—	—	—	—	—	—	—	12/0	12/0
Atropine	—	—	—	—	—	—	—	—	—	12/0	12/0
Demerol	—	—	—	—	—	—	—	—	—	6/0	6/0

64 SP-368, p. 33.

310 Apollo by the Numbers

Apollo Medical Kits[65]

	Apollo 7	Apollo 8	Apollo 9	Apollo 10	Apollo 11	Apollo 12	Apollo 13	Apollo 14	Apollo 15	Apollo 16	Apollo 17
Apollo Medical Accessories Kit											
Constant Wear Garment Harness Plug	—	—	—	—	—	—	—	—	3	3	3
ECG Sponge Packages	—	—	—	—	—	—	—	—	14	14	14
Electrode Bag	1	1	1	1	1	1	1	1	1	1	1
Electrode Attachment Assembly	12	12	12	12	20	20	20	20	100	100	100
Micropore Disc	12	12	12	12	20	20	20	20	50	50	50
Sternal Harness	1	1	1	1	3	3	3	3	3	3	3
Axillary Harness	1	1	1	1	1	1	1	1	1	1	1
Electrode Paste	1	1	1	1	1	1	1	1	1	1	1
Oral Thermometer	1	1	1	1	1	1	1	1	1	1	1
pH Paper	1	1	1	1	1	1	1	1	1	None	None
Urine Collection and Transfer Assembly Roll-On Cuffs	3	3	6	6	6	6	6	6	6	6	6
Lunar Module Medical Kit[66]											
Rucksack	—	—	—	—	—	1					
Stimulant Pills (Dexedrine®)	—	—	—	—	—	4					
Pain Pills (Darvon®)	—	—	—	—	—	4					
Decongestant Pills (Actifed®)	—	—	—	—	—	8					
Diarrhea Pills (Lomotil®)	—	—	—	—	—	12					
Aspirin	—	—	—	—	—	12					
Band-Aids	—	—	—	—	—	6					
Compress Bandages	—	—	—	—	—	2					
Eye Drops (Methylcellulose)	—	—	—	—	—	1					
Antibiotic Ointment (Neosporin®)	—	—	—	—	—	1					
Sleeping Pills (Seconal®)	—	—	—	—	—	6					
Anesthetic Eye Drops	—	—	—	—	—	1					
Nose Drops (Afrin®)	—	—	—	—	—	1					
Urine Collection and Transfer Assembly Roll-On Cuffs	—	—	—	—	—	6					
Pronestyl	—	—	—	—	—	12					
Injectable Drug Kit											
Injectable Drug Kit Rucksack	—	—	—	—	—	1					
Lidocaine (cardiac)	—	—	—	—	—	8					
Atropine (cardiac)	—	—	—	—	—	4					
Demerol (pain)	—	—	—	—	—	2					

65 SP-368, P. 33.

66 Typical quantities and items; there was no "standard" lunar module medical kit. The adequacy of the kits was reviewed after each mission and appropriate modifications were made for the next mission.

Crew Weight History (kg)[67]

Mission	Crew member	30 Days Before Launch	30-Day Average	Launch	Recovery
Apollo 7	Schirra	87.1	87.8	88.0	86.1
	Eisele	69.4	69.5	71.2	66.7
	Cunningham	69.4	70.7	70.8	67.8
Apollo 8	Borman	76.2	76.6	76.6	72.8
	Lovell	76.4	76.8	78.0	74.4
	Anders	66.0	66.4	64.4	62.6
Apollo 9	McDivitt	73.5	73.0	72.1	69.6
	Scott	82.8	82.0	80.7	78.2
	Schweickart	74.7	74.3	71.2	69.4
Apollo 10	Stafford	80.1	79.6	77.6	76.4
	Young	76.6	76.8	74.8	72.3
	Cernan	79.4	79.4	78.5	73.9
Apollo 11	Armstrong	78.0	78.4	78.0	74.4
	Collins	74.4	75.6	75.3	72.1
	Aldrin	77.6	78.1	75.7	75.3
Apollo 12	Conrad	66.2	66.6	67.7	65.8
	Gordon	71.0	70.7	70.4	67.1
	Bean	69.4	69.9	69.1	63.5
Apollo 13	Lovell	79.8	78.7	80.5	74.2
	Swigert	89.1	89.4	89.3	84.4
	Haise	71.0	70.8	70.8	67.8
Apollo 14	Shepard	78.0	78.4	76.2	76.6
	Roosa	74.2	75.3	74.8	69.4
	Mitchell	83.5	83.2	79.8	80.3
Apollo 15	Scott	80.5	81.1	80.2	78.9
	Worden	73.7	73.6	73.5	72.1
	Irwin	74.3	74.3	73.2	70.8
Apollo 16	Young	80.8	80.1	78.9	75.5
	Mattingly	63.2	62.6	61.5	58.5
	Duke	73.1	73.2	73.0	70.5
Apollo 17	Cernan	81.0	80.7	80.3	76.1
	Evans	78.2	77.3	75.7	74.6
	Schmitt	76.0	76.0	74.8	72.9

67 *Biomedical Results of Apollo*, SP-368, pps. 76-77. Note that on Apollo 14, Shepard and Mitchell each gained weight.

312 Apollo by the Numbers

Inflight Medical Problems in Apollo Crews[68]

Symptom/Finding	Etiology	Cases
Barotitis	Barotrauma	1
Cardiac arrhythmia	Undetermined, possibly linked with potassium deficit	2
Dehydration	Reduced water intake during emergency	2
Dysbarism (bends)[69]	Undetermined	1
Excoriation, urethral meatus	Prolonged wearing of urine collection device	2
Eye irritation	Spacecraft atmosphere	4
	Fiberglass	1
Flatulence	Undetermined	3
Genitourinary infection with prostatic congestion	Pseudomonas aeruginosa	1
Head cold	Undetermined	3
Headache	Spacecraft environment	1
Nasal stuffiness	Zero gravity	2
Nausea, vomiting	Labyrinthine	1
	Undetermined (possibly virus-related)	1
Pharyngitis	Undetermined	1
Rash, facial, recurrent inguinal	Contact dermatitis	1
	Prolonged wearing of urine collection device	11
Respiratory irrigation	Fiberglass	1
Rhinitis	Oxygen, low relative humidity	2
Seborrhea	Activated by spacecraft environment	2
Shoulder strain	Lunar core drilling	1
Skin irrigation	Biosensor sites	11
	Fiberglass	2
	Undetermined	1
Stomach awareness	Labyrinthine	6
Stomatitis	Aphthous ulcers	1
Subungual hemorrhages	Glove fit	5
Urinary tract infection	Undetermined	1

[68] *Biomedical Results of Apollo*, SP-368.

[69] Also occurred during Gemini 10; later incidences were reported by the same crew member five years after his Apollo mission.

Postflight Medical Problems in Apollo Crews[70]

Diagnosis	Etiology	Cases
Barotitis media	Eustachian tube blockage	7
Folliculitis, right anterior chest	Bacterial	1
Gastroenteritis	Bacterial	1
Herpetic lesion, lip	Herpes virus	1
Influenza syndrome	Influenza B virus	1
	Undetermined	1
	Influenza A virus	1
Laceration of the forehead	Trauma	1
Rhinorrhea, mild	Fiberglass particle	1
Papular lesions, parasacral	Bacteria	1
Prostatitis	Undetermined	2
Pulpitis, tooth 7		1
Pustules, eyelids		1
Rhinitis	Viral	3
Acute maxillary sinusitis	Bacterial	1
Ligamentous strain, right shoulder		1
Urinary tract infection	Pseudomonas	1
Vestibular dysfunction, mild		1
Rhinitis and pharyngitis	Influenza B virus	1
Rhinitis and secondary bronchitis	Beta-streptococcus (not group A)	1
Contact dermatitis	Fiberglass	1
	Beta cloth	1
	Micropore tape	6
Subungual hemorrhages, finger nails	Trauma	3

70 *Biomedical Results of Apollo*, SP-368.

NASA Photo Numbers for Crew Portraits and Mission Emblems

Event	NASA Photo Number
Apollo 1 Mission Emblem	S66-36742
Portrait of Apollo 1 Prime Crew	S66-30236
Apollo 7 Mission Emblem	S68-26668
Portrait of Apollo 7 Prime Crew	S68-33744
Apollo 8 Mission Emblem	S68-51093
Portrait of Apollo 8 Prime Crew	S68-50265
Apollo 9 Mission Emblem	S69-19974
Portrait of Apollo 9 Prime Crew	S69-17590
Apollo 10 Mission Emblem	S69-31959
Portrait of Apollo 10 Prime Crew	S69-32616
Apollo 11 Mission Emblem	S69-34875
Portrait of Apollo 11 Prime Crew	S69-31739
Apollo 12 Mission Emblem	S69-52336
Portrait of Apollo 12 Prime Crew	S69-38852
Apollo 13 Mission Emblem	S69-60662
Portrait of Apollo 13 Original Prime Crew	S69-62224
Portrait of Apollo 13 Flight Crew	S70-36485
Apollo 14 Mission Emblem	S70-17851
Portrait of Apollo 14 Prime Crew	S70-55635
Apollo 15 Mission Emblem	S71-30463
Portrait of Apollo 15 Prime Crew	S71-37963
Apollo 16 Mission Emblem	S71-56246
Portrait of Apollo 16 Prime Crew	S72-16660
Apollo 17 Mission Emblem	S72-49079
Portrait of Apollo 17 Prime Crew	S72-50438

Bibliography

Akens, Davis S, editor, *Saturn Illustrated Chronology: Saturn's First Ten Years, April 1957 Through April 1967*, MHR-5, August 1, 1968, George C. Marshall Space Flight Center, National Aeronautics and Space Administration

Apollo 7 Mission Commentary, Prepared by Public Affairs office, NASA Johnson Space Center, October, 1968

Apollo 7 Mission Report, Prepared by Apollo 7 Mission Evaluation Team, National Aeronautics and Space Administration, Manned Spacecraft Center, Houston, Texas, December 1968 (MSC-PA-R-68-15)

Apollo 7 Press Kit, Release #68-168K, National Aeronautics and Space Administration, Washington, DC, October 6, 1968

Apollo 8 Mission Report, Prepared by Mission Evaluation Team, National Aeronautics and Space Administration, Manned Spacecraft Center, Houston, Texas, February 1969 (MSC-PA-R-69-1)

Apollo 8 Press Kit, Release #68-208, National Aeronautics and Space Administration, Washington, DC, December 15, 1968

Apollo 9 Mission Report (MSC-PA-R-69-2 May 1969) (NTIS-34370)

Apollo 9 Press Kit, Release #69-29, February 23, 1969

Apollo 10 Mission Report (MSC-00126 November 1969)

Apollo 10 Press Kit, Release #69-68, May 7, 1969

Apollo 11 Mission Report (MSC-00171 November 1969/NASA-TM-X-62633) (NTIS N70-17401)

Apollo 11 Preliminary Science Report, Scientific and Technical Information Division, National Aeronautics and Space Administration, Washington, DC, 1969 (NASA SP-214)

Apollo 11 Press Kit, Release #69-83K, July 6, 1969

Apollo 12 Mission Report (MSC-01855 March 1970)

Apollo 12 Preliminary Science Report, Scientific and Technical Information Division, office of Technology Utilization, National Aeronautics and Space Administration, Washington, DC, 1970 (NASA SP-235)

Apollo 12 Press Kit, Release #69-148, November 5, 1969

Apollo 13 Mission Report (MSC-02680 September 1970/NASA-TM-X-66449) (NTIS N71-13037)

Apollo 13 Press Kit, Release #70-50K, April 2, 1970

Apollo 14 Mission Report (MSC-04112 May 1971)

Apollo 14 Preliminary Science Report, Scientific and Technical Information office, National Aeronautics and Space Administration, Washington, DC, 1971 (NASA SP-272)

Apollo 14 Press Kit, Release #71-3K, January 21, 1971

Bibliography

Apollo 15 Mission Report (MSC-05161 December 1971/NASA-TM-X-68394) (NTIS N72-28832)

Apollo 15 Preliminary Science Report, Scientific and Technical Information office, National Aeronautics and Space Administration, Washington, DC, 1972 (NASA SP-289)

Apollo 15 Press Kit, Release #71-119K, July 15, 1971

Apollo 16 Mission Report (MSC-07230 August 1972/NASA-TM-X-68635) (NTIS N72-33777)

Apollo 16 Preliminary Science Report, Scientific and Technical Information office, National Aeronautics and Space Administration, Washington, DC, 1972 (NASA SP-315)

Apollo 16 Press Kit, Release #72-64K, April 6, 1972

Apollo 17 Mission Report (MSC-07904 March 1973) (NTIS N73-23844)

Apollo 17 Preliminary Science Report, Scientific and Technical Information office, National Aeronautics and Space Administration, Washington, DC, 1973 (NASA SP-330)

Apollo 17 Press Kit, Release #72-220K, November 26, 1972

Apollo Lunar Surface Experiments Package (ALSEP): Five Years of Lunar Science and Still Going Strong, Bendix Aerospace

Apollo Program Summary Report, National Aeronautics and Space Administration, Lyndon B. Johnson Space Center, Houston, Texas, April 1975 (JSC-09423) (NTIS N75-21314)

Apollo/Saturn Postflight Trajectory (AS-503), Boeing Corporation Space Division, February 19, 1969 (D-5-15794), (NASA-CR-127240) (NTIS N92-70422)

Apollo/Saturn Postflight Trajectory (AS-504), Boeing Corporation Space Division, May 2, 1969 (D-5-15560-4), (NASA-CR-105771) (NTIS N69-77056)

Apollo/Saturn Postflight Trajectory (AS-505), Boeing Corporation Space Division, July 17, 1969 (D-5-15560-5), (NASA-CR-105770) (NTIS N69-77049)

Apollo/Saturn Postflight Trajectory (AS-506), Boeing Corporation Space Division, October 6, 1969 (D-5-15560-6), (NASA-CR-102306) (NTIS N92-70425)

Apollo/Saturn Postflight Trajectory (AS-507), Boeing Corporation Space Division, January 13, 1970 (D-5-15560-7), (NASA-CR-102476) (NTIS N92-70420)

Apollo/Saturn Postflight Trajectory (AS-508), Boeing Corporation Space Division, June 10, 1970 (D-5-15560-8), (NASA-CR-102792) (NTIS N92-70437)

Apollo/Saturn Postflight Trajectory (AS-509), Boeing Corporation Space Division, June 30, 1971 (D-5-15560-9), (NASA-CR-119870) (NTIS N92-70433)

Apollo/Saturn Postflight Trajectory (AS-510), Boeing Corporation Space Division, November 23, 1971 (D-5-15560-10), (NASA-CR-120464) (NTIS N74-77459)

Bibliography

Apollo/Saturn Postflight Trajectory (AS-511), Boeing Corporation Space Division, August 9, 1972 (D-5-15560-11), (NASA-CR-124129) (NTIS N73-72531)

Apollo/Saturn Postflight Trajectory (AS-512), Boeing Corporation Space Division, April 11, 1973 (D-5-15560-12), (NASA-CR-144080) (NTIS N76-19199)

Apollo/Skylab, ASTP and Shuttle Orbiter Major End Items, Final Report (JSC-03600), March 1978

Bilstein, Roger E., *Stages To Saturn: A Technological History of the Apollo/Saturn Launch Vehicles*, Scientific and Technical Information Branch, National Aeronautics and Space Administration, November, 1980 (NASA SP-4206)

Boeing Company, *Final Flight Evaluation Report: Apollo 10 Mission*, for the office of Manned Space Flight, National Aeronautics and Space Administration, (D-2-117017-7/NASA-TM-X-62548) (NTIS N70-34252)

Boeing Company, *Final Flight Evaluation Report: Apollo 7 Mission*, for the office of Manned Space Flight, National Aeronautics and Space Administration, February 1969 (D-2-117017-4 Revision A)

Boeing Company, *Final Flight Evaluation Report: Apollo 8 Mission*, for the office of Manned Space Flight, National Aeronautics and Space Administration, April 1969 (D-2-117017-5)

Boeing Company, *Final Flight Evaluation Report: Apollo 9 Mission*, for the office of Manned Space Flight, National Aeronautics and Space Administration, (D-2-117017-6/NASA-TM-X-62316)

Brooks, Courtney G., James M. Grimwood, and Loyd S. Swenson, Jr., *Chariots For Apollo: A History of Manned Lunar Spacecraft*, The NASA History Series, Scientific and Technical Information Branch, National Aeronautics and Space Administration, Washington, DC, 1979 (NASA SP-4205) (NTIS N79-28203)

Cassutt, Michael, *Who's Who in Space: The International Edition*, MacMillan Publishing Company, New York, 1993.

Compton, William David, *Where No Man Has Gone Before: A History of Apollo Lunar Exploration Missions*, The NASA History Series, Office of Management, Scientific and Technical Information Division, National Aeronautics and Space Administration, Washington, DC, 1989 (NASA SP-4214)

Cortright, Edgar, Chairman, *Report of the Apollo 13 Review Board*, National Aeronautics and Space Administration, June 15, 1970

Davies and Colvin, *Journal of Geophysical Research*, Vol. 105, American Geophysical Union, Washington, DC, September 2000

Ertel, Ivan D., and Roland W. Newkirk, *The Apollo Spacecraft: A Chronology*, Volume IV, January 21, 1966 - July 14, 1974, Scientific and Technical Information office, National Aeronautics and Space Administration, Washington, DC, 1978 (NASA SP-4009)

Ezell, Linda Neuman, *NASA Historical Data Book, Volume II, Programs and Projects 1958-1968*, The NASA Historical Series, Scientific and Technical Information Division, National Aeronautics and Space Administration, Washington, DC, 1988 (NASA SP-4012)

Ezell, Linda Neuman, *NASA Historical Data Book, Volume III, Programs and Projects 1958-1968*, The NASA Historical Series, Scientific and Technical Information Division, National Aeronautics and Space Administration, Washington, DC, 1988 (NASA SP-4012)

Bibliography

First Americans In Space: Mercury to Apollo-Soyuz, National Aeronautics and Space Administration (undated)

Johnson, Dale L., *Summary of Atmospheric Data Observations For 155 Flights of MSFC/ABMA Related Aerospace Vehicles,* NASA George C. Marshall Space Flight Center, Alabama, December 5, 1973 (NASA-TM-X-64796) (NTIS N74-13312)

Johnston, Richard S., Lawrence F. Dietlein, M.D., and Charles A. Berry, M.D., *Biomedical Results of Apollo,* Scientific and Technical Information office, National Aeronautics and Space Administration, Washington, DC, 1975 (NASA SP-368)

Jones, Dr. Eric, *Apollo Lunar Surface Journal,* Internet: http://www.hq.nasa.gov/office/pao/History/alsj/, National Aeronautics and Space Administration, 1996.

Kaplan, Judith and Robert Muniz, *Space Patches From Mercury to the Space Shuttle,* Sterling Publishing Co., New York, 1986

King-Hele, D. G., D. M. C. Walker, J. A. Pilkington, A. N. Winterbottom, H. Hiller, and G. E. Perry, *R. A. E. Table of Earth Satellites 1957-1986,* Stockton Press, New York, NY, 1987

Lattimer, Dick, *Astronaut Mission Patches and Spacecraft Callsigns,* unpublished draft, July 4, 1979, Lyndon B. Johnson Space Center History office

McFarlan, Donald and Norris D. McWhirter et al, editors, *1990 Guinness Book of World Records,* Sterling Publishing Co., New York

NASA Facts: Apollo 7 Mission (E-4814)

NASA Facts: Apollo 8 Mission

NASA Facts: Apollo 9 Mission

NASA Facts: Apollo 10 Mission

NASA Facts: Apollo 11 Mission

NASA Facts: Apollo 12 Mission

NASA Facts: Apollo 13 Mission

NASA Facts: Apollo 14 Mission

NASA Facts: Apollo 15 Mission

NASA Facts: Apollo 16 Mission

NASA Facts: Apollo 17 Mission

NASA Information Summaries, Major NASA Launches, PMS 031 (KSC), National Aeronautics and Space Administration, November 1985

Bibliography

NASA Information Summaries, PM 001 (KSC), National Aeronautics and Space Administration, November 1985

National Aeronautics and Space Administration Mission Report: Apollo 9 (MR-3)

National Aeronautics and Space Administration Mission Report: Apollo 10 (MR-4)

National Aeronautics and Space Administration Mission Report: Apollo 11 (MR-5)

National Aeronautics and Space Administration Mission Report: Apollo 12 (MR-8)

National Aeronautics and Space Administration Mission Report: Apollo 13 (MR-7)

National Aeronautics and Space Administration Mission Report: Apollo 14 (MR-9)

National Aeronautics and Space Administration Mission Report: Apollo 15 (MR-10)

National Aeronautics and Space Administration Mission Report: Apollo 16 (MR-11)

National Aeronautics and Space Administration Mission Report: Apollo 17 (MR-12)

Newkirk, Roland W., and Ivan D. Ertel with Courtney G. Brooks, *Skylab: A Chronology,* Scientific and Technical Information office, National Aeronautics and Space Administration, Washington, DC (NASA SP-4011)

Nicogossian, Arnauld E., M.D., and James F. Parker, Jr., Ph.D., *Space Physiology and Medicine,* (SP-447), National Aeronautics and Space Administration, 1982

Project Apollo: Manned Exploration of the Moon, Educational Data Sheet #306, NASA Ames Research Center, Moffett Field, California, Revised May, 1974

Project Apollo: NASA Facts, National Aeronautics and Space Administration

Report of Apollo 13 Review Board, National Aeronautics and Space Administration, June 15, 1970

Saturn AS-205/CSM-101 Postflight Trajectory, Chrysler Corporation Space Division (TN-AP-68-369) (NASA CR-98345) (NTIS N92-70426)

Saturn IB Flight Evaluation Working Group, Results of the Fifth Saturn IB Launch Vehicle Test Flight AS-205 (Apollo 7 Mission), NASA George C. Marshall Space Flight Center, Alabama, January 25, 1969 (MPR-SAT-FE-68-4)

Saturn V Flight Evaluation Working Group, Saturn V Launch Vehicle Flight Evaluation Report AS-503: Apollo 8 Mission, NASA George C. Marshall Space Flight Center, Alabama, February 20, 1969 (MPR-SAT-FE-69-1)

Saturn V Launch Vehicle Flight Evaluation Report AS-504: Apollo 9 Mission, NASA George C. Marshall Space Flight Center, Alabama, (MPR-SAT-FE-69-4/NASA-TM-X-62545) (NTIS 69X-77591)

Saturn V Launch Vehicle Flight Evaluation Report AS-505: Apollo 10 Mission, NASA George C. Marshall Space Flight Center, Alabama, (MPR-SAT-FE-69-7/NASA-TM-X-62548) (NTIS 69X-77668)

Bibliography

Saturn V Launch Vehicle Flight Evaluation Report AS-506: Apollo 11 Mission, NASA George C. Marshall Space Flight Center, Alabama, (MPR-SAT-FE-69-9/NASA-TM-X-62558) (NTIS 90N-70431/70X-10801)

Saturn V Launch Vehicle Flight Evaluation Report AS-507: Apollo 12 Mission, NASA George C. Marshall Space Flight Center, Alabama, (MPR-SAT-FE-70-1/NASA-TM-X-62644) (NTIS 70X-12182)

Saturn V Launch Vehicle Flight Evaluation Report AS-508: Apollo 13 Mission, NASA George C. Marshall Space Flight Center, Alabama, (MPR-SAT-FE-70-2/NASA-TM-X-64422) (NTIS 90N-70432/70X-16774)

Saturn V Launch Vehicle Flight Evaluation Report AS-509: Apollo 14 Mission, NASA George C. Marshall Space Flight Center, Alabama, (MPR-SAT-FE-71-1/NASA-TM-X-69536) (NTIS N73-33824)

Saturn V Launch Vehicle Flight Evaluation Report AS-510: Apollo 15 Mission, NASA George C. Marshall Space Flight Center, Alabama, (MPR-SAT-FE-71-2/NASA-TM-X-69539) (NTIS N73-33819)

Saturn V Launch Vehicle Flight Evaluation Report AS-511: Apollo 16 Mission, NASA George C. Marshall Space Flight Center, Alabama, (MPR-SAT-FE-72-1/NASA-TM-X-69535) (NTIS N73-33823)

Saturn V Launch Vehicle Flight Evaluation Report AS-512: Apollo 17 Mission, NASA George C. Marshall Space Flight Center, Alabama, (MPR-SAT-FE-73-1/NASA-TM-X-69534) (NTIS N73-33822)

The Early Years: Mercury to Apollo-Soyuz, PM 001 (KSC), NASA Information Summaries, National Aeronautics and Space Administration, November 1985

Thompson, Floyd, Chairman, *Report of the Apollo 204 Review Board*, National Aeronautics and Space Administration, April 5, 1967

Toksoz, M.N., Dainty, A.M., Solomon, S.C., Anderson, K.R.; *Structure of the Moon published in* Reviews of Geophysics and Space Physics, Vol. 12, No. 4, American Geohpysical Union, Washington, DC, November 1974

Trajectory Reconstruction Unit, *Saturn AS/205/CSM-101 Postflight Trajectory*, Aerospace Physics Branch, Chrysler Corporation Space Division, December 1968

Photo Credits

To the author's knowledge, all images in this work originated with the National Aeronautics and Space Administration. Some were scanned from original NASA photographs by the author, but most were acquired via the Internet from either the NASA Johnson Space Center Digital Image Collection Web site *(http://images.jsc.nasa.gov/iams/html/pao/apollo.htm)*; Dr. Eric Jones' Apollo Lunar Surface Journal Web site *(http://www.hq.nasa.gov/alsj/)*;[1] or Kipp Teague's The Project Apollo Archive Web site *(http://www.apolloarchive.com/apollo_gallery.html)*[2] and are used with permission. Much of Teague's work also appears in the Apollo Lunar Surface Journal. The author has resized, or cropped some images to fit the needs of this work and he is solely responsible for the results.

Except where noted, images for Apollo 1, Apollo 7, Apollo 8, Apollo 9, Apollo 10 and Apollo 13 were downloaded from the Johnson Space Center Web site or scanned and edited by the author. Lunar surface images not listed below are from Apollo Lunar Surface Journal Web site. Other images not listed below (particularly launch, recovery and post-mission images) for Apollo 11, Apollo 12, Apollo 14, Apollo 15 and Apollo 17 are also from the Johnson Space Center Web site. The remaining images are noted below with appropriate credits and are listed in order of mission and by NASA image numbers.

NASA photos reproduced from this work should include photo credit to "NASA" or "National Aeronautics and Space Administration" and should include scanning credit to the appropriate individual as noted below, to whom the author extends special thanks.

Apollo 1
J. L. Pickering: 67HC21 from The Project Apollo Archive.

Apollo 8
Ed Hengeveld: S68-53187 from Apollo Lunar Surface Journal.

Kipp Teague: S68-56050.

Apollo 9
Kipp Teague: AS09-19-2919; AS09-19-2994; and AS09-21-3236.

Apollo 10
Kipp Teague: S69-34385.

Apollo 11
Kipp Teague: AS11-36-5390; AS11-37-5528; AS11-40-5869; AS11-40-5877; AS11-40-5886; AS11-40-5899; AS11-40-5927; AS11-40-5942; AS11-40-5964; AS11-44-6574; AS11-44-6642; AS11-44-6667; S69-21365; S69-31740; S69-39526; S69-40308.

Apollo 12
Kipp Teague: AS12-46-6716; AS12-46-6728; AS12-46-6729; AS12-46-6790; AS12-47-6897; AS12-47-6988; AS12-48-7071; AS12-48-7110; AS12-48-7133; AS12-49-7278; AS12-49-7286; AS12-51-7507; S69-38852.

Apollo 13
Kipp Teague: 70-H-724; AS13-59-8500; AS13-59-8562; AS13-62-9004; KSC-70PC-0130; S69-62224; S70-15511; S70-34853; S70-35145; S70-35632.

Apollo 14
Kipp Teague: AS14-64-9089; AS14-64-9135; AS14-66-9344; AS14-68-9414; S70-55387; S71-18398; S71-18753.

Apollo 15
David Harland: AS15-82-11057; AS15-85-11471; AS15-85-11514; AS15-86-11603; AS15-87-11748; AS15-87-11847 from Apollo Lunar Surface Journal.

Kipp Teague: AS15-88-11866; AS15-88-11894; AS15-88-11901; AS15-88-11972; AS15-88-11980; S71-37963; S71-41356.

---

[1] Apollo Lunar Surface Journal Web site, Copyright © 1995-2000, edited by Eric M. Jones. All rights reserved.

[2] The Project Apollo Archive Web site, Copyright © 2000, Kipp Teague.

Apollo 16

John Pfannerstill: Apollo 16 Pan Camera frame 4623 from Apollo Lunar Surface Journal.

David Harland: AS16-106-17413; AS16-109-17804; S72-37002 from Apollo Lunar Surface Journal.

Kipp Teague: AS16-107-17436; AS16-107-17446; AS16-108-17629; AS16-108-17670; AS16-108-17701; AS16-110-18020; AS16-113-18340; AS16-113-18359; AS16-114-18423; AS16-114-18439; AS16-117-18826; AS16-117-18841; AS16-117-18841; AS16-118-18885; AS16-118-18894; AS16-122-19533; KSC-72PC-176; S72-16660

Ricardo Saleme: AS16-107-17442; AS16-116-18579 from The Project Apollo Archive.

Apollo 17.

Scott Cornish: AS17-134-20380 from The Project Apollo Archive.

Kipp Teague: AS17-134-20384; AS17-134-20425; AS17-134-20435; AS17-134-20469; AS17-134-20482; AS17-137-20979; AS17-137-20990; AS17-140-21493; AS17-140-21496; AS17-145-22165; AS17-145-22257; AS17-147-22465; AS17-147-22526; AS17-147-22527; AS17-148-22695; AS17-148-22726; AS17-149-22857; AS17-162-24149; S72-50438; S72-55482; S72-55834.

The NASA History Series

Reference Works, NASA SP-4000:

Grimwood, James M. *Project Mercury: A Chronology.* (NASA SP-4001, 1963).

Grimwood, James M., and Hacker, Barton C., with Vorzimmer, Peter J. *Project Gemini Technology and Operations: A Chronology.* (NASA SP-4002, 1969).

Link, Mae Mills. *Space Medicine in Project Mercury.* (NASA SP-4003, 1965).

Astronautics and Aeronautics, *1963: Chronology of Science, Technology, and Policy.* (NASA SP-4004, 1964).

Astronautics and Aeronautics, *1964: Chronology of Science, Technology, and Policy.* (NASA SP-4005, 1965).

Astronautics and Aeronautics, *1965: Chronology of Science, Technology, and Policy.* (NASA SP-4006, 1966).

Astronautics and Aeronautics, *1966: Chronology of Science, Technology, and Policy.* (NASA SP-4007, 1967).

Astronautics and Aeronautics, *1967: Chronology of Science, Technology, and Policy.* (NASA SP-4008, 1968).

Ertel, Ivan D., and Morse, Mary Louise. *The Apollo Spacecraft: A Chronology, Volume I, Through November 7, 1962.* (NASA SP-4009, 1969).

Morse, Mary Louise, and Bays, Jean Kernahan. *The Apollo Spacecraft: A Chronology, Volume II, November 8, 1962-September 30, 1964.* (NASA SP-4009, 1973).

Brooks, Courtney G., and Ertel, Ivan D. *The Apollo Spacecraft: A Chronology, Volume III, October 1, 1964-January 20, 1966.* (NASA SP-4009, 1973).

Ertel, Ivan D., and Newkirk, Roland W., with Brooks, Courtney G. *The Apollo Spacecraft: A Chronology, Volume IV, January 21, 1966-July 13, 1974.* (NASA SP-4009, 1978).

Astronautics and Aeronautics, *1968: Chronology of Science, Technology, and Policy.* (NASA SP-4010, 1969).

Newkirk, Roland W., and Ertel, Ivan D., with Brooks, Courtney G. *Skylab: A Chronology.* (NASA SP-4011, 1977).

Van Nimmen, Jane, and Bruno, Leonard C., with Rosholt, Robert L. *NASA Historical Data Book, Volume I: NASA Resources, 1958-1968.* (NASA SP-4012, 1976, rep. ed. 1988).

Ezell, Linda Neuman. *NASA Historical Data Book, Volume II: Programs and Projects, 1958-1968.* (NASA SP-4012, 1988).

Ezell, Linda Neuman. *NASA Historical Data Book, Volume III: Programs and Projects, 1969-1978.* (NASA SP-4012, 1988).

Gawdiak, Ihor Y., with Fedor, Helen. Compilers. *NASA Historical Data Book, Volume IV: NASA Resources, 1969-1978.* (NASA SP-4012, 1994).

Rumerman, Judy A. Compiler. *NASA Historical Data Book, 1979-1988: Volume V, NASA Launch Systems, Space Transportation, Human Spaceflight, and Space Science.* (NASA SP-4012, 1999).

Rumerman, Judy A. Compiler. *NASA Historical Data Book, Volume VI: NASA Space Applications, Aeronautics and Space Research and Technology, Tracking and Data Acquisition/Space Operations, Commercial Programs, and Resources, 1979-1988* (NASA SP-2000-4012, 2000).

Astronautics and Aeronautics, *1969: Chronology of Science, Technology, and Policy.* (NASA SP-4014, 1970).

Astronautics and Aeronautics, *1970: Chronology of Science, Technology, and Policy.* (NASA SP-4015, 1972).

Astronautics and Aeronautics, *1971: Chronology of Science, Technology, and Policy.* (NASA SP-4016, 1972).

Astronautics and Aeronautics, *1972: Chronology of Science, Technology, and Policy.* (NASA SP-4017, 1974).

Astronautics and Aeronautics, *1973: Chronology of Science, Technology, and Policy.* (NASA SP-4018, 1975).

Astronautics and Aeronautics, *1974: Chronology of Science, Technology, and Policy.* (NASA SP-4019, 1977).

Astronautics and Aeronautics, *1975: Chronology of Science, Technology, and Policy.* (NASA SP-4020, 1979).

Astronautics and Aeronautics, *1976: Chronology of Science, Technology, and Policy.* (NASA SP-4021, 1984).

Astronautics and Aeronautics, *1977: Chronology of Science, Technology, and Policy.* (NASA SP-4022, 1986).

Astronautics and Aeronautics, *1978: Chronology of Science, Technology, and Policy.* (NASA SP-4023, 1986).

Astronautics and Aeronautics, *1979-1984: Chronology of Science, Technology, and Policy.* (NASA SP-4024, 1988).

Astronautics and Aeronautics, *1985: Chronology of Science, Technology, and Policy.* (NASA SP-4025, 1990).

Noordung, Hermann. *The Problem of Space Travel: The Rocket Motor.* Stuhlinger, Ernst, and Hunley, J.D., with Garland, Jennifer. Editor. (NASA SP-4026, 1995).

Astronautics and Aeronautics, *1986-1990: A Chronology.* (NASA SP-4027, 1997).

Astronautics and Aeronautics, *1990-1995: A Chronology.* (NASA SP-2000-4028, 2000).

Management Histories, NASA SP-4100:

Rosholt, Robert L. *An Administrative History of NASA, 1958-1963.* (NASA SP-4101, 1966).

Levine, Arnold S. *Managing NASA in the Apollo Era.* (NASA SP-4102, 1982).

Roland, Alex. *Model Research: The National Advisory Committee for Aeronautics, 1915-1958.* (NASA SP-4103, 1985).

Fries, Sylvia D. *NASA Engineers and the Age of Apollo.* (NASA SP-4104, 1992).

Glennan, T. Keith. *The Birth of NASA: The Diary of T. Keith Glennan.* Hunley, J.D. Editor. (NASA SP-4105, 1993).

Seamans, Robert C., Jr. *Aiming at Targets: The Autobiography of Robert C. Seamans, Jr.* (NASA SP-4106, 1996)

Project Histories, NASA SP-4200:

Swenson, Loyd S., Jr., Grimwood, James M., and Alexander, Charles C. *This New Ocean: A History of Project Mercury.* (NASA SP-4201, 1966; rep. ed. 1998).

Green, Constance McL., and Lomask, Milton. *Vanguard: A History.* (NASA SP-4202, 1970; rep. ed. Smithsonian Institution Press, 1971).

Hacker, Barton C., and Grimwood, James M. *On Shoulders of Titans: A History of Project Gemini.* (NASA SP-4203, 1977).

Benson, Charles D. and Faherty, William Barnaby. *Moonport: A History of Apollo Launch Facilities and Operations.* (NASA SP-4204, 1978).

Brooks, Courtney G., Grimwood, James M., and Swenson, Loyd S., Jr. *Chariots for Apollo: A History of Manned Lunar Spacecraft.* (NASA SP-4205, 1979).

Bilstein, Roger E. *Stages to Saturn: A Technological History of the Apollo/Saturn Launch Vehicles.* (NASA SP-4206, 1980, rep. ed. 1997).

SP-4207 not published.

Compton, W. David, and Benson, Charles D. *Living and Working in Space: A History of Skylab.* (NASA SP-4208, 1983).

Ezell, Edward Clinton, and Ezell, Linda Neuman. *The Partnership: A History of the Apollo-Soyuz Test Project.* (NASA SP-4209, 1978).

Hall, R. Cargill. *Lunar Impact: A History of Project Ranger.* (NASA SP-4210, 1977).

Newell, Homer E. *Beyond the Atmosphere: Early Years of Space Science.* (NASA SP-4211, 1980).

Ezell, Edward Clinton, and Ezell, Linda Neuman. *On Mars: Exploration of the Red Planet, 1958-1978.* (NASA SP-4212, 1984).

Pitts, John A. *The Human Factor: Biomedicine in the Manned Space Program to 1980.* (NASA SP-4213, 1985).

Compton, W. David. *Where No Man Has Gone Before: A History of Apollo Lunar Exploration Missions.* (NASA SP-4214, 1989).

Naugle, John E. *First Among Equals: The Selection of NASA Space Science Experiments.* (NASA SP-4215, 1991).

Wallace, Lane E. *Airborne Trailblazer: Two Decades with NASA Langley's Boeing 737 Flying Laboratory.* (NASA SP-4216, 1994).

Butrica, Andrew J. Editor. *Beyond the Ionosphere: Fifty Years of Satellite Communication.* (NASA SP-4217, 1997).

Butrica, Andrews J. *To See the Unseen: A History of Planetary Radar Astronomy.* (NASA SP-4218, 1996).

Mack, Pamela E. Editor. *From Engineering Science to Big Science: The NACA and NASA Collier Trophy Research Project Winners.* (NASA SP-4219, 1998).

Reed, R. Dale. With Lister, Darlene. *Wingless Flight: The Lifting Body Story.* (NASA SP-4220, 1997).

Heppenheimer, T.A. *The Space Shuttle Decision: NASA's Search for a Reusable Space Vehicle.* (NASA SP-4221, 1999).

Hunley, J.D. Editor. *Toward Mach 2: The Douglas D-558 Program.* (NASA SP-4222, 1999).

Swanson, Glen E. Editor. *"Before this Decade is Out...": Personal Reflections on the Apollo Program* (NASA SP-4223, 1999).

Tomayko, James E. *Computers Take Flight: A History of NASA's Pioneering Digital Fly-by-Wire Project.* (NASA SP-2000-4224, 2000).

Center Histories, NASA SP-4300:

Rosenthal, Alfred. *Venture into Space: Early Years of Goddard Space Flight Center.* (NASA SP-4301, 1985).

Hartman, Edwin, P. *Adventures in Research: A History of Ames Research Center, 1940-1965.* (NASA SP-4302, 1970).

Hallion, Richard P. *On the Frontier: Flight Research at Dryden, 1946-1981.* (NASA SP- 4303, 1984).

Muenger, Elizabeth A. *Searching the Horizon: A History of Ames Research Center, 1940-1976.* (NASA SP-4304, 1985).

Hansen, James R. *Engineer in Charge: A History of the Langley Aeronautical Laboratory, 1917-1958.* (NASA SP-4305, 1987).

Dawson, Virginia P. *Engines and Innovation: Lewis Laboratory and American Propulsion Technology.* (NASA SP-4306, 1991).

Dethloff, Henry C. *"Suddenly Tomorrow Came...": A History of the Johnson Space Center.* (NASA SP-4307, 1993).

Hansen, James R. *Spaceflight Revolution: NASA Langley Research Center from Sputnik to Apollo.* (NASA SP-4308, 1995).

Wallace, Lane E. *Flights of Discovery: 50 Years at the NASA Dryden Flight Research Center.* (NASA SP-4309, 1996).

Herring, Mack R. *Way Station to Space: A History of the John C. Stennis Space Center.* (NASA SP-4310, 1997).

Wallace, Harold D., Jr. *Wallops Station and the Creation of the American Space Program.* (NASA SP-4311, 1997).

Wallace, Lane E. *Dreams, Hopes, Realities: NASA's Goddard Space Flight Center, The First Forty Years* (NASA SP-4312, 1999).

Dunar, Andrew J., and Stephen P. Waring. *Power to Explore: A History of the Marshall Space Flight Center* (NASA SP-4313, 1999).

Bugos, Glenn E. *Atmosphere of Freedom: Sixty Years at the NASA Ames Research Center Astronautics and Aeronautics, 1986-1990: A Chronology.* (NASA SP-2000-4314, 2000).

General Histories, NASA SP-4400:

Corliss, William R. *NASA Sounding Rockets, 1958-1968: A Historical Summary.* (NASA SP-4401, 1971).

Wells, Helen T., Whiteley, Susan H., and Karegeannes, Carrie. *Origins of NASA Names.* (NASA SP-4402, 1976).

Anderson, Frank W., Jr. *Orders of Magnitude: A History of NACA and NASA, 1915-1980.* (NASA SP-4403, 1981).

Sloop, John L. *Liquid Hydrogen as a Propulsion Fuel, 1945-1959.* (NASA SP-4404, 1978).

Roland, Alex. *A Spacefaring People: Perspectives on Early Spaceflight.* (NASA SP-4405, 1985).

Bilstein, Roger E. *Orders of Magnitude: A History of the NACA and NASA, 1915-1990.* (NASA SP-4406, 1989).

Logsdon, John M. Editor. With Lear, Linda J., Warren-Findley, Jannelle, Williamson, Ray A., and Day, Dwayne A. *Exploring the Unknown: Selected Documents in the History of the U.S. Civil Space Program, Volume I, Organizing for Exploration.* (NASA SP-4407, 1995).

Logsdon, John M. Editor. With Day, Dwayne A., and Launius, Roger D. *Exploring the Unknown: Selected Documents in the History of the U.S. Civil Space Program, Volume II, Relations with Other Organizations.* (NASA SP-4407, 1996).

Logsdon, John M. Editor. With Launius, Roger D., Onkst, David H., and Garber, Stephen E. *Exploring the Unknown: Selected Documents in the History of the U.S. Civil Space Program, Volume III, Using Space.* (NASA SP-4407, 1998).

Logsdon, John M. General Editor. With Ray A. Williamson, Roger D. Launius, Russell J. Acker, Stephen J. Garber, and Jonathan L. Friedman. *Exploring the Unknown: Selected Documents in the History of the U.S. Civil Space Program, Volume IV, Accessing Space.* (NASA SP-4407, 1999).

Siddiqi, Asif A. *Challenge to Apollo: The Soviet Union and the Space Race, 1945-1974.* (NASA SP-2000-4408, 2000).

Index

A

N

O

P

Q

R

S

T

U

W

Y

www.ingramcontent.com/pod-product-compliance
Lightning Source LLC
Chambersburg PA
CBHW082136210326
41599CB00031B/6001